CRC Handbook of Antibiotic Compounds

Author

János Bérdy
Senior Research Fellow
Institute of Drug Research
Budapest, Hungary

Contributors

Adorjan Aszalos
Head of Biochemistry, Chemotherapy
Frederick Cancer Research Center
Frederick, Maryland

Melvin Bostian
Manager of Information Systems Department
Frederick Cancer Research Center
Frederick, Maryland

Karen L. McNitt
Senior Programmer
Frederick Cancer Research Center
Frederick, Maryland

Volume I
Carbohydrate Antibiotics

Volume II
Macrocyclic Lactone (Lactam) Antibiotics

Volume III
Quinone and Similar Antibiotics

Volume IV Part 1
Amino Acid and Peptide Antibiotics

Volume IV Part 2
Peptolide and Macromolecular Antibiotics

Volume V
Heterocyclic Antibiotics

Volume VI
Alicyclic, Aromatic, and Aliphatic Antibiotics

Volume VII
Miscellaneous Antibiotics with Unknown Chemical Structure

Volume VIII Parts 1 and 2
Antibiotics from Higher Forms of Life: Higher Plants

Volume IX
Antibiotics from Higher Forms of Life: Lichens, Algae, and Animal Organisms

Volume X
General Indexes

Volume XI, Part 1 and 2
Microbial Metabolites

Volume XII
Antibiotics from Higher Forms of Life

CRC Handbook of Antibiotic Compounds

Volume XI
Microbial Metabolites, Part 1

Author

János Bérdy
Senior Research Fellow
Institute of Drug Research
Budapest, Hungary

Contributors

Adorjan Aszalos
Head of Biochemistry, Chemotherapy
Frederick Cancer Research Center
Frederick, Maryland

Karen L. McNitt
Senior Programmer
Frederick Cancer Research Center
Frederick, Maryland

CRC Press, Inc.
Boca Raton, Florida

Library of Congress Cataloging in Publication Data
(Revised for volumes 11 and 12)

Bérdy, János.
CRC Handbook of antibiotic compounds.

Title on spine: Handbook of antibiotic compounds.
Includes bibliographical references and indexes.
Contents: v. 1. Carbohydrate antibiotics. — v. 2. Macrocyclic lactone (lactam) antibiotics. — [etc.] — v. 11. Microbial metabolites. 2. v. — v. 12. Antibiotics from higher forms of life.
1. Antibiotics — Handbooks, manuals, etc. 2. Chemistry, Pharmaceutical — Handbooks, manuals, etc. 3. Antibiotics. I. Handbook of antibiotic compounds. [DNLM: QV 350.3 H236]
RS431.A6B47 615'.329 78-31428
ISBN 0-8493-3450-0 (set)
ISBN 0-8493-3463-2 (Volume XI, Part 1)
ISBN 0-8493-3464-0 (Volume XI, Part 2)

Direct all inquiries to CRC Press, Inc., 2000 Corporate Blvd., N.W., Boca Raton, Florida, 33431.

International Standard Book Number 0-8493-3450-0 (Complete Set)
International Standard Book Number 0-8493-3463-2 (Volume XI, Part 1)
International Standard Book Number 0-8493-3464-0 (Volume XI, Part 2)

Library of Congress Card Number 78-31428
Printed in the United States

HANDBOOK OF ANTIBIOTIC COMPOUNDS, VOLUME XI

INTRODUCTION

During the years that have passed since the publication of the Main Volumes (I through X) of the *CRC Handbook of Antibiotic Compounds*, about 2200 new antibiotics were published in the scientific literature. The Main Volumes covered 6700 antibiotic compounds (according to their sequence numbers), while these volumes cover compounds with sequence numbers 6700 through 8872.

In this cumulative volume about 1200 new microbial metabolites, which were published mainly during 1979 to 1981, are collected. In the same period, numerous important additional results appeared in the literature for about 1000 already known (old) antibiotics. These new data are included in the first section of this volume, while the novel compounds will be listed in the second section. The new plant and animal products and the new data for their old derivatives will be published in Volume XII (the Second Supplement).

Some recently recognized new types of chemical structures and the newly discovered or modified structures of old ones require the completion of the Antibiotic Classification System. Among new entries some 700 compounds have known structures and out of these more than a hundred have new types of structures.

In most cases the changes in the Antibiotic Numbers are minimal, such as the more precise classification within the antibiotic families or groups, but occasionally several entries have been moved and now appear under a new family or group. For the newly discovered chemical structures, including the plant and animal products also, more than 50 new Antibiotic Code Numbers were constructed to complete the classification system. One Antibiotic Number (44130, Taitomycin type) must be deleted because the compounds listed here proved to belong to the thiostrepton type antibiotics. These changes are summarized in Table 1, which includes the new Antibiotic Numbers of the Plant and Animal products (marked by asterisks).

János Bérdy
Budapest
March, 1982

Table 1
NEW ANTIBIOTIC TYPES

AN	CT	Representative
12224	Aminoglycoside	S-11-A, SU-2
12225	Aminoglycoside, Combimicin t.	Combimicins, I-SK-A1
1235	Aminoglycoside 1.	X-14847
21224	Macrolide	Mycinamycin II
21234	Macrolide	Mycinamycin IV
2324	Macrolactone	Izumenolide
2333	Dilactone, Elaiophillin t.	Azalomycin B
23513	Macrolactone, Tetrocarcin t.	Antlermicin, Kijanmycin
23514*	Macrolactone	Latrunculins
2355	Macrolactone, Nargenicin t.	Nargenicins, Nodusmicin
3124	Anthracyclinone l.	Tetracenomycins
3138	Anthraquinone l.	Viocrystin
32227	Naphtoquinone derivatives	Guanacin
32228		
3315	Benzoquinone derivatives	Sarubicin, U-58431
3325	Saframycin t.	Saframycins, Chloracarcin
3326	Dnactin t.	Dnactins
34133	Semiquinone	Cercosporin
3414*	Semiquinone	Obtusaquinone
41125	Amino Acid	Forphenicine
41216	Beta lactam, Monobactam t.	Monobactams, Sulfazecin
4214	Oligopeptide l., Indole derivatives	CC-1065, Rachelmycin
4227	Peptide, Leucinamycin t.	P-168, Trichopolins
4426	Peptolide	A-21978
4445	Peptolide	Lipopeptins
4537	Cell-Wall Component	CP, 60-F, CW-5
51134	Tetramic acid derivatives, Oligopeptide l.	Capsimycin, Ikarugamycin
51223	Pyridone derivatives	G-1549, BN-227
5141	Azepine derivatives	Ophiocordin
5224*	N-Heterocyclic derivatives	Azaskatole, Spinaceamine
5235	Indole glycoside	Sydomycin
5236		Oxanosine
5311*	Alkaloid (Terpene alkaloids)	Norcassidine, aconitine
5312*	Alkaloid (Colchicine t.)	Demecolcine, colchicine
5313*	Alkaloid (Aliphatic alkaloids)	Sphaerophysine
5321*	Alkaloid (Pyrrole, imidazole derivative)	Chacsine
5322*	Alkaloid (Pyridine, peperidine derivative)	Anabasin, Visanin
5331*	Alkaloid (Harmane t.)	Gramine, Chanthin-6-one
5332*	Alkaloid (Ellipticine t.)	Olivacine, guatambuine
5333*	Alkaloid (Ibogaine alkaloids)	Coronaridin, apparicine
5334*	Alkaloid (Vinca alkaloids)	Vinchristine, gabunine
5335*	Alkaloid (Other indole alkaloids)	Caracurine
5341*	Alkaloid (Bisbenzylisoquinoline t.)	Tetradrine, curine
5342*	Alkaloid (Benzphenanthridine t.)	Nitidine, sanguinarine
5343*	Alkaloid (Protoberberin t.)	Berberin, coralyne
5344*	Alkaloid (Glaucine t.)	Liriodenine, annonaine
5345*	Alkaloid (Other isoquinoline alkaloids)	Emetine, protopine
5346*	Alkaloid (Quinoline t.)	Quinine, Casimiroin
534*	Alkaloid	Vasicin, febrifugin
5351*	Alkaloid (Pyrrolizidine t.)	Indicine, senecionine
5352*	Alkaloid (Quinolizidine t.)	Tylophorine, lycorine
5353*	Alkaloid (Cephalotaxine t.)	Harringtonine
5354*	Alkaloid (Camptothecin t.)	Camptothecins

Table 1 (continued)
NEW ANTIBIOTIC TYPES

AN	CT	Representative
5355*	Alkaloid	Sparteine, matrine
5361*	Alkaloid (O-heterocyclic alkaloids)	Acronycin, pteleatin
5362*	Alkaloid (S-heterocyclic alkaloids)	Gerrardin, cassipourine
5371*	Alkaloid	
5431	Thiazol deriv.	Myxothiazol
5432*		Cycloalliin
6213*		Pederin
6234		Diplosporin
6316*	Flavan t.	7-Hydroxyflavan
6346*	Chromane	2,2-Dimethylchromane
6351*		Demethoxyageratochomen
6352*		Lathodoratin
6417*		Malyngolide
65111	Polyether, Monensin t.	Monensins, Laidlomycin
65112	Polyether	CP-47433
6523	Polyether	M-139603
72143*	Sesterterpene	Desacetylscalaradial
7216*	Carotenoides	Retinal, Retinoids
73114*	Limonoid	Aphanostatin
73135*	Physalin t.	Physalin-B
7324*	Limonoid	Hispidins
8313*	Tropolone derivative	Hainanolide
8412	Aromatic ether	
8413*	Aromatic ether	
84221	Glycosidic antibiotic, Chartreusin t.	Chartreusin, G-261-B
84222	Glycosidic antibiotic, Chartreusin l.	Gilvocarcins, Toromycin
8424	Glycosidic antibiotic, Trioxacarcin t.	Trioxacarcins
8425*		Amygdalin

Deleted Antibiotic Type

AN	CT	Representative
4413[a]	Peptide l., Taitomycin t.	Taitomycin, RP-9671

[a] Compounds moved to type: 43211

THE AUTHOR

János Bérdy, Ph.D., is a Senior Research Fellow of the Institute of Drug Research (formerly the Research Institute for Pharmaceutical Chemistry), Budapest.

Dr. Bérdy graduated in 1958 from Eötvös Loránd University, Budapest, and received his Ph.D. degree (summa cum laude) in organic chemistry in 1961 from Kossuth Lajos University, Debrecen. He was qualified as a Pharmaceutical Chemistry Engineer in 1969 at the Technical University, Budapest. He is a member of the Hungarian Chemical Society and many other scientific associations.

Dr. Bérdy's research interests include the isolation of new antibiotics, the development of industrial production, and classification and identification problems of antibiotics, as well as the theoretical problems of antibiotics research.

CONTRIBUTORS

Adorjan Aszalos, Ph.D., is Head of Biochemistry, Chemotherapy, Frederick Cancer Research Center, Frederick, Maryland, and is Lecturer in Biochemistry at Hood College, Frederick, Maryland. He recently joined the Bureau of Drugs, Food and Drug Administration, in Washington, D.C. He previously held positions at the Squibb Institute for Medical Research and at Princeton University.

Dr. Aszalos graduated from Technical University of Budapest, Hungary with B.S. and M.S. degrees in chemical engineering and biochemistry. He received his Ph.D. in bioorganic chemistry from Technical University of Vienna, Austria in 1961. Subsequently, he was Post-Doctoral Fellow at Rutgers University.

Dr. Aszalos is a member of the American Chemical Society, Interscience Foundation, and New York Academy of Sciences. In the latter society, he served as Vice Chairman of the Biophysics Section in 1973 to 1975. Dr. Aszalos received, among other awards, the Austrian Industrial Research Award and the Army Post-Doctoral Research Award.

Dr. Aszalos has presented over 30 lectures at National and International meetings and published over 60 research papers and several review articles and chapters. His current major interest is antibiotics and enzymes in chemotherapy.

Karen L. McNitt is a Senior Programmer at the Frederick Cancer Research Center, Frederick, Maryland. In this capacity she was responsible for the programming and implementation of the data base system used to collect the information on the antibiotic compounds. She also directed the data entry and validation of the compound information.

Ms. McNitt has a degree in computer science and specializes in the analysis of scientific data.

TABLE OF CONTENTS

SECTION I
NEW AND CORRECTED DATA FOR COMPOUNDS INCLUDED IN VOLUMES I THROUGH IV OF THE *CRC HANDBOOK OF ANTIBIOTIC COMPOUND*

SECTION II
NEW COMPOUNDS

SELECTION OF COMPOUNDS INCLUDED

The guiding principles in selection of material to include in the Handbook are as follows:

1. The compounds listed in this book are derived from the whole living world, including all types of prokaryotes and eukaryotes, namely microorganisms, lichens, fungi, mosses, algae, higher plants, protozoa, molluscs, sponges, worms, insects, and vertebrates.
2. An essential requirement is the in vitro, or perhaps only in vivo, antimicrobial (at least at a concentration of 500 μg/mℓ) activity or some antitumor, cytotoxic, antiprotozoal, or antiviral (antiphage) effect, regardless that this activity is observable in a specific medium or circumstance only.
3. Every chemical entity, e.g., stereoisomer, forms a separate entry. Components of antibiotic complexes, when they are separated and when some of their properties are determined, are listed individually.
4. The unresolved antibiotic complexes (components are detected by chromatography only) form a single entry. These complexes in many instances differ only by proportions of the same components (e.g., streptothricin or heptaene antibiotic complexes) and are designated by their own name.
5. Crude antibiotic extracts, characterized by some properties such as UV spectra, stability, or others, possessing interesting activity, especially those originating from uncommon sources, also form separate entries.
6. Derivatives of antibiotics made by chemical methods are not listed, unless they are produced by biosynthetic or enzymatic methods also. The products of directed and conversion-type fermentations or mutational biosynthetic processes employing precursor-like compounds incorporated into the active products are included.
7. Alkaloids, stress metabolites, insecticides, anthelminthics with some antimicrobial or antitumor activity, and mycotoxins without significant antimicrobial effect but with high (cyto)toxicity are included. Phytotoxins, enzyme inhibitors, plant growth regulators, animal growth promoters, and other physiologically active metabolites without any antimicrobial, antitumor, antiviral, or cytotoxic activities are excluded from this Handbook.

Consequently this work includes all antibiotically active natural products (antibacterial, antifungal, antiprotozoal, antitumor, antiviral, and occasionally anthelminthic or insecticide agents) discovered, having one or more of the characteristic properties described, although many compounds have not been isolated in pure state and their structures are unknown. After all, the number of entries is not exactly identical with the number of presently existing antibiotic compounds. It is very likely that numerous identities are undetermined and numerous components are unresolved yet.

This Handbook series contains more than 6000 entries, of which about 4500 represent the antibiotics prepared by the fermentation of microorganisms. Approximately 3000 antibiotics are derived from different *Actinomycetales* species, of which about 88 to 90% originate from *Streptomyces* species. It must be noted that in this decade about 20% of *Actinomycetales* antibiotics were derived from non-*Streptomyces* species. Almost 1000 antibiotics come from different fungi, and 500 to 600 come from various bacterial strains (including *Pseudomonales*).

The total number of antibiotics with known chemical structure is about 2500 (nearly 2000 are microbial antibiotics), and about 400 compounds are synthetized. Additionally, there are about 1500 antibiotics about which we have satisfactory knowledge regarding their chemical structure (degradation products, skeleton, principal moieties, etc.). Numerous compounds might be classified on the basis of physical, chemical, and microbiological similarities (e.g., cross-resistance) to the known type compounds. After all, about 85% of the antibiotics have more or less known chemical structural features.

HOW TO USE THIS HANDBOOK

Although this Handbook details vastly different types of compounds, an effort has been made to present the material according to a general format. All compounds (antibiotic entries) have a specific *compound number*, which serves as a title to a group of entries and as a unique numerical identifier. This number consists of two parts. The first element is, in fact, identical to our previously reported[9] *antibiotic code number* (without the separation by commas), which is characteristic of the chemical type of the compound. The second element of the compound number, separated by a hyphen, is a simple *sequence number* assigned individually to any compound according to its addition to the data base. The complete compound number provides access to that compound through the indices for any compound for which no name is listed.

Most of the compounds in this Handbook have been arranged according to our previously reported, continuously revised and completed chemical classification system.[9] This system follows the formal chemical classification but not in the strictest sense. Since this is merely a superficial classification, taking into account some biogenetic and other points of view, it is obvious that the same compound may belong to more than one class. To avoid these duplications, we selected nine basic chemical moieties (principal constituents) most characteristic of the compound, and the primary classification was done accordingly.

Assignment to antibiotic families is performed according to the following principal constituents

1. Sugar
2. Macrocyclic lactone ring (more than eight members)
3. Quinone (or quinone-like) skeleton
4. Amino acid
5. Nitrogen-containing heterocyclic system
6. Oxygen-containing heterocyclic system
7. Alicyclic skeleton
8. Aromatic skeleton
9. Aliphatic chain

The construction of some more or less arbitrary class of compounds seems to be justified. The formation of a family for the macrocyclic lactones and the separation of the quinones and quinone-like compounds from the aromatic (mainly phenolics) compounds was unavoidable. Beyond their frequent occurrence and great importance, their complete new biological properties, different from those of normal aliphatic and aromatic antibiotics, justifies listing them as a separate family of antibiotics. Moreover, the limitation of the carbohydrate (sugar) family of compounds to the mostly sugar-containing structures, excluding most of the glycosides (macrolide-, anthracycline-, peptide-, purine-pyrimidine-, and aromatic-glycosides), which are classified on the basis of their diversified aglycones, surely contributes to the logical classification. In the course of detailed systemization, some further arbitrary decisions became necessary. The grouping of streptothricines among the carbohydrates was permitted because of their properties and activities similar to other water-soluble basic antibiotics. The tetracyclines are grouped together with anthracycline quinones in the family of quinone compounds. Again, all glutarimides were grouped together as alicyclic compounds, rather than grouping them as heterocyclic, aromatic (actiphenol), or aliphatic (streptimidone) compounds. Alkaloids having antimicrobial or antitumor activity (except steroid alkaloids) were grouped as N-heterocyclic compounds. The terpenes were distributed according to their structures into the alicyclic, aromatic, or aliphatic families. The skeleton of this system includes only the families, subfamilies, and groups shown in Table 1.

Table 1

CLASSIFICATION OF ANTIBIOTIC COMPOUNDS

AN	Family, subfamily, group	Important representatives
1	Carbohydrate antibiotics	
11	Pure saccharides	
111	Mono and oligosaccharides	Streptozotocin, nojirimycin
112	Polysaccharides	Glucans, soedomycin
12	Aminoglycoside antibiotics	
121	Streptamine derivatives	Streptomycins, bluensomycin
122	2-Deoxystreptamine derivatives	Neomycin, gentamicin, etc.
123	Inositol-inoseamine derivatives	Kasugamycin, validamycin
124	Other aminocyclitols	Fortimicin
125	Aminohexitols	Sorbistin
13	Other glycosides	
131	Streptothricin group	Streptolin, racemomycin
132	Glycopeptides, C-glycosides	Vancomycin, chromomycin
14	Sugar derivatives	
141	Sugar esters, amides	Everninomicin, lincomycin
142	Sugar lipids	Moenomycin, labilomycin
2	Macrocyclic lactone (lactam) antibiotics	
21	Macrolide antibiotics	
211	Small (12-, 14-membered) macrolide	Erythromycin, picromycin
212	16-membered macrolides	Leucomycin, tylosin
213	Other macrolides	Borrelidin, lankacidin
22	Polyene antibiotics	
221	Trienes	Mycotrienine, proticin
222	Tetraenes	Nystatin, rimocidin
223	Pentaenes	Eurocidin, filipin
224	Hexaenes	Candihexin, mediocidin
225	Heptaenes	Candicidin, amphotericin B
226	Octaenes	Ochramycin
227	Oxo-polyenes	Flavofungin, dermostatin
228	Mixed polyenes	Tetrahexin
23	Macrocyclic lactone antibiotics	
231	Macrolide-like antibiotics	Oligomycin, primycin
232	Simple lactones	Albocyclin, A-26771 B
233	Dilactones	Antimycin, boromycin
234	Polylactones	Nonactin, tetranactin
235	Condensed macrolactones	Chlorothricin, cytochalasin
24	Macrolactam antibiotics	
241	Ansamycin group	Rifamycin, tolypomycin
242	Ansa-lactams (maytanosides)	Ansamitocin, maytansin
243	Lactone-lactams	Viridenomycin
3	Quinone and similar antibiotics	
31	Tetracyclic compounds and anthraquinones	
311	Tetracyclines	Tetracycline, chlorotetracycline
312	Anthracyclines	Adriamycin, rhodomycin
313	Anthraquinone derivatives	Ayamycin, hedamycin
32	Naphtoquinones	
321	Simple naphtoquinones	Javanicin, juglomycin
322	Condensed naphtoquinones	Granaticin, rubromycin
33	Benzoquinones	
331	Simple benzoquinones	Spinulosin, oosporein
332	Condensed benzoquinones	Mitomycin, streptonigrin
34	Quinone-like compounds	
341	Semiquinones	Resistomycin, maytenin
342	Other quinone-like compounds	Epoxidon, aeroplysinin

Table 1 (continued) CLASSIFICATION OF ANTIBIOTIC COMPOUNDS

AN	Family, subfamily, group	Important representatives
4	Amino acid, peptide antibiotics	
41	Amino acid derivatives	
411	Simple amino acids	Cycloserine, alanosin
412	Amino acid derivatives	Penicillin, aureothricin
413	Diketopiperazine derivatives	Gliotoxin, chaetocin
42	Homopeptides	
421	Oligopeptides	Netropsin, negamycin
422	Linear homopeptides	Gramicidin, alamethicin
423	Cyclic homopeptides	Tyrocidin, bacitracin, viomycin
43	Heteromer peptides	
431	Cyclic lipopeptides	Polymyxin, amphomycin, iturin
432	Thiapeptides	Thiostrepton, althiamycin
433	Chelate-forming peptides	Bleomycin, sideromycins
44	Peptolides	
441	Chromopeptolides	Actinomycin, quinomycin
442	Lipopeptolides	Enduracidin, surfactin
443	Heteropeptolides	Etamycin, ostreogrycin B
444	Simple peptolides	Telomycin, grisellimycin
445	Depsipeptides	Valinomycin, ostreogrycin A
45	Macromolecular peptides	
451	Polypeptides	Nisin, licheniformin
452	Proteins	Neocarzinostatin, pacibilin
453	Proteids (chromo-, gluco-, nucleo-)	Asparaginase, bacteriocins
5	Nitrogen (or S) containing heterocyclic antibiotics	
51	Single heterocycles	
511	Five-membered ring	Pyrrolnitrin, azomycin
512	Six-membered ring	Mocimycin, abikoviromycin
513	Pyrimidine glycosides	Amicetin, polyoxin, blasticidins
52	Condensed heterocycles	
521	Aromatic fused compounds	Albofungin, pyocyanine
522	Fused heterocycles	Anthramycin, fervenulin
523	Purine glycosides	Puromycin, tubercidin
53	Alkaloids	
54	S-containing heterocycles	
6	Oxygen-containing heterocyclic antibiotics	
61	Furan derivatives	
611	Simple furans	Botriodiploidin
612	Condensed furans	Usnic acid, aflatoxins
613	Benzofurans	Furasterin
62	Pyran derivatives	
621	Simple pyrans	Aucubin, plumericin
622	α-Pyrones	Phomalactone, asperline
623	γ-Pyrones	Distacin, kojic acid
63	Benzpyran derivatives	
631	Flavonoids	Chloroflavonin, eupafolin
632	Isoflavonoids	Pisatin, pterocarpan
633	Neoflavones	Dalbergione
634	Other benzopyran derivatives	Radicinin, morellin
64	Small lactones	
641	Simple lactones	Acetomycin, penicillic acid
642	Condensed lactones (coumarins)	Actinobolin, mycophenolic acid

Table 1 (continued)
CLASSIFICATION OF ANTIBIOTIC COMPOUNDS

AN	Family, subfamily, group	Important representatives
65	Polyether antibiotics	
651	Saturated polyethers	Monensin, nigericin
652	Unsaturated polyethers	Narasin, salinomycin
653	Aromatic polyethers	Lasalocid
654	Polyether-like antibiotics	A-23187
7	Alicyclic antibiotics	
71	Cycloalkane derivatives	
711	Cyclopentane derivatives	Sarcomycin, pentanenomycin
712	Cyclohexane derivatives	Fumagillin, ketomycin
713	Glutarimide antibiotics	Cycloheximide, streptimidone
72	Small terpenes	
721	Simple mono, sesqui, and diterpenes	Coriolin, cyathin, siccanin
722	Terpene lactones	Vernolepin, enmein, quassin
73	Oligoterpenes	
731	Steroids	Fusidic acid, viridin
732	Triterpenes	Saponins, cardenolides, etc.
733	Terpenoides (Scirpene derivatives)	Trichotecin, verrucarins
8	Aromatic antibiotics	
81	Benzene derivatives	
811	Monocyclic derivatives	Flavipin, versicolin
812	Alkyl-benzene derivatives	Chloramphenicol, ascochlorin
813	Polycyclic benzene derivatives	Xanthocyllin, alternariol
82	Condensed aromatic compounds	
821	Spiro compounds (Grisans)	Griseofulvin, geodin
822	Naphtalene derivatives	Gossypol, carzinophyllin
823	Anthracene-phenantrene derivatives	Thermorubin, orchinol
83	Nonbenzoid aromatic compounds	
831	Tropolones	Puberulic acid
832	Azulene	Lactaroviolin
84	Other aromatic derivatives	
841	Aromatic ethers	Zinninol, bifuhalol
842	Glycosidic antibiotics	Novobiocin, hygromycin A
843	Aromatic esters	Nidulin, phlorizin
9	Aliphatic antibiotics	
91	Alkane derivatives	
911	Saturated alkane derivatives	Elaiomycin, lipoxamycin
912	Polyines	Marasin, mycomycin
92	Carboxylic acid derivatives	
921	Small carboxylic acid derivatives	Enteromycin, cellocidin
922	Fatty acid derivatives	Eulicin, myriocin
93	Sulfur- and phosphor-containing aliphatic compounds	
931	S-containing compounds	Allicin, fluopsin
932	P-containing compounds	Phosphonomycin

The most characteristic feature of this classification system is the utilization of the previously mentioned *antibiotic code number* (AN). This number carries information about the structure or structural type of the compound. The first member of this five-digit number indicates the nine large antibiotic families to which the compound belongs. The second, third, fourth, and occasionally the fifth digits indicate the subfamilies, groups, types, and subtypes, respectively, e.g., 12222 represents the gentamicin subtype among the 4,5-disubstituted (1222) deoxystreptamine (122) derivatives of aminoglycoside antibiotics (12) in the family of carbohydrate antibiotics (1). The less well-known agents receive only the first few figures, indicating the large group to where the compound can surely be ranged. Thus, 12000 indicates an aminoglycoside antibiotic with unknown type. The zero means lack of information; thus the compounds listed throughout the tables without the antibiotic code number (00000) are those for which structural information has not been established as yet. They represent about 600 (10% of all) compounds which are listed in separate volume(s) divided into sections according to type of producing organisms, namely, the compounds produced by Actinomycetales, fungi, and bacteria as well as plants and animals have been arranged alphabetically according to their producing genus and species.

Another identifier, *chemical type,* is a short description of the structural type (aminoglycoside, ansamycin, purine glycoside, etc.) and/or the specification of a peculiar type (neomycin type, oligomycin type, cycloheximide type, etc.) of compound. While the compound always bears one antibiotic code number, sometimes two designations are attached to it. One is characteristic of the larger group, while the other refers to the specific type, e.g., aminoglycoside, neomycin type. Occasionally a compound may bear the antibiotic code number without any chemical type designation. This usually belongs to the newer groups or types of compounds with only a few representatives.

In some cases compounds have been included in the most probable group, even though insufficient data are available to verify the final grouping, hoping that these entries will promote further work in a class of these compounds. Close chemical and microbiological properties will certainly suggest to investigators to do further work on these compounds.

SELECTION OF DATA INCLUDED

The data included in this book were determined by the content of our original card index file system. The selection of data for maintenance in the file system and in the computerized data bank was decided on the basis of their usefulness in screening work searching for new antibiotics. Consequently, the properties which are characteristic for the crude substances or active extracts, isolated in the early phase of research, were emphasized. This is one reason why some properties such as melting points or NMR spectral data are excluded.

The following data, when available, are included for each antibiotic compound:

1. Name, alternate names, and trade name	Name
2. Identical with	Identical
3. Producing organism(s)	PO
4. Chemical type, chemical nature	CT
5. Molecular formula	Formula
6. Elemental analysis	EA
7. Molecular/equivalent weight	MW/EW
8. Color, appearance: Physical characteristics	PC
9. Optical rotation	OR
10. Ultraviolet spectra, solvent(s)	UV
11. Solubility	
Good	SOL-Good
Fair	SOL-Fair
Poor	SOL-Poor
12. Qualitative chemical reactions	Qual

13. Stability	Stab
14. Antimicrobial activity, test organisms	TO
15. Toxicity	LD_{50}
16. Antitumor and/or antiviral activity	TV
17. Isolation methods employed	
Filtration	IS-Fil
Extraction	IS-Ext
Ion exchange	IS-Ion
Absorption	IS-Abs
Chromatography	IS-Chr
Crystallization	IS-Cry
18. Practical application, utility	Utility
19. Structural formula	Structure
20. References	References

The presentation of data involved some compromises to meet the special requirements of computer programming and to maintain the easy usefulness of the cross indices. For example, all letters are capital, and, of course, it would be impossible to use subscript and superscript symbols.

Some data are given slightly differently than they stand in the literature. The molecular weights and elemental analysis values are given in rounded numerals to facilitate the key back to the compounds using the indices. The solubility and stability data are given, sometimes after intelligent transcription of the terminology found in the original papers. All other data reported have been extracted from the literature and presented in their original values. The accuracy of values (±), when available, is indicated by a vertical bar (|). The data presented (solubility, optical rotation) are related to room temperature.

1. Name, Alternate Names, and Trade Name

The *basic name* (title) of any antibiotic compound is generally the trivial, nonproprietary name (underlined). Alternate names (synonyms), such as other chemical and patent names, experimental drug codes (e.g., NSC numbers), and occasionally systematic names, as far as they came to our attention, all have been listed after the basic name, without any differentiation. The trade names (s), if they occur, appear on a separate line. Specific names evidently derived from a related well-known compound, such as 14-hydroxydaunomycin (adriamycin), have also been included. If a compound is not designated specifically, the letter and/or number designations (in most cases the experimental drug codes) were used as the basic name of the compound without the terms "antibiotic" or "number". For example, PA-616 and not Antibiotic PA-616, or 2230-C and not Antibiotic number 2230-C, were used. In a few instances, lacking other identification, exotic names have been used, e.g., "Landy substance" and "bromine-rich marine antibiotics", and are given in quotation marks.

Naturally occurring compounds derived from other trivially named compounds by modifying adjectives such as allo-, methyl-, dihydro-, etc. are listed separately, but in the name index they are listed as the derivatives of the parent compound, sometimes upholding the name in the original format also. Dihydrostreptomycin appears in the index as both Dihydrostreptomycin and Streptomycin, dihyro-. Stereochemical descriptors such as D and L or + and − are incorporated into the names wherever possible, but in the name index these compounds are alphabetized according to the general name.

No attempt has been made to supply systematic names, except when these were obvious or the compound had no trivial or other alternate name. The systematic names sometimes appear slightly modified because of the impossibility of always applying the proper signs, i.e., double parenthesis, commas, or apostrophes, as a consequence of the restrictions of the computer programming. The following symbols are used for Greek letters: A″ = α (alpha), B″ = β (beta), G″ = γ (gamma), D″ = δ (delta), E″ = ε (epsilon), H″ = η (eta), O″ = ω (omega), and X″ = ξ (xi).

Special care has been taken to eliminate duplicates from this Handbook, that is the multiple description of the same compounds referred to by different names by the same authors (quintomycin-lividomycin, marcomycin-hygromycin B, etc.) in patent and the subsequent article.

Compounds unnamed by the discoverers and compounds for which no structures are given or where the structure is too complicated to give a short, meaningful systematic name have been titled simply by their antibiotic code number and/or their sequence number, e.g., 11210-0023 or 00000-4138.

2. Identical With

Very often, a single antibiotic has been isolated independently by several authors from several different sources and thus different names have been given to the same compound. In those cases where the identity has been proved only later, these antibiotics are listed as separate entries, but the identity has been noted under "identical with". In such instances where the first publication is included, the verification of the identity of the newly isolated compound with another formerly known substance — or where this identity has been known for a long time (carbomycin-magnamycin, novobiocin-streptonivicin, etc.) — does not form an individual entry. The specific name (if given) of the newly isolated compound is listed as a synonym of the old compound only, together with its specific data. The possible alternate names of the identical compounds are not listed. All identities are always noted mutually.

When it is not known whether the products are identical, due to the lack of sufficient data, they are treated as separate entries. It is very likely that careful perusal of the data included in the Handbook will indicate that some further compounds reported in one study may be identical to those reported by others. Those questions remain to be solved in the future.

3. Producing Organisms

The genus and species names have been given as stated in the original publications but the obvious faults are corrected and the style is standardized. The unidentified species are indicated by the abbreviation sp. In a few reports the genus name is not given; it is designated by the family or other specification of the type of producing organism given in inverted commas. An attempt has been made to list all organisms which are able to produce the compound in question. In most cases the variants or subspecies are also indicated and separated by a hyphen. *Streptomyces* species are frequently referred as *Actinomyces* according to Krassilnikov's systematization. They are transcribed to *Streptomyces* species, upholding the *Actinomyces* designation.

The + sign after the name of a microorganism followed by a special term signifies the addition of the indicated precursors or other substances to the medium for obtaining the compound desired (directed fermentations, mutational biosynthesis with idiothropes). The + sign between the name of a plant organism and another organism (generally fungi) indicates the production of phytoalexin (stress metabolite) by the host caused by infection.

It is hoped that the classification of compounds through the cross index by taxonomical origin will be of value to the taxonomists and biogeneticists who are able, by means of the data presented, to trace the occurrence of particular structural types through the family of organisms.

4. Chemical Type, Chemical Nature

The identifier chemical type gives a short description of the structural type (aminoglycoside, ansamycin, purine glycoside, etc.) and/or the specification of a peculiar type (neomycin type, oligomycin type, cycloheximide type, etc.) of compound. While the compound always bears one antibiotic code number, sometimes two designations are attached to it. One is characteristic of the larger group, while the other refers to the specific type, e.g., aminoglycoside, neomycin type. Occasionally a compound may bear the antibiotic code number without any chemical type designation. This usually belongs to the newer groups of types of compounds with only a few representatives. The terms used throughout the Handbook to characterize the compounds are listed in Table 2, including the antibiotic code numbers.

The identifier chemical nature indicates the acid-base character of the compounds, distinguising the acidic, basic, amphoteric and neutral characters. In the case of a compound listed as both acidic and amphoteric, an amphoteric compound is meant with a more distinct acidic character. When this information is not given directly in the original paper, but the chemical nature of the compound can evidently be concluded from the properties or structure published, it is stated and listed.

5. Molecular Formula

Chemical formulas are always listed in the sequence of C, H, N, O, S, Hlg, and other elements. When the formulae are uncertain, the most probable average values are given, e.g., $C_{22-24}H_{44-46}NO_4$ becomes $C_{23}H_{45}NO_4$ rather than $C_{23\pm1}H_{45\pm1}NO_4$. If in the literature the formula of only some simple salt is provided (e.g., $C_{40}H_{78}O_{11}Na$), the free form is calculated ($C_{40}H_{79}O_{11}$) and the compound listed accordingly. The formulas do not contain subscript symbols. Note that the symbols O (oh) and 0 (zero) are very similar.

6. Elemental Analysis

When the molecular formula is unknown, the elemental analyses have been given in percent to the nearest whole value. For all compounds independent of the known formula or structure, the percentage of nitrogen, sulfur, halogen, phosphor, and other rare elements is given in rounded whole numbers. These values, if given for simple salts or their derivatives only, are calculated for the free form of compounds. The S, Cl, Na, etc. elements occurring in the simple salt-forming radicals (sulfates, hydrochlorides, sodium salts, etc.) are excluded from the coded elemental analysis. When the molecular formula is established and no data about found percentage composition are provided, these values are calculated from the published formula and listed among experimental values.

7. Molecular and/or Equivalent Weight

These are as a rule experimentally found values. Data are given in whole numbers. When these data are not available, they are calculated from the reported molecular formula. All equivalent weights are experimentally found values.

8. Color and Appearance, Physical Characteristics

The colors of substances are given by the author's original description. The different colors listed mean the transitional ones, e.g., white, yellow means a yellowish white. Only the crystalline and solid or liquid state are distinguished by crystalline, powder, oil, liquid, syrup, etc., respectively. No crystal form or other physical properties are given.

9. Optical Rotation

Rotation values in degrees and solvent employed, separated by a comma, are listed. Solvents are abbreviated according to the generally accepted abbreviations (see list of

Table 2
KEY TO THE ANTIBIOTIC TYPES AND SUBTYPES

AN	Chemical type designations	AN	Chemical type designations
	Volume I: Carbohydrate Antibiotics		
11111	Sugar, monosaccharide	1224	Aminoglycoside, apramycin-type
11112	Sugar, disaccharide	1225	Aminoglycoside, neamine-type
11121	Amino sugar	1231	Aminoglycoside-like, validamycin-type
11122	Aminodisaccharide	1232	Aminoglycoside-like, kasugamycin-type
11131	Oligosaccharide	1233	Aminoglycoside-like
11132	Aminooligosaccharide	1234	Aminoglycoside-like
1114	Sugar derivative	1241	Aminoglycoside-like, fortimicin-type
112	Polysaccharide	1251	Aminoglycoside-like, aminohexitol derivatives
1121	Polysaccharaide, glucan		
1122	Polysaccharaide-protein complex	131	Streptothricin-type
1123	Lipopolysaccharide	1311	Streptothricin-type
1124	Polysaccharide (hemicellulose)	1312	Streptothricin-like
12	Aminoglycoside	1313	Streptothricin-like
1211	Aminoglycoside, streptomycin-type	132	Glycopeptide
1212	Aminoglycoside	1321	Glycopeptide, ristocetin-type
1213	Aminoglycoside, spectinomycin-type	1322	Glycopeptide, vancomycin-type
12211	Aminoglycoside, neomycin-type	1323	Chromomycin-type
12212	Aminoglycoside, ribostamycin-type	1411	Everninomicin-type
12221	Aminoglycoside, kanamycin-type	1412	Lincomycin-type
12222	Aminoglycoside, gentamicin-type	1413	
12223	Aminoglycoside, seldomycin-type	1421	Moenomycin-type
1223	Aminoglycoside, hygromycin B-type	1422	Glycolipid
		1423	Sugar derivatives
	Volume II: Macrocyclic Lactone (Lactam) Antibiotics		
21	Macrolide	223	Pentaene
2111	Macrolide, methymycin-type	2231	Pentaene, methylpentaene
21121	Macrolide, picromycin-type	2232	Pentaene, eurocidin-type
21122	Macrolide, erythromycin-type	2233	Pentaene, capacidin-type
21123	Macrolide, megalomycin-type	2234	Pentaene
21124	Macrolide, lankamycin-type	2241	Hexaene, candihexin-type
21211	Macrolide, leucomycin-type	225	Heptaene
21212	Macrolide, spiramycin-type	22511	Aromatic heptaene, candicidin-type
21221	Macrolide, carbomycin-type	22512	Aromatic heptaene
21222	Macrolide, angolamycin-type	22513	Aromatic heptaene
21223	Macrolide, cirramycin-type	2252	Nonaromatic heptaene, amphotericin B-type
21231	Macrolide. carbomycin B-type		
21232	Macrolide, tylosin-type	2261	Octaene
21233	Macrolide, juvenimycin B-type	2271	Oxo-pentaene, flavofungin-type
21241	Macrolide, maridomycin-type	2272	Oxo-hexaene
21242	Macrolide, cirramycin-type	2281	Polyene, tetra + hexaene
21251	Macrolide, neutramycin-type	23	(azalomycin F-type)
21252	Macrolide, aldgamycin E-type	231	Macrolide-like
21311	Macrolide-like, bundlin-type	2311	Macrolide-like, oligomycin-type
21312	Macrolide-like	2312	Macrolide-like, venturicidin-type
2132	Macrolide-like	2313	Macrolide-like
22	Polyene	2314	Macrolide-like
2211	Triene, trienine-type	2315	Macrolide-like, melanosporin-type
2212	Triene	2316	Macrolide-like, humidin-type
222	Tetraene	2317	Macrolide-like, blasticidin A-type
2221	Tetraene, pimaricin-type	2321	Simple lactone
2222	Tetraene, rimocidin-type	2322	Macrolactone
2223	Tetraene, nystatin-type	2323	Macrolactone

Table 2 (continued)
KEY TO THE ANTIBIOTIC TYPES AND SUBTYPES

Volume II: Macrocyclic Lactone (Lactam) Antibiotics (continued)

AN	Chemical type designations
2331	Antimycin-type
2332	Dilactone
2341	Cyclopolylactone, nonactin-type
2342	Cyclopolylactone-like
23511	Macrolactone, chlorothricin-type
23512	Macrolactone, milbemycin-type
2352	Cytochalasin-type
2353	Brefeldin-type
2354	Zygosporin-type
2411	Ansamycin, rifamycin-type
2412	Ansamycin, streptovaricin-type
2413	Ansamycin
2414	Ansamycin
2421	Ansa-macrolactam, maytansin-type
2422	Ansa-macrolactam, rubradirin-type
243	Lactone-lactam

Volume III: Quinone and Similar Antibiotics

AN	Chemical type designations
3	Quinone-type
31111	Tetracycline-type
31112	Tetracycline-type
3112	Tetracycline-like
3113	Tetracycline-like
312	Anthracycline-like
31211	Anthracycline, rhodomycin-type
31212	Anthracycline, cinerubin-type
31213	Anthracycline, aklavin-type
31214	Anthracycline, daunomycin-type
31221	Anthracycline, nogalamycin-type
31222	Anthracycline, steffimycin-type
31223	Anthracycline, quinocycline-type
3123	Anthracyclinone
313	Anthraquinone derivatives
3131	Anthraquinone
3132	Dianthraquinone
3133	Benzanthraquinone, tetrangomycin-type
3134	Anthraquinone derivatives
3135	Pluramycin-type
3136	Phenanthrenequinone
3137	Anthraquinone derivatives
3211	o-Naphtoquinone
3212	p-Naphtoquinone
32211	Naphtoquinone derivatives, cervicarcin-type
32212	Naphtoquinone derivatives, julimycin-type
32221	Naphtoquinone derivatives, actinorhodin-type
32222	Naphtoquinone derivatives, rubromycin-type
32223	Naphtoquinone derivates, granaticin-type
32224	Naphtoquinone derivates, naphtazarin
32225	Naphtoquinone derivatives, kalafungin-type
32226	Naphtoquinone derivatives, pyranonaphtoquinone
3223	Naphtoquinone derivatives
3231	Luteomycin-type
3232	Xanthomycin-type
3311	Benzoquinone
3312	Benzoquinone derivates
3313	Benzoquinone derivatives, bibenzoquinones
3314	Benzoquinone derivates
33211	Mitomycin-type
33212	Mitomycin-like
3322	Streptonigrin-type
3323[a]	Benzoquinone derivatives
3324	Benzoquinone derivatives, saframycin-type
3411	Semiquinone, quinone-methide
3412[a]	Semiquinone, quinone-methide
34131	Semiquinone, resistomycin-type
34132	Semiquinone, herqueinone-type
3421	Quinone-like, epoxydone-type
3422[a]	Quinone-like
3423[a]	Quinone-like

Volume IV Part 1: Amino Acid and Peptide Antibiotics

AN	Chemical type designations
4	Peptide/polypeptide
41	Amino acid derivatives
4111	Azaamino acid
41121	Amino acid
41122	Amino acid
41123	Amino acid
41124	Amino acid
4113	Amino acid analog
41211	Beta lactam, penicillin-type
41212	Beta lactam, cephalosporin-type
41213	Beta lactam, nocardicin-type
41214	Beta lactam, thienamycin-type
41215	Beta lactam, clavulanic acid-type
4122	Pyrrothine-type
4123	Actithiazic acid-type
4131	Diketopiperazine derivatives
41321	Epidiketooligothiapiperazine, gliotoxin-type
41322	Epidiketooligothiapiperazine, chaetocin-type
41323	Epidiketooligothiapiperazine, sporidesmin-type
41324	Epidiketooligothiapiperazine, hyalodendrin-type

Table 2 (continued)
KEY TO THE ANTIBIOTIC TYPES AND SUBTYPES

AN	Chemical type designations	AN	Chemical type designations
	Volume IV Part 1: Amino Acid and Peptide Antibiotics (continued)		
41325	Epidiketooligothiapiperazine, sirodesmin-type	4311	Lipopeptide, amphomycin-type
41326	Epidiketooligothiapiperazine	43121	Lipopeptide, polymyxin-type
4133	Aspergillic acid-type	43122	Lipopeptide, octapeptin-type
4211	Oligopeptide, netropsin-like	43123	Lipopeptide, polypeptin-type
42111	Oligopeptide, netropsin-type	43124	Lipopeptide, tridecaptin-type
42112	Oligopeptide, noformicin-type	4313	Peptide, echinocandin-type
42113	Oligopeptide, kikumycin-type	4314	Peptide, bacillomycin-type
4212	Oligopeptide	4315	Peptide
4213	Cyclic oligopeptide-like, diketopiperazine derivatives	432	Peptide
4221	Peptide, gramicidin-type	43211	Thiazolyl-peptide, thiostrepton-type
4222	Peptide, edein-type	43212	Thiazolyl-peptide, althiomycin-type
4223	Peptide	43213	Thiazolyl-peptide
4224	Peptide	43214	Thiazolyl-peptide, micrococcin-type
4225	Peptide, peptaibophol-type	4322	Peptide, bottromycin-type
4226	Peptide, cerexin-type	4323	Peptide, berninamycin-type
423	Cyclopeptide	4324	Peptide, leucinamycin-type
4231	Cyclopeptide, tyrocidine-type	43311	Sideromycin, albomycin-type
4232	Cyclopeptide, bacitracin-type	43312	Sideromycin, ferrimycin-type
4233	Cyclopeptide, viomycin-type	43313	Sideromycin, succinimycin-type
4234	Cyclopeptide, ilamycin-type	43314	Pseudosideromycin
4235	Cyclopeptide, cyclosporin-type	43315	Sideramine
431	Lipopeptide	4332	Glycopeptide, bleomycin-type
		4333	Peptide-like, chelate-forming
	Volume IV Part 2: Peptolide and Macromolecular Antibiotics		
4411	Chromopeptolide, actinomycin-type	4452	Depsipeptide, serratamolide, type
4412	Quinoxaline-peptide, echinomycin-type	4453	Depsipeptide, ostreogrycin-A-type
4413	Peptide like, taitomycin-type	4454	Depsipeptide
4421	Peptolide, enduracidin-type	45	Polypeptide, protein, macromolecular
4422	Peptolide, stendomycin-type	4511	Polypeptide
4423	Peptolide, peptidolipid	4512	Polypeptide
4424	Peptolide	4513	Polypeptide
4425	Peptolide	4514	Polypeptide, nisin-type
44311	Peptolide, virginiamycin-type	45211	Acidic protein
44312	Peptolide, etamycin-type	45212	Basic protein
44313	Peptolide, pyridomycin-type	45213	Amphoteric protein
4432	Peptolide, monamycin-type	4522	Protein
4441	Peptolide, telomycin-type	4523	Lipoprotein
4442	Peptolide, grisellimycin-type	4531	Chromoproteid
4443	Peptolide	4532	Glycoproteid
4444	Peptolide	4533	Nucleoproteid
4451	Depsipeptide, valinomycin-type	4534	Enzyme-like
		4535	Bacteriocin
		4536	
	Volume V: Heterocyclic Antibiotics		
5111	Pyrrole derivatives, pyrrolnitrin-type	5117[a]	Pyrrole derivatives, chlorophyll derivative
5112	Pyrrole derivatives, prodigiosin-type	5121	Pyridine derivatives
51131	Tetramic acid derivatives	51221	Pyridone derivatives, mocimycin-type
51132	Tetramic acid derivatives, streptolydigin-type	51222	Pyridone derivatives
51133	Tetramic acid derivatives, oleficin-type	5123	Piperidine derivatives
5114	Pyrrolidine derivatives	5124	Piperidine derivatives
5115	Imidazole derivatives	5125	Pyrimidine derivatives
5116	Pyrrole	5126	Pyrazine derivatives
		51311	Cytosine glycoside, amicetin-type
		51312	Cytosine glycoside, blasticidin-S-type

Table 2 (continued)
KEY TO THE ANTIBIOTIC TYPES AND SUBTYPES

AN	Chemical type designations
	Volume V: Heterocyclic Antibiotics (continued)
51313	Cytosine glycoside, gougerotin-type
51314	Azacytosine glycoside
51315	Pyrimidine glycoside, ezomycin-type
51321	Uracil glycoside, polyoxin-type
51322	Uracil glycoside, mycospocidin-type
51323	Azauracil glycoside
51324	Uracil glycoside
51325	Thymine glycoside
51331	Imidazole glycoside
51341	N-Heterocyclic C-glycoside
51342	N-Heterocyclic C-glycoside
5211	Indole derivatives
5212	Quinoline (quinoxaline) derivatives
5213	Phenazine derivatives
5214	Phenoxazine derivatives
5215	Albofungin-type
5216	N-Heterocyclic derivatives
5217[a]	Alkaloid-like, carbazole derivatives
5221	Benzdiazepine, anthramycin-type
5222	Purine derivatives
5223	Pyrimidotriazine derivatives, fervenulin-type
523	Adenine glycoside-like
52311	Purine glycoside
52312	Adenine glycoside
52313	Adenine derivative
52314	Deazaadenine glycoside
52315	Adenine analog
52316	2-Aminopurine glycoside
52321	Pyrazolopyrimidine C-glycoside, formycin-type
52322	Homopurine glycoside, coformycin-type
5233	Guanine glycoside
5234	Nucleotide
53	Alkaloid
5411	Cyclic polysulfide
5421	Tiophene derivatives
6111	Furan derivatives
6112[a]	Furan derivatives, lignan-like
6121	Aflatoxin-type
6122[a]	Furan derivatives
6131[a]	Dibenzofuran derivatives
6132	Dibenzofuran derivatives, usnic acid-type
6133	Benzofuran derivatives
6134	Benzofuran derivatives
6211	Pyran derivatives
6212[a]	Pyran derivatives
6221	Alpha pyrone, asperline-type
6222	Alpha pyrone
6223[a]	Alpha pyrone derivatives
6224	Alpha pyrone derivatives
6225	Alpha pyrone derivatives
6226	Alpha pyrone derivatives
6231	Gamma pyrone derivatives, aureothin-type
6232	Gamma pyrone derivatives
6233	Gamma pyrone derivatives
6241	Beta pyrone
631[a]	Flavonoid
6311[a]	Flavone-type
6312	Flavonol-type
6313[a]	Flavanone-type
6314	Anthocyanide
6315[a]	Chalcone derivatives
6321[a]	Isoflavone
6322[a]	Pterocarpan-type
6323[a]	Isoflavanone
6324[a]	Isoflavan
6331[a]	Neoflavone-type
6341	Xanthone derivatives
6342[a]	Xanthone derivatives, morellin-type
6343	Condensed gamma pyrone derivatives
63441	Bisnaphtopyran derivatives, viridotoxin-type
63442	Bisnaphtopyran derivatives, cephalochromin-type
6345	Xanthone derivative, ergochrome type
6411	Beta lactone derivatives
6412	Gamma lactone derivatives
6413[a]	Lichesteric acid-type
6414	Tetronic acid
6415	Dilactone derivatives
6416	Dilactone derivatives
64211[a]	Podophyllotoxin-type
64212	Condensed small lactone
6422	Coumarone derivatives
6423	Coumarin derivatives
6424[a]	Coumarin derivatives
6425	Isocoumarin derivatives
65	Polyether
6511	Polyether, monensin-type
65121	Polyether, nigericin-type
65122	Polyether, septamycin-type
6513	Polyether, alborixin-type
6514	Polyether, lysocellin-type
6521	Polyether, narasin-type
6522	Polyether, dianemycin-type
6531	Polyether, lasalocid-type
6532	Polyether
6541	Polyether-like, calcimycin-type
6551	Polyether-like, ambrutycin-type
	Volume VI: Alicyclic, Aromatic, and Aliphatic Antibiotics
7111	Cyclopentane derivatives
7112	Cyclopentane derivatives
7113	Cyclopentane derivatives
7114	Cyclopentane derivatives
7121	Cyclohexane derivatives, fumagillin-type
7122	Cycloohexenone derivatives
7123	Cyclohexene derivatives
7124	Cyclohexanol derivatives

Table 2 (continued)
KEY TO THE ANTIBIOTIC TYPES AND SUBTYPES

Volume VI: Alicyclic, Aromatic, and Aliphatic Antibiotics (continued)

AN	Chemical type designations	AN	Chemical type designations
713	Glutarimide-like	8134	Polycyclic benzene derivatives
7131	Glutarimide, cycloheximide-type	8211	Grisan derivatives, griseofulvin-type
7132	Glutarimide, actiphenol-type	8212	Grisan derivatives, geodin-type
7133	Glutarimide, streptimidone-type	8221	Naphtalene derivatives
7141	Cyclobutane derivatives	8222	Naphtalene derivatives, naphtalonone
72	Terpene-like	8223[a]	Naphtalene derivatives, gossypol-type
7211[a]	Monoterpene	8224[a]	Naphtalene derivatives
7212	Sesquiterpene	8231	Anthracene derivatives
7213	Diterpene	8232[a]	Phenanthrene derivatives
72141	Sesterterpene, ophiobolin-type	8232[a]	Phenanthrene derivatives
72142[a]	Sesterterpene	8241[a]	Indane derivatives, pterosin-type
7215	Terpene glycoside	8311	Tropolone
7221[a]	Sesquiterpene lactone	8312	Tropolone
7222[a]	Diterpene lactone	8321	Azulene
7223[a]	Terpene lactone derivatives	8411	Aromatic ether
7224[a]	Glaucarubin-type, terpene derivatives	8421	Glycosidic antibiotic, hygromycin A-type
7225	Nonadrine	8422	Glycosidic antibiotic, chartreusin-type
7311	Steroid	84231	Glycosidic antibiotic, novobiocin-type
73111	Fusidic acid-type	84232	Glyyosidic antibiotic, coumermycin-type
73112	Polyporenic acid-type	84311	Depsidone, nidulin-type
73113[a]	Cucurbitacin-type	84312[a]	Depsidone
7312[a]	Steroid alkaloid	8432[a]	Depside
73131[a]	Cardenolide	8433	Aromatic ester
73132[a]	Bufadienolide	8434	Aromatic ester
73133[a]	Withanolide	91	Aliphatic
73134[a]	Sterol glycoside	9111	Aliphatic derivatives
7314	Viridin-type	9112[a]	Aliphatic derivatives
7315	Azasteroid	9113	Aliphatic derivatives, elaiomycin-type
7321[a]	Triterpene	9114	Aliphatic derivatives
7322[a]	Triterpene glycoside	9115	Aliphatic derivatives
7323[a]	Saponine	912	Polyine
73311	Scirpene derivatives, trichotecin-type	9121	Polyine
73312	Scirpene derivatives, trichodermin-type	9122	Polyine
7332	Scirpene derivatives, macrolactone	9123	Polyine
8111	Benzene derivatives	92111	Acrylic acid derivatives, enteromycin-type
8112	Benzene derivatives	92112	Acrylic acid derivatives
8113	Benzene derivatives	9212	Acetylene derivatives
8114	Benzene derivatives	9213	Simple carboxylic acid derivatives
81211	Chloramphenicol-type	922	Fatty acid-like
81212	Benzene derivatives	9221	Fatty acid
8122	Aromatic terpene derivatives, ascochlorin-type	9222	Fatty acid
		9223	Fatty acid
8123	Benzene-aliphatic derivatives	9224	Fatty acid derivatives
8124	Benzene-aliphatic derivatives	9225	Fatty acid derivatives
8125	Benzene-aliphatic derivatives	9226	Fatty acid derivatives
8126[a]	Aromatic terpene derivatives	9311	Thioformine derivatives
8131	Polycyclic benzene derivatives, xanthocillin-type	9312[a]	Aliphatic sulnoxide
		93131[a]	Isothiocyanate derivatives
8132[a]	Polycyclic benzene derivatives, stilbene-type	93132	Isothiocyanate derivatives
		9314[a]	Simple aliphatic thio derivatives
8133	Polycyclic benzene derivatives, diphenyl-type	932	Phosphonomycin-type

Table 2 (continued)
KEY TO THE ANTIBIOTIC TYPES AND SUBTYPES

Volume VII: Miscellaneous Antibiotics with Unknown Chemical Structure	
00000	Unclassified antibiotics (no antibiotic number)

[a] Only plant and animal products exist in this type.

abbreviations). Generally, all the values found in different solvents are listed. No exact temperature and concentration of the compounds is given; the values are regarded to near room temperature. If the magnitude of rotation is unknown, just a + or − sign is coded.

10. Ultraviolet Spectra

The wavelength of all observed maxima (in nanometers) and the corresponding extinction values ($E^{1\%}_{1\,cm}$ and/or molecular extinction) are listed. Values are referred to for any solvents determined. Solvents are abbreviated according to the listed abbreviations. If the spectrum is taken at a particular pH (in water or buffer), the number instead of the solvent indicates these values. Combination solvents are separated by hyphens, e.g., MeOH-HCl means generally 0.1 *N* methanolic hydrochloric acid. UV-MeOH: (235,, 35000) = λ_{max}: 235 nm, ε: 35000 in methanolic solution. UV-8: (240,422,) = λ_{max}: 240 nm, $E^{1\%}_{1\,cm}$: 422 in a pH 8 buffer. UV- : (200,,) means end absorption.

11. Solubility

Data about solubility range from good to fair to poor (insoluble), according to the author's original statement. Solvents are abbreviated according to the general list. The following organic solvents are considered with special emphasis: methanol, ethanol, butanol, acetone, ethyl acetate, chloroform, benzene, ether, and hexane (petrolether). For each compound when two solvents are given which do not appear consecutively in the above list, then include those solvents which are enlisted between these two solvents. For example, the methanol, ether pair in the category solubility-good means that the compound is well soluble in all organic solvents listed before, with the exception of hexane. The acetone, hexane pair in the solubility-poor category means the insolubility of the compound in acetone, ethyl acetate, chloroform, ether, benzene, and hexane.

12. Qualitative Chemical Reactions

The name of the selected eight reactions, listed below with any abbreviations used, and the result (positive, +, and negative, −), separated by a comma, are listed.

Ninhidrine	Ninh.
Sakaguchi	Saka.
Fehling	Fehl.
Ferrichloride	$FeCl_3$
Ehrlich	Ehrl.
2,4-Dinitrophenyl hydrazone	DNPH
Biuret	
Pauly	

13. Stability

The condition (acid, base, heat, light) and the result (stable, +, or unstable, −), separated by a comma, are listed. These data are more or less uncertain because the coding is obviously arbitrary.

14. Antimicrobial Activity

The data, if known, for the following ten test organisms are given for each compound:

Staphylococcus aureus	*S. aureus*
Sarcina lutea	*S. lutea*
Bacillus subtilis	*B. subtilis*
Escherichia coli	*E. coli*
Shigella species	*Shyg.* sp.
Pseudomonas aeruginosa	*Ps. aer.*
Proteus vulgaris	*P. vulg.*
Klebsiella pneumoniae	*K. pneum.*
Saccharomyces cerevisiae	*S. cerev.*
Candida albicans	*C. alb.*

The overall activity on the following microorganism types is also listed, especially when it is outstandingly characteristic of the compound or if specification of the activity is not detailed.

Gram-positive bacteria	G. pos.
Gram-negative bacteria	G. neg.
Mycobacterium species	*Mycob.* sp.
Phytopathogen bacteria (*Xanthomonas oryzae*)	Phyt. bact.
Phytopathogen fungi (*Piricularia oryzae*)	Phyt. fungi
Fungi (excluding yeasts)	
Protozoa	

Other specific test organisms are listed only when the compound is ineffective (or no data about the activity are available) against the formerly listed microorganisms or the activity against these specific test organisms, e.g., *B. mycoides, Corynebacterium* species, *Cryptococcus* species, *Trichophyton* species, *Mycoplasma* species, etc., is very significant.

The microorganisms (abbreviated as before) and the MIC values — if known — (in micrograms per milliliter) are listed and separated by commas. Data of the most sensitive strains (among the identical species or types) are taken into account. If no specific organism is known, the terms antibacterial, antifungal, antimicrobial, etc. are listed. Specific activities are coded as follows: anthelminthic, herbicide, insecticide, nematocide, etc.

15. Toxicity

All data listed are acute LD_{50} values in mice. The abbreviation of the method of administration follows the value, given in milligrams per kilogram, e.g., LD_{50}: (10, i.v.). If no particular dose level is known, the terms toxic or nontoxic are used freely, usually after the author's original statement. The phytotoxicities are also frequently noted.

16. Antitumor and/or Antiviral Activity

The following antitumor (cytotoxic) activities are listed:

Adenocarcinoma 755, mouse	CA-755
Other carcinomas	CA
Croecker sarcoma, mouse	Croecker
Ehrlich ascites or solid carcinoma, mouse	Ehrlich
Guerin carcinoma, rat	Guerin
Lymphoid leukemia L-1210, mouse	L-1210
Lewis lung carcinoma, mouse	Lewis
Human epidermoid carcinomas	H-1, H-2
Other lymphosarcomas	LS
Melanomas (B-16)	Melanoma
Myelosarcomas	Myelo
Novikoff hepatoma, rat	Novikoff
Lymphocytic leukemia P388, mouse	P-388
Leukemia P-815, mouse	P-815
Leukemia P-1534, mouse	P-1534
Sarcoma 37, mouse	S-37
Sarcoma 45, rat	S-45
Sarcoma 180, mouse	S-180
Leukemia SN-36, mouse	SN-36
Walker carcinosarcoma 256, rat	Walker 256
Yoshida sarcoma, rat	Yoshida
Cell lines	
HeLa human carcinoma, cell culture	HeLa
Portio carcinoma, cell culture	Earle
Human epidermoid carcinoma of the nasopharynx	KB
Németh-Kellner lymphoma, cell culture	NK-Ly

The following antiviral activities are listed:

Columbia SK virus	Columbia
Coxsackie virus	Coxsackie
Influenza virus, PR-8	Influenza
Newcastle disease virus	NDV
Poliovirus	Polio
Rhinovirus	Rhino
Herpesvirus	Herpes
Plant viruses	Plant virus

Tobacco mosaic virus	TMV
Vaccinia virus, pox	Vaccinia

For both antitumor and antiviral activity, no data about the specific circumstances or effective doses are provided; only the existence of the effects is listed. If the specific activity is unknown, or the compound is active only in test(s) excluding the above list, the terms antitumor, cytotoxic, antiviral, or antiphage are listed.

17. Isolation Methods Employed

This information is not listed in the case of plant and animal products.

A. Filtration

The value of the pH when the cultural broth has been filtered is listed. The term original pH means the filtration at the original pH of the broth.

B. Extraction

The solvents of with what, at what pH, and from what, separated by commas, are listed. For example: (EtOAc, 3, filtrate) means the extraction of the active compounds from the cultural filtrate at pH 3. The free places mean the unknown data, e.g., (BuOH,,filtrate) means the extraction with butanol from the filtrate at an unknown pH.

C. Ion Exchange

The resins or other ion-exchange materials and the eluting solvents or solvent systems employed (without quantitative composition), separated by comma, are listed. For example, (IRC-50-Na, HCl) means the absorption of the active compound on the Amberlite IRC-50 resin in sodium form, followed by the elution with diluted hydrochloric acid.

D. Absorption

The absorbent and the eluting solvents are listed, as before.

E. Chromatography

The adsorbent and the eluting solvents or solvent systems (the solvents are listed always according to increasing polarity) are listed as before. The components of the eluting system are separated by hyphens and no quantitative composition is given. The ion-exchange chromatographic methods are included here, e.g., (SILG., CHL-MeOH) means silica gel chromatography with chloroform-methanol mixtures.

F. Crystallization

This identifier summarizes the final purification steps of the isolation process, according to the following:

Isolation-crystallization: Crystallized from what solvent or solvent system.

Isolation-precipitation: Precipitated from what and with what. Sometimes it means the first step of the isolation. Examples: (Prec., Acet, Et_2O) means precipitation with ether from acetone; (Prec., Filt., HCl) means that the active compound was precipitated from the cultural filtrate by acidification with hydrochloric acid.

Isolation-dry: Evaporated to dry from what solvent.

Isolation-lyophilization: Lyophilized from water.

Examples

Is-fil: 2

Is-ext: (BuOAc, 8, filt.) (w, 2, BuOAc) (CHL, 7, w) (MeOH,,mic.)

Is-chr: (AL, benz-MeOH)

Is-cry: (prec., CHL, hex) (cryst., acet-benz)

The meaning of this coded information is the following: the cultural broth was filtered at pH 2 and the filtrate was extracted at pH 8 with butyl acetate. Some additional active substance was extracted from the mycelium with methanol. The active substance(s) was transferred at pH 2 to water and, after neutralization to pH 7, this substance(s) was reextracted with chloroform. The crude substance — after evaporation of the final chloroformic extract — was precipitated by hexane, then was chromatographed on an aluminum oxide column with benzene-methanol mixtures. Finally, the active compound(s) — after evaporation of the pooled active fractions — was crystallized from the acetone-benzene mixture.

18. Utility

In this identifier the practical utilization or potential usefulness of the compounds has been summarized. The following areas of utility are listed:

1. Antibacterial drug	Human drugs (1–5)	Commercialized compounds (1–10)
2. Antifungal drug		
3. Antiprotozoal drug		
4. Antitumor drug		
5. Antiviral drug		
6. Veterinary drug (including anthelminthics, coccidiostatics)		
7. Feed additive		
8. Food preservative		
9. Plant protecting agent		
10. Biochemical reagent		
11. On clinical trial — potential drug		
12. No longer available as a commercial product (have only some historical importance)		

The human drugs include the so-called "pre-compounds" such as cephalosporin C or rifamycin B which are only practically and historically important.

19. Structural Formula

The chemical structures of compounds or occasionally partial structures appear as a part of or following the introductory material. The related structures always have been given in summarized form as derivatives of a basic skeleton, as far as it enhances understanding the relationship between the compounds. These collective structural formulae are provided within the scope of the introduction to the group, type, or subtype, preceding the listing of compounds. The unique chemical structures generally are given following the introductory material.

An attempt has been made to depict all structures with the most modern stereochemical representations. Spatial drawings are used where appropriate. The standard convention of heavy and dotted lines is used to demonstrate the spatial arrangements of bonds.

20. References

References are given annexed to the introductory material (reviews) and the individual compounds (original papers, patents, etc.) as well. Referencing is not exhaustive, but much attention has been given to selecting the useful references. Our intent is to give a concise but not full reference story of any compound. In general, an attempt has been made to cite the first publications including the important properties of compounds and the recent papers or reviews, to work on the subject which, by its own bibliography, will include the earlier literature. In the references we made an attempt at completeness in the case of newer, less-known but potentially useful or interesting agents. The well-known antibiotics are referred to in reviews and monographs.

Special attention has been given to the patent literature, which is particularly quick in reporting antibiotics. Some antibiotics have only been published in the patent literature. *Chemical Abstracts* references have been used liberally, particularly when the original journal or patent is unlikely to be readily available. On certain topics the literature is vast and it is impossible, and it is not our aim, to cite all publications. If someone wants to know everything about penicillins or tetracyclines, many excellent monographs and reviews are at hand.

The periodicals and patents in this Handbook are abbreviated unlike the usual citation (owing to the requirements of computer programming); moreover, we refer only to the volume (or year), page, and year of publication of the periodicals and to the nationality and number of patents. The books and monographs (from review journals) are referred to by the name of the author(s) and/or editor, title, and year of publication.

The following lists include the most important books, periodicals, and other publications (proceedings, abstracts, reports, etc.) which are devoted exclusively or partly to antibiotics. These listings cover — together with the information obtained from the patent literature — the sources of data listed in this Handbook. The lists of periodicals include the abbreviations used in the reference part of this Handbook.

List 1

HANDBOOKS, TEXTBOOKS, AND PERIODICALS DEVOTED EXCLUSIVELY TO ANTIBIOTICS

A. Most Useful and Complete Handbooks

1. **Umezawa, H., Ed.,** *Index of Antibiotics from Actinomycetes,* University of Tokyo Press, Tokyo, 1967.
2. **Korzybski, T., Kowszyk-Gindifer, Z., and Kurylowicz, W.,** *Antibiotics: Origin, Nature and Properties,* Pergamon Press, Oxford, and Polish Scientific Publishers, Warsaw, 1967.
3. **Gottlieb, D. and Shaw, P. D., Eds.,** *Antibiotics,* Vol. 1 and 2, Springer-Verlag, Berlin, 1967.
4. **Corcoran, J. W. and Hahn, F. E., Eds.,** *Antibiotics,* Vol. 3, Springer-Verlag, Berlin, 1975.
5. **Glasby, J. S.,** *Encyclopedia of Antibiotics,* John Wiley & Sons, London, 1976.
6. **Shemyakin, M. M., Khokhlov, A. S., Kolosov, M. N., Bergelson, L. D., and Antonov, V. K.,** *Chemistry of Antibiotics,* 3rd ed., Izdanja Akademii Nauk SSSR, Moscow, 1961.
7. **Sevcik, V.,** *Antibiotika aus Actinomyceten,* VEB Gustav Fischer Verlag, Jena, 1963.
8. **Brunner, R. and Machek, G.,** *Die Antibiotica,* Hans Carl Verlag, Nurnberg, 1962.
9. **Waksman, S. A. and Lechevalier, H. A.,** *The Actinomycetes,* Vol. 3, Williams & Wilkins, Baltimore, 1962.

B. Outdated Handbooks

1. **Spector, W. S., Porter, J. N., DeMallo, and G. C., Eds.,** *Handbook of Toxicology,* Vol. 2, W. B. Saunders, Philadelphia, 1957.
2. **Florey, H. W., Chain, E., Heatley, N. G., Jennings, M. A., Sanders, A. G., Abraham, E. P., and Florey, M. E.,** *Antibiotics,* Oxford University Press, Oxford, 1949.
3. **Karel, L. and Roach, E. S.,** *A Dictionary of Antibiosis,* Columbia University Press, New York, 1951.
4. **Baron, A. L.,** *Handbook of Antibiotics,* Reinhold, New York, 1950.
5. **Klosa, J.,** *Antibiotika,* Verlag Technik, Berlin, 1952.
6. **Robinson, F. A.,** *Antibiotics,* Pitman & Sons, London, 1953.
7. **Werner, G. E.,** *Antibiotica Codex,* Wissenschaftliche Verlag, Stuttgart, 1963.

C. Textbooks

1. General (Chemistry, Biochemistry)

1. **Umezawa, H.,** *Recent Advances in Chemistry and Biochemistry of Antibiotics,* Microbial Chemistry Research Foundation, Tokyo, 1964.
2. **Hash, J. H., Ed.,** *Methods in Enzymology,* Vol. 43, Academic Press, New York, 1975.
3. **Sammes, P. G., Ed.,** *Topics in Antibiotic Chemistry,* Vol. 1, Ellis Horwood Ltd., Chichester, 1977.
4. **Mitsuhashi, S., Ed.,** *Drug Action and Drug Resistance on Bacteria,* Vol. 1 and 2, University of Tokyo Press, Tokyo, 1975.
5. **Perlman, D., Ed.,** *Structure-Activity Relationships Among the Semisynthetic Antibiotics,* Academic Press, New York, 1977.
6. **Perlman, D.,** *Antibiotics,* Rand McNally, Chicago, 1970.
7. **Evans, R. M.,** *Chemistry of the Antibiotics Used in Medicine,* Pergamon Press, London, 1965.
8. **Goldberg, H. S., Ed.,** *Antibiotics, Their Chemistry and Non-medical Uses,* D. Van Nostrand, Princeton, 1959.
9. **Prescott, S. C. and Dunn, C. G.,** *Antibiotics, Industrial Microbiology,* McGraw-Hill, New York, 1959.
10. **Waksman, S. A. and Lechevalier, H. A.,** *Actinomycetes and Their Antibiotics,* Williams & Wilkins, Baltimore, 1953.
11. **Gause, G. F.,** *The Search for New Antibiotics,* Yale University Press, New Haven, 1960.

2. General (Biosynthesis, Mechanism of Action)

12. **Snell, J. F.,** *Biosynthesis of Antibiotics,* Academic Press, New York, 1966.
13. **Vanek, Z. and Hostalek, Z., Eds.,** *Biogenesis of Antibiotic Substances,* Academic Press, New York, 1966.
14. **Gale, E. F., Cundliffe, E., Reynolds, P. E., Richmond, M. H. and Waring, M. J.,** *The Molecular Basis of Antibiotic Action,* John Wiley & Sons, London, 1972.
15. **Franklin, T. J. and Snow, G. A.,** *Biochemistry of Antimicrobial Action,* Chapman and Hall, London, 1971.
16. **Zähner, H.,** *Biologie der Antibiotica,* Springer-Verlag, Berlin, 1965.
17. **Newton, B. A. and Reynolds, P. E., Eds.,** *Biochemical Studies of Antimicrobial Drugs,* Cambridge University Press, Cambridge 1966.
18. **Mitsuhashi, S.,** *Transferable Drug Resistance Factor R,* University Park Press, Baltimore, 1971.
19. **Barber, M. and Garrod, L. P.,** *Antibiotic and Chemotherapy,* E & S Livingstone, London, 1963.
20. **Garrod, L. P. and O'Grady, F.,** *Antibiotic and Chemotherapy,* 3rd ed., E & S Livingstone, Edinburgh, 1971.

3. Special (Assay, Physical, Applications)

21. **Grove, D. C. and Randall, W. A.,** *Assay Methods of Antibiotics. A Laboratory Manual,* Medical Encyclopedia, New York, 1955.
22. **Kavanagh, F., Ed.,** *Analytical Microbiology,* Academic Press, New York, 1963.
23. **Wagman, G. H. and Weinstein, M. J.,** *Chromatography of Antibiotics,* Elsevier, Amsterdam, 1973.
24. **Blinov, N. O. and Khokhlov, A. S.,** *Paper Chromatography of Antibiotics,* Izdanja Akademii Nauk SSSR, Moscow, 1970.
25. **Abraham, E. P.,** *Biochemistry of Some Peptide and Steroid Antibiotics,* John Wiley & Sons, London, 1957.
26. **Maeda, K.,** *Streptomyces Products Inhibiting Mycobacteria,* John Wiley & Sons, New York, 1965.
27. **Rinehart, K. L., Jr.,** *The Neomycins and Related Antibiotics,* John Wiley & Sons, New York, 1964.
28. **Woodbine, M., Ed.,** *Antibiotics in Agriculture,* Butterworths, London, 1962.
29. **Bücher, T. and Sies, H., Eds.,** *Inhibitors: Tools in Cell Research,* Springer-Verlag, Berlin, 1969.
30. **Jukes, T. H.,** *Antibiotics in Nutrition,* Medical Encyclopedia, New York, 1955.
31. **Bilai, V. I.,** *Antibiotic Producing Microscopic Fungi,* Elsevier, Amsterdam, 1963.
32. **Flynn, E. H., Ed.,** *Cephalosporins and Penicillins; Chemistry and Biology,* Academic Press, New York, 1972.
33. **Barker, B. M. and Prescott, F.,** *Antimicrobial Agents in Medicine,* Blackwell, Oxford, 1973.
34. **Sermonti, G.,** *Genetics of Antibiotic-Producing Microorganisms,* John Wiley & Sons, London, 1969.

D. Journals

D. Journals	Abbreviation
1. *Journal of Antibiotics* (formerly *Journal of Antibiotics, Series A*)	*JA*
2. *Japanese Journal of Antibiotics* (in Japanese) (formerly *Journal of Antibiotics, Series B*)	*Jap. J. Ant.*
3. *Antimicrobial Agents and Chemotherapy* (1971—)	*AAC*

No.	Title	Abbreviation
4.	*Antibiotiki (Moscow)* (in Russian)	*Antib.*
5.	*Hindustan Antibiotic Bulletin*	*HAB*
6.	*Antibiotics & Chemotherapy* (1951—1962)	*Ant. & Chem.*
7.	*Revista Instituto de Antibioticos (Recife)*	*Rev. Inst. Antib.*
8.	*Information Bulletin, International Center of Information on Antibiotics*	*ICIA*

E. Other Periodicals

No.	Title	Abbreviation
1.	*Antibiotics Annual* (1953—54 to 1959—60)	*Ant. An.*
2.	*Antimicrobial Agents Annual 1960*	*Ant. A. An.*
3.	*Antimicrobial Agents and Chemotherapy 1961—1970* (Proc. Interscience Conf. Antimicrobial Agents and Chemotherapy)	*AAC year*
4.	*Abstracts Interscience Conf. Antimicrobial Agents and Chemotherapy* (1971—)	*Abst. AAC*
5.	*Antibiotica et Chemotherapia* (S. Karger, Basel) Vol. 1 to 24 (1954—1978)	
6.	*Progress in Antimicrobial and Anticancer Chemotherapy,* Proc. 6th Int. Congr. Chemotherapy, Tokyo, 1969, University of Tokyo Press, Tokyo, 1970)	*Progr. AAC*
7.	*Advances in Antimicrobial and Antineoplastic Chemotherapy,* Proc. 7th Int. Congr. Chemotherapy, Prague, 1971, Urban & Schwarzenberg Verlag, Munich, 1972	*Adv. AAC*
8.	*Progress in Chemotherapy (Antimicrobial, Antiviral, Antineoplastic),* Proc. 8th Int. Congr. Chemotherapy, Athens, 1973, Hellenic Society of Chemotherapy, Athens, 1974	
9.	*Antibiotics. Advances in Research, Production and Clinical Use,* Proc. Congr. Antibiotics, Prague, 1964, Butterworths, London, 1966	
10.	*Biochemistry of Antibiotics,* Proc. 4th Int. Congr. Biochemistry, Vienna, 1958, Pergamon Press, London, 1959	
11.	*Antibiotics and Mould Metabolites,* Symp. Chem. Soc., Nottingham, 1956, Chemical Society, London, 1956	
12.	*Antibiotics. Their Production, Utilization and Mode of Action,* Symp. Hindustan Antibiotics Ltd., Pimpri, 1956, Council of Science and Industrial Research, New Delhi, 1958	
13.	*Symposyum on Antibiotics,* Quebec, 1971, Butterworths, London, 1971	

List 2

HANDBOOKS, TEXTBOOKS, AND PERIODICALS DEVOTED PARTLY TO ANTIBIOTICS

A. General Handbooks

1. **Laskin, A. I. and Lechevalier, H. A., Eds.,** *CRC Handbook of Microbiology,* Vol. 3, CRC Press, Cleveland, 1973.
2. **Miller, M. W.,** *The Pfizer Handbook of Microbial Metabolites,* McGraw-Hill, New York, 1961.
3. **Turner, W. B.,** *Fungal Metabolites,* Academic Press, New York, 1971.
4. **Stecher, P. G., Ed.,** *The Merck Index,* 9th ed., Merck & Company, Rahway, N.J., 1977.
5. **Shibata, S., Natori, S., and Udagawa, S.,** *List of Fungal Products,* Charles C Thomas, Springfield, Ill., 1964.
6. **Devon, T. K. and Scott, A. I.,** *Handbook of Naturally Occurring Compounds,* Academic Press, New York, 1972.
7. **Ciegler, A., Kadis, S., and Ajl, S. J., Eds.,** *Microbial Toxins,* Vol. 1 to 7, Academic Press, New York.
8. **Foster, J. W.,** *Chemical Activities of Fungi,* Academic Press, New York, 1949.
9. **Thompson, R. H.,** *Naturally Occurring Quinones,* Academic Press, New York, 1971.
10. **Dean, F. M.,** *Naturally Occurring Oxygen Ring Compounds,* Butterworths, London, 1963.
11. **Culberson, C. F.,** *Chemical and Botanical Guide to Lichen Products,* University of North Carolina Press, Chapel Hill, 1969.

B. Special Textbooks

1. Chemistry

1. **Nakanishi, K., Goto, T., Ito, S., Natori, S., and Nozoe, S., Eds.,** *Natural Products Chemistry,* Vol. 1 and 2, Kodansha Ltd. and Academic Press, Tokyo-New York, 1975.

2. **Coffey, S., Ed.,** *Rodd's Chemistry of Carbon Compounds,* Elsevier, Amstedam, 1967.
3. **Geissman, T. A. and Crout, D. H. G.,** *Organic Chemistry of Secondary Plant Metabolism,* Freeman, Cooper & Company, San Francisco, 1969.
4. **Ollis, W. D., Ed.,** *Recent Developments in the Chemistry of Natural Phenolic Compounds,* Pergamon Press, Oxford, 1961.
5. **Pigman, W. and Horton, D.,** *The Carbohydrates. Chemistry and Biochemistry,* 2nd ed., Academic Press, New York, 1970.
6. **Jeanloz, R. W., Ed.,** *The Amino Sugars,* Academic Press, New York, 1969.
7. **Asahina, Y. and Shibata, S.,** *Chemistry of Lichen Substances,* Japan Society for Promotion of Science, Tokyo, 1954.

2. Biosynthesis and Microbiology

8. **Bu'Lock, J. D.,** *The Biosynthesis of Natural Products,* McGraw-Hill, London, 1965.
9. **Bu'Lock, J. D.,** *Essays in Biosynthesis and Microbial Development,* John Wiley & Sons, London, 1967.
10. **Bernfeld, P., Ed.,** *Biogenesis of Natural Compounds,* Pergamon Press, Oxford, 1967.
11. **Grisebach, H.,** *Biosynthetic Patterns in Microorganisms and Higher Plants,* John Wiley & Sons, New York, 1967.
12. **Sykes, G. and Skinner, F. A.,** *Actinomycetales: Characteristics and Practical Importance,* Academic Press, New York, 1973.
13. **Rainbow, C. and Rose, A. H.,** *Biochemistry of Industrial Microorganisms,* Academic Press, New York, 1963.

3. Mechanism of Action

14. **Sexton, W. A.,** *Chemical Constitution and Biological Activity,* E & FN Spon Ltd., London, 1963.
15. **Blood, F., Ed.,** *Essays in Toxicology,* Academic Press, New York, 1970.
16. **Hochster, R. M., and Quastel, J. H., Eds.,** *Metabolic Inhibitors,* Academic Press, New York, 1964.
17. **Goodman, L. S. and Gilman, A.,** *The Pharmacological Basis of Therapeutics,* 3rd ed., MacMillan, New York, 1965.
18. **Meynell, G. G.,** *Bacterial Plasmids,* MIT Press, Cambridge, Mass., 1973.

C. Periodicals (Review Journals)

	Abbreviations
1. *Zechmeister's Forschritte der Organische Chemie Naturstoffe* (Herz, W., Grisebach, H., and Kirby, G. W., Eds., Springer-Verlag, Vienna)	*Forschr.*
2. *Advances in Applied Microbiology* (Perlman, D., Ed., Academic Press, New York)	*Adv. Appl. Microb.*
3. *Progress in Industrial Microbiology* (Hockenhull, D. J. D., Ed., Churchil Livingstone, Edinburgh)	*Progr. Ind. Microb.*
4. *Advances in Carbohydrate Chemistry and Biochemistry* (Tipson, R. S. and Horton, D., Eds., Academic Press, New York)	*Adv. Carb. Chem.*
5. *Annual Reviews in Biochemistry*	*An. Rev. Bioch.*
6. *Annual Reviews in Microbiology*	*An. Rev. Microb.*
7. *Progress in Medicinal Chemistry* (Ellis, G. P. and West, G. B., Eds., Elsevier, Amsterdam)	*Progr. Med. Chem.*

D. Journals

1. General

Annales da Academia Brasileira de Ciencias	*An. Acad. Brasil.*
Annals of the New York Academy of Sciences	*An. N.Y. Acad. Sci.*
Canadian Journal of Research, Section E: Medical Science	*Can. J. Res. Sect. E*
Comptes Rendus Hebdomadaires des Seances de l' Academie des Sciences, Serie D: Sciences Naturelles	*CR Ser. D*
Current Science	*Curr. Sci.*
Doklady Akademii Nauk SSSR	*Dokl.*
Experientia	*Exp.*
Izvestiya Akademii Nauk SSSR, Seriya Biologicheskaya	*Izv. Ser. Biol.*
Izvestiya Akademii Nauk SSSR, Seriya Khimicheskaya	*Izv. Ser. Khim.*
Journal of Scientific and Industrial Research, Section C: Biological Sciences	*J. Sci. Ind. Res. Sect. C*

Nature (London)	*Nature*
Naturwissenschaften	*Naturwiss.*
Pakistan Journal of Scientific and Industrial Research	*Pak. J. Sci. Ind. Res.*
Proceedings of the Japan Academy	*Proc. Jap. Acad.*
Proceedings of the National Academy of Sciences of the United States of America	*Proc. Nat. Acad. Sci.*
Science	*Sci.*
Scientia Sinica (Hua Hsueh Pao)	*Sci. Sinica*

2. Chemistry

Acta Chemica Scandinavica	*Acta Chem. Scand.*
Acta Chimca Sinica	*Acta Chim. Sinica*
Analytical Chemistry	*Anal. Chem.*
Angewandte Chemie	*Angew.*
Arkiv fUr Kemi	*Ark. Kemi*
Australian Journal of Chemistry	*Aust. J. Chem.*
Bulletin de la Societe Chimique de Belgique	*Bull. Soc. Chim. Belg.*
Bulletin de la Societe Chimique de France	*Bull. Soc. Chim. Fr.*
Bulletin of the Chemical Society of Japan	*Bull. Ch. Soc. Jap.*
Canadian Journal of Chemistry	*Can. J. Chem.*
Carbohydrate Research	*Carb. Res.*
Chemical Communications — Journal of the Chemical Society, Series D (formerly Proceedings of the Chemical Society, to 1969)	*CC*
Chemical Letters	*Chem. Lett.*
Chemicke Zvesti	*Chem. Zv.*
Chemische Berichte	*Ber.*
Chemistry & Industry (London)	*Chem. & Ind.*
Chimia (Basel)	*Chim. (Basel)*
Collection of Czechoslovak Chemical Communications	*Coll.*
Gazzetta Chimica Italiana	*Gaz.*
Helvetica Chimica Acta	*Helv.*
Heterocycles	*Heterocycl.*
Indian Journal of Chemistry	*Ind. J. Chem.*
Journal of the American Chemical Society	*JACS*
Journal of the Chemical Society (London)	*JCS*
Journal of the Chemical Society, Section C: Organic	*JCSC*
Journal of the Chemical Society, Perkin Transactions I: Organic and Bio-Organic Chemistry	*JCS Perkin I*
Journal of Chromatography	*J. Chrom.*
Journal of Heterocyclic Chemistry	*J. Heterocycl. Chem.*
Journal of the Indian Chemical Society	*J. Ind. Ch. Soc.*
Journal of Organic Chemistry	*JOC*
Liebig's Annalen der Chemie	*Liebigs Ann.*
Monatshefte fUr Chemie	*Monatsh.*
Recueil des Travaux Chimiques des Pays-Bas	*Rec.*
Suomen Kemistilehti	*Suomen Kem.*
Svensk Kemisk Tidskrift	*Svensk Kem. Tid.*
Tetrahedron	*Tetr.*
Tetrahedron Letters	*TL*
Zeitschrift fUr Chemie	*Z. Chem.*

3. Microbiology, Bacteriology, Pathology, Fermentation

Acta Microbiologica Hungarica	*Acta Micr. Hung.*
Acta Microbiologica Polonica	*Acta Micr. Pol.*
Acta Microbiologica Sinica (Wei Sheng Wu Hsueh Pao)	*Acta Micr. Sinica*
Biotechnology and Bioengineering	*Biotech. Bioeng.*
Annales de l'Institute Pasteur (Paris)	*Ann. Pasteur*
Annali di Microbiologia et Enzymologia	*Ann. Micr. Enzym.*
Antoine van Leeuwenhoek, Journal of Microbiology and Serology	*J. Leeuwenhoek*
Applied and Environmental Microbiology (formerly Applied Microbiology)	*Appl. Micr.*
Archiv fUr Mikrobiologie	*Arch. Mikr.*
Bacteriological Proceedings	*Bact. Proc.*

Bacteriological Reviews	*Bact. Rev.*
British Journal of Experimental Pathology	*Brit. J. Exp. Path.*
Canadian Journal of Microbiology	*Can. J. Micr.*
Developments in Industrial Microbiology	*Dev. Ind. Micr.*
European Journal of Applied Microbiology	*Eur. J. Appl. Micr.*
Folia Microbiologica	*Folia Micr.*
Giornale di Microbiologia	*Giorn. Micr.*
Japanese Journal of Bacteriology (Nippon Saikug. Zashi)	*Jap. J. Bact.*
Japanese Journal of Microbiology	*Jap. J. Micr.*
Journal of Applied Bacteriology	*J. Appl. Bact.*
Journal of Bacteriology	*J. Bact.*
Journal of Fermentation Technology (Hakko Kagaku Kaishi)	*J. Ferm. Techn.*
Journal of General Microbiology	*J. Gen. Micr.*
Medical Microbiology and Immunology	*Med. Microb.*
Mikrobiologichnii Zhurnal (Kiev)	*Mikrob. Zh.*
Mikrobiologiya (Moscow)	*Mikrob.*
Mycologia	*Mycol.*
Mycopathologia & Mycologia Applicata	*Mycopath.*
Postepy Higieny i Medycyny Doswiadczalnej	*Med. Dosw.*
Prikladnaya Biokhimiya i Mikrobiologiya	*Prikl. Biokh. Mikr.*
Transactions of the British Mycological Society	*Trans. Brit. Mycol. Soc.*
Zentralblatt für Bakteriologie, Parasitenkunde, Infectionkrankheiten und Hygiene, Abteilung: Originale	*Zbl. Bakt. Parasit.*
Zertschrift für Allgemeine Mikrobiologie	*Z. Allg. Mikr.*
Zhurnal Microbiologii, Epidemiologii i Immunologii	*Zh. Micr. Epid. Imm.*

4. Pharmaceutical Chemistry, Pharmacology, and Natural Products

Annales Pharmaceutiques Francoises	*Ann. Farm. Franc.*
Archiv für Pharmazei	*Arch. Pharm.*
Arzneimittelforschung	*Arzn. Forsch.*
Bioorganic Chemistry	*Bioorg. Chem.*
Bioorganicheskaya Khimiya	*Bioorg. Khim.*
Chemical Pharmaceutical Bulletin	*Chem. Ph. Bull.*
Dissertation Pharmacy	*Diss. Pharm.*
Il Farmaco, Edizione Scientifica (Pavia)	*Farmaco, Sci.*
Il Farmaco, Edizione Practica (Pavia)	*Farmaco, Pract.*
Indian Journal of Pharmacy	*Ind. J. Pharm.*
Journal of Medicinal Chemistry	*J. Med. Chem.*
Journal of Pharmaceutical Sciences	*J. Pharm. Sci.*
Journal of Pharmaceutical Society Japan (Yakugaku Zasshi)	*J. Ph. Soc. Jap.*
Journal of Pharmacy and Pharmacology	*J. Pharm. Pharmacol.*
Khimicheskaya Promyslennost, Khimiko-Farmatsevticheskii Zhurnal	*Khim. Prom.*
Khimiya Prirodnykh Soedinenii	*Khim. Prir. Soed.*
Lloydia (Journal of Natural Products)	*Lloydia*
Phytochemistry	*Phytoch.*
Polish Journal Pharmacy and Pharmacology	*Pol. J. Pharm. Pharmacol.*
(Die) Pharmazie	*Pharm.*
Zeitschrift für Naturforschung Teil B: Inorganic and Organic Chemistry	*Z. Naturforsch. Ser. B*
Zeitschrift für Naturforschung Teil C: Biosciences	*Z. Naturforsch. Ser. C*

5. Biochemistry, Physiology, Biology

Agricultural and Biological Chemistry	*Agr. Biol. Ch.*
Anais do Sociedade de Biologia da Pernambuco	*Anais Biol. Pernambuco*
Annals of Applied Biology	*Ann. Appl. Biol.*
Archives of Biochemistry and Biophysics	*ABB*
Biochimica and Biophysica Acta	*BBA*
Biochemical and Biophysical Research Communications	*BBRC*
Biochemical Journal (London)	*Bioch. J.*
Biochemical Pharmacology	*Bioch. Pharm.*
Biochemical Society Transactions	*Bioch. Soc. Trans.*

Biochemistry	*Biochem.*
Biochemische Zeitschrift	*Bioch. Z.*
Biologica (Bratislava)	*Biol. (Bratislava)*
Biologicheskie Nauki (Moscow)	*Biol. Nauk.*
Bolletino della Societa Italiane di Biologica Sperimentale	*Boll. Soc. Ital. Biol.*
Bulletin de la Societe de Chimie Biologique	*Bull. Soc. Chim. Biol.*
Comptes Rendus des Seances de la Societe de Biologie et de Ses Filiales	*CR Soc. Biol.*
European Journal of Biochemistry	*Eur. J. Bioch.*
Federation Proceedings	*Fed. Proc.*
FEBS Letters	*FEBS Lett.*
Hoppe Seyler's Zeitschrift für Physiologische Chemie	*Hoppe Seyler*
Indian Journal of Biochemistry	*Ind. J. Bioch.*
Indian Journal of Experimental Biology	*Ind. J. Exp. Biol.*
Journal of Biochemistry (Tokyo)	*J. Bioch. (Tokyo)*
Journal of Biological Chemistry	*J. Biol. Chem.*
Journal of Cell Physiology	*J. Cell Physiol.*
Life Sciences	*Life Sci.*
Marine Biology	*Marine Biol.*
Molecular Biology	*Mol. Biol.*
Molecular Pharmacology	*Mol. Pharm.*
Process Biochemistry	*Proc. Bioch.*
Rivista de Biologia	*Riv. Biol.*

6. Chemotherapy, Clinical

American Review of Tuberculosis	*Am. Rev. Tub.*
Archivum Immunologie et Therapiae Experimentalis (Wroclaw)	*Arch. Immun.*
Cancer Chemotherapy Reports from 1976	*Canc. Chemoth. Rep.*
Cancer Treatment Reports	*Canc. Tmt. Rep.*
Cancer Research	*Cancer Res.*
Chemotherapy (Basel)	*Chemother.*
Chemotherapy (Tokyo)	*Chemother. (Tokyo)*
Gann	*Gann*
Indian Journal of Medical Research	*Ind. J. Med. Res.*
Japanese Journal of Experimental Medicine	*Jap. J. Exp. Med.*
Japanese Journal of Medical Science & Biology	*Jap. J. Med. Sci. Biol.*
Japan Medical Gazetta	*Jap. Med. Gaz.*
Japanese Medical Journal	*Jap. Med. J.*
Journal of the American Medical Association	*JAMA*
Journal of Clinical Investigations	*J. Clin. Invest.*
Journal of Experimental Medicine	*J. Exp. Med.*
Journal of Infectious Diseases	*J. Inf. Dis.*
Presse Medica	*Presse Med.*
Proceedings of the Society for Experimental Biology & Medicine	*Proc. Soc. Exp. B. M.*
Rassagne Medica	*Rass. Med.*
Revue Internationale d'Oceanographie Medicale	*Rev. Int. Oceanogr.*

7. Botanical, Agriculture

Annales Phytopathological Society Japan	*Ann. Phytop. Soc. Jap.*
Botanical Gazetta	*Bot. Gaz.*
Canadian Journal of Botany	*Can. J. Bot.*
Indian Journal of Phytopathology	*Ind. J. Phyt.*
Japanese Journal of Botany	*Jap. J. Bot.*
Journal of Agricultural Chemical Society of Japan	*J. Agr. Chem. Soc. Jap.*
Journal of Agricultural and Food Chemistry	*J. Agr. Food Chem.*
Physiologica Plantarum	*Physiol. Plant.*
Physiological Plant Pathology	*Physiol. Plant Path.*
Phytopathologische Zeitschrift	*Phytopath. Z.*
Phytopathology	*Phytopath.*
Plant Disease Reporter	*Plant Dis. Rep.*
Plant Physiology	*Plant Phys.*
Plant Science Letters	*Plant Sci. Lett.*
Planta Medica	*Planta Med.*

Rastitel'nye Resursy	*Rast. Res.*
Science & Culture	*Sci. & Cult.*
South African Journal of Agricultural Sciences	*S. Afr. J. Agr. Sci.*

8. Report Journals

Annual Report Takeda Research Laboratories (Takeda Kenkyusho Ho)	*An. Rep. Takeda*
Annual Reports Sankyo Co. (Sankyo Kenkyusho Nempo)	*An. Rep. Sankyo*
Annual Reports Shionogi Co. (Shionogi Kenkyusho Nempo)	*An. Rep. Shionogi*
Annual Reports, Institute of Food Microbiology, Chiba University (Chiba)	*An. Rep. Chiba Univ.*
Bulletin Faculty Meiji University	*Bull. Fac. Meiji*
Journal of the National Cancer Institute	*J. Nat. Cancer Inst.*
Kitasato Archives of Experimental Medicine	*Kitasato Arch.*
Scientific Reports, Meiji Pharmaceutical Co. (Meiji Seika Kenkyusho Nempo)	*Sci. Rep. Meiji*
Tanabe Seiyaku Kenkyu Nempo	*Tanabe Seiyaku*
Tohoku Journal of Experimental Medicine	*Tohoku J. Exp. Med.*

9. Abstract Journals

Biological Abstracts	*BA*
Chemical Abstracts	*CA*
Dissertation Abstracts, International Section B	*Diss. Abst.*
Microbiological Abstracts	*Micr. Abst.*

E. Patents

Belgian patent	*Belg. P*
British patent	*BP*
Canadian patent	*Can. P*
Czechoslovakian patent	*Cz. P*
Dutch (Holland) patent	*Holl. P (year/number)*
East German patent	*DDR P*
European patent	*EP*
French patent	*Fr. P*
German patent	*DT*
Hungarian patent	*Hung. P*
Indian patent	*Ind. P*
Japanese Patent (kokai)	*JP (year/number)*
Polish patent	*Pol. P*
Soviet (USSR) patent	*SU P*
Swiss patent	*Swiss P*
USA patent	*USP*

USING THE INDICES

Each volume will contain a general name index, including "identical with" and trade names, and an index of producing organisms. A separate index volume (Volume X) covering name, formula, producing organisms, molecular weight, elemental analysis, optical rotation, chemical type, antitumor/antiviral activity and the utility of the compounds will also be published.

Each listing in the indices directs the reader to the *compound number*. The compound number covers the antibiotic code number and the sequence number, separated by a hyphen, and this numerical identifier refers to the volumes in which the compounds are arranged according to this number, except the plant and animal products, which are listed separately in Volumes VIII and IX. The sequence number is assigned individually to any compound according to the addition of a new entry to the data base. Compounds with the same antibiotic code number are listed in numerical order by sequence number.

Similar types of compounds are listed in the same volume. If someone knows the chemical type of a compound, he will find it accordingly among its relatives on the basis of the chemical type key (Table 2) or code number index. When only the trivial, patent, trade, or chemical name is known, or when the name is restricted in common local use (in a particular region of the world), then it may easily be located in the alphabetical cross index. In this index almost 7000 names are listed.

If someone is interested in the active metabolic products of a given organism (species or genus), he only has to turn to the index of the producing organisms, which contains the full reference.

If a "new" compound was isolated and there are sufficient data known about this compound, e.g., mass spectrometric molecular weight, UV spectra, rotation, some activities, etc., on the basis of these indices it is relatively easy to recognize the similar or perhaps identical compounds.

This Handbook, as is evident from the preceding explanation, has a strong chemical character. It seemed to be very complicated, in contrast to the exact physical and chemical data, to formulate the special microbiological, taxonomic, chemotherapeutic, pharmacological, and clinical data in a standardized and computer-searchable format.

During the compilation and the editing of this work great care has been taken to assure the accuracy of the information; however there is a possibility that some mistakes do exist in the values. The Editor and the Publisher cannot be responsible for errors in the original publications.

It is recognized, however, that all data are the subject of continuous revision; therefore special care has been taken to collect and carefully select all new information related to any known compounds and add or occasionally replace them in the existing data.

The work of editing was finished at the end of 1977, but some important new data have been included recently (1979).

LIST OF ABBREVIATIONS

General

Simple chemicals are abbreviated by the formula (NaCl, HCl, NaOH, NH_4OH, NH_4Cl, HCOOH, CH_2Cl_2, etc.). Generally the *Chemical Abstracts* abbreviations are used.

Solvents

W	Water	Acet	Acetone
Pyr	Pyridine	Benz	Benzene
		Tol	Toluene
MeOH	Methanol	Et_2O	Diethyl ether
EtOH	Ethanol	Hex	Hexane, petroleter
PrOH	n-Propanol	AcOH	Acetic acid (glacial)
BuOH	n-Butanol	AcCN	Acetonitrile
AmOH	Amyl alcohol	DMSO	Dimethylsulfoxide
EtOAc	Ethyl acetate	DMF	Dimethylformamide
CHL	Chloroform	THF	Tetrahydrofurane

Other Chemicals

i-	iso-	Me-	Methyl-
i-PrOH	Isopropanol	Et-	Ethyl- as in Me-Et-Ketone
t-	tertier	Bu-	Butyl- as in di-Bu-ether
t-BuOH	tertier-Butanol		
c-	cyclo-	NH_4	Ammonium
Ac	Acetyl, acetate	PTSA	*p*-Toluene sulfonic acid
NH_4OAc	Ammonium acetate		

Absorbents

Cel	Cellulose
Pap.	Paper
SILG	Silica gel, SiO_2-xH_2O
AL	Aluminum oxide, Al_2O_3
Carbon	Carbon (Norit A)
Diatom	Kieselguhr, etc.
XAD-2	Amberlite XAD-2

IRC-50-H	Amberlite IRC-50 resin in hydrogene form
IR-120-Na	Amberlite IR-120 resin in sodium form
$CG\text{-}50\text{-}NH_4$	Amberlite CG-50 resin in ammonium form
XE-	Amberlite XE resins
Dowex-1-OH	Dowex-1 resin in hyroxyl form
| (vertical bar)	±
+ (in producing organism)	Addition of precursors or other compounds to the medium (directed fermentations, mutational biosynthesis with idiothrops)

Special

Some abbreviations are listed in "How to Use This Handbook" (qualitative reactions, antimicrobial activity and antitumor, antiviral activity)

Producing Organisms

S.	*Streptomyces*
Act.	*Actinomyces*
Noc.	*Nocardia*
Mic.	*Micromonospora*
Stv.	*Streptoverticillium*
B.	*Bacillus*
Ps.	*Pseudomonas*
P.	*Penicillium*
Asp.	*Aspergillus*
Fus.	*Fusarium*
Cep.	*Cephalosporium*
sp.	*species*
PL	*Plant products*
AN	*Animal products*

Chemical Type

Deriv.	Derivative
t.	Type
l.	like

Color and Appearance

Pow.	Powder, amorphous substance
Wh.	White
Cryst.	Crystalline

Toxicity

IV	Intravenous
IP	Intraperitoneal
SC	Subcutaneous
IM	Intramuscular
PEROS	Per os

Isolation Methods

Filt.	Filtered fermentation broth
Orig.	Original pH
Mic.	Mycelium
WB.	Whole broth
Evap.	Evaporated
Cryst.	Crystallization
Prec.	Precipitation
Liof.	Lyophilization

Microbial Metabolites

Section I
New and Corrected Data for Compounds Existing in Volumes I Through VII

SECTION I

NEW DATA FOR OLD COMPOUNDS

In this section compounds having either new data or new references are arranged in the same form as in the Main Volumes. The compounds listed are headed by their Compound Number (Antibiotic Number and Sequence Number) and Basic Name. The new data covered in this section are (if any):

1. New antibiotic code number (AN)
2. New names, trade names. Identities (Identical with:)
3. New producing species
4. Some important new physical, chemical and/or biological data. New practical applications (Utility)
5. New or missing references
6. New or corrected chemical structures

Any changes in the existing data and data missing from the Main Volumes are usually included here. Errata have been inserted for errors as if they were new data. Data to be deleted are indicated in the same line after the correct data between slashes (/.../).

Compounds with changed Antibiotic Code Number are marked by an asterisk and are listed according to the new classification number. Old Antibiotic Numbers are always given in reference to the Main Volumes. A few compounds were moved to another family, which would mean moving to another Volume. In most cases compounds were moved from the miscellaneous group of antibiotics (Volume VII) to another well defined Family according to their recently discovered, or modified, structures.

Attempts have been made to list all new names, synonym names and trade names of compounds as they come to our attention. Special care has been taken to indicate the identity of all newly proved structures with the compounds listed in the second part of this supplement. In some cases the basic name of a compound is also changed and is indicated by underlining. The names of compounds and new sources of antibiotics — the new producing species — are always listed. In some cases wrong spellings are corrected.

Efforts have been made to correct the originally published empirical formulae and/or molecular weights according to recent publications. Sometimes, if only approximate experimental values were reported, the molecular weight data were calculated. Some important newly published physical and or biological data (OR, UV, TO, TV, LD_{50}), particularly if these data were not included in the Main Volumes, are also enlisted. New practical applications are always noted.

Special care has been taken to list all new references concerning old compounds. Some mistakes or missing data occurring in the Main Volumes are compensated here. Changes in journal names and the list of titles and abbreviations of some new periodicals and monographs are covered in Table 2.

The newly recognized or modified chemical structures (altogether 116 new structures are listed) are indicated in the text after the listing of new references by STR-NEW: (new structure), STR-CORR: (corrected structure). The drawings of structures are inserted before the subfamilies of compounds and are titled by their basic names or sequence numbers. The structural corrections are frequently only stereochemical alterations. These minor changes are not listed in all known cases. Besides the structural corrections proposed in the literature under STR-CORR: some mistakes occurring in the Main Volumes are also repaired here.

Table 1
HANDBOOKS AND TEXTBOOKS

1. **Umezawa, H., Ed.,** *Index of Antibiotics from Actinomycetes,* Vol. 2, Japan Scientific Societies Press, Tokyo, 1978.
2. **Korzybski, T., Kowszyk-Gindifer, Z., and Kurylowitz, W.,** *Antibiotics: Origin, Nature and Properties,* Vols. 1, 2, and 3, American Microbiological Society, Washington, 1980.
3. **Hahn, F. E., Ed.,** *Antibiotics,* Vol. 5. Parts 1 and 2, Springer Verlag, Berlin, 1979.
4. **Glasby, J. S.,** *Encyclopedia of Antibiotics,* 2nd ed., John Wiley & Sons, London, 1980.
5. **Sammes, P. G., Ed.,** *Topics in Antibiotic Chemistry,* Vols. 2, 3, 4, and 5, Elis Horwood Ltd., Chicester, 1978—1980.
6. **Weinstein, M. J. and Wagman, G. H., Eds.,** *Antibiotics: Isolation, Separation and Purification,* J. Chrom. Library, Vol. 15, Elsevier, Amsterdam, 1978.
7. **Gialdroni Grassi, G. and Sabath, L. D., Eds.,** *New Trends in Antibiotics: Research and Therapy,* Elsevier, Amsterdam, 1981.
8. **Aszalos, A., Ed.,** *Antitumor Compounds of Natural Origin,* Vols. 1 and 2, CRC Press, Boca Raton, Fla., 1981.
9. **Cassady, J. M. and Douros, J. D., Eds.,** *Anticancer Agents Based on Natural Product Models,* Academic Press, New York, 1980.
10. **Rinehart, K. L. and Tsuami, T., Eds.,** *Aminocyclitol Antibiotics,* ACS Symp. Ser. 125, American Chemical Society, Washington, 1980.
11. **Remers, W. A.,** *The Chemistry of Antitumor Antibiotics,* John Wiley & Sons, New York, 1980.
12. **Crooke, S. T. and Reich, S. D., Eds.,** *Anthracyclines: Current Status and New Developments,* Academic Press, New York, 1980.
13. **Petit, G. R. and Grogg, G. M.,** *Biosynthetic Products for Cancer Chemotherapy,* Vol. 2, Plenum Press, New York, 1978.
14. **Petit, G. R. and Ode, R. H.,** *Biosynthetic Products for Cancer Chemotherapy,* Vol. 3, Plenum Press, New York, 1979.

Table 2
PERIODICALS AND JOURNALS

Periodicals	
Annual Reports on Fermentation Processes (Ed.: Perlman, D., Tsao, G. T., Academic Press, New York)	*Ann. Rep. Ferm. Proc.*
Microbiology — 1979 (Ed: Schlesinger, D., Am. Soc. Microbiol., Washington D.C.)	*Microbiol.*
Recent Results in Cancer Research (Springer Verlag, Berlin)	*Rec. Res. Cancer Res.*
Pure and Applied Chemistry (Pergamon Press, Oxford)	*Pure Appl. Chem.*
Drugs and Food from the Sea (Proc. Food-Drugs from the Sea, Symposium Series, Marine Techn. Soc., Washington DC)	*Drugs-Food Sea*
Cancer and Chemotherapy (Academic Press, New York)	*Cancer Chemoth.*
Journals	
Biomedical Mass Spectroscopy	*Biomed. Mass Spectr.*
Biotechnological Letters	*Biotechn. Lett.*
Cancer Chemotherapy and Pharmacology	*Cancer Chem. Pharm.*
Chung Ts'ao Yao	
Current Microbiology	*Curr. Microbiol.*
Drugs under Experimental Clinical Research	*Drugs Exp. Clin. Res.*
Egypt Journal of Microbiology	*Egypt J. Micr.*
Journal of Antimicrobial Chemotherapy	*J. Antimicr. Chem.*
Journal of Chemical Technology and Biotechnology	*J. Chem. Techn. Biotech.*
Journal of Liquid Chromatography	*J. Liquid Chrom.*
Microbiological Reviews (formerly Bacteriological Reviews)	*Microb. Rev.*
Microbios	
Neoplasma	
Toxicon	
Yao Hsueh Hsueh Pao	
Yao Hsueh T'ung Pao	

Carbohydrate Antibiotics

11
PURE SACCHARIDES

111 Mono- and Oligosaccharide Types
112 Polysaccharides

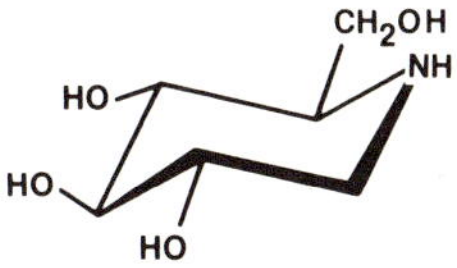

Nojirimycin-B

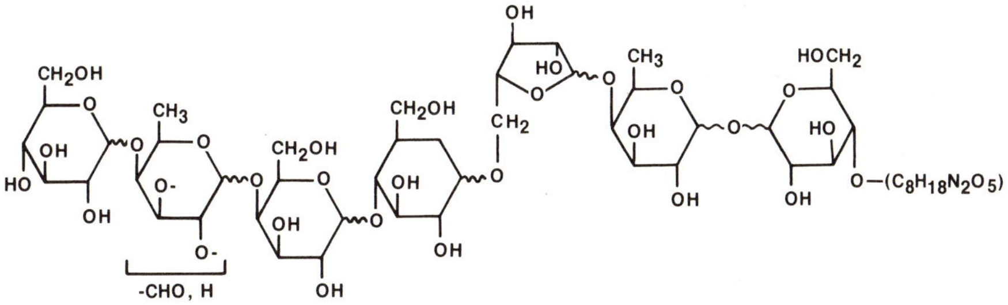

K-52-B

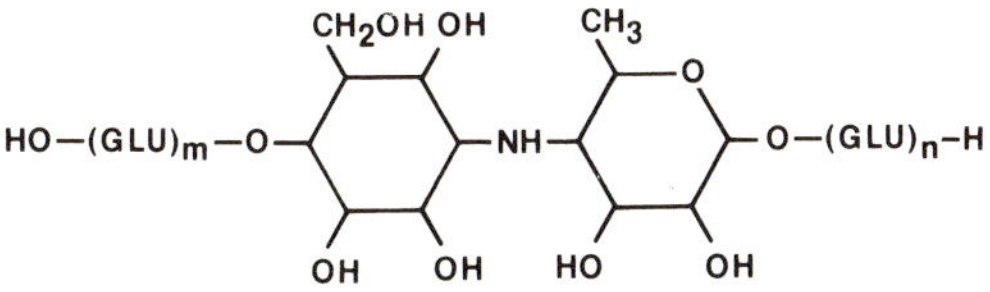

m + n

Oligostatin E
Oligostatin D

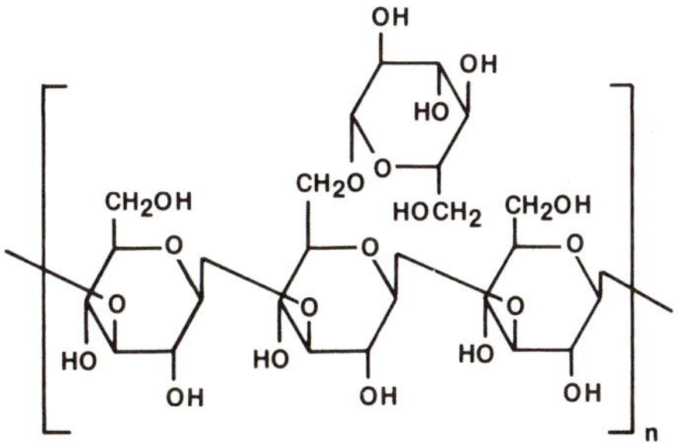

Schysophyllan

11121-5564

NAME: NOJIRIMYCIN-B
IDENTICAL: SG-1
PO: S.LAVENDULAE-TREHALOSTATICUS
FORMULA: C6H13NO4 /C6H13NO15/
MW: 164
STR-NEW:
REFERENCES:

Agr. Biol. Ch., 44, 219, 1980; *Tetr.*, 23, 2125, 1968; *Naturwiss.*, 66, 584, 1979; Belg. P 862165

11132-5843

NAME: K-52-B
MW: 1180 /770/
UTILITY: ON CLINICAL TRIAL
STR-NEW:
REFERENCES:

Agr. Biol. Ch., 44, 1863, 2073, 1980, 45, 641, 645, 1981

11132-6178

! NAME: OLIGOSTATIN-E
STR-CORR:
REFERENCES:

DT 3035193; *JA*, 34, 1424, 1429, 1981

11132-6179

! NAME: OLIGOSTATIN-D
STR-CORR:
REFERENCES:

DT 3035193; *JA*, 34, 1424, 1429, 1981

11140-0012

NAME: STREPTOZOTOCIN
TRADE NAMES: ZANOSAR
UV: MEOH: (228, , 6360)(380, , 136)
REFERENCES:

JA, 32, 379, 1979; *JOC*, 44, 9, 1979; *Abst. AAC*, 18, 168, 1978

11140-0013

NAME: PRUMYCIN
PO: B.CEREUS
TV: P-388, CA
REFERENCES:
Carb. Res. 60, 75, 1978; *JA*, 32, 347, 1979

11210-0019

NAME: LENTINAN
REFERENCES:
Exp., 35, 409, 1979

11210-0020

NAME: G-2
REFERENCES:
Acta Med. Univ., Kagoshima, 20, 1, 1973; 25, 209, 1978

11210-0023

NAME: NO NAME
REFERENCES:

11210-0025

NAME: SCHYSOPHYLLAN, SCHYZOPHYLLAN
UTILITY: ANTITUMOR DRUG
STR-NEW:
REFERENCES:
Carb. Res., 89, 121, 1981; *Showa Igakku Zasshi*, 40, 193, 1980; *CA*, 94, 41360

11210-6181

NAME: GU-1 (GU-2, GU-3)
REFERENCES:
Mie Med. J., 23, 117, 1973

11210-6182

NAME: GU-4
REFERENCES:
Mie Med. J., 23, 117, 1973

11220-0044

NAME: KRESTIN, POLYSACCHARIDE-K
REFERENCES:
Gann, 66, 365, 1975; 69, 255, 699, 1978; *Proc. Jap. Cancer Ass.*, 32, 282, 1973; DT 2659808; Belg. P 852987

11220-6485

NAME: KS-2-A, "PEPTIDOMANNAN"
MW: 9000|300 /8000|3000/
LD50: (875, IV)
TV: INFL.
UTILITY: ON CLINICAL TRIAL
REFERENCES:
JP 80/27101; *CA*, 93, 236923

* 11240-4084 /11200/

NAME: H-2609, H-2075 /SOEDOMYCIN/
IDENTICAL: SOEDOMYCIN
REFERENCES:
Nippon Acta Radiol., 26, 1483, 1492, 1967; *Boei Eisei*, 18, 5, 1971; *Proc. 9th Int. Cancer Congr.*, 1966, p. 332

12
AMINOGLYCOSIDE ANTIBIOTICS

121 Streptamine Derivatives
122 2-Deoxystreptamine Derivatives
123 Inositol- Inoseamine Derivatives
124 Other Aminocyclitols
125 Aminohexitols

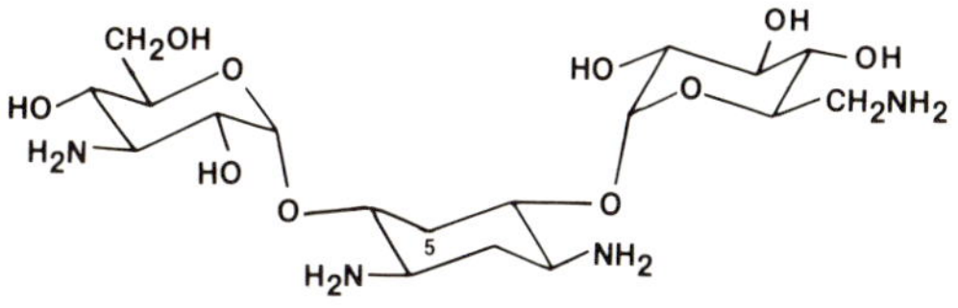

5-Deoxykanamycin

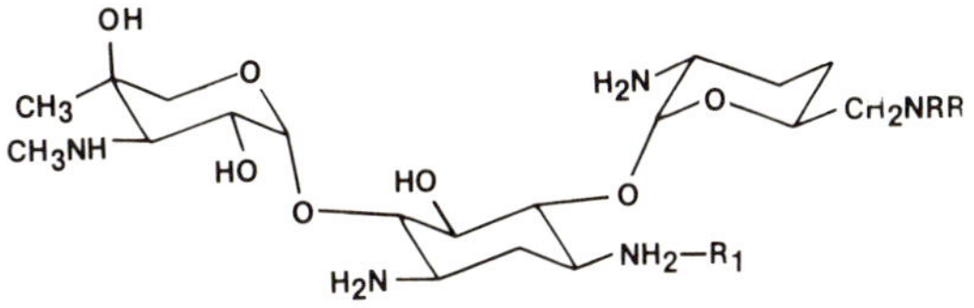

	R	**R_1**
XK-62-3	H	CH_3
XK-62-4	CH_3	H

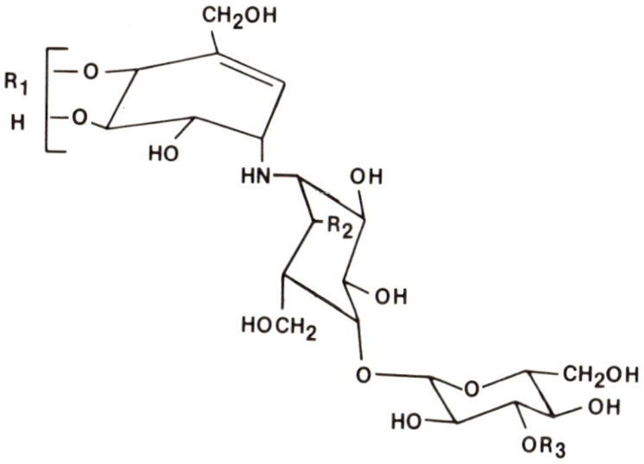

	R_1	**R_2**	**R_3**	
Validamycin A	H	H	H	
Validamycin B	H	OH	H	glu: D-glucose
Validamycin C	glu	H	H	
Validamycin D	H	H	glu	

	n
Myomycin B	2
LL-BM-782 α_2	3
LL-BM-782 α_1	4
LL-BM-782 α_{1a}	5

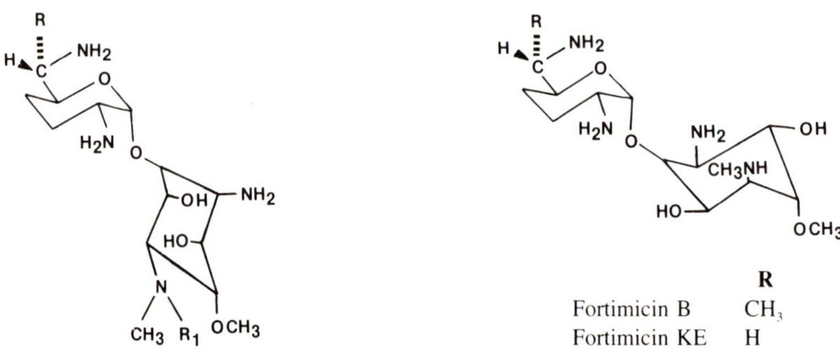

LL-BM-123α

	R
Fortimicin B	CH_3
Fortimicin KE	H

	R	R_1
Fortimicin A	CH_3	$COCH_2NH_2$
Fortimicin C	CH_3	$COCH_2NHCONH_2$
Fortimicin D	H	$COCH_2NH_2$
SF-1854 (Dactimicin-B)	CH_3	$COCH_2NHCHO$

12110-0076

NAME: DIHYDROSTREPTOMYCIN
REFERENCES:
JA, 31, 1233, 1978

12130-0090

NAME: SPECTINOMYCIN
TRADE NAMES: SPECTAM, SPECTOGARD
REFERENCES:
JACS, 101, 5839, 1979; 102, 6817, 1980; *JOC*, 43, 4355, 1978; *Lloydia*, 42, 691, 1979; *Planta Med.*, 34, 345, 1978

* 12130-4043 /00000/

NAME: B-2847-A"
CT: AMINOGLYCOSIDE, SPECTINOMYCIN T.
REFERENCES:

12211-0096

NAME: NEOMYCIN-B
REFERENCES:
JACS, 102, 857, 1980; *Egypt J. Microb.*, 14, 103, 1979; *J. Chrom.*, 211, 223, 1981

12211-0097

NAME: NEOMYCIN-C
REFERENCES:
Bull. Ch. Soc. Jap., 53, 3259, 1980

12212-0119

NAME: RIBOSTAMYCIN
TRADE NAMES: IBISTACIN, RIBOMYCINE, RIBAMICINA, RIBOSTAMIN
IDENTICAL: H-60
PO: MORAXELLA SP., PS.SP.
REFERENCES:
JP 79/84095; *CA*, 91, 156037; JP 81/88800; *CA*, 95, 167132; *JACS*, 103, 5614, 1981; *JOC*, 46, 4298, 1981

12212-0124

NAME: XYLOSTATIN, XYLOSTACIN, XYLOSTAMYCIN
REFERENCES:

12212-0126

NAME: BUTIROSIN-A
REFERENCES:
JA, 32, 18, 891, 1979; *Carb. Res.*, 60, 289, 1979

12212-0127

NAME: BUTIROSIN-B
REFERENCES:
JA, 32, 18, 891, 1979; *Carb. Res.*, 60, 289, 1979

12212-6129

! NAME: 6′-N-METHYLBUTIROSIN-A, /6-N-METHYLBUTIROSIN-A/
REFERENCES:

12212-6346

NAME: 3′.4′-DIDEOXY-6′-C-METHYLBUTIROSIN, DMB-A
/3′.4′-DIDEOXY-6′-C-METHYLBUTIROSIN-B, DMB-A/
REFERENCES:
JA, 31, 247, 1978

12212-6347

NAME: 6′-N-METHYLBUTIROSIN-B
REFERENCES:
JA, 31, 247, 1978

12221-0149

NAME: NEBRAMYCIN COMPLEX
REFERENCES:
DT 2921022; *JA*, 31, 503, 1978; Hung. P 172951; 174315, 176103

12221-0154

NAME: TOBRAMYCIN, 3′-DEOXYKANAMYCIN-B /3-DEOXYKANAMYCIN-B/
TRADE NAMES: GERNEBCIN, OBRACIN, TOBRACIN
REFERENCES:
Antib., 60, 1979; 891, 1980; USP 4032404, 4234685; Hung. P 172951; 174315; 176103

12221-6595

NAME: 5-DEOXYKANAMYCIN, 5-DEOXYKANAMYCIN-A
REFERENCES:
STR-NEW: 5

* 12221-6596 /12211/

NAME: NO NAME

12222-0156

NAME: GENTAMICIN-C1
TRADE NAMES: REFOBACIN, SULMYCIN
PO: MIC.SCALBATINA-RUBRA, MIC.SCALBATINA-PALLIDA, MIC.SAGAMIENSIS-NONREDUCTANS, MIC.PURPUREA-VIOLACEUS, MIC.ECHINOSPORA-PALLIDA, MIC.LONGISPOROFLAVUS
REFERENCES:
Agr. Biol. Ch., 44, 2507, 1980; *Biotech. Bioeng.*, 21, 2045, 1979; *Antib.*, 488, 1981; *JCS Perkin I*, 2137, 2151, 2168, 1981; *Chemother. (Tokyo)*, 29, 482, 1981; BP 2036719; Hung. P 168778; JP 79/160795; *CA*, 92, 162179; 95, 78387

12222-0157

NAME: GENTAMICIN-C2
TRADE NAMES: REFOBACIN, SULMYCIN
PO: MIC.SCALBATINA-RUBRA, MIC.SCALBATINA-PALLIDA, MIC.SAGAMIENSIS-NONREDUCTANS, MIC.PURPUREA-VIOLACEUS, MIC.ECHINOSPORA-PALLIDA, MIC.LONGISPOROFLAVUS
REFERENCES:
Agr. Biol. Ch., 44, 2507, 1980; *Biotech. Bioeng.*, 21, 2045, 1979; *Antib.*, 488, 1981; *JCS Perkin I*, 2137, 2151, 2168, 1981; *Chemother. (Tokyo)*, 29, 482, 1981; BP 2036719; Hung. P 168778; JP 79/160795; *CA*, 92, 162179; 95, 78387

12222-0158

NAME: GENTAMICIN-C1A
TRADE NAMES: REFOBACIN, SULMYCIN
PO: DACTYLSPORANGIUM THAILANDENSE, MIC.SCALBATINA-PALLIDA, MIC.SAGAMIENSIS-NONREDUCTANS, MIC.PURPUREA-VIOLACEUS, MIC.ECHINOSPORA-PALLIDA, MIC.LONGISPOROFLAVUS
REFERENCES:

Agr. Biol. Ch., 44, 2507, 1980; *Biotech. Bioeng.*, 21, 2045, 1979; *Antib.*, 488, 1981; *JCS Perkin I*, 2137, 2151, 2168, 1981; *Chemother. (Tokyo)*, 29, 482, 1981; BP 2036719; Hung. P 168778; JP 79/160795; *CA*, 92, 162179; 95, 78387; JP 80/156592; 81/1892; *CA*, 94, 137804

12222-0159

NAME: SAGAMICIN
TRADE NAMES: MICRONOMYCIN
UTILITY: ANTIBACTERIAL DRUG /ON CLINICAL TRIAL/
REFERENCES:

References for 12222-0159 and 12222-0160 in Volume I should be transposed.

12222-0160

NAME: GENTAMICIN-A
REFERENCES:

JA, 33, 1380, 1980

12222-0163

NAME: GENTAMYCIN-B
REFERENCES:

Arzn. Forsch., 29, 1695, 1979; *CA*, 95, 78387

12222-0165

NAME: GENTMICIN-X2
REFERENCES:

Ber., 114, 322, 1981

12222-0166

NAME: G-418
REFERENCES:

Arzn. Forsch., 29, 1695, 1979; *Am. J. Trop. Med. Hyg.*, 5 (Suppl.), 1089, 1980; *BBRC*, 101, 1031, 1981; *CA*, 94, 96942

12222-0167

NAME: SISOMICIN
TRADE NAMES: BAYMICINE,SISOMINE, SISOLINE
IDENTICAL: G-52-2
PO: MIC.ZINONENSIS, MIC.DANUBIENSIS, MIC.ROSEA, MIC.CYANEOGRAMA, DACTYLOSPORANGIUM THAILANDENSE
REFERENCES:
Holl. P. 80/623; JP 79/62392; 80/104895; 80/156593; Esp. P 461247; BP 2078714; Hung. P T21541; *JCS Perkin I*, 2168, 1981; *CA*, 91, 122150; 93, 184285, 236939

12222-0168

NAME: 66-40-B
REFERENCES:
Ber., 114, 843, 1980

12222-0169

NAME: 66-40-D
REFERENCES:
Ber., 114, 843, 1980

12222-0170

NAME: VERDAMICIN-I
PO: MIC.CYANEOGRANULATA
REFERENCES:
USP 3997524; JP 79/62393; 79/160796; *CA*, 91, 122151

12222-0171

NAME: G-52
PO: MIC.SAGAMIENSIS
REFERENCES:
JP 79/140794; *CA*, 92, 126909

12222-3875

NAME: MUTAMICIN-6
REFERENCES:
AAC, 17, 798, 1979; *Abst. AAC*, 18, 768, 1978; USP 4053591

12222-5309

NAME: 2-HYDROXYGENTAMICIN-C1, WIN-42122-2
TRADE NAMES: IPROMICIN
UTILITY: ON CLINICAL TRIAL
REFERENCES:
AAC, 16, 813, 1979; *Abst. AAC*, 19, 761, 1979

12222-5394

NAME: 2-HYDROXYGENTAMICIN-C2, WIN-42122-2
TRADE NAMES: IPROMICIN
UTILITY: ON CLINICAL TRIAL
REFERENCES:
AAC, 16, 813, 1979; *Abst. AAC*, 19, 761, 1979

12222-6196

NAME: 66-40-G, 3"-N-DEMETHYLSISOMICIN /3"-DE-N-METHYLSISOMICIN/
REFERENCES:

12222-6597

NAME: XK-62-4, 6′-N-METHYLSAGAMICIN
STR-NEW:
REFERENCES:

12222-6598

NAME: XK-62-3, 1-N-METHYLGENTAMICIN-C1A
STR-NEW:
REFERENCES:

12223-4756

NAME: SELDOMYCIN-5
REFERENCES:
USP 4189569, 4214077; Holl. P 77/8324

12230-0173

NAME: DESTOMYCIN-A
IDENTICAL: XK-33-F1
REFERENCES:
Bull. Ch. Soc. Jap., 54, 2147, 1981

12230-0176

NAME: A-396-I
IDENTICAL: A-16316-A
REFERENCES:

12240-0178

NAME: APRAMYCIN
PO: SACCHAROPOLYSPORA HIRSUTA
MW: 539
REFERENCES:
Folia Micr., 16, 205, 1971; *Can. J. Chem.*, 57, 924, 1979; *Eur. J. Bioch.*, 99, 623, 1979; JP 80/102397; Hung. P 172951, 176103, 174315; *CA*, 93, 236935

12240-3877

! NAME: OXYAPRAMYCIN
REFERENCES:

12250-5835

NAME: GARAMINE
REFERENCES:
Hung. P 170416; *CA*, 95, 78387

12250-5837

NAME: GENTAMINE-C2
REFERENCES:
JA, 32, 1357, 1979

12310-0179

NAME: VALIDAMYCIN-A
TRADE NAMES: VALIDACIN
PO: S.HYGROSCOPICUS-LIMONEUS
MW: 497
STR-CORR:
REFERENCES:
JA, 33, 98, 764, 1575, 1980; *Agr. Biol. Ch.*, 44, 143, 1980

12310-5830

NAME: JINGGANGMYCIN-II
REFERENCES:
CA, 93, 236830, 236835

12320-0187

NAME: MINOSAMINOMYCIN
PO: S.AUREOMONOPLODIALES
MW: 618
REFERENCES:

12330-0189

NAME: MYOMYCIN-B
PO: CORYNEBACTERIUM SP.
STR-CORR:
REFERENCES:
JOC, 46, 792, 1981; *CA*, 90, 101945; 93, 232522

12340-0420

NAME: LL-BM-123-A"
STR-CORR:
REFERENCES:
CC, 1134, 1980

12410-0135

NAME: FORTIMICIN-B
STR-CORR:
REFERENCES:

12410-3861

NAME: FORTIMICIN-A, ABBOT-44747, KW-1070
TRADE NAMES: ASTROMICIN
UTILITY: ON CLINICAL TRIAL
STR-CORR:
REFERENCES:
JA, 32, 371, 868, 1979; 33, 1071, 1289, 1980; 34, 1360, 1981; *AAC*, 16, 823, 1979; 18, 761, 766, 773, 1980; *Carb. Res.*, 79, 71, 1980; 85, 61, 1980; 92, 207, 1981; *Chem. Lett.*, 1125, 1978; *Bull. Ch. Soc. Jap.*, 52, 2727, 1979; *J. Chrom.*, 208, 257, 1981; DT 2748530; JP 78/6487; *CA*, 88, 150600

* 12410-5435 /12000/

NAME: SF-1854, N-FORMYLFORTIMICIN-A
IDENTICAL: DACTIMICIN-B
PO: MIC.OLIVOASTEROSPORA
CT: AMINOGLYCOSIDE L., FORTIMICIN T.
FORMULA: C18H35N5O7
MW: 433 /500|50/
STR-CORR:
REFERENCES:
JA, 33, 510, 1980

12410-6133

NAME: FORTIMICIN-D, 6′-DEMETHYLFORTIMICIN-A
STR-CORR:
REFERENCES:
JA, 32, 868, 1273, 1979; DT 2748530

12410-6134

NAME: FORTIMICIN-KE, 6′-DEMETHYLFORTIMICIN-B
STR-CORR:
REFERENCES:
JA, 32, 868, 1273, 1979

12410-6486

NAME: SPORARICIN-A, KA-6606-I /KA-6606F-I/
PO: SACCHAROPOLYSPORA HIRSUTA-KOBENSIS
/SACCHAROPOLYSPORA HIRSUTA/
UTILITY: ON CLINICAL TRIAL
REFERENCES:
JA, 32, 173, 180, 187, 1137, 1979; *AAC*, 17, 337, 1980; DT 2942194;
JP 79/66603

12410-6487

NAME: SPORARICIN-B
REFERENCES:
DT 2942194; JP 79/66603; *JA*, 34, 811, 1981

12410-6488

! NAME: SPORARICIN-C
REFERENCES:
DT 2942194; JP 79/66603; *JA*, 34, 811, 1981

12410-6489

! NAME: SPORARICIN-D

REFERENCES:

DT 2942194; JP 79/66603; *JA*, 34, 811, 1981

12510-5156

NAME: SORBISTIN-A1

IDENTICAL: GL-A1

REFERENCES:

J. Ferm. Techn., 56, 193, 1978; *Chem. Ph. Bull.*, 27, 65, 1979; *Abst. AAC*, 19, 509, 1979; *Tetr.*, 36, 2727, 1980

12510-5158

NAME: SORBISTIN-B

IDENTICAL: GL-A2

REFERENCES:

12510-5439

NAME: P-2563-A

IDENTICAL: GL-A2

REFERENCES:

Chem. Ph. Bull., 26, 1083, 1091, 1978; *J. Ferm. Techn.*, 56, 193, 1978

12510-5440

NAME: P-2563-B

IDENTICAL: GL-A1

REFERENCES:

* 12000-4105 /10000/

NAME: CORALLINOMYCIN

REFERENCES:

12000-5306

NAME: FA-252-C

REFERENCES:

JP 76/35496

13
OTHER GLYCOSIDES

131 Streptothricin Group
132 Glycopeptides, C-Glycosides

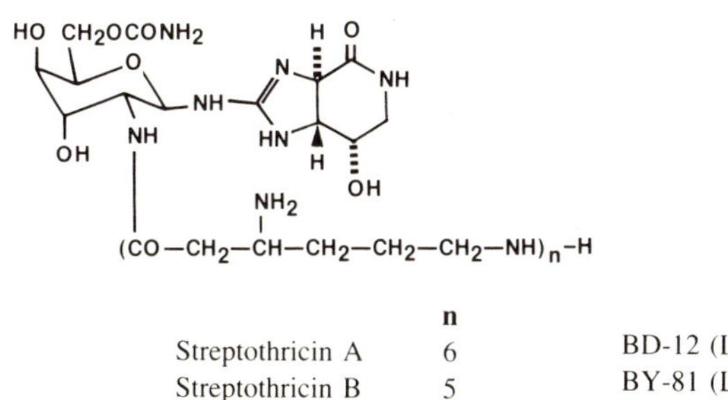

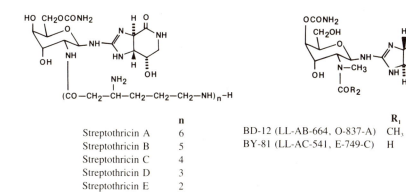

	n
Streptothricin A	6
Streptothricin B	5
Streptothricin C	4
Streptothricin D	3
Streptothricin E	2
Streptothricin F	1

	R_1	R_2
BD-12 (LL-AB-664, O-837-A)	CH_3	$COCH_2NHCH=NH$
BY-81 (LL-AC-541, E-749-C)	H	$COCH_2NHCH=NH$

Ristocetin-A, Ristomycin-A	
Ristocetin-B, Ristomycin-B	X:H

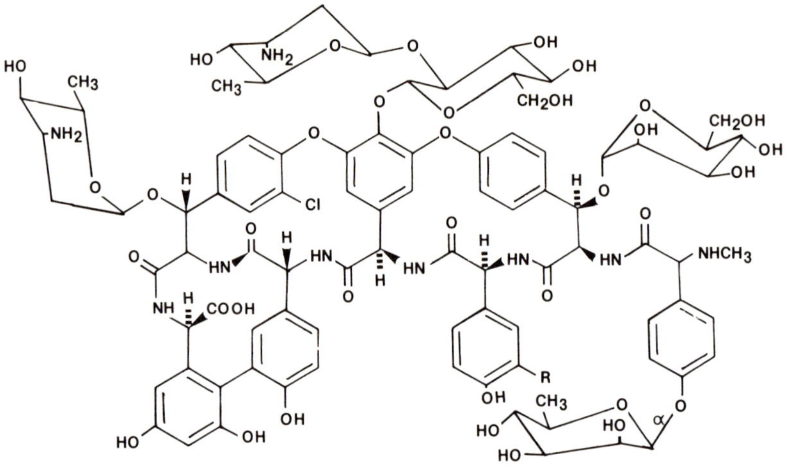

Avoparcin-α: R = H
Avoparcin-β: R = Cl

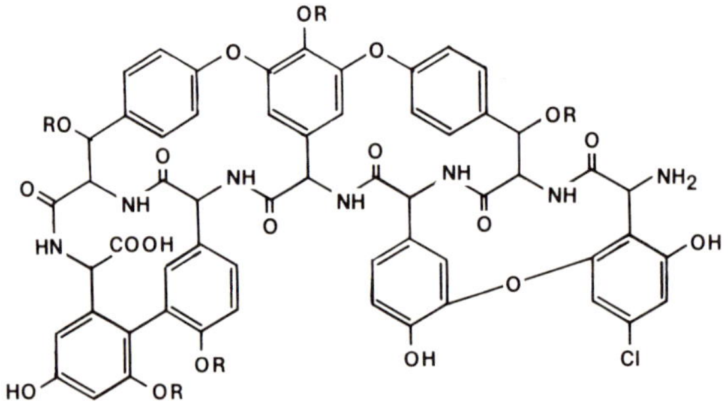

A-35512-B

R: glucose, fucose, mannose, rhamnose,

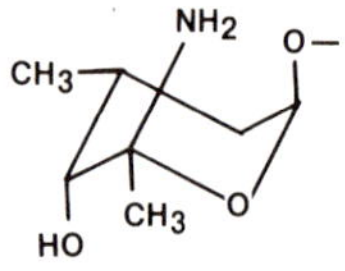

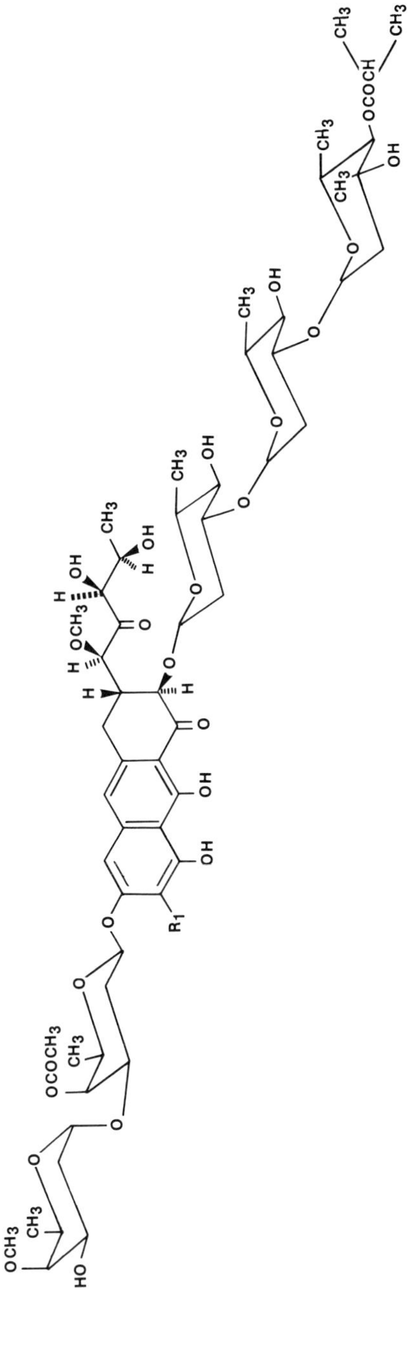

	R_1
Chromomycin-A_3	CH_3
Olivomycin-A	H

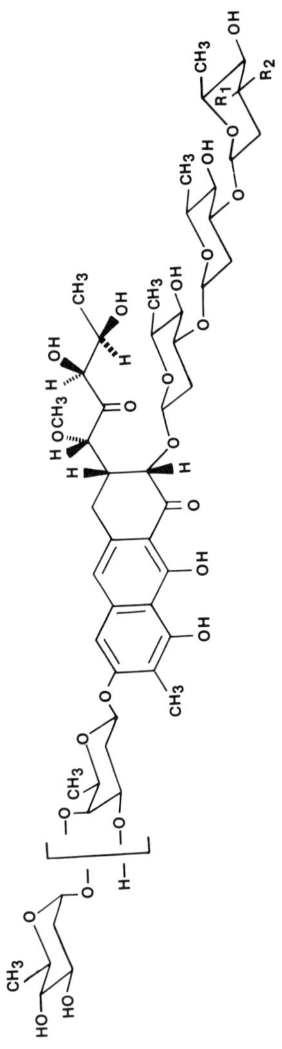

	R_1	R_2
Mithramycin	OH	CH_3
Variamycin	OCH_3	H

13110-0190

NAME: STREPTOTHRICIN-F
STR-CORR:
REFERENCES:
Chem. Ph. Bull., 26, 885, 1147, 1978; 27, 230, 1979; 28, 2884, 1980; 29, 580, 1981; *J. Ferm. Techn.*, 56, 15, 1978; *J. Chrom. Libr.*, 15, 617—713, 1978; *JA*, 33, 636, 1980; *JACS*, 103, 2871, 1981; *Chem. Lett.*, 1317, 1981

13110-0205

NAME: RACEMOMYCIN-B
IDENTICAL: PRACTOMYCIN-C

* 13110-0271 /13130/

NAME: BOSEIMYCIN-III

13110-0278

NAME: S-15-1-A
REFERENCES:
JA, 34, 292, 1981

13110-0437

NAME: AMYCIN
REFERENCES:
Ann. Microbiol. Enzymol., 10, 115, 1960; BP 922952; *CA*, 55, 15606; 59, 1053

13110-1884

NAME: PRACTOMYCIN-C
IDENTICAL: RACEMOMYCIN-B, STREPTOTHRICIN-D
CT: STREPTOTHRICIN T.
REFERENCES:

13110-4757

NAME: Y-U17W-C1
EA: (C, 35)(H, 7)(N, 16) /(N,)/
OR: (-47, W)
REFERENCES:
Yamanochi Cent. Res. Lab., 2, 164, 1974

13120-0259

NAME: BY-81
STR-CORR:
REFERENCES:
Jap. J. Antib., 32, 720, 1979; *JA*, 34, 921, 1981

13120-0261

NAME: BD-12
STR-CORR:
REFERENCES:
Jap. J. Antib., 32, 720, 1979; *JA*, 34, 921, 1981

13120-0262

NAME: LL-AC-541
STR-CORR:
REFERENCES:
Jap. J. Antib., 32, 720, 1979; *JA*, 34, 921, 1981

13120-0264

NAME: LL-AB-664
STR-CORR:
REFERENCES:
Jap. J. Antib., 32, 720, 1979; *JA*, 34, 921, 1981

13130-0436

NAME: 24010-B1
REFERENCES:
JA, 25, 604, 1972

13210-0280

NAME: RISTOCETIN-A
FORMULA: C95H110N8044 /C100H133N9051/
MW: 2066 /2500/
STR-CORR:
REFERENCES:
TL, 3905, 1979; 4187, 1980; *CC*, 906, 1979; *JCS Perkin I*, 1483, 1981; *JACS*, 102, 897, 7093, 1980

13210-0281

NAME: RISTOCETIN-B
FORMULA: C78H76N8030
MW: 1604 /2200/
STR-CORR:
REFERENCES:

13210-0282

NAME: RISTOMYCIN-A
FORMULA: C95H110N8044 /C100H133N9051/
REFERENCES:

TL, 3906, 1979; 2983, 1980; *JOC*, 44, 1009, 1979; *JCS Perkin I*, 787, 1979; *Antib.*, 179, 1979; 581, 1981; *Khim. Prir. Soed.*, 742, 1979; *JA*, 32, 446, 1979; *JACS*, 102, 7093, 1980; *JCS Perkin II*, 201, 1981

13210-0283

NAME: ACTINOIDIN-A
FORMULA: C83H93N9031CL
EA: (CL, 2) /(CL,)/
MW: 1747 /2100/
REFERENCES:

TL, 2801, 1979

13210-0284

NAME: ACTINOIDIN-B
FORMULA: C83H92N9031CL2
EA: (CL, 4) /(CL,)/
MW: 1781 /2100/
REFERENCES:

13220-0286

NAME: VANCOMYCIN
FORMULA: C66H77N9024CL2 /C64H71N9022CL2/
MW: 1450 /1388, 1560/
REFERENCES:

CC, 153, 1979; *JCS Perkin II*, 201, 1981; *JACS*, 103, 6530, 1981

13220-0289

NAME: AVOPARCIN, AVOPARCIN-B"
FORMULA: C89H100N9036CL2
MW: 1942 /3000/
STR-NEW:
REFERENCES:
JACS, 101, 2237, 1979; 102, 1671, 1980; 103, 6522, 1981; *Analyst*, 104, 1075, 1979; DT 2154633

* 13220-5378 /13200/

NAME: TEICHOMYCIN-A2
REFERENCES:
USP 4239751

13220-5815

NAME: A-35512-B
FORMULA: C90H101N8039CL.2HCL /C98H103N8.5047CL/
EA: (N, 6)(CL, 2) /(N, 5) (CL, 1.5)/
STR-NEW:
REFERENCES:
JA, 33, 1397, 1407, 1980; 34, 469, 1981; *JOC*, 45, 4685, 1980; USP 4122168

13230-0293

NAME: CHROMOMYCIN-A3
STR-CORR:
REFERENCES:
JCS Perkin II, 1331, 1979

13230-0305

NAME: OLIVOMYCIN-A
STR-CORR:
REFERENCES:
Tetr., 37, 551, 1981

13230-0311

NAME: MITHRAMYCIN
STR-CORR:
REFERENCES:
Tetr., 37, 551, 1981

13230-0317

NAME: VARIAMYCIN
PO: S.OLIVOVARIABLIS
STR-CORR:
REFERENCES:

Carb. Res., 72, C6, 1979; *JCS Perkin I*, 1800, 1980

13230-4864

NAME: S-2449, C-2449
LD50: (4.1, IP)
TV: ANTITUMOR
REFERENCES:

14
SUGAR DERIVATIVES

141 Sugar Esters, Sugar Amides
142 Sugar Lipids

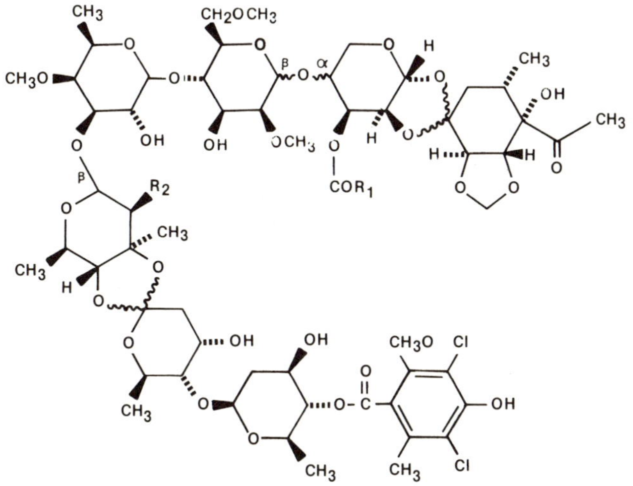

	R_1	R_2
Avilamycin A	H	$CH(CH_3)_2$
Curamycin A	H	CH_3
Flambamycin	OH	$CH(CH_3)_2$

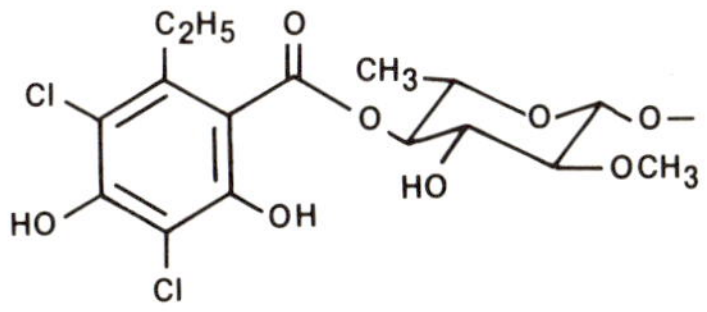

Lipiarmycin (partial structure)

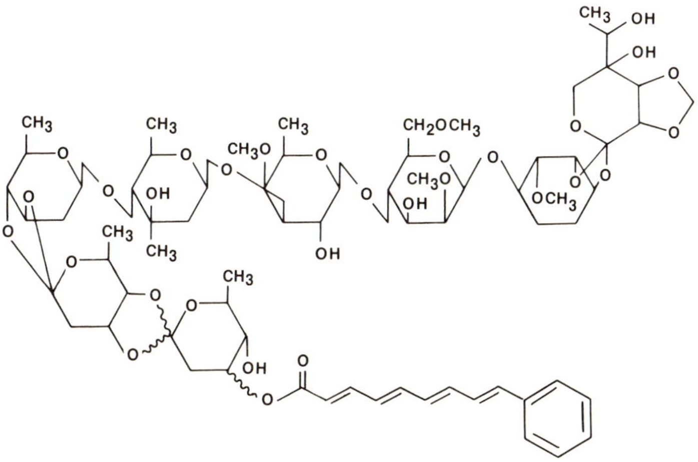

SE-73B

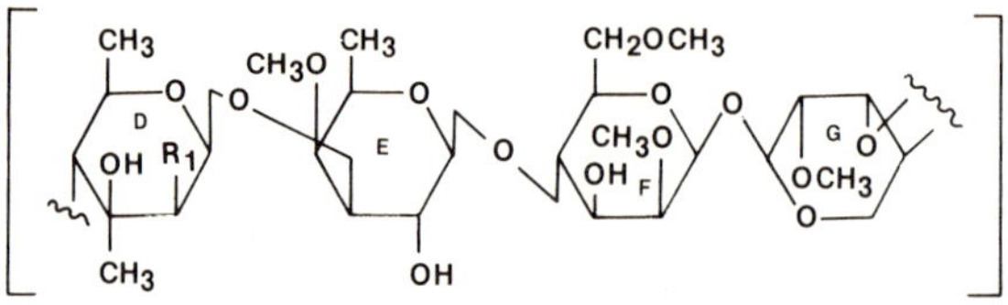

Everinomycin-D
(partial structure)

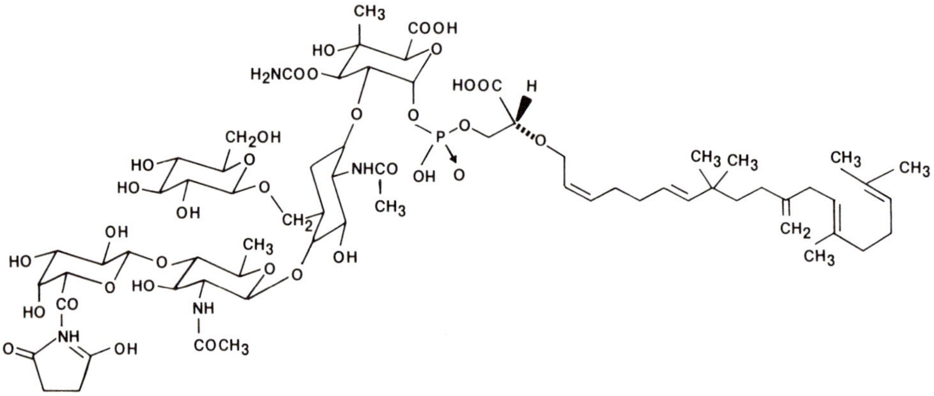

Moenomycin-A

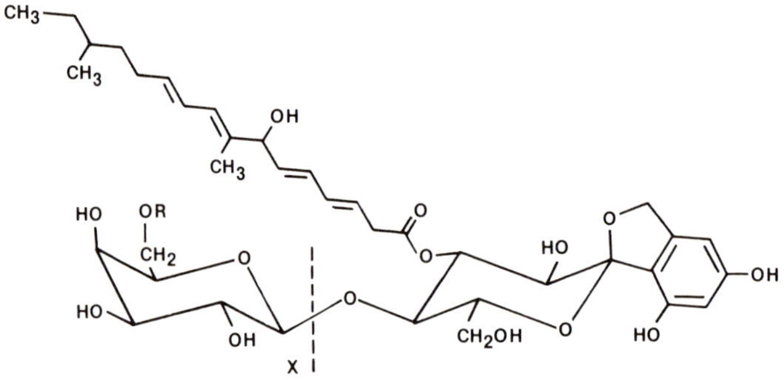

Papulacandin B R =

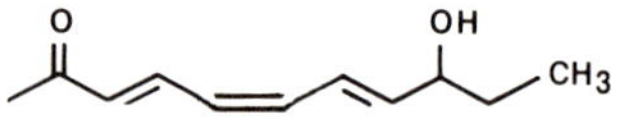

Papulacandin A R =

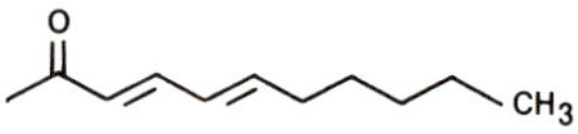

Papulacandin C R =

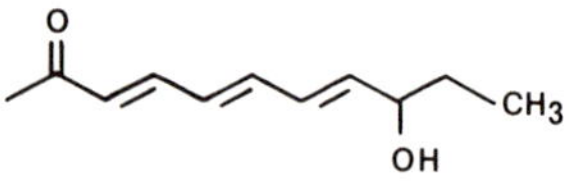

Papulacandin D X = H

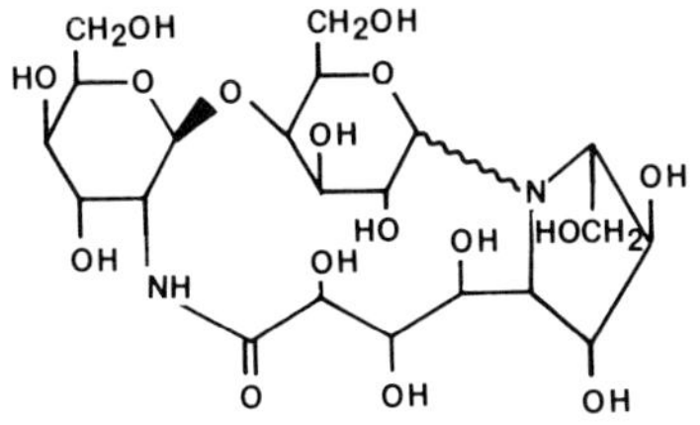

84-B-3

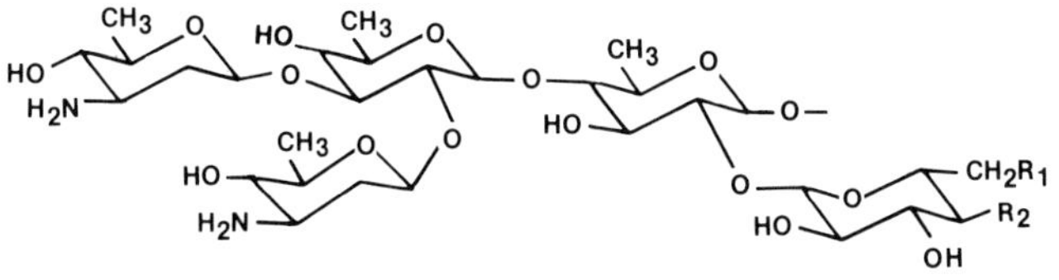

Sporaviridin (partial structure)
R_1 = H, OH
R_2 = OH, NH_2
X = aglycone (tentative)

14110-0325

NAME: EVERNINOMYCIN-D
STR-CORR:
REFERENCES:
CC, 56, 1980; *Carb. Res.*, 90, 329, 1981; *Helv.*, 62, 1, 1979; USP 4129720

14110-0332

NAME: AVILAMYCIN-A
FORMULA: C61H88O32CL2 /C63H94O35CL2/
MW: 1403 /1324/
IS-EXT: (ETOAC, 7, FILT.)
IS-CHR: (SILG, CHL-MEOH)
STR-CORR:
REFERENCES:
Helv., 63, 1141, 1980

14110-0333

NAME: CURAMYCIN, CURAMYCIN-A
FORMULA: C59H84O32CL2 /C54H84O33CL2/
MW: 1375 /1360/
STR-NEW:
REFERENCES:
JA, 32, 1213, 1979; *Heterocycl.*, 15, 1621, 1981

14110-0334

NAME: FLAMBAMYCIN
MW: 1419 /1400/
STR-CORR:
REFERENCES:
Tetr., 35, 993, 1003, 1979; *Helv.*, 63, 1141, 1980

14110-4001

NAME: LIPIARMYCIN
UV: MEOH: (232, 354,)(268, 214,)(315, 108,)
TV: ANTIPHAGE
STR-NEW:
REFERENCES:
Ann. Rev. Microb., 33, 389, 1979; *J. Virol.*, 33, 945, 1980; USP 3978211

* 14110-5312 /00000/

NAME: SE-73B
CT: EVERNINOMYCIN T.
STR-NEW:
REFERENCES:
DT 2848793

14210-0354

NAME: MOENOMYCIN-A
STR-NEW:
REFERENCES:
TL, 2119, 1978; 3493, 1979; *Tetr.*, 37, 97, 105, 113, 1981; *Angew.*, 93, 130, 1981; *CA*, 95, 25449

14210-4760

NAME: ENSANCHOMYCIN
REFERENCES:
Abst. AAC, 19, 233, 1979

14130-5386

NAME: PAPULACANDIN-A
STR-NEW:
REFERENCES:
JA, 33, 967, 1980; *Exp.*, 34, 1667, 1978; *Eur. J. Bioch.*, 97, 345, 1979; *Helv.*, 64, 1533, 1981

14130-5387

NAME: PAPULACANDIN-B
STR-NEW:
REFERENCES:
JA, 33, 967, 1980; *Exp.*, 34, 1667, 1978; *Eur. J. Bioch.*, 97, 345, 1979; *Helv.*, 64, 1533, 1981

14130-5388

NAME: PAPULACANDIN-D
STR-NEW:
REFERENCES:
JA, 33, 967, 1980; *Exp.*, 34, 1667, 1978; *Eur. J. Bioch.*, 97, 345, 1979; *Helv.*, 64, 1533, 1981

14130-5525

NAME: PAPULACANDIN-C
STR-NEW:
REFERENCES:
JA, 33, 967, 1980; *Exp.*, 34, 1667, 1978; *Eur. J. Bioch.*, 97, 345, 1979; *Helv.*, 64, 1533, 1981

14220-0380

NAME: CORD FACTOR
REFERENCES:
Chem. Phys. Lipids, 21, 97, 1978

14230-0421

! NAME: CINODINE-B"1, BM-123-B"1
REFERENCES:
JOC, 44, 1166, 1979; *JAMA*, 241, 402, 1979; *AAC*, 20, 424, 1981; USP 4154925

14230-0422

! NAME: CINODINE-G"1, BM-123-G"1
REFERENCES:
JOC, 44, 1166, 1979; *JAMA*, 241, 402, 1979; *AAC*, 20, 424, 1981; USP 4154925

14230-4070

NAME: PULVOMYCIN
IDENTICAL: 1063-Z
PO: STV.MOBARANSE
FORMULA: C23H34O8 /C41H56NO12/
MW: 438
REFERENCES:
Proc. Nat. Acad. Sci., 75, 5324, 1978

* 14230-4088 /00000/

NAME: 84-B-3
CT: SUGAR DERIV.
FORMULA: C21H36N2016
EA: (N, 4)
MW: 572
STR-NEW:
REFERENCES:

Abst. 8th Int. Symp. Carbohydr. Chem., Kyoto, 1976, 2A-II; JP 71/19593, 23279, 23280

14230-5400

! NAME: CINODINE-G"2, BM-123-G"2
REFERENCES:

JOC, 44, 1166, 1979; USP 4154925; *AAC*, 20, 424, 1981

14230-5401

! NAME: CINODINE-B"2, BM-123-B"2
REFERENCES:

JOC, 44, 1166, 1979; USP 4154925; *AAC*, 20, 424, 1981

* 10000-3594 /84210/

NAME: SPORAVIRIDIN
CT: /GLYCOSIDE ANTIBIOTIC, HYGROMYCIN-A T./
STR-NEW:
REFERENCES:

Carb. Res., 75, C-17, 1979; *TL*, 3965, 1979; *Heterocycl.*, 13(Spec.), 145, 1979

Macrocyclic Lactone (Lactam) Antibiotics

21
MACROLIDE ANTIBIOTICS

211 Small Macrolides
212 16-Membered Macrolides
213 Other Macrolides

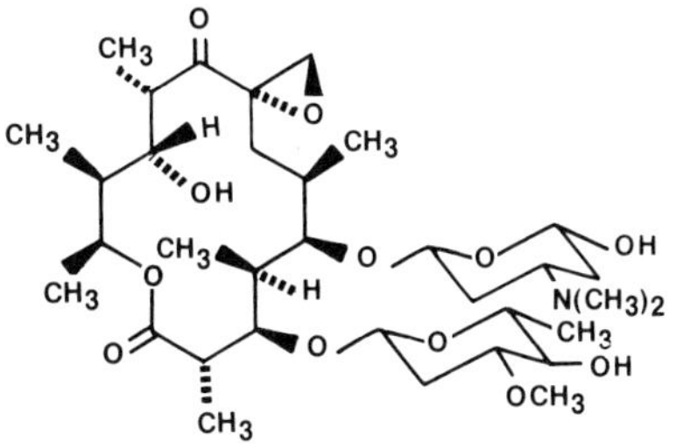

Oleandomycin

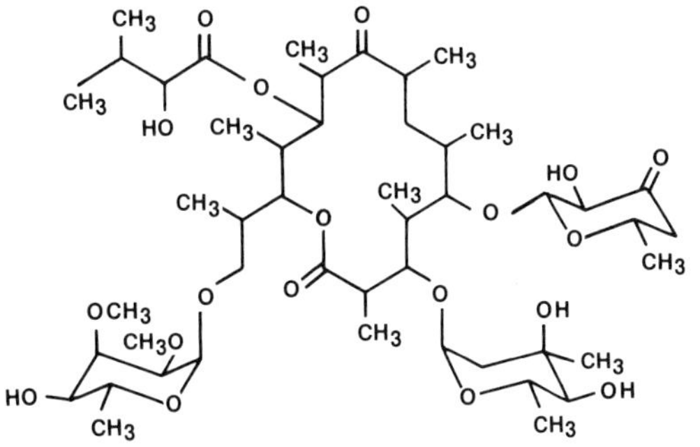

RP-23672

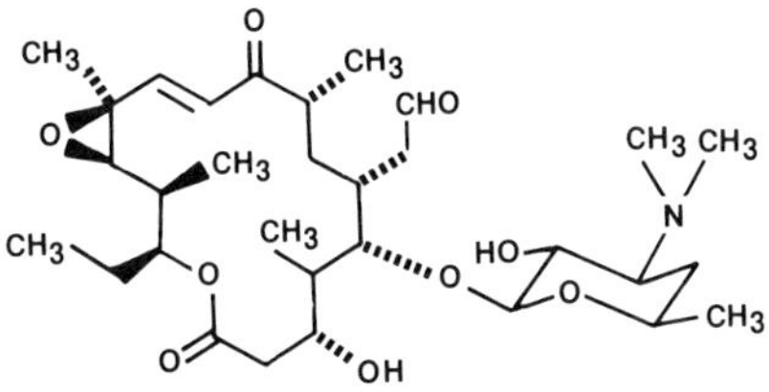

Rosamycin

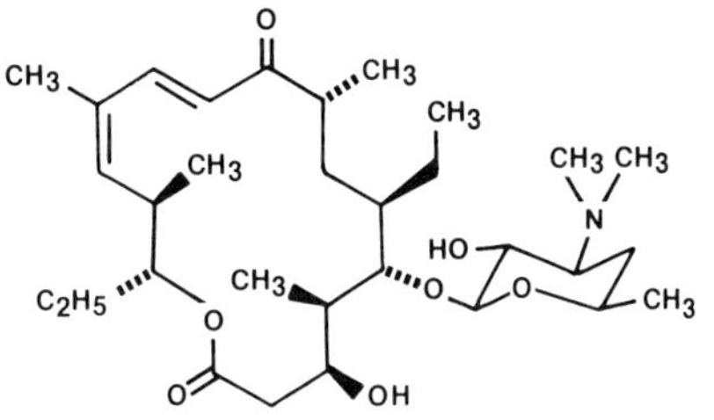

M-4365-G_2

21110-0439

NAME: METHYMYCIN
REFERENCES:
Chem. Lett., 1021, 1979

21110-0441

NAME: NA-181
PO: S.FUNGICIDICUS-CHIUSAENSIS /S.FUNGICIDICUS/
FORMULA: C25H43NO7 /C25H42NO7/
REFERENCES:

21122-0457

NAME: ERYTHROMYCIN-A
TRADE NAMES: ANAMYCIN, EROMERZIN
REFERENCES:
JACS, 103, 3210, 3213, 3215, 1981

21122-0461

NAME: OLEANDOMYCIN
TRADE NAMES: OLEANDOCIN
REFERENCES:
STR-CORR: 25

21122-5027

NAME: 1745-A-X
REFERENCES:
USP 4152424

21123-0465

NAME: MEGALOMYCIN-A
REFERENCES:
JCS Perkin I, 1600, 1979

* 21124-4082 /00000/

NAME: RP-23671
CT: MACROLIDE, LANKAMYCIN T., NEUTRAL
REFERENCES:
JACS, 102, 3605, 1980

* 21124-4083 /00000/

NAME: RP-23672, BF-2126208
CT: MACROLIDE, LANKAMYCIN T., NEUTRAL
FORMULA: C48H82O20
MW: 979
STR-NEW:
REFERENCES:
Proc. IUPAC Conf. Nat. Prod., Varna, 1978, 2, 247, 1979; *JACS*, 102, 3605, 1980

21211-0475

NAME: MYDECAMYCIN
TRADE NAMES: MIDICACIN, MIDECAMINE, RUBIMYCIN
IDENTICAL: MAIDIMEISU
REFERENCES:
Acta Micr. Sinica, 17, 311, 1977; *CA*, 88, 85705; 93, 161821

21211-0478

NAME: LEUCOMYCIN-A3
REFERENCES:
JA, 32, 1055, 1979; *AAC*, 15, 738, 1979; *TL*, 2837, 1980; *JACS*, 103, 1224, 1981; *Agr. Biol. Ch.*, 43, 1331, 1979; Can. P 1046438

21211-0480

NAME: LEUCOMYCIN-A5
REFERENCES:
JA, 32, 78, 1979

21211-0485

NAME: LEUCOMYCIN-U, LEUCOMYCIN-A10
IDENTICAL: DHA

21211-0486

NAME: LEUCOMYCIN-V, LEUCOMYCIN-A11

21211-0492

NAME: JOSAMYCIN
TRADE NAMES: JOSACIN
REFERENCES:
TL, 2837, 1980

21211-0495

NAME: PLATENOMYCIN-A1
REFERENCES:
Heterocycl, 15, 1123, 1981

21211-0503

NAME: ESPINOMYCIN-A2
REFERENCES:
JP 79/998; *CA*, 90, 136243

21211-0506

NAME: TURIMYCIN COMPLEX
IDENTICAL: LEUCOMYCIN COMPLEX
TO: (MYCOPLASMA SP., .1)
LD50: (320, IV)
REFERENCES:
JA, 33, 566, 574, 663, 1980; *Arch. Toxicol., Suppl. I*, 235, 1978; *Pol. J. Pharm. Pharmacol.*, 32, 211, 217, 1980; *Pharm.*, 34, 338, 1979; *CA*, 91, 163095

21211-0575

NAME: DHH
IDENTICAL: TURIMYCIN-HO
REFERENCES:

21212-0514

NAME: SPIRAMYCIN-I
REFERENCES:
Chromatogr., 12, 294, 1979; *Chem. Ph. Bull.*, 27, 176, 1979; *JA*, 33, 911, 1980

21221-0524

NAME: CARBOMYCIN-A
REFERENCES:
J. Chem. Techn. Biotechn., 31, 241, 1981; *CA*, 95, 78393

21221-5317

NAME: DELTAMYCIN-A1
REFERENCES:
JA, 32, 878, 1979; 33, 284, 1980

* 21221-5954 /21222/

NAME: DELTAMYCIN-X

21223-0529

NAME: ROSAMYCIN
UTILITY: NO LONGER AVAILABLE /ON CLINICAL TRIAL/
STR-CORR:
REFERENCES:
JA, 32, 915, 920, 926, 1979; *TL*, 4699, 1980; USP 4279896

21223-0535

NAME: CIRRAMYCIN-A1
IDENTICAL: A-6888-F
PO: S.FLOCCULUS
REFERENCES:

21223-0536

NAME: CIRRAMYCIN-B
IDENTICAL: A-6888-A
PO: S.FLOCCULUS
REFERENCES:
Abst. AAC, 19, 1026, 1979; USP 4252898; JP 80/154994; *CA*, 93, 166021; 94, 137798

21223-0537

NAME: ACUMYCIN
IDENTICAL: A-6888-A
REFERENCES:
21223-0537 ACUMYCIN
Tetr., 37(Suppl. 1), 91, 1981

21231-0539

NAME: CARBOMYCIN-B
REFERENCES:
TL, 2837, 1980; *Zbl. Bakt. Parasit.*, Abt. 2, 135, 541, 1980; *J. Chem. Techn. Biotechn.*, 31, 241, 1981

21232-0541

NAME: RELOMYCIN, S-1, 20-DIHYDROTYLOSIN
PO: S.AMBOFACIENS + TYLOSIN + CERULENIN
REFERENCES:
JA, 33, 911, 1980

21232-0549

NAME: TYLOSIN
REFERENCES:
JOC, 44, 2050, 1979; *JA*, 33, 915, 1980; *AAC*, 19, 209, 1981; *Chem. Ph. Bull.*, 28, 1963, 1980; *J. Ferm. Techn.*, 59, 235, 1981; SU P 755837; *CA*, 93, 236815

21232-5849

NAME: 3-ACETYLTYLOSIN
REFERENCES:
JA, 33, 1300, 1309, 1980; 32, 542, 1979; USP 4092473; EP 1841

21232-5850

NAME: 3-PROPYONYLTYLOSIN
REFERENCES:
JA, 33, 1300, 1309, 1980; 32, 542, 1979; USP 4092473; EP 1841

21232-5851

NAME: 4"-BUTYRLTYLOSIN
REFERENCES:
JA, 33, 1300, 1309, 1980; 32, 542, 1979; USP 4092473; EP 1841

21232-5852

NAME: 4"-ISOVALERYLTYLOSIN
REFERENCES:
JA, 33, 1300, 1309, 1980; 32, 542, 1979; USP 4092473; EP 1841

21233-5174

NAME: M-4365-G1, DESOSAMINYL-PROTYLONOLIDE
PO: S.SP + PROTYLONOLIDE
TO: (MYCOPLASMA SP.,)
REFERENCES:
JA, 33, 1570, 1980; 34, 719, 1981

21233-5175

NAME: M-4365-G2
TO: (MYCOPLASMA SP., .004)
STR-CORR:
REFERENCES:
JA, 34, 719, 1981

21241-0555

NAME: MARIDOMYCIN-I
REFERENCES:
Chem. Ph. Bull., 26, 2718, 1978; *Agr. Biol. Ch.*, 43, 847, 1103, 1111, 1979

21241-0556

NAME: MARIDOMYCIN-II
REFERENCES:
Chem. Ph. Bull., 26, 2718, 1978; *Agr. Biol. Ch.*, 43, 847, 1103, 1111, 1979

21241-0557

NAME: MARIDOMYCIN-III
REFERENCES:
Chem. Ph. Bull., 26, 2718, 1978; *Agr. Biol. Ch.*, 43, 847, 1103, 1111, 1979

21241-0558

NAME: MARIDOMYCIN-IV
REFERENCES:
Chem. Ph. Bull., 26, 2718, 1978; *Agr. Biol. Ch.*, 43, 847, 1103, 1111, 1979

21241-0559

NAME: MARIDOMYCIN-V

REFERENCES:

Chem. Ph. Bull., 26, 2718, 1978; *Agr. Biol. Ch.*, 43, 847, 1103, 1111, 1979

21241-0561

NAME: MARIDOMYCIN-VI

REFERENCES:

Chem. Ph. Bull., 26, 2718, 1978; *Agr. Biol. Ch.*, 43, 847, 1103, 1111, 1979

22
POLYENE ANTIBIOTICS

221	**Trienes**
222	**Tetraenes**
223	**Pentaenes**
224	**Hexaenes**
225	**Heptaenes**
226, 227, 228	**Other Polyenes**

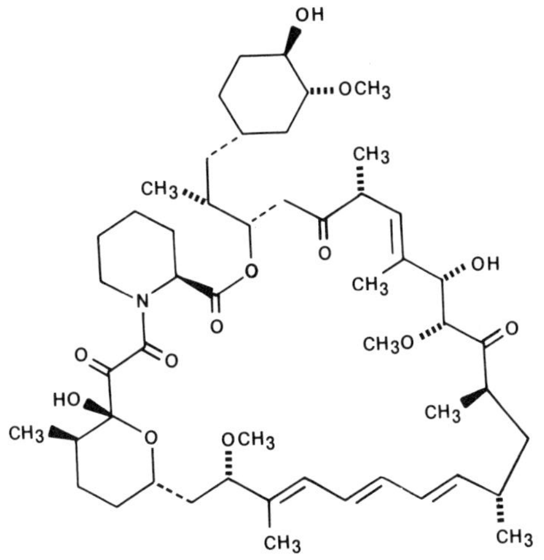

Rapamycin

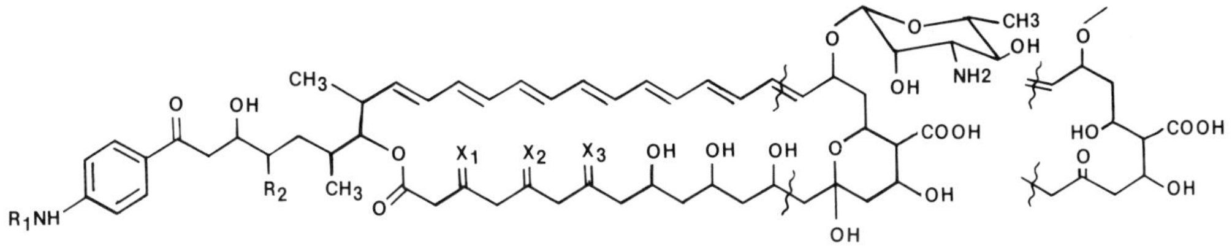

	R_1	R_2	X_1	X_2	X_3
Candicidin-D_1	H	CH_3	=O	H, H	=O
Trichomycin-B	H	H	=O	H, H	=O
Levorin-A_2	H	CH_3	=O	H, H	=O
Levorin-A_3	H	CH_3	OH, H	OH, H	=O
Aureofungin-A	CH_3	H	OH, H	OH, H	=O
Partricin-A	CH_3	H	OH, H	=O	OH, H
Partricin-B	H	H	OH, H	=O	OH, H
Vacidin-A	H	H	=O	OH, H	OH, H
Candimycin	CH_3	H	OH, H	OH, H	=O

22110-0595

NAME: RAPAMYCIN
STR-CORR:
REFERENCES:
JA, 32, 630, 1979; *Can. J. Chem.*, 58, 579, 1980

22210-0603

NAME: ETRUSCOMYCIN
IDENTICAL: 3605-5, 3608-5, 4995-35
REFERENCES:

22210-0608

NAME: TETRAMYCIN
FORMULA: C35H53NO13 /C34H53NO14/
MW: 695 /731/
REFERENCES:
Tetr., 35, 1851, 1979

22230-0618

NAME: NYSTATIN-A2
REFERENCES:
Antib., 423, 519, 1981; *Rec.*, 91, 780, 1972

22230-0619

NAME: NYSTATIN-A3
REFERENCES:
JA, 32, 565, 1979

22230-5853

NAME: POLYFUNGIN-B
REFERENCES:
JA, 32, 565, 1979; Pol. P 87664; *CA*, 90, 85273

22200-0633

NAME: AC2-435
REFERENCES:
Zbl. Bakt. Parasit., Abt. 2, 136, 207, 1981; *CA*, 95, 128948

22200-5294

NAME: BARODAMYCIN
REFERENCES:
Ind. P 126592; *CA*, 92, 109107

22310-0649

NAME: FUNGICHROMIN
REFERENCES:

22310-0650

NAME: LAGOSIN
REFERENCES:

22310-0651

NAME: MOLDICIDIN-B
REFERENCES:

22310-0652

NAME: PENTAMYCIN
REFERENCES:

22310-0653

NAME: COGOMYCIN
REFERENCES:

22310-0659

NAME: FILIPIN COMPLEX, U-5956
UTILITY: ON CLINICAL TRAIL
REFERENCES:
Microbios Lett., 10, 115, 1980

22310-6204

NAME: FCRC-21, NSC-277813
IDENTICAL: PENTAMYCIN, MOLDCIDIN-B
TV: KB, MELANOMA
REFERENCES:
Biomed. Mass Spectr., 7, 93, 1980

22310-6602

NAME: LYMPHOSARCIN, NSC-208642
REFERENCES:

22330-0703

NAME: LIENOMYCIN, 2995
PO: S.VANDERGENSIS, ACT.VANDERGENSIS /ACT.SP./
REFERENCES:
JA, 33, 980, 989, 1980

22511-0726

NAME: DJ-400-B1
IDENTICAL: AUREOFUNGIN, CANDIMYCIN
FORMULA: C65H86N2020 /C66H97N2021/
MW: 1240 /1253/
REFERENCES:
JA, 33, 591, 1980

22511-0727

NAME: DJ-400-B2
FORMULA: C58H86N2020 /C65H95N2021/
MW: 1130
REFERENCES:
JA, 33, 591, 1980

22511-0728

NAME: CANDICIDIN, CANDICIDIN COMPLEX
IDENTICAL: HEPTAMYCIN
REFERENCES:
JA, 33, 591, 1980; *TL*, 1791, 1979

22511-0729

NAME: HAMYCIN
FORMULA: /C64H110N2O21, C60H95N2O20/
MW: 1114
REFERENCES:
JA, 33, 591, 1980

22511-0732

NAME: CANDICIDIN-D1, CANDICIDIN-D
PC: POW.
UV: MEOH: (342, ,)(363, ,)(384, ,)(401, ,)
STR-CORR:
REFERENCES:
Proc. IUPAC Conf. Nat. Prod., Varna, 1978, 2, 251; *JA*, 33, 591, 1980

22511-0733

NAME: TRICHOMYCIN-A
FORMULA: C58H84N2O18 /C61H86N2O21/
MW: 1096 /1094/
STR-NEW:
REFERENCES:
JA, 33, 591, 1980; *Proc. IUPAC Conf. Nat. Prod., Varna*, 1978, 2, 251

22511-0734

NAME: TRICHOMYCIN-B
FORMULA: C58H82N2O18
MW: 1094
STR-CORR:
REFERENCES:
Proc. IUPAC Conf. Nat. Prod., Varna, 1978, 2, 251; *Proc. Antib. Symp., Weimar*, 1979, B-15

22511-0735

NAME: LEVORIN-A0
FORMULA: C59H84N2O17
MW: 1092
REFERENCES:
Proc. Antib. Symp., Weimar, 1979, B-16

22511-0736

NAME: LEVORIN-A1
FORMULA: C59H82N2O17
MW: 1090
REFERENCES:
Proc. Antib. Symp., Weimar, 1979, B-16; *JA*, 33, 591, 1980

22511-0737

NAME: LEVORIN-A2
FORMULA: C59H84N2O18 /C49H84N2O18/
STR-CORR:
REFERENCES:
JA, 33, 591, 1980; *Antib.*, 566, 1981; SU P 251764

22511-0738

NAME: LEVORIN-A3
FORMULA: C59H86N2O18 /C59H93N2O22/
MW: 1110 /1100/
STR-CORR:
REFERENCES:
Proc. Antib. Symp., Weimar, 1979, B-16

22511-0741

NAME: LEVORIN, LEVORIN COMPLEX
IDENTICAL: HEPTAMYCIN
REFERENCES:
JA, 33, 591, 1980, SU P 251764

22511-0747

NAME: HEPTAFUNGIN
REFERENCES:
Hung. P 155813

22511-0769

NAME: <u>AUREOFUNGIN-A</u>
IDENTICAL: CANDIMYCIN-A, DJ-400-B1
FORMULA: C59H86N2O19 /C62H87N2O18/
MW: 1126
REFERENCES:
JA, 33, 591, 1980

22511-0771

NAME: <u>CANDIMYCIN</u>
IDENTICAL: AUREOFUNGIN-A
FORMULA: C59H86N2O19
MW: 1126
STR-NEW:
REFERENCES:
JA, 33, 591, 1980

* 22511-0777 /22520/

NAME: <u>PARTRICIN</u>, PARTRICIN-A, SN-654
TRADE NAMES: ORAFUNGIN, TRICANDINE (METHYL ESTER)
CT: AROMATIC HEPTAENE, CANDICIDIN T. /NONAROMATIC HEPTAENE, AMPHOTERICIN-B T./
FORMULA: C59H86N2O19
MW: 1126
UTILITY: ANTIFUNGAL DRUG
STR-NEW:
REFERENCES:
JA, 33, 591, 904, 1980; *Farmaco, Sci.*, 34, 188, 1979

* 22511-0806 /22500/

NAME: <u>HEPTAMYCIN</u>
IDENTICAL: CANDICIDIN, ASCOSIN, LEVORIN
CT: AROMATIC HEPTAENE, CANDICIDIN T., /HEPTAENE/
REFERENCES:
JA, 33, 591, 1981

22511-4767

NAME: VACIDIN-A, AUREOFACIN-A /AUREOFACIN/
IDENTICAL: AYFACTIN, AYF
CT: CANDICIDIN T.
FORMULA: C58H84N2O19
EA: (N, 3) /(N,)/
MW: 1112
STR-CORR:
REFERENCES:

JA, 33, 904, 1980; 34, 884, 1981; *Abst. AAC*, 15, 428, 1975

* 22511-5854 /22500/

NAME: NO NAME
IDENTICAL: 67-121 COMPLEX, HAMYCIN
CT: AROMATIC HEPTAENE
REFERENCES:

* 22511-6205 /22500/

NAME: GLOBORUBERMYCIN, GLOBOROSEOMYCIN
CT: AROMATIC HEPTAENE, CANDIDIDIN T. /HEPTAENE/
TO: (C.ALB., .5)(FUNGI, .4) /(C.ALB,) (FUNGI,)/
LD50: (139, IV)
REFERENCES:

Chin. Med. J., 92, 443, 1979; *CA*, 93, 37520

22520-0776

NAME: AMPHOTERICIN-B
REFERENCES:

Tetr., 1421, 1981; *TL*, 2847, 1981; *BBRC*, 101, 853, 1981; *CA*, 90, 99959

22520-0786

NAME: CANDIDININ
FORMULA: C53H79NO20 /C47H72NO17/
MW: 1049
REFERENCES:

JA, 32, 565, 1979; Pers. comm. from E. Borowsky

22520-0787

NAME: MYCOHEPTIN, MYCOHEPTIN-A2
MW: 921
LD50: (5.25, IP)
REFERENCES:
Antib., 499, 1980

* 22520-4058 /22511/

NAME: HYDROHEPTIN
CT: NONAROMATIC HEPTAENE, AMPHOTERICIN-B T., NEUTRAL /AROMATIC HEPTAENE, CANDICIDIN T./
PC: ORANGE
UV: DMSO: (365, 540,)(388, 830,)(410, 900,)
UV: MEOH: (341, 320,)(359, 560,)(379, 950,) (402, 1080,)
UV: W: (361, 480,)(381, 810,)(404, 920,)
SOL-GOOD: MEOH, ETOH, DMSO
SOL-POOR: ACET, HEX
STAB: (HEAT, -)(LIGHT, -)
TO: (C.ALB., 1)(S.CEREV., 1)(FUNGI, .5) /(C.ALB.,) (S.CEREV.,) (FUNGI,)/
IS-FIL: 7.5
IS-EXT: (BUOH, 7.5, FILT.)
REFERENCES:
JA, 32, 1223, 1230, 1979

22500-0791

NAME: NEOHEPTAENE
REFERENCES:
HAB, 11, 38, 1968

* 22600-5266 /22500/

NAME: EPIRODIN-A
CT: OCTAENE /HEPTAENE/
REFERENCES:
J. Agr. Food Chem., 24, 555, 1976; *JA*, 31, 159, 1978

22710-0814

NAME: FLAVOFUNGIN
IDENTICAL: 464
REFERENCES:
Bioorg. Khim., 4, 1244, 1978

22710-0817

! NAME: FLAVOMYCOIN, ROFLAMYCIN
FORMULA: C40H66O12 /C40H60O12/
MW: /721/
REFERENCES:
JA, 34, 122, 1981; DDR P 67208

22710-0818

NAME: MYCOTICIN-A
IDENTICAL: 464
REFERENCES:
Bioorg. Khim., 4, 1244, 1978

22710-0820

NAME: BRUNEOFUNGIN
REFERENCES:
Bioorg. Khim., 4, 1244, 1978

22710-6207

NAME: FLAVOPENTIN
FORMULA: C41H66O10 /C44H66O10/
UV: MEOH: (260, ,)(368, ,)
REFERENCES:
Antib., 736, 1979; *CA*, 90, 54795

22720-0822

NAME: DERMOSTATIN-A
REFERENCES:
HAB, 22, 47, 1980

23
MACROCYCLIC LACTONE ANTIBIOTICS

231 Macrolide-Like Antibiotics
232 Simple Lactones
233 Dilactones
234 Cyclopolylactones
235 Condensed Macrolactones

A-26771-B

Asposterol (Aspochalasin-B)

Azalomycin F_{4a}

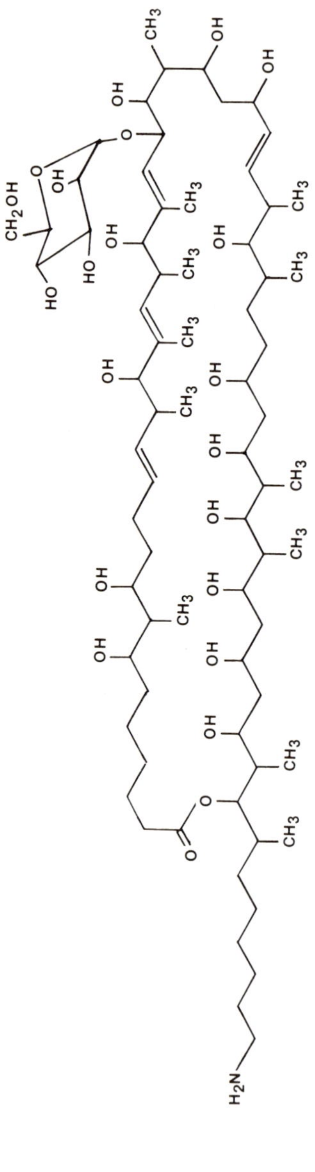

Monazomycin

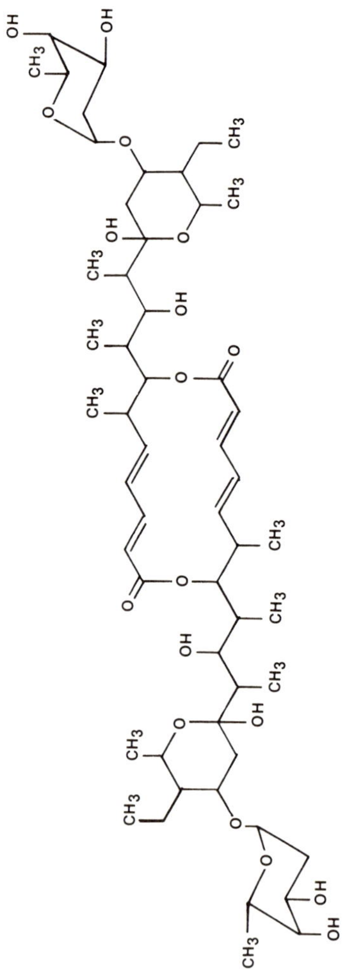

Elaiophylin-A, Azalomycin-B

23110-0829

NAME: OLIGOMYCIN-A
IDENTICAL: 178
PO: S.CHIBAENSIS
UV: ETOH: (226, 490,)(232, 460,)(241, 280,)
REFERENCES:
JA, 33, 514, 1980

23110-4987

NAME: 560
CT: MACROLIDE L., OLIGOMYCIN T.
REFERENCES:

* 23120-0838 /23110/

NAME: BOTRYCIDIN
IDENTICAL: VENTURICIDIN-X
CT: VENTURICIDIN T. /OLIGOMYCIN T./
REFERENCES:

23120-0843

NAME: AABOMYCIN, AABOMYCIN-A
PO: S.HYGROSCOPICUS-AABOMYCETICUS /S.HYGROSCOPICUS/
REFERENCES:
USP 3657422

23120-0844

NAME: A-150-A /TETRAESIN/

23120-0845

NAME: VENTURICIDIN-X
IDENTICAL: BOTRYCIDIN
REFERENCES:

23120-5034

NAME: AABOMYCIN-S
REFERENCES:
JP 74/11051

23130-0846

NAME: AXENOMYCIN-A
REFERENCES:

Z. Naturforsch. Ser. C, 35, 936, 1980; *Acta Chem. Scand.*, 29B, 507, 1975

23130-0848

NAME: AXENOMYCIN-D
REFERENCES:

Z. Naturforsch. Ser. C, 35, 936, 1980; *Acta Chem. Scand.*, 29B, 507, 1975

23140-0849

NAME: PRIMYCIN
PO: MIC.GALERIENSIS, THERMOPOLYSPORA SP.
PC: CRYST. /MICROCRYST./
TO: (MYCOB.SP., .06)(C.ALB., 2)(FUNGI, 1)
IS-CRY: (CRYST., DMFA-MEOH)(CRYST., DMSO-MEOH)
REFERENCES:

JA, 32, 408, 1979; 33, 523, 1980

* 23140-4002 /00000/

NAME: MONAZOMYCIN
CT: MACROLIDE L.
FORMULA: C72H133NO22 /C62H119NO20/
MW: 1364 /1051/
OR: (+16.3, MEOH)
STR-NEW:
REFERENCES:

TL, 5217, 1981

* 23140-4003 /00000/

NAME: TAKACIDIN
IDENTICAL: MONAZOMYCIN
CT: MACROLIDE L.
REFERENCES:

* 23140-4004 /00000/

NAME: LL-A-491
CT: MACROLIDE L.
REFERENCES:

23150-0952

NAME: SCOPAFUNGIN
REFERENCES:
Abst. AAC, 21, 188, 1981

23150-3835

NAME: P-6226
REFERENCES:
Antib. Res. Mezh. Sborn., 5, 22, 1970

23150-3836

NAME: AZALOMYCIN-F
STR-NEW:
REFERENCES:
AAC, 13, 454, 1978; *Abst. 23rd Symp. Chem. Nat. Prod., Nagoya*, 1980, 600

* 23160-0956 /23100/

NAME: SCOPATHRICIN, SCOPATHRICIN-I
IDENTICAL: SCOPAMYCIN-A (TENTATIVE)
CT: HUMIDIN T.
MW: 842 /860/
UV: MEOH: (245, , 37000)(286, , 7800)

23160-5940

NAME: SF-1540, SF-1540-A
FORMULA: C36H55NO10 /C34H50NO9/
MW: 661 /600/
UV: MEOH-HCL: (247, 505,)
UV: MEOH-NAOH: (249, 620,)

23160-5958

NAME: SF-1540-B
FORMULA: C35H56O10
MW: 636 /600/
PC: YELLOW
REFERENCES:

23210-0866

NAME: RECIFEOLIDE
REFERENCES:
Tetr., 37, 4059, 1981

23220-4774

NAME: A-26771-B
STR-CORR:
REFERENCES:
TL, 2633, 1979; 1479, 4611, 1980; *Finn Chem. Lett.*, 24, 1979; *CA*, 90, 186322

23230-0444

NAME: ALBOCYCLIN
REFERENCES:
JA, 31, 319, 1978

23310-0870

NAME: ANTIMYCIN-A3
REFERENCES:
Bull. Ch. Soc. Jap., 52, 198, 1979; *Tanagawa Daigaku Kenkyu Ho*, 20, 79, 1980

23320-0884

NAME: PYRENOPHORIN
REFERENCES:
JOC, 46, 3137, 1981; *CA*, 56, 9234; TL, 159, 1982

23320-0885

NAME: VERMICULIN
REFERENCES:
Biol. (Bratislava), 34, 735, 1979; *JOC*, 46, 3137, 1981; Cz P 4083/1977

* 23330-0953 /23100/

NAME: ELAIOPHYLIN-A, ELAIOPHYLIN /255-E/
IDENTICAL: 255-E
PO: S.VIOLACEONIGER
FORMULA: C54H88018 /C54H90018/
MW: 1024
STR-NEW:
REFERENCES:
Helv., 64, 407, 1981; *JA*, 34, 1107, 1981; *Giorn. Micr.*, 7, 207, 1959

* 23330-0954 /23100/

NAME: AZALOMYCIN-B
IDENTICAL: SAPROMYCETIN-A, 255-E
PO: S.VIOLACEONIGER
FORMULA: C54H88018 /C56H92019/
MW: 1024 /713, 1032/
STR-NEW:
REFERENCES:
Helv., 64, 407, 1981

* 23330-0955 /23100/

NAME: SAPROMYCETIN-A
IDENTICAL: AZALOMYCIN-B, ELAIOPHYLIN-A
REFERENCES:

23410-0890

NAME: TETRANACTIN
PO: S.ROSEOCHROMOGENES
REFERENCES:
Chem. Ph. Bull., 28, 745, 1980

23410-0891

NAME: MACROTETROLIDE B
PO: S.ROSEOCHROMOGENES
REFERENCES:
JA, 28, 1000, 1975; *Arch. Mikr.*, 85, 233, 239, 1972

23410-0892

NAME: MACROTETROLIDE C
PO: S.ROSEOCHROMOGENES
REFERENCES:
JA, 28, 1000, 1975; *Arch. Mikr.*, 85, 233, 239, 1972

23410-0893

NAME: MACROTETROLIDE D
PO: S.ROSEOCHROMOGENES
REFERENCES:
JA, 28, 1000, 1975; *Arch. Mikr.*, 85, 233, 239, 1972

23410-0894

NAME: MACROTETROLIDE G
PO: S.ROSEOCHROMOGENES
REFERENCES:
JA, 28, 1000, 1975; *Arch. Mikr.*, 85, 233, 239, 1972

23420-0883

NAME: BOROMYCIN
REFERENCES:
TL, 3123, 1981; USP 4102997

23420-5322

NAME: APLASMOMYCIN-A
REFERENCES:
JACS, 101, 5826, 1979; *JA*, 33, 1316, 1980; EP 2893

23511-0915

NAME: CHLOROTHRICIN
REFERENCES:
Lloydia, 42, 455, 679, 1979; *Biochem.*, 20, 919, 1981

23512-0850

NAME: MILBEMYCIN-A"10
REFERENCES:
JA, 33, 1120, 1980; USP 3950360, 4093629, 4144352

23512-0851

NAME: MILBEMYCIN-A"9
REFERENCES:
JA, 33, 1120, 1980; USP 3950360, 4093629, 4144352

23512-0852

NAME: MILBEMYCIN-A"3
REFERENCES:
JA, 33, 1120, 1980; USP 3950360, 4093629, 4144352

23512-0853

NAME: MILBEMYCIN-A"6
REFERENCES:
JA, 33, 1120, 1980; USP 3950360, 4093629, 4144352

23512-0854

NAME: MILBEMYCIN-B"1
REFERENCES:
JA, 33, 1120, 1980; USP 3950360, 4093629, 4144352

23512-0855

NAME: MILBEMYCIN-A"5
REFERENCES:
JA, 33, 1120, 1980; USP 3950360, 4093629, 4144352

23512-0856

NAME: MILBEMYCIN-A"1
REFERENCES:
JA, 33, 1120, 1980; USP 3950360, 4093629, 4144352

23512-0857

NAME: MILBEMYCIN-A"2
REFERENCES:
JA, 33, 1120, 1980; USP 3950360, 4093629, 4144352

23512-0858

NAME: MILBEMYCIN-A"4
REFERENCES:
JA, 33, 1120, 1980; USP 3950360, 4093629, 4144352

23512-0859

NAME: MILBEMYCIN-B"2

REFERENCES:

JA, 33, 1120, 1980; USP 3950360, 4093629, 4144352

23512-0860

NAME: MILBEMYCIN-B"3

REFERENCES:

JA, 33, 1120, 1980; USP 3950360, 4093629, 4144352

23512-6142

NAME: AVERMECTIN-A1A

TRADE NAMES: IVERMECTIN

UTILITY: VETERINARY DRUG

REFERENCES:

AAC, 15, 361, 368, 372, 1979; *J. Med. Chem.*, 23, 1134, 1980; *Proc. Nat. Acad. Sci.*, 76, 2062, 1979; *Brit. Veter. J.*, 136, 88, 1980; *JACS*, 103, 4216, 4221, 1981; *Exp.*, 37, 963, 1981; USP 4160083, 4160084, 4160861, 4161583, 4172940, USP 4285963; EP 1688, 1689

23512-6143

NAME: AVERMECTIN-A1B

TRADE NAMES: IVERMECTIN

UTILITY: VETERINARY DRUG

REFERENCES:

AAC, 15, 361, 368, 372, 1979; *J. Med. Chem.*, 23, 1134, 1980; *Proc. Nat. Acad. Sci.*, 76, 2062, 1979; *Brit. Veter. J.*, 136, 88, 1980; *JACS*, 103, 4216, 4221, 1981; *Exp.*, 37, 963, 1981; USP 4160083, 4160084, 4160861, 4161583, 4172940, USP 4285963; EP 1688, 1689

23512-6144

NAME: AVERMECTIN-A2A

TRADE NAMES: IVERMECTIN

UTILITY: VETERINARY DRUG

REFERENCES:

AAC, 15, 361, 368, 372, 1979; *J. Med. Chem.*, 23, 1134, 1980; *Proc. Nat. Acad. Sci.*, 76, 2062, 1979; *Brit. Veter. J.*, 136, 88, 1980; *JACS*, 103, 4216, 4221, 1981; *Exp.*, 37, 963, 1981; USP 4160083, 4160084, 4160861, 4161583, 4172940, USP 4285963; EP 1688, 1689

23512-6145

NAME: AVERMECTIN-A2B
TRADE NAMES: IVERMECTIN
UTILITY: VETERINARY DRUG
REFERENCES:
AAC, 15, 361, 368, 372, 1979; *J. Med. Chem.*, 23, 1134, 1980; *Proc. Nat. Acad. Sci.*, 76, 2062, 1979; *Brit. Veter. J.*, 136, 88, 1980; *JACS*, 103, 4216, 4221, 1981; *Exp.*, 37, 963, 1981; USP 4160083, 4160084, 4160861, 4161583, 4172940, USP 4285963; EP 1688, 1689

23512-6146

NAME: AVERMECTIN-B1A
TRADE NAMES: IVERMECTIN
UTILITY: VETERINARY DRUG
REFERENCES:
AAC, 15, 361, 368, 372, 1979; *J. Med. Chem.*, 23, 1134, 1980; *Proc. Nat. Acad. Sci.*, 76, 2062, 1979; *Brit. Veter. J.*, 136, 88, 1980; *JACS*, 103, 4216, 4221, 1981; *Exp.*, 37, 963, 1981; USP 4160083, 4160084, 4160861, 4161583, 4172940, USP 4285963; EP 1688, 1689

23512-6147

NAME: AVERMECTIN-B1B
TRADE NAMES: IVERMECTIN
UTILITY: VETERINARY DRUG
REFERENCES:
AAC, 15, 361, 368, 372, 1979; *J. Med. Chem.*, 23, 1134, 1980; *Proc. Nat. Acad. Sci.*, 76, 2062, 1979; *Brit. Veter. J.*, 136, 88, 1980; *JACS*, 103, 4216, 4221, 1981; *Exp.*, 37, 963, 1981; USP 4160083, 4160084, 4160861, 4161583, 4172940, USP 4285963; EP 1688, 1689

23512-6148

NAME: AVERMECTIN-B2A
TRADE NAMES: IVERMECTIN
UTILITY: VETERINARY DRUG
REFERENCES:
AAC, 15, 361, 368, 372, 1979; *J. Med. Chem.*, 23, 1134, 1980; *Proc. Nat. Acad. Sci.*, 76, 2062, 1979; *Brit. Veter. J.*, 136, 88, 1980; *JACS*, 103, 4216, 4221, 1981; *Exp.*, 37, 963, 1981; USP 4160083, 4160084, 4160861, 4161583, 4172940, USP 4285963; EP 1688, 1689

23512-6149

NAME: AVERMECTIN-B2B
TRADE NAMES: IVERMECTIN
UTILITY: VETERINARY DRUG
REFERENCES:

AAC, 15, 361, 368, 372, 1979; *J. Med. Chem.*, 23, 1134, 1980; *Proc. Nat. Acad. Sci.*, 76, 2062, 1979; *Brit. Veter. J.*, 136, 88, 1980; *JACS*, 103, 4216, 4221, 1981; *Exp.*, 37, 963, 1981; USP 4160083, 4160084, 4160861, 4161583, 4172940, USP 4285963; EP 1688, 1689

23520-0922

NAME: CYTOCHALASIN-E
PO: ASP.TERREUS
REFERENCES:

J. Ferm. Techn., 57, 15, 1979; JP 79/86696; *CA*, 90, 99762, 119689; 91, 156041

23540-0932

NAME: CYTOCHALASIN-D
REFERENCES:

JP 80/88699; *CA*, 93, 166105

23540-0940

NAME: CHAETOGLOBOSIN-A
PO: CHAETOMIUM SUBAFFINE, CHAETOMIUM RECTUM, CHAETOMIUM MOLLIPILUM
REFERENCES:

Can. J. Micr., 24, 1082, 1978; 25, 170, 1979

23540-0941

NAME: CHAETOGLOBOSIN-B
PO: CHAETOMIUM SUBAFFINE, CHAETOMIUM RECTUM, CHAETOMIUM MOLLIPILUM
REFERENCES:

Can. J. Micr., 24, 1082, 1978; 25, 170, 1979

23540-3325

NAME: ASPOSTEROL
IDENTICAL: ASPOCHALASIN-B
STR-NEW:
REFERENCES:
Helv., 62, 1501, 1979

23540-5035

NAME: CHAETOGLOBOSIN-C
PO: CHAETOMIUM SUBAFFINE, CHAETOMIUM RECTUM, CHAETOMIUM MOLLIPILUM
REFERENCES:
Can. J. Micr., 24, 1082, 1978; 25, 170, 1979

23540-5036

NAME: CHAETOGLOBOSIN-D
PO: CHAETOMIUM SUBAFFINE, CHAETOMIUM RECTUM, CHAETOMIUM MOLLIPILUM
REFERENCES:
Can. J. Micr., 24, 1082, 1978; 25, 170, 1979

23540-5037

NAME: CHAETOGLOBOSIN-E
PO: CHAETOMIUM SUBAFFINE, CHAETOMIUM RECTUM, CHAETOMIUM MOLLIPILUM
REFERENCES:
Can. J. Micr., 24, 1082, 1978; 25, 170, 1979

23540-5038

NAME: CHAETOGLOBOSIN-F
PO: CHAETOMIUM SUBAFFINE, CHAETOMIUM RECTUM, CHAETOMIUM MOLLIPILUM
REFERENCES:
Can. J. Micr., 24, 1082, 1978; 25, 170, 1979

23530-0925

NAME: BREFELDIN-A
REFERENCES:
Folia Micr., 22, 43, 1977; *TL*, 3021, 1979; 2679, 1981; *JACS*, 102, 7583, 1980; *Ind. J. Exp. Biol.*, 18, 208, 1980

23530-0929

NAME: RADICICOL
PO: P.LUTEO-AURANTIUM, DIHETEROSPHORA CHLAMYDOSPORA
REFERENCES:
Lloydia, 42, 374, 1979; Belg. P 873855, 873856

23530-0930

NAME: MONORDEN
PO: MONOCILLIUM NORDINII
MW: 365 /323, 332, 337/
REFERENCES:
Can. J. Micr., 26, 766, 1980

* 23000-2344 /45000/

NAME: MUSARIN
REFERENCES:

* 23000-2345 /45000/

NAME: HYGROSTATIN, Y-5443
REFERENCES:
DT 2805197

* 23000-4028 /00000/

NAME: COPIAMYCIN
CT: AZALOMYCIN-F T.
REFERENCES:
AAC, 13, 454, 1978

24
MACROLACTAM ANTIBIOTICS

241 Ansamycins
242 Ansa-Lactams (Maytanosides)
243 Lactone-Lactams

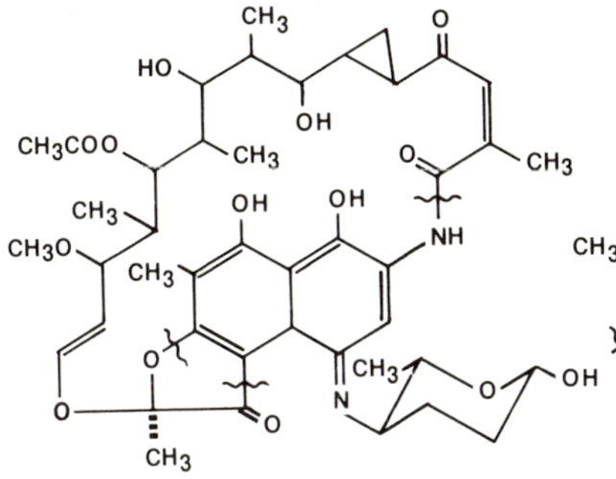

Tolypomycin R

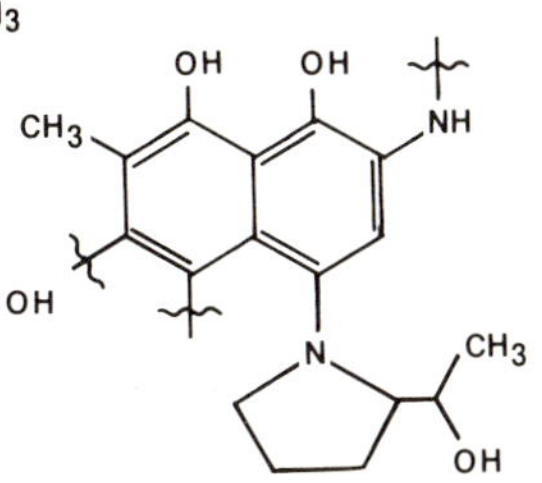

Tolypomycin RB

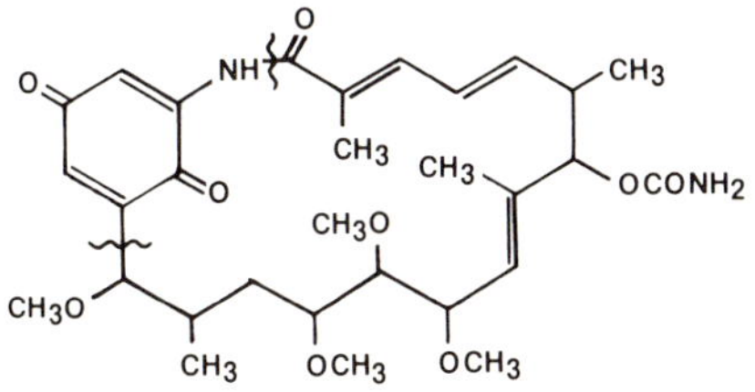

Macbecin I

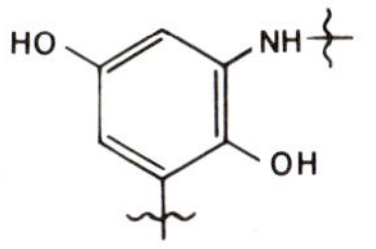

Macbecin II

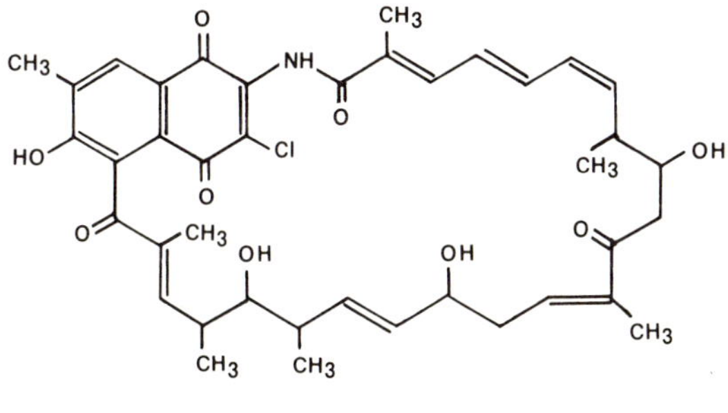

Naphtomycin

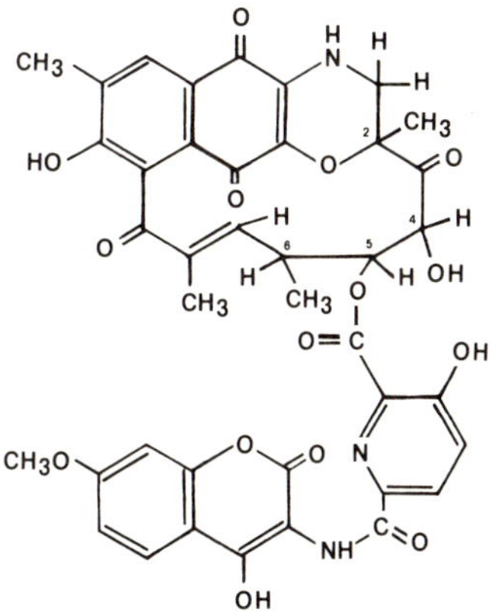

Rubradirin B
Absolute configuration: 2S, 4S, 5R, 6S

24110-0957

NAME: RIFAMYCIN-B
PO: S.ALBOBIOLACEUS, NOC.LURIDA
REFERENCES:

JA, 25, 11, 1972; *Appl. Micr.*, 13, 600, 1965; *Farmaco, Sci.*, 15, 228, 1960; 16, 165, 1961; *Exp.*, 20, 336, 339, 343, 1964; 23, 508, 1967; *Nature*, 236, 163, 166, 1972; *An. Rev. Microb.*, 26, 199, 1972; *Helv.*, 56, 2287, 1973; *Tetr.*, 30, 3087, 1974; *TL*, 1364, 1974; *Proc. Nat. Acad. Sci.*, 71, 3260, 1974; *JA*, 31, 202, 215, 1978; 33, 847, 1980; 34, 58, 965, 1981; *AAC*, 19, 134, 1981; *TL*, 2317, 1979; Belg. P 805923; JP 79/110392; 79/110393; Rum. P 65096; *CA*, 92, 20566, 56812, 56813

24110-0966

NAME: RIFAMYCIN-S
PO: NOC.SP.
REFERENCES:

JACS, 102, 7965, 1980; 103, 5553, 1981; *JA*, 34, 965, 971, 1981; EP 14181; Pol. P 99294; JP 79/110390, 79/110391

24110-0981

NAME: TOLYPOMYCIN-R
STR-NEW:
REFERENCES:

J. Med. Chem., 20, 1287, 1977; *JA*, 31, 1195, 1978

24110-0982

NAME: TOLYPOMYCIN-RB
FORMULA: C43H56N2O14
STR-NEW:
REFERENCES:

J. Med. Chem., 20, 1287, 1977; *JA*, 31, 1195, 1978

* 24110-4113 /00000/

NAME: B-2847-A"L
REFERENCES:

24110-5039

NAME: RIFAMYCIN-P
REFERENCES:

JA, 33, 842, 1980; *Tetr.*, 36, 1415, 1980

24110-5040

NAME: RIFAMYCIN-Q
REFERENCES:
JA, 33, 842, 1980; *Tetr.*, 36, 1415, 1980

24110-5041

NAME: RIFAMYCIN-R
REFERENCES:
JA, 31, 949, 1978

24110-6150

NAME: RIFAMYCIN-VERDE
REFERENCES:
JA, 33, 842, 1980; *Tetr.*, 36, 1415, 1980

24120-5044

NAME: DAMAVARICIN-C
REFERENCES:
JA, 32, 545, 1979

24120-5045

NAME: DAMAVARICIN-D
REFERENCES:
JA, 32, 545, 1979

24130-0990

NAME: GELDANAMYCIN
PO: S.HYGROSCOPICUS-DUAMYCETICUS
REFERENCES:
JA, 32, 849, 1979; JP 79/135300; *CA*, 92, 126908

24130-6493

NAME: MACBECIN-I
STR-NEW:
REFERENCES:
JA, 33, 199, 205, 1980; *TL*, 309, 1980; *Tetr.*, 37, 1123, 1981; *CA*, 92, 159321

24130-6494

NAME: MACBECIN-II
STR-NEW:
REFERENCES:
JA, 33, 199, 205, 1980; *TL*, 309, 1980; *Tetr.*, 37, 1123, 1981; *CA*, 92, 159321

24140-1289

NAME: NAPHTOMYCIN, RO-7-7961
STR-CORR:
REFERENCES:

24210-5961

NAME: ANSAMITOCIN-P2, C-15003-P2
REFERENCES:
Tetr., 35, 1079, 1979; *JA*, 33, 192, 1980; JP 79/76859

24210-5962

NAME: ANSAMITOCIN-P1, C-15003-P1
REFERENCES:
Tetr., 35, 1079, 1979; *JA*, 33, 192, 1980

24210-5963

NAME: ANSAMITOCIN-P3′, C-15003-P3′
TRADE NAMES: ANSACRIN
UTILITY: ON CLINICAL TRIAL
REFERENCES:
Tetr., 35, 1079, 1979; *JA*, 33, 192, 1980; 34, 489, 496, 1981; *J. Gen. Micr.*, 118, 411, 1980; *AAC*, 16, 101, 1979; DT 2849696, 2849666; USP 4162940

24210-5964

NAME: ANSAMITOCIN-P3, C-15003-P3
TRADE NAMES: ANSACRIN
UTILITY: ON CLINICAL TRIAL
REFERENCES:
Tetr., 35, 1079, 1979; *JA*, 33, 192, 1980; 34, 489, 496, 1981; *J. Gen. Micr.*, 118, 411, 1980; *AAC*, 16, 101, 1979; DT 2849696, 2849666; USP 4162940

24210-5965

NAME: ANSAMITOCIN-P4, C-15003-P4
TRADE NAMES: ANSACRIN
UTILITY: ON CLINICAL TRIAL
REFERENCES:
Tetr., 35, 1079, 1979; *JA*, 33, 192, 1980; 34, 489, 496, 1981; *J. Gen. Micr.*, 118, 411, 1980; *AAC*, 16, 101, 1979; DT 2849696, 2849666; USP 4162940

24220-1307

NAME: RUBRADIRIN, RUBRADIRIN-A
REFERENCES:
JA, 32, 771, 773, 1186, 1979; *Biochem.*, 12, 1136, 1973; *JOC*, 46, 2426, 1981; *TL*, 3381, 1981; DT 2816082

24220-6351

NAME: RUBRADIRIN-B
STR-CORR:
REFERENCES:
DT 2810264; *CA*, 90, 53129

Quinone and Similar Antibiotics

31
TETRACYCLIC COMPOUNDS AND ANTHRAQUINONES

311 Tetracyclines
312 Anthracyclines*
313 Anthraquinone Derivatives

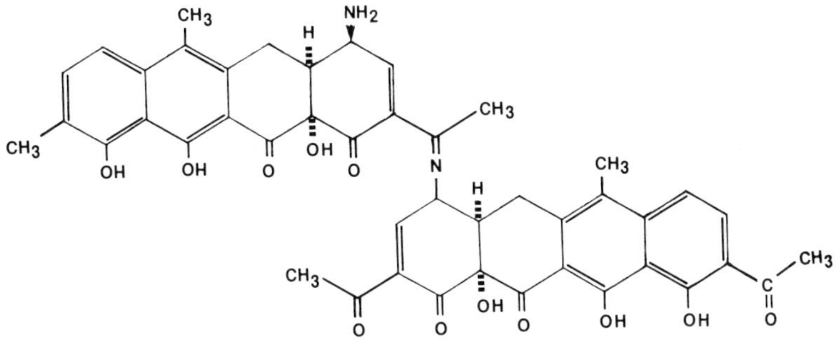

Isochelocardin

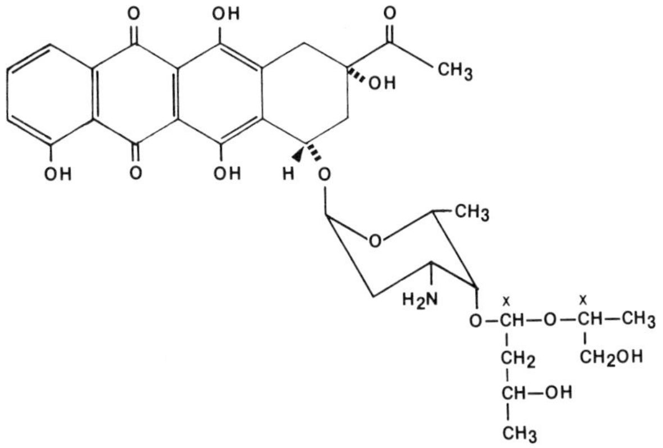

Carminomycin 2
Carminomycin 3

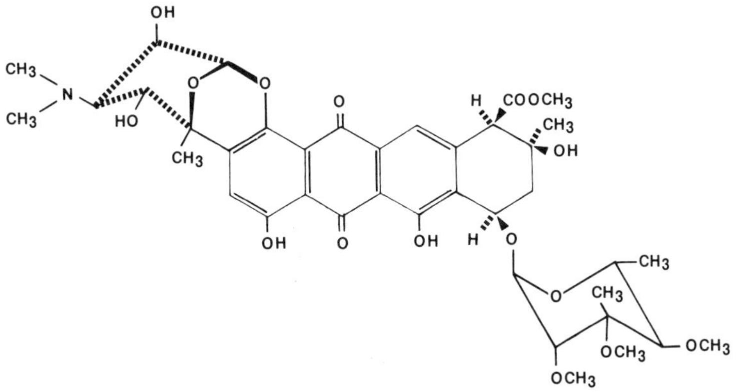

Nogalamycin

* The sterochemistry of anthracyclines and their aglycones should be inverted at C-7, C-9, and C-10 on pages 62 to 63 and 77 in Volume III.

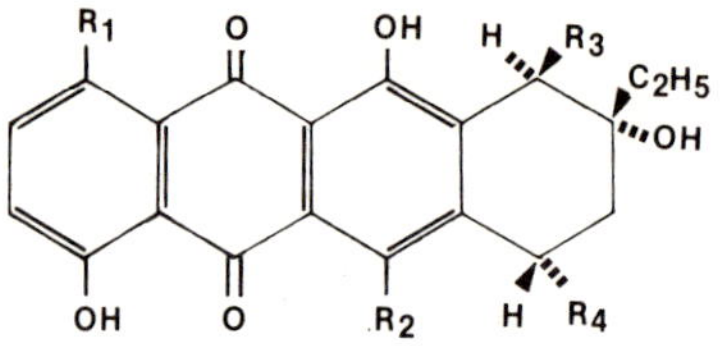

	R_1	R_2	R_3	R_4
Rhodomycin A (β-rhodomycin II)	H	OH	O–Roa	O–Roa
Rhodomycin B (β-rhodomycin I)	H	OH	OH	O–Roa
γ-Rhodomycin I	H	OH	O–Roa	H
Violamycin A_1	OH	OH	OH	O–Roa
Violamycin A_2	OH	H	OH	O–Roa
γ-Rhodomycin II	H	OH	O–Roa–Roa	H

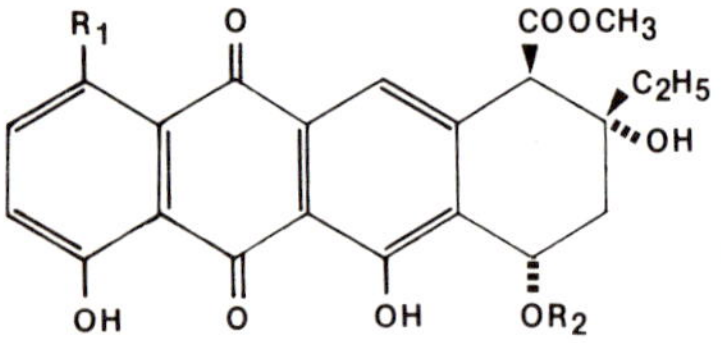

	R_1	R_2
Pyrromycin	OH	O–Roa
Aklavin	H	O–Roa
Cinerubin A	OH	O–Roa–dF–LcinA
Rudolphomycin	OH	O–Roa–dF–Red

Red=L-rednose:

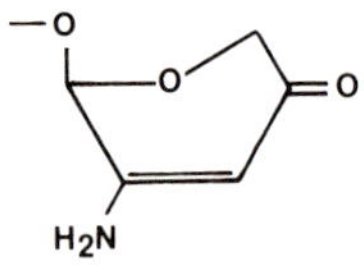

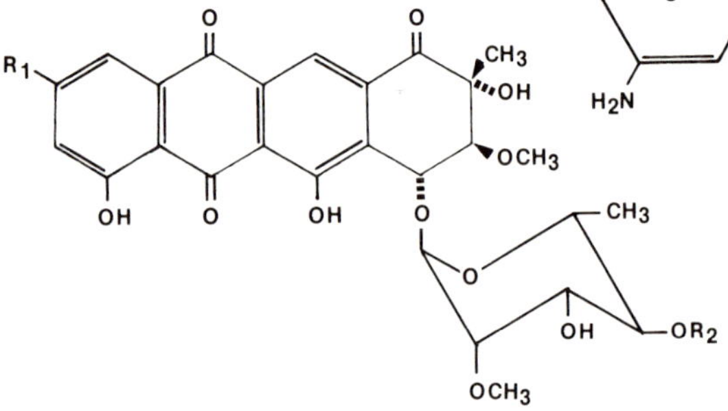

	R_1	R_2
Steffimycin	OCH_3	H
Aranciamycin	H	H
Steffimycin B	OCH_3	CH_3

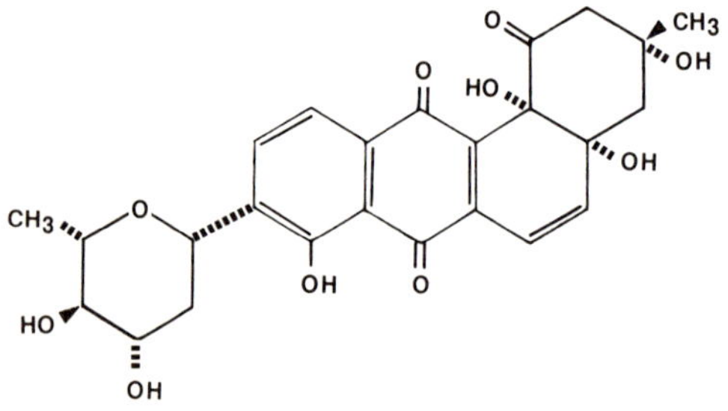

Aquayamycin

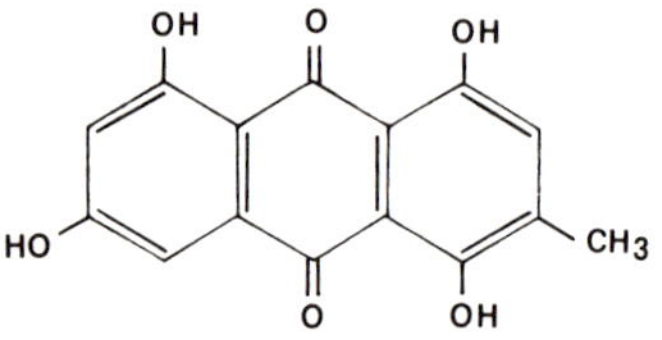

Catenarin

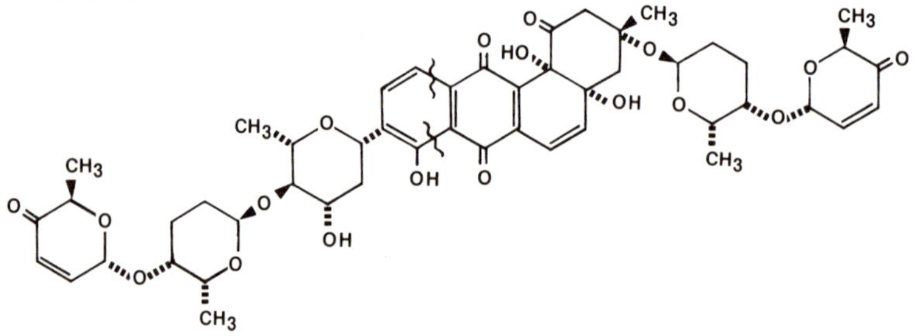

Vineomycin A_1 (OS-4742-A_1)

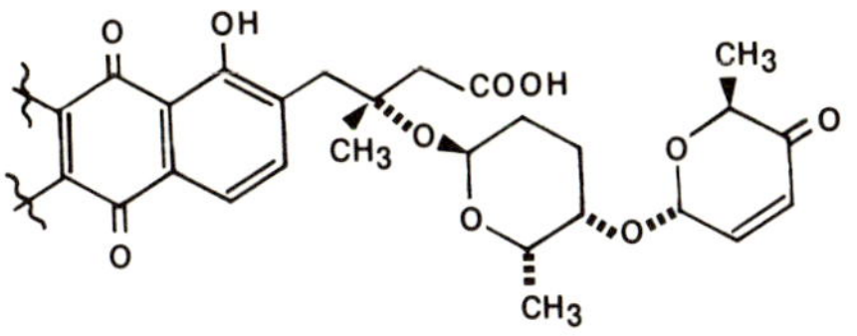

Vineomycin B_2 (OS-4742-B_2)

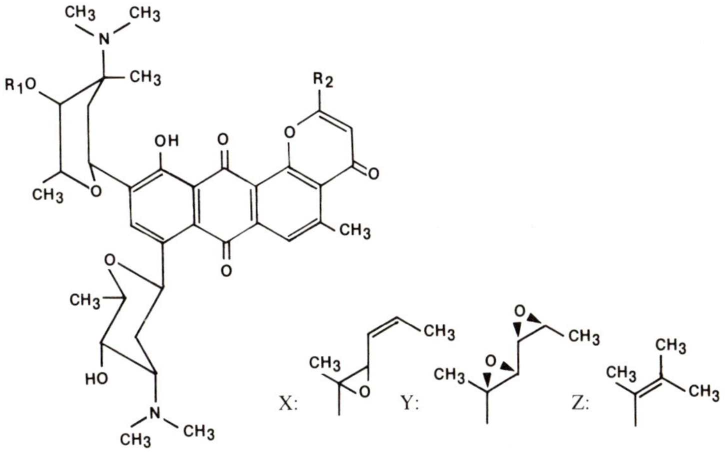

X: Y: Z:

	R_1	R_2
Pluramycin A	$COCH_3$	X
Kidamycin-F (Rubiflavin-B)	H	Z
Hedamycin	H	Y
Rubiflavin-A (Desacetylpluramycin-A)	H	X
Neopluramycin	$COCH_3$	Z

31111-5979

NAME: DOXYCYCLINE
REFERENCES:
JACS, 85, 2643, 1963; *BBRC*, 18, 325, 1965; *Proc. Antib. Symp., Weimar*, 1979, C-23; Fr. P 2187301

31112-1011

NAME: CHELOCARDIN
REFERENCES:
Can. J. Chem., 56, 1059, 1978; USP 4104306

31112-6491

! NAME: ISOCHELOCARDIN
STR-NEW:
REFERENCES:

31120-1012

NAME: PILLAROMYCIN-A
REFERENCES:
JACS, 97, 6250, 1975; *Can. J. Chem.*, 58, 2694, 1980; *JOC*, 45, 4241, 1980

31211-1035

NAME: G"-RHODOMYCIN-I
IDENTICAL: IREMYCIN
STR-CORR:
REFERENCES:
JA, 34, 1457, 1981

31211-1036

NAME: G"-RHODOMYCIN-II
IDENTICAL: ROSEORUBICIN-B
STR-CORR:
REFERENCES:
JA, 32, 420, 1979

31211-1039

NAME: RHODOMYCIN-Y
IDENTICAL: G"-RHODOMYCIN-ROA.DEOFUC.ROD
MW: 760
REFERENCES:

31211-1041

NAME: G"-RHODOMYCIN-ROA2.ROD
PO: S.PURPURASCENS
REFERENCES:

31211-1043

NAME: RETAMYCIN, RETAMICINA
TV: HERPES
REFERENCES:
Rev. Inst. Antib., 15, 25, 1975

31211-1047

NAME: B-5794
IDENTICAL: G"-RHODOMYCIN COMPLEX /G"-RHODOMYCINONES/
REFERENCES:

31211-1048

! NAME: VIOLAMYCIN-A, VIOLAMYCIN-A1 /VIOLAMYCIN-A1-A6/
IDENTICAL: ISORHODOMYCIN-B
TV: P-388, HERPES
STR-CORR:
REFERENCES:
Proc. Antib. Symp., Weimar, 1979, A-16, A-22

31211-1049

! NAME: VIOLAMYCIN-B1 /VIOLAMYCIN-B1-B6/
TV: P-388,HERPES
REFERENCES:
Proc. Antib. Symp., Weimar, 1979, A-17, A-18, A-19, A-20, A-21,
A-23, A-25

31212-1058

NAME: CINERUBIN-A
IDENTICAL: G-4, 5888-III
STR-CORR:
REFERENCES:
JA, 33, 49, 1331, 1341, 1980; *Z. Naturforsch. Ser. C*, 34, 1024, 1979

31212-5668

NAME: MUSETTAMYCIN, C-36145, NSC-219941, NSC-284671
IDENTICAL: PYRROCYCLINE-A
REFERENCES:
Lloydia, 40, 611, 1978; 42, 242, 1980; USP 4064014

31212-5669

NAME: MARCELLOMYCIN
IDENTICAL: PYRROCYCLINE-B
UTILITY: ON CLINICAL TRIAL
REFERENCES:
Lloydia, 43, 242, 1980; *Mol. Pharmacol.*, 14, 290, 1978; *Cancer & Chemoth.*, 1, 313, 1980

31212-5816

NAME: RHODIRUBIN-A
REFERENCES:
JA, 32, 472, 1979; DT 2738656

31212-5876

NAME: RHODIRUBIN-B
REFERENCES:
JA, 32, 472, 1979; DT 2738656

31212-5975

NAME: RHODIRUBIN-D
IDENTICAL: MUSETTAMYCIN,MA-144-S2 /MARCELLOMYCIN, MA-144-U2/
REFERENCES:
JA, 32, 472, 1979; DT 2738656

31212-6603

NAME: RUDOLPHOMYCIN, NSC-267695, NSC-293858
IDENTICAL: PYRROCYCLINE-C
PO: S.SP.
REFERENCES:

Lloydia, 40, 611, 1977; 43, 242, 1980; *JACS*, 101, 7041, 1979; USP 4123608, 4162938; JP 81/25183; *CA*, 95, 95467

31213-1064

NAME: AKLAVIN
IDENTICAL: 5888-II
REFERENCES:

JACS, 103, 4247, 4248, 4251, 1981

31213-4778

NAME: ACLACINOMYCIN-A, NSC-208734
TRADE NAMES: ACLARUBICIN, ACLACINON
IDENTICAL: G-1
PO: S.GALILAEUS-SIVENENSIS
REFERENCES:

JA, 31, 1149, 1978; 33, 1323, 1980; *Jap. J. Ant.*, 33, 618, 1980; *Gann*, 70, 403, 411, 1979; *Acta Micr. Sinica*, 19, 365, 1979; *JA*, 32, 791, 801, 1979; 33, 1331, 1980; 34, 47, 1981; USP 3988315, 4219622; JP 76/15690, 76/19701, 76/34915

31213-5855

NAME: MA-144-T1
REFERENCES:

USP 4219622 DT 2715255; JP 79/38104; 80/120786; *CA*, 93, 24457

31213-5857

NAME: MA-144-L1
REFERENCES:

USP 4219622 DT 2715255; JP 79/38104; 80/120786; *CA*, 93, 24457

31213-5858

NAME: MA-144-M1
REFERENCES:

USP 4219622 DT 2715255; JP 79/38104; 80/120786; *CA*, 93, 24457

31213-5860

NAME: MA-144-G1

REFERENCES:

USP 4219622 DT 2715255; JP 79/38104; 80/120786; *CA*, 93, 24457

31213-5861

NAME: MA-144-N1

REFERENCES:

USP 4219622 DT 2715255; JP 79/38104; 80/120786; *CA*, 93, 24457

31213-5862

NAME: MA-144-S1

REFERENCES:

USP 4219622 DT 2715255; JP 79/38104; 80/120786; *CA*, 93, 24457

31213-6062

! NAME: ACLACINOMYCIN-Y

REFERENCES:

JA, 32, 472, 791, 801, 1979; USP 4207313

31214-1069

NAME: DAUNOMYCIN

REFERENCES:

JACS, 100, 6188, 1978; 102, 5881, 1980; *JA*, 31, 336, 1978; 32, 223, 1038, 1979; *J. Liquid Chrom.*, 2, 533, 1979; *Folia Micr.*, 23, 249, 1978; 24, 117, 1979; 22, 275, 1977; *J. Chrom.*, 198, 407, 1980

31214-1070

NAME: ADRIAMYCIN

REFERENCES:

JA, 31, 336, 1978; 33, 1038, 1980; 34, 1229, 1981; *JACS*, 100, 6188, 1978; *J. Pharm. Sci.*, 70, 265, 1981; USP 3803124

31214-1085

NAME: CARMINOMYCIN-1,CARMINOMYCIN

REFERENCES:

JA, 27, 254, 1974; *JACS*, 100, 3635, 1978; *Biotech. Lett.*, 1, 471, 1979; *Z. Allg. Mikr.*, 20, 219, 1980; USP 4189568

31214-1086

NAME: CARMINOMYCIN-2, CARMINOMYCIN-II, 4-HYDROXYBAUMYCIN-A2
IDENTICAL: D-326-III, RUBEOMYCIN-A, DF-4466-A
FORMULA: C33H41NO13
MW: 659
STR-NEW:
REFERENCES:

Antib., 488, 492, 563, 1980; *JA*, 34, 774, 1981

31214-1087

NAME: CARMINOMYCIN-3, CARMINOMYCIN-III, 4-HYDROXYBAUMYCIN-A1
IDENTICAL: D-326-IV, RUBEOMYCIN-A1, DF-4466-B
FORMULA: C33H41NO13
MW: 659
STR-NEW:
REFERENCES:

Antib., 488, 492, 563, 1980; *JA*, 34, 774, 1981

31214-5323

NAME: DIHYDROCARMINOMYCIN
PO: ACTINOMADURA CARMINATA
REFERENCES:

Antib., 492, 1980; Swiss P 614974; *CA*, 93, 112306

31214-5818

NAME: BAUMYCIN-A1, BAUNOMYCIN
REFERENCES:

JP 80/19254; EP 26849; *CA*, 93, 68658

31221-1090

NAME: NOGALAMYCIN
UTILITY: ON CLINICAL TRIAL (NOGAMYCIN)
STR-CORR:
REFERENCES:

JOC, 43, 3457, 1978; 44, 4030, 1979; *Lloydia*, 42, 568, 1979; *TL*, 1153, 1979; *Cancer Res.*, 39, 4816, 1979; *Recent Results Canc. Res.*, 70, 21, 1980

31221-1098

NAME: BEROMYCIN-B
REFERENCES:
Folia Micr., 25, 207, 214, 1980

31221-1099

NAME: BEROMYCIN-C
REFERENCES:
Folia Micr., 25, 207, 214, 1980

31222-1121

NAME: ARANCIAMYCIN
IDENTICAL: SM-173-A
PO: S.CHROMOFUSCUS
FORMULA: C27H28012 /C27H25012/
STR-CORR:
REFERENCES:
CA, 93, 130479

31222-1123

NAME: STEFFIMYCIN
MW: 558 /574, 568/
STR-CORR:
REFERENCES:
JOC, 43, 3457, 1978; USP 4077844

31222-6211

NAME: SM-173-B
CT: ANTHRACYCLINE,STEFFIMYCIN T. /ANTHRACYCLINONE/
REFERENCES:
DT 2804495; *CA*, 92, 144895; 93, 130479

* 31200-1423 /30000/

NAME: 54
CT: ANTHRACYCLINE /QUINONE T./
EA: (C, 57)(H, 5)
PC: RED
OR: (+80.5, CHL)
UV: MEOH: (234, 748,)(255, 383,)(492, 177,)
UV: MEOH-HCL: (235, 880,)(258, 460,)(492, 247,)
UV: MEOH-NAOH: (242, 830,)(567, 305,)
LD50: (125, IV)
IS-EXT: (ETOAC, 9.5, FILT.)(ACET, , MIC,)
REFERENCES:

31310-1141

NAME: ENDOCROCIN
IDENTICAL: CLAVOXANTHIN
PO: CLAVICEPS PURPUREA, ASP.SP., P.SP.
MW: 314
UV: MEOH: (274, , 21000)(287, , 15000)(311, , 8300)(442, , 10600)
REFERENCES:
JA, 32, 1256, 1979

31310-4780

NAME: EMODIN
PO: PYRENOCHAETA TERRESTRIS
TO: (B.SUBT.,)(P.VULG.,)
TV: EHRLICH
REFERENCES:
JA, 32, 1256, 1979; *Arch. Mikr.*, 126, 223, 231, 1980; *CA*, 93, 88550

31310-5568

NAME: CYNODONTHIN
PO: HELMINTHOSPORIUM SP.
UV: ETOH: (241, ,)(295, ,)(483, ,)(539, ,)(552, ,)
REFERENCES:

31310-6063

NAME: CATENARIN
PO: ASP.CRISTATUS
TO: (B.SUBT.,)(P.VULG.,)
TV: EHRLICH
STR-NEW:
REFERENCES:
Arch. Mikr., 126, 223, 231, 1980; *Liebigs Ann.*, 2247, 1981

31310-6604

NAME: PHYSCION
IDENTICAL: CHRYSOROBIN, PARIETHIN
PO: P.HERQUEI, ASP.CRISTATUS
TO: (FUNGI,)
REFERENCES:
J. Agr. Food Chem., 28, 1139, 1980; *Arch. Mikr.*, 126, 223, 231, 1980

31320-1144

NAME: LUTEOSKYRIN
REFERENCES:
J. Agr. Food Chem., 24, 964, 1976; *Tetr.*, 29, 3703, 1973

31320-5978

NAME: RUGULIN
REFERENCES:
Cz. P 187049; *CA*, 95, 202076

31330-1149

NAME: AQUAYAMYCIN
STR-CORR:
REFERENCES:
Chem. Ph. Bull., 29, 1788, 1981

31330-5329

NAME: 792
REFERENCES:
Antib., 3, 1977

31330-5555

NAME: <u>YORONOMYCIN</u>, SEN-136-A
REFERENCES:
Abst. Ann. Meet. Agr. Ch. Soc. Jap., Kyoto, 1976, 21-8

31330-5966

! NAME: <u>VINENOMYCIN-A1</u>, OS-4742-A1
IDENTICAL: P-1894-B
STR-NEW:
REFERENCES:
Chem. Ph. Bull., 29, 1788, 1981; *JA*, 34, 1517, 1981

31330-5967

NAME: <u>VINENOMYCIN-A2</u>, OS-4742-A2
REFERENCES:

31330-5968

! NAME: <u>VINENOMYCIN-B1</u>, OS-4742-B1
REFERENCES:

31330-5969

NAME: <u>VINENOMYCIN-B2</u>, OS-4742-B2
FORMULA: C49H58018 /C65H54017/
STR-NEW:
REFERENCES:

31350-1157

NAME: <u>PLURAMYCIN-A</u>
REFERENCES:
Helv., 61, 2241, 1978

31350-1160

NAME: <u>RUBIFLAVIN</u>, RUBIFLAVIN-A, DESACETYLPLURAMYCIN-A
FORMULA: C41H50N2010 /C25H30NO5/
MW: 730 /412/
STR-NEW:
REFERENCES:
Helv., 63, 2446, 1980

31350-1162

NAME: HEDAMYCIN
STR-CORR:
REFERENCES:
Tetr., 34, 3623, 1978; *TL*, 3703, 1979

31350-1170

NAME: KIDAMYCIN-F, KIDAMYCIN
IDENTICAL: RUBIFLAVIN-B
STR-CORR:
REFERENCES:
Tetr., 43, 3623, 1979; *Helv.*, 63, 2446, 1980

31350-5234

NAME: GRISEORUBIN COMPLEX
PC: VIOLET
TO: (MYCOPLASMA SP., .31)(PROTOZOA, .1)
TV: WALKER-256
REFERENCES:
JA, 33, 1, 9, 1980

31300-4685

NAME: MADURAMYCIN
REFERENCES:
Proc. Antib. Symp., Weimar, 1979, B-32, D-10; *Zbl. Bakt. Parasit., Abt. I, Suppl.*, 377, 1976; *CA*, 88, 168436

32
NAPHTOQUINONE DERIVATIVES

321 Simple Naphtoquinones
322 Condensed Naphtoquinones

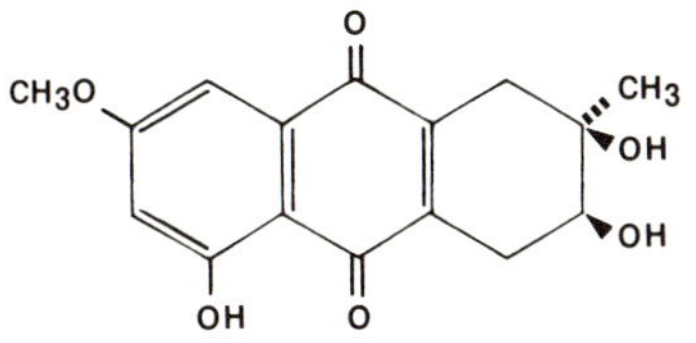

Dactylarin (Altersolanol B)

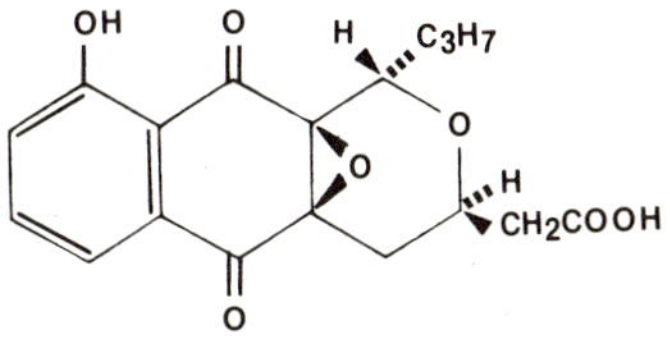

Frenolicin

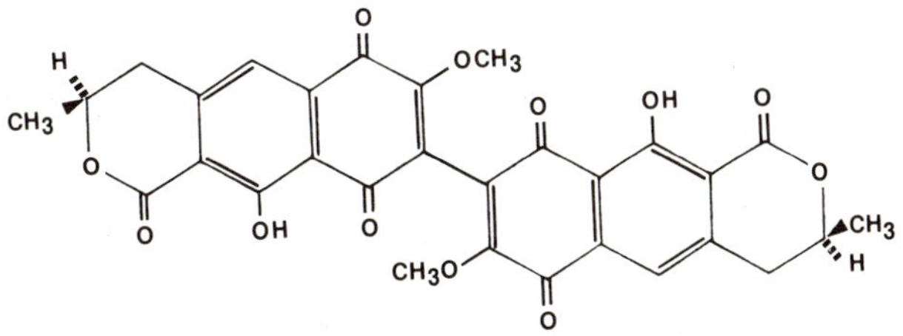

Xanthomegnin

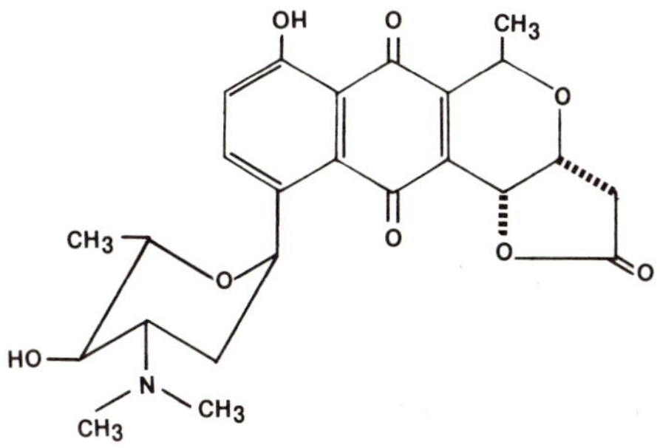

Medermycin

32120-1186

NAME: JUGLOMYCIN-A
REFERENCES:
JCS Perkin I, 2091, 1981

32120-1187

NAME: JUGLOMYCIN-B
REFERENCES:
JCS Perkin I, 2091, 1981

32120-1199

NAME: FUSARUBIN
PO: FUS.JAVANICUM, FUS.MARTII, NEOCOSMOSPORA VASINFECTA
PC: ORANGE
UV: MEOH: (301, , 6800)(472, , 5100)(488, , 5700) (525, , 4200)
TO: (S.LUTEA, 21)(S.AUREUS, 50)(E.COLI, 75)(C, ALB., 30)(S.CEREV., 100)
LD50: PHYTOTOXIC
TV: ANTITUMOR
REFERENCES:
J. Chrom., 133, 291, 1977; *Ann. Phytopathol.*, 10, 327, 1978; *JA*, 32, 679, 685, 1979; 33, 1376, 1980

32120-1200

NAME: JAVANICIN
PO: NEOCOSMOSPORA VASINFECTA
REFERENCES:
Ann. Phytopathol., 10, 327, 1978

32120-1201

NAME: NOVARUBIN
PO: NEOCOSMOSPORA VASINFECTA
REFERENCES:
Ann. Phytopathol., 10, 327, 1978

32120-1202

NAME: NORJAVANICIN
PO: NEOCOSMOSPORA VASINFECTA
REFERENCES:
Ann. Phytopathol., 10, 327, 1978

32211-1213

NAME: BOSTRYCIN
PO: ALT.EICHORNIAE, ARTHRINIUM PHAEOSPERMUM
REFERENCES:
Phytoch., 18, 1579, 1979; *J. Chem. Res. (S)*, 306, 1978; *CA*, 92, 106982

32211-3539

NAME: DACTYLARIN
STR-CORR:
REFERENCES:
JA, 34, 708, 1981

32211-6114

NAME: ALTERSOLANOL-B
PO: ALT.PORRI
UV: MEOH: (217, , 3700)(265, , 9700)(284, , 14200)
(422, , 18600)
STR-CORR:
REFERENCES:
Can. J. Chem., 50, 122, 1972; *TL*, 4309, 1978; 2481, 1979; *Agr. Biol. Ch.*, 42, 1801, 1978

32221-1234

NAME: ACTINORHODIN
REFERENCES:
Lloydia, 42, 691, 1979; *JOC*, 46, 455, 1981

32222-1239

NAME: PURPUROMYCIN
IDENTICAL: 4041-I
PO: ACTINOPLANES IANTHOGENES-OCTAMYCINI
REFERENCES:
Antib., 563, 582, 1979

32222-1242

NAME: G"-NAPHTOCYCLINONE
REFERENCES:
JOC, 43, 1438, 1978; *Planta Med.*, 34, 345, 1978

32222-1243

NAME: B"-NAPHTOCYCLINONE

REFERENCES:

JOC, 43, 1438, 1978; *Planta Med.*, 34, 345, 1978

32222-1244

NAME: A"-NAPHTOCYCLINONE

REFERENCES:

JOC, 43, 1438, 1978; *Planta Med.*, 34, 345, 1978

32222-1247

NAME: GRISEORHODIN-A

REFERENCES:

Tetr., 34, 2693, 1978; 35, 1621, 1979

32222-1248

NAME: GRISEORHODIN-B

REFERENCES:

Proc. IUPAC Conf. Nat. Prod., Varna, 1978, 1, 217, 1979

32222-1249

NAME: GRISEORHODIN-C

REFERENCES:

JA, 31, 970, 1978; 32, 197, 1979; DDR P 133056

32222-6115

NAME: GRISEORHODIN-C2

REFERENCES:

Proc. IUPAC Conf. Nat. Prod., Varna, 1978, 1, 217, 1979

32222-6565

NAME: GRISEORHODIN-G, FCRC-57G
PO: S.GRISEUS /S.SP./
MW: 510 /526/
UV: MEOH: (233, , 59900)(315, , 8880)(355, , 6950)
(505, , 7350)(545, , 5100)
TO: (B.SUBT.,)(S.AUREUS,)
TV: KB
IS-EXT: (ETOAC-HCL, , MIC.)
IS-CHR: (SILG, CHL-ACOH-MEOH)
REFERENCES:
Proc. 12th Region. Meet. ACS, 1978, OR-37

32222-6566

NAME: FCRC-57U
PO: S.GRISEUS /S.SP./
MW: 526 /510/
UV: MEOH: (230, , 5880)(275, , 12000)(310, , 7700)
(360, , 6000)(505, , 6150)(540, , 4000)
TO: (B.SUBT.,)(S.AUREUS,)
TV: KB
REFERENCES:

32223-1261

NAME: GRANATICIN
IDENTICAL: GRANATOMYCIN-C
PO: S.THERMOVIOLACEUS, S.LATERITUS
TO: (P.VULG.,)(PS.ARE.,)(PROTOZOA, 1)
TV: P-388
REFERENCES:
JACS, 101, 7018, 1978; *Lloydia*, 42, 627, 1979; *Planta Med.*, 34, 345, 178; *Z. Allg. Mikr.*, 20, 543, 1980; DDR P 147115

32223-5410

NAME: DIHYDROGRANATICIN
IDENTICAL: GRANATOMYCIN-D
PO: S.LATERITUS
UV: ETOH: (224, ,)(284, ,)(484, ,)(523, ,)
(556, ,)
TO: (P.VULG.,)(PS.AER.,)
REFERENCES:
Z. Allg. Mikr., 20, 543, 1980; *Helv.*, 62, 30, 1979; DDR P 147115

32224-1268

NAME: BIKAVERIN
TV: CYTOTOXIC
REFERENCES:
JA, 31, 615, 1978; *Chem. Ph. Bull.*, 26, 209, 1978; *Heterocycl.*, 12, 1, 1979

32224-1269

NAME: NORBIKAVERIN
TV: CYTOTOXIC
REFERENCES:
Chem. Ph. Bull., 26, 209, 1978

32224-1277

NAME: MARTICIN
PO: NEOCOSMOSPORA VASINFECTA
REFERENCES:
Ann. Phytopath., 10, 327, 1978

32224-1278

NAME: ISOMARTICIN
PO: NEOCOSMOSPORA VASINFECTA
REFERENCES:
Ann. Phytopath., 10, 327, 1978

32225-1267

NAME: KALAFUNGIN
REFERENCES:
JACS, 100, 6263, 1978; *JOC*, 43, 4923, 1978

32225-1271

NAME: DESOXYFRENOLICIN, DEOXYFRENOLICIN
REFERENCES:
JACS, 88, 4109, 1966; *JA*, 31, 959, 1978; JP 79/110396; *CA*, 92, 20565

32225-1272

NAME: FRENOLICIN
STR-CORR:
REFERENCES:
TL, 4469, 1980

32225-3849

NAME: NANAOMYCIN-A
REFERENCES:
JACS, 100, 6263, 1978; *TL*, 4469, 1980; *Jap. J. Ant.*, 33, 728, 1980; *JA*, 33, 711, 1980; *JCS Perkin I*, 1197, 1981; *J. Bioch. (Tokyo)*, 90, 355, 1981

32225-3850

NAME: NANAOMYCIN-B
REFERENCES:
JP 77/128230; *CA*, 93, 46466

32225-5048

NAME: NANAOMYCIN-D
REFERENCES:
JA, 33, 711, 1980; JP 78/35591; *CA*, 88, 168489

32225-5049

NAME: GRISEUSIN-A
REFERENCES:
Heterocycl., 16, 1659, 1981

32225-6352

NAME: FRENOLICIN-B
REFERENCES:
EP 4128; *CA*, 92, 20563

32226-1270

NAME: XANTHOMEGNIN
PO: P.SP., ASP.SP., NANIZZIA CAJETANI, MICROSPORUM COOKEI
UV: MEOH-NAOH: (265, ,)(398, ,)(540, ,)
TO: (B.SUBT.,) /ANTIBACTERIAL/
STR-CORR:
REFERENCES:

Phytoch., 4, 505, 1965; *Exp.*, 26, 803, 1970; *Appl. Micr.*, 33, 351, 1977; 42, 446, 1981; *Ber.*, 112, 957, 1979; *Coll.*, 46, 1210, 1981; *Diss. Abst.*, 40, 1107; *CA*, 91, 204976

32226-1279

NAME: LAMBERTELLIN
REFERENCES:

Phytopath., 52, 753, 1962; *Lloydia*, 28, 359, 1965; *JCS*, 5927, 1965; *JCSC*, 109, 1970; *Trans. Brit. Mycol. Soc.*, 36, 109, 1953; *Phytoch.*, 17, 1804, 1978; USP 3438998

32230-1284

NAME: BOSTRYCOIDIN
REFERENCES:

TL, 5089, 1980

32310-1421

NAME: MEDERMYCIN
FORMULA: C24H27NO8 /C23H29NO8/
MW: 457 /496/
STR-NEW:
REFERENCES:

Mass Spectr., 27, 97, 1979

32320-1300

NAME: XANTHOMYCIN
REFERENCES:

JA, 34, 856, 1981; Hung. P 148963

32320-5549

NAME: SF-1739
REFERENCES:

DT 2904628; BP 2017079; *CA*, 88, 119253

33
BENZOQUINONE DERIVATIVES

331 Simple Benzoquinones
332 Condensed Benzoquinones

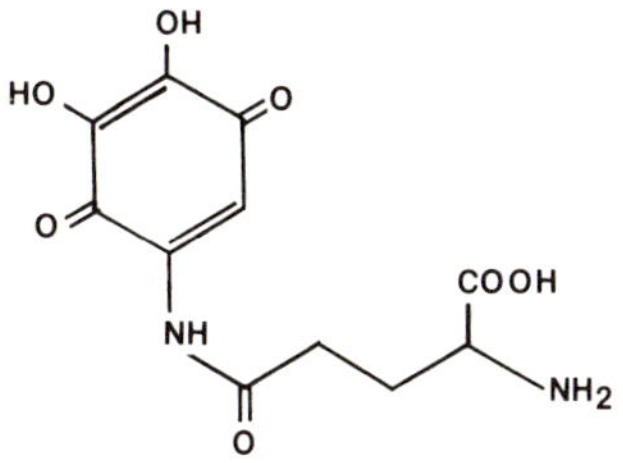

490-quinone

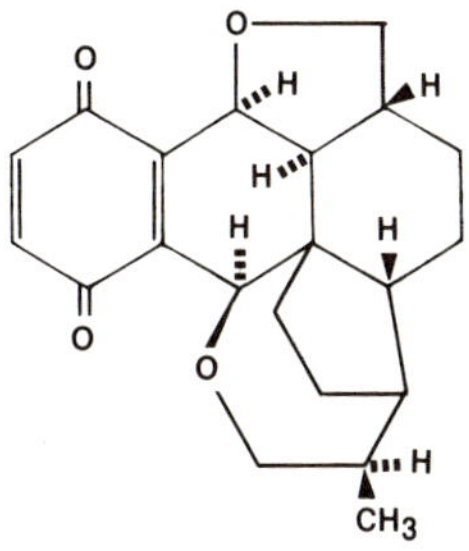

Pleurotin

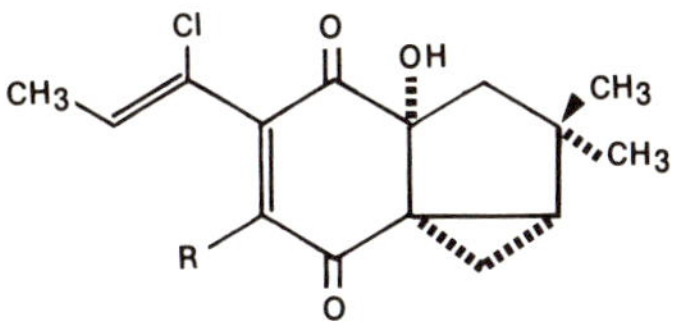

Mycorrhizin R = H
Chloromycorrhizin R = Cl

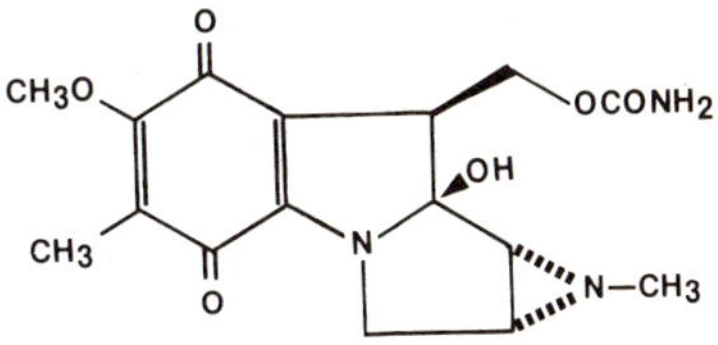

Mitomycin-B

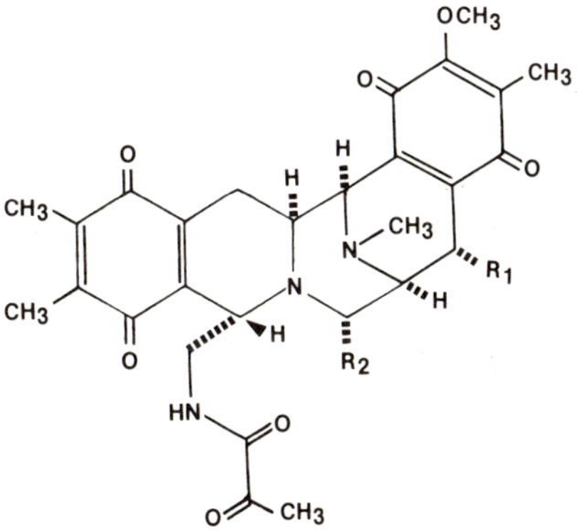

	R_1	R_2
Saframycin A	H	CN
Saframycin B	H	H
Saframycin C	OCH_3	H

33110-6492

NAME: 490-QUINONE
FORMULA: C11H12N2O6
EA: (N, 10) /(N,)/
MW: 268
STR-NEW:
REFERENCES:

Proc. Nat. Acad. Sci., 67, 1050, 1970; *J. Biol. Chem.*, 246, 2010, 2015, 1971; *Am. J. Pathol.*, 76, 165, 1974; 78, 33, 1975; *Cancer Res.*, 37, 436, 1133, 1977

33140-1334

NAME: PLEUROTIN
IDENTICAL: COMPOUND (NO NAME) 6663, P1
STR-CORR:
REFERENCES:

Arzn. Forsch., 31, 293, 1981; *CA*, 91, 173367

33140-5327

NAME: ASTERRIQUINONE
REFERENCES:

Chem. Ph. Bull., 29, 961, 991, 1005, 1981

33140-6116

NAME: MYCORRHIZIN
PO: MONOTROPA HYPOPITYS, GILMANIELLA HUMICOLA
STR-CORR:
REFERENCES:

Phytoch., 17, 1359, 1978; *TL*, 2915, 1981

33140-6117

NAME: CHLOROMYCORRHIZIN
PO: MONOTROPA HYPOPITYS, GILMANIELLA HUMICOLA
STR-CORR:
REFERENCES:

Phytoch., 17, 1359, 1978; *TL*, 2915, 1981

33211-1336

NAME: MITOMYCIN-B
STR-CORR:
REFERENCES:

33211-1337

NAME: MITOMYCIN-C
PO: MIC.SP.
REFERENCES:
TL, 3207, 1978; *Ind. J. Technol.*, 15, 85, 1977; *Farmaco, Sci.*, 33, 651, 1978; *Heterocycl.*, 9, 11, 1978; 13 (Spec.), 373, 411, 1979; *Lloydia*, 42, 549, 1979; *Bull. Ch. Soc. Jap.*, 52, 2334, 1979; *JA*, 33, 804, 1980; EP 8021; JP 78/107487; *CA*, 88, 4686; 90, 53134; 93, 43923

33211-1344

NAME: PORFIROMYCIN
PO: MIC.SP.
REFERENCES:
JP 78/107487

33220-1350

NAME: STREPTONIGRIN
REFERENCES:
TL, 3207, 1978; *JACS*, 103, 1271, 1981

33240-5252

NAME: MIMOSAMYCIN
REFERENCES:
Chem. Ph. Bull., 26, 2175, 1978; *JA*, 31, 847, 1978; JP 77/122301; *CA*, 88, 49009

* 33250-5253 /33240/

NAME: CHLOROCARCIN-A
REFERENCES:

* 33250-5254 /33240/

NAME: CHLOROCARCIN-B
REFERENCES:

* 33250-5255 /33240/

NAME: CHLOROCARCIN-C
REFERENCES:

* 33250-5970 /33240/

NAME: SAFRAMYCIN-A, 21-CYANOSAFRAMYCIN-B
UV: MEOH: (370, ,)
LD50: (12, IV)(40, IP)
TV: P-388, L-1210, EHRLICH
IS-EXT: (CHL, 8, FILT.,)(W, 1, ETOAC)
IS-CHR: (SILG, ETOAC-MEOH)(SEPHADEX LH-20)
IS-CRY: (CRYST., ET20)
REFERENCES:

Exp., 36, 1025, 1980; *Biochem.*, 17, 2545, 1978; *Gann,* 71, 790, 1980; *JA*, 33, 951, 1980; *Can. J. Chem.*, 59, 2945, 1981; *CA*, 92, 142076; 93, 93466

* 33250-5971 /33240/

NAME: SAFRAMYCIN-B
LD50: (250, IP)
TV: L-1210, ANTITUMOR
REFERENCES:

Exp., 36, 1025, 1980; *TL*, 2355, 1979; *Chiba Igakku Zasshi,* 56, 337, 1980; *CA*, 94, 150362

* 33250-5972 /33240/

NAME: SAFRAMYCIN-C
LD50: (250, IP)
TV: L-1210, P-388, EHRLICH
REFERENCES:

Gann, 71, 790, 1980; *TL*, 2355, 1979; *Can. J. Chem.*, 59, 2945, 1981

* 33250-5973 /33240/

NAME: SAFRAMYCIN-D
REFERENCES:

* 33250-5974 /33240/

NAME: SAFRAMYCIN-E
REFERENCES:

34
QUINONE-LIKE COMPOUNDS

341 Semiquinones
342 Other Quinone-Like Compounds

Ascochytin

34110-1363

NAME: ASCOCHYTIN
STR-CORR:
REFERENCES:
JCS Perkin I, 675, 1980

34110-1367

NAME: CITRININ
REFERENCES:
CC, 1055, 1979; *Lloydia*, 42, 423, 1979; DDR P 149082

34132-1388

NAME: ATROVENTIN
REFERENCES:
TL, 1051, 1980; *Agr. Biol. Ch.*, 44, 1333, 1980

34132-1389

NAME: HERQUEICHRYSIN
REFERENCES:
JCS Perkin I, 1233, 1979

34210-1392

NAME: EPOXYDON
REFERENCES:
Bioch. J., 182, 445, 1979; *Agr. Biol. Ch.*, 42, 1308, 1978

34210-1396

NAME: EPOFORMIN, DESOXYEPOXYDON
REFERENCES:
Agr. Biol. Ch., 42, 2421, 1979

34210-1397

NAME: TERREIC ACID
IDENTICAL: Y-8980
OR: (-23.9,CHL)
UV: W: (206,42,)(252,26,)(311,39,)(378,19,)
TO: (S.LUTEA,75)(C.ALB.,200)(PHYT.BACT.,12.5)
LD50: (75,IP)
TV: EHRLICH
REFERENCES:
Jap. J. Ant., 33, 320, 1980; *CA*, 89, 159850; 93, 43852

34210-1399

NAME: TERREMUTIN
REFERENCES:
Chem. Ph. Bull., 28, 920, 1980

34210-1400

NAME: PHYLLOSTIN, PHYLLOSTINE
PO: P.URTICAE
REFERENCES:
Biochem., 17, 1785, 1978

34210-1401

NAME: PANEPOXYDONE
REFERENCES:
Bioch. J., 182, 445, 1979

34210-1402

NAME: PANEPOXYDIONE
REFERENCES:
Bioch. J., 182, 445, 1979

34210-6118

NAME: ANTIPHENICOL
QUAL: (NINH.,+)
TO: (S.AUREUS,25)(B.SUBT.,50)(K.PNEUM.,50)
(PS.AER.,25)
REFERENCES:
Agr. Biol. Ch., 43, 2431, 2437, 1979; JP 79/52001

30
LESS KNOWN QUINONE ANTIBIOTICS

30000-1453

NAME: 768
FORMULA: C18H23N5O6
OR: (-315, NAOH)
UV: MEOH: (218, 180,)(258, 790,)(492, 760,)
REFERENCES:

30000-3892

NAME: CRATERIFERMYCIN, CRATERIFERMYCIN COMPLEX
PO: S.CRATERIFER-ANTIBIOTICUS /S.SP./
UV: MEOH: (250, ,)(334, ,)
LD50: (22, IP)
REFERENCES:

Amino Acid and Peptide Antibiotics

41
AMINO ACID DERIVATIVES

411 Simple Amino Acids
412 Cyclic Amino Acid Derivatives
413 Diketopiperazine Derivatives

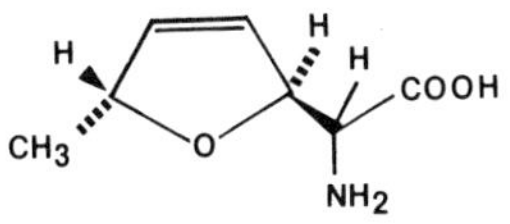

Furanomycin

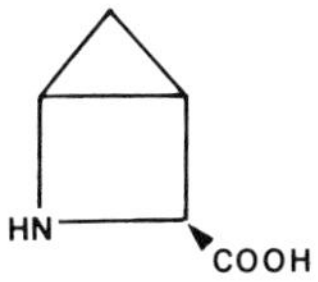

SF-1836

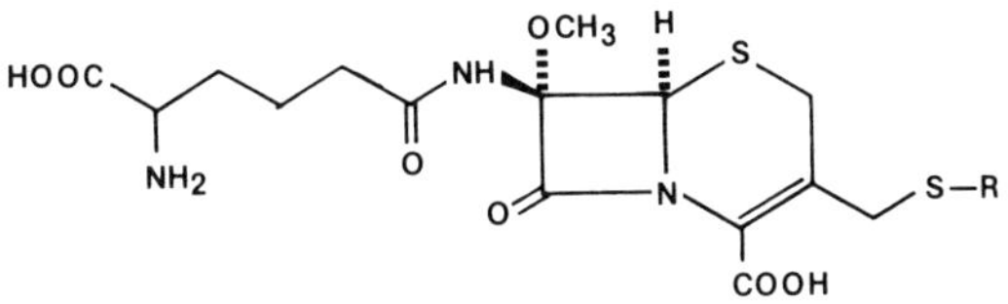

Oganomycin F R =

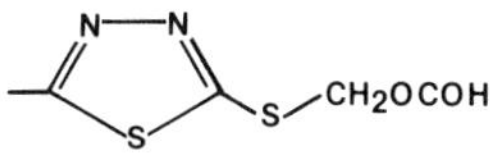

Oganomycin G R =

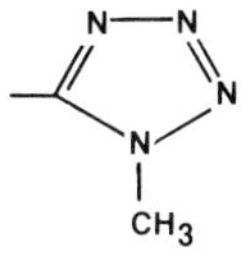

Oganomycin H R =

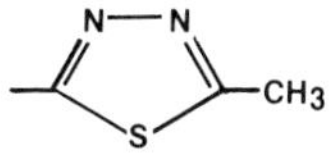

Oganomycin I R =

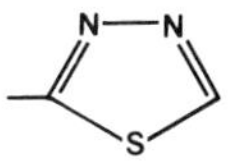

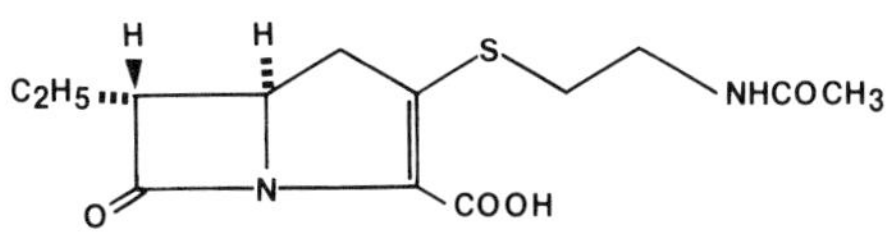

PS-5

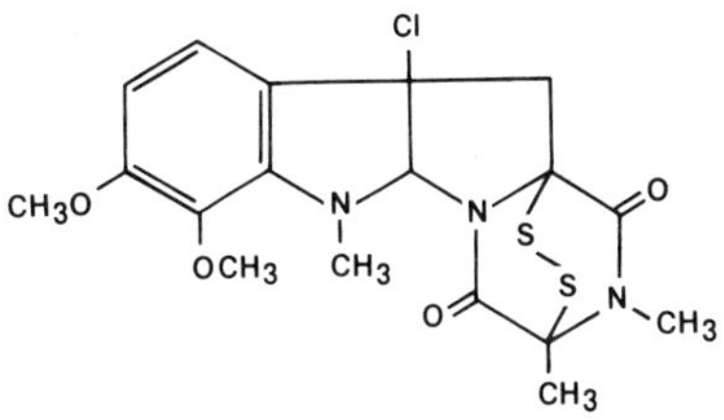

Sporidesmin H

41110-1454

NAME: AZASERIN, AZASERINE, 40816
PO: ACTINOPLANES DECCANENSIS-AZASERINUS
UV: W: (250,,)
UTILITY: NO LONGER AVAILABLE
REFERENCES:
JOC, 43, 4666, 1978; *An. Rep. Sankyo*, 30, 84, 1978; *Canc. Tmt. Rep.*, 63, 1031, 1033, 1979; *CA*, 90, 150248

41110-1455

NAME: DON
REFERENCES:
Canc. Tmt. Rep., 63, 1031, 1033, 1979

41110-1457

NAME: DUAZOMYCIN-B
UTILITY: NO LONGER AVAILABLE
REFERENCES:
Canc. Tmt. Rep., 63, 1033, 1979

41121-1460

NAME: ALANOSINE, L-ALANOSINE, NSC-153353
UTILITY: ON CLINICAL TRIAL
REFERENCES:
J. Med. Chem., 16, 289, 1973

41121-1464

NAME: ARMENTOMYCIN
REFERENCES:
Helv., 64, 1379, 1981

41121-1474

NAME: L-2-AMINO-4-METHOXY-TRANS-3-BUTENOIC ACID, L-2-AMINO-4-METHOXY-TRANS-BUT-3-ENOIC ACID
OR: (+123,W)
REFERENCES:
JOC, 43, 3711, 1978; *Plant Physiol.*, 48, 1, 1971

41122-1466

NAME: L-2-AMINO-4-2-AMINOETHOXY-3-BUTENOIC ACID
REFERENCES:
JOC, 43, 3713, 1978; *Plant Physiol.*, 48, 1, 1971

41122-1477

NAME: L-4-OXALYSINE
IDENTICAL: I-677
REFERENCES:

41123-1483

NAME: L-2.5-DIHYDROPHENYLALANINE, DHPA /U-15738/
REFERENCES:
JACS, 90, 2992; 1968; *AAC*, 7, 601, 1975; *Biochem.*, 17, 3054, 1978; USP 4029548

41124-1486

NAME: L-AZETIDINE-2-CARBOXYLIC ACID
PO: ACTINOPLANES FERRUGINEUS
OR: (-108, W)
REFERENCES:
J. Syst. Bacteriol., 29, 51, 1979

41124-1488

! NAME: ACIVICIN, U-42126
TV: EHRLICH
REFERENCES:
Proc. Am. Ass. Canc. Res., 19, 40, 1978; *CC*, 795, 1976; *JACS*, 101, 1054, 1979; 103, 942, 1981; *Canc. Tmt. Rep.*, 63, 473, 1109, 1979; *JAMA*, 243, 788, 1980; *TL*, 229, 1980; *JA*, 34, 459, 1981; *Cancer Clin. Trials*, 4, 327, 1981; USP 4188324

41124-1491

NAME: CYCLOSERINE
REFERENCES:
Arch. Mikr., 129, 210, 1981

41124-1492

! NAME: FURANOMYCIN
PO: S.THREOMYCETICUS /S.SP./
OR: (+164, HCL)
STR-CORR:
REFERENCES:
Bull. Ch. Soc. Jap., 48, 491, 1975; *Chem. Lett.*, 625, 1975; *JACS*, 102, 887, 7505, 1980; *CC*, 375, 1980

41124-6221

NAME: SF-1836, SF-1835
PO: S.ZAOMYCETICUS /S.ZAOMYCELICUS/
STR-NEW:
REFERENCES:
Ann. Phytop. Soc. Jap., 45, 192, 1979; *Agr. Biol. Ch.*, 43, 2279, 1979; 44, 73, 1980

41125-6119

NAME: FORPHENICINE
TRADE NAMES: FORPHENICOL
UTILITY: ON CLINICAL TRIAL
REFERENCES:
JP 79/73190; *CA*, 91, 138855

41211-1497

NAME: 6-AMINOPENICILLANIC ACID
PO: TRICHODERMA VIRIDAE, TRICHODERMA LIGNORUM, GYMNOASCUS SP. /GYMNOASAUS SP./
REFERENCES:
J. Antimicr. Chem., 5, 7, 1979; JP 80/39712; *CA*, 89, 143023

41211-1506

NAME: PENICILLIN-N
REFERENCES:
JA, 34, 567, 1981

41211-1507

NAME: ISOPENICILLIN-N
REFERENCES:
Bioch. J., 184, 427, 1979; *JA*, 33, 722, 1980

41212-1508

NAME: CEPHALOSPORIN-C
PO: S.INOSITOVORUS
REFERENCES:
Lloydia, 40, 519, 1977; *JA*, 32, 855, 1979; DT 2748659, 2852596; JP 80/37136; 80/144896; *CA*, 92, 126910

41212-1510

NAME: DESACETYLCEPHALOSPORIN-C
REFERENCES:
JP 79/123885; 80/37136; DT 3027380; Belg. P 884385

41212-1511

NAME: CEPHAMYCIN-A
PO: S.CHARTREUSIS
REFERENCES:
USP 4103083; JP 75/121489

41212-1512

NAME: CEPHAMYCIN-B
PO: S.CHARTREUSIS
REFERENCES:
USP 4103083; JP 75/121489

41212-1513

NAME: CEPHAMYCIN-C, S-3907C-2
IDENTICAL: SF-1584
PO: S.TODOROMINENSIS, S.LACTAMGENES, S.INOSITOVORUS
UTILITY: ANTIBACTERIAL DERIVATIVES /ANTIBACTERIAL/
REFERENCES:
Heterocycl., 8, 719, 1977; *J. Chem. Techn. Biotechn.*, 31, 127, 1981; *JA*, 33, 585, 1980; USP 3886044, 3895742, 4103083, 4137405; DT 2908848,3022250; BP 2052502; EP 28511; JP 76/110097; 79/14800; 80/3750; *CA*, 92, 126910, 162183

41212-1518

NAME: WS-3442-D
REFERENCES:
USP 4283492

41212-1519

NAME: WS-3442-E
REFERENCES:
USP 4283492

41212-5054

NAME: OGANOMYCIN-D3, Y-G19Z-D3 /7-METHOXYDEACETYLCEPHALOSPORIN-C/
IDENTICAL: 7-METHOXYDEACETYLCEPHALOSPORIN-C /Y-G19Z-D3/
REFERENCES:

41212-5278

NAME: 7-METHOXYDEACETYLCEPHALOSPORIN-C
IDENTICAL: OGANOMYCIN-D3
REFERENCES:

41212-5671

! NAME: OGANOMYCIN-F, Y-G19Z-F
STR-NEW:
REFERENCES:
JA, 33, 1074, 1980

41212-5672

! NAME: OGANOMYCIN-G, Y-G19Z-G
TO: (B.SUBT., 6.25)(S.LUTEA, 6.25)(S.AUREUS, 50)
(E.COLI, 3.13)(K.PNEUM., 3.13)(P.VULG., 3.13)
(SHYG., 3.13)(PS.AER., 100) /(E.COLI,)
(P.VULG.,)(K.PNEUM.,)/
STR-NEW:
REFERENCES:
JA, 33, 1074, 1980; DT 2657599; JP 77/83777

41212-5673

! NAME: OGANOMYCIN-H, Y-G19Z-H
TO: (B.SUBT., 12.5)(S.LUTEA, 25)(S.AUREUS, 25)
(E.COLI, 5.25)(K.PNEUM., 6.25)(P.VULG., 3.13)
(SHYG., 6.25)(PS.AER., 50) /(E.COLI,)(P.VULG.)
(K.PNEUM.)/
STR-NEW:
REFERENCES:
JA, 33, 1074, 1980; DT 2657599

41212-5674

! NAME: OGANOMYCIN-I, Y-G19Z-I
TO: (B.SUBT., 6.25)(S.LUTEA, 6.25)(S.AUREUS, 50)
(E.COLI, 3.13)(K.PNEUM., 6.25)(SHYG., 6.25)
(PS.AER., 100) /(E.COLI,)(K.PNEUM.,)/
STR-NEW:
REFERENCES:
JA, 33, 1074, 1980; DT 2657599

41212-5874

! NAME: OGANOMYCIN-D2
REFERENCES:

41213-5186

NAME: NOCARDICIN-A
PO: S.ALCALOPHYLUS, NOCARDIOPSIS ATRA, NOC.SP.
REFERENCES:
JACS, 100, 6780, 1978; 103, 2873, 4582, 1981; *TL*, 5119, 1978; *CC*, 770, 1980; USP 4212860, 4212944; DT 2758867; JP 79/151196; 80/34030, 80/45327; *CA*, 92, 126911; 93, 184266

41213-5675

NAME: NOCARDICIN-E
REFERENCES:
Swiss P 622824

41213-5676

NAME: NOCARDICIN-F
REFERENCES:
Swiss P 622824

41213-5875

NAME: NOCARDICIN-C
REFERENCES:
Swiss P 622824

41213-5876

NAME: NOCARDICIN-D
REFERENCES:

41213-5980

NAME: NOCARDICIN-G
REFERENCES:
Swiss P 622824

41214-2262

NAME:	MM-4550, "OLIVANIC ACID"
PO:	S.FLAVUS, S.LIPMANI, S.FLAVOVIRENS, S.SIOYAENSIS
FORMULA:	C13H16N2O9S2 /C13H14N2O9S2/
MW:	/364/
UV:	7: (287, 268,)
UV:	HCL: (293, ,)
SOL-GOOD:	MEOH, DMF, DMSO
SOL-POOR:	BUOH, HEX
TO:	(B.SUBT., 3.1)(S.AUREUS, 25)(E.COLI, 6.2) (P.VULG., 3.1)(K.PNEUM., 12)(PS.AER., 500) /(S.AUREUS,)(K.PNEUM.,)/

REFERENCES:
JA, 32, 287, 295, 961, 1979; 33, 878, 1980; DT 2808563

41214-4253

NAME: MC696-SY2-A
REFERENCES:
Fr. P 2396078

41214-4800

NAME: MM-13902
IDENTICAL: EPITHIENAMYCIN-E
PO: S.FLAVUS, S.LIPMANI, S.FLAVOVIRENS, S.SIOYAENSIS, S.ARGENTEOLUS, S.GEDAENSIS /ALTEROMONAS LUTEO-VIOLACEUS/
EA: (N, 7)(S, 16) /(N,)(S,)/
MW: 392
UV: HCL: (319, ,)
UTILITY: ON CLINICAL TRIAL
REFERENCES:
JA, 32, 287, 295, 1979; 33, 878, 1980; 34, 600, 628, 637, 1981; *CC*, 1083, 1980; USP 4206202

41214-4902

! NAME: EPITHIENAMYCIN-A, MSD-890-A1 /N-ACETYL-EPI-THIENAMYCIN-A/
IDENTICAL: MM-22380, 17927-A1
TO: (B.SUBT., .5)(S.LUTEA,)(S.AUREUS, .39)(E.COLI, .2)(K.PNEUM., .39)(P.VULG., 1.56) /(G.POS.,) (G.NEG.,)/
REFERENCES:
Abst. AAC, 17, 80, 81, 1977; *Heterocycl.*, 16, 65, 1981; *JCS Perkin I*, 2282, 1981; *JA*, 34, 628, 1981; USP 4162324, 4235967; DT 2808563

41214-5056

NAME: THIENAMYCIN
MW: /290/
UV: 7: (297, 272,)
REFERENCES:
JACS, 100, 6491, 8004, 1978; 102, 6161, 1980; *JOC*, 46, 2208, 1981; 45, 1130, 1139, 1142, 1980; *JA*, 32, 1, 1979; *AAC*, 14, 436, 1978; 15, 518, 1979; 19, 114, 201, 1981; *Abst. AAC*, 19, 592—597, 1979; *TL*, 4359, 1979; 2783, 1980; *J. Antimicr. Chem.*, 6, 601, 1980; *Biotech. Bioeng.*, 23, 1255, 1981; *J. Chem. Techn. Biotechn.*, 31, 127, 1981: USP 4081548, 4172140, 4198338; DT 2808636, 2819453; EP 931

41214-5370

NAME: MM-17880
IDENTICAL: EPITHIENAMYCIN-F
PO: S.FLAVUS, S.LIPMANI, S.FLAVOVIRENS, S.SIOYAENSIS, S.ARGENTEOLUS, S.GEDAENSIS
EA: (N, 7)(S, 16) /(N, 6)(S, 13)/
UV: HCL: (310, ,)
UV: W: (298, 192,)
SOL-GOOD: MEOH, DMF, DMSO
SOL-POOR: BUOH, HEX
LD50: NONTOXIC
REFERENCES:
JA, 32, 287, 295, 1979; 33, 878, 1980; 34, 600, 628, 1981; USP 4206202; DT 2808563

41214-5677

NAME: N-ACETYL-THIENAMYCIN
UV: W: (301, 208,)(301, 270,)
TO: (S.AUREUS, .082)(B.SUBT., .027)(E.COLI, .74) (P.VULG., .25) /(S.AUREUS,)(B.SUBT.,)(E.COLI,)/
REFERENCES:
JA, 34, 628, 637, 1981; USP 4135978, 4229534

41214-5982

! NAME: EPITHIENAMYCIN-B, MSD-890-A2
IDENTICAL: MM-22382, 17927-A2 /N-ACETYL-EPI-THIENAMYCIN-B/
TO: (S.AUREUS, .12)(B.SUBT., .014)(E.COLI, .12) (P.VULG., .12) /(S.AUREUS,)(B.SUBT.,)(E.COLI,)(P.VULG.,)/
REFERENCES:
Abst. AAC, 17, 80, 81, 1977; *Heterocycl.*, 16, 65, 1981; *CC*, 1085, 1980 *JCS Perkin I*, 2282, 1981; *JA*, 34, 628, 637, 1981; USP 4141986, 4162324; DT 2816608, 2808563; JP 78/109997

41214-5983

! NAME: EPITHIENAMYCIN-D, MSD-980-A5
IDENTICAL: MM-22383, PS-4 /N-ACETYL-EPI-THIENAMYCIN-D/
REFERENCES:
Abst. AAC, 17, 80, 81, 1977; *Heterocycl.*, 16, 65, 1981; *CC*, 1085, 1980; *JCS Perkin I*, 2282, 1981; *JA*, 34, 628, 637, 1981; USP 4141986, 4162324; DT 2816608, 2808563; JP 78/109997

41214-6080

NAME: PS-5
PO: S.CREMEUS-AURATILIS, S.FULVOVIRIDIS
OR: (+77.3, PH8 PUFF) /(+1.23, PH8 PUFF)/
STR-CORR:
REFERENCES:

JA, 32, 262, 272, 280, 1979; 33, 293, 543, 796, 1128, 1980; 34, 341, 1981; *J. Ferm. Techn.*, 57, 265, 1979; *CC*, 1083, 1084, 1980; *Heterocycl.*, 14, 1967, 1980; *JCS Perkin I*, 2228, 1981; EP 1567, 2058; JP 78/121702; 79/30195; 80/72191; *CA*, 94, 3920

41214-6120

! NAME: EPITHIENAMYCIN-E, MSD-890-A9
FORMULA: C13H16N2O8S2 /C13H16N2O8S2/
MW: 392 /456/
REFERENCES:

JA, 34, 628, 637, 1981; USP 4264735; DT 2805724

41214-6121

NAME: EPITHIENAMYCIN-F, MSD-890-A10
MW: 394 /458/
REFERENCES:

JA, 34, 628, 637, 1981; USP 4264734, 4264736

41214-6223

! NAME: MM-22380 /N-ACETYL-EPI-THIENAMYCIN-A, M-22380/
IDENTICAL: EPITHIENAMYCIN-A, 17927-A1
REFERENCES:

JA, 32, 961, 1239, 1979; 33, 878, 1980; 34, 600, 628, 637, 1981; *CC*, 1083, 1085, 1980; *Heterocycl.*, 16, 65, 1981; *JCS Perkin I*, 2282, 1981; DT 2718782; USP 4141986

41214-6224

! NAME: MM-22381 /N-ACETYL-8-EPI-THIENAMYCIN-C, M-22381/
IDENTICAL: EPITHIENAMYCIN-C, PS-3
REFERENCES:

JA, 32, 961, 1239, 1979; 33, 878, 1980; 34, 600, 628, 637, 1981; *CC*, 1083, 1085, 1980; *Heterocycl.*, 16, 65, 1981; *JCS Perkin I*, 2282, 1981; DT 2718782; USP 4141986

41214-6225

! NAME: MM-22382 /N-ACETYL-EPI-THIENAMYCIN-B, M22382/
IDENTICAL: EPITHIENAMYCIN-B, 17927-A2
REFERENCES:

JA, 32, 961, 1239, 1979; 33, 878, 1980; 34, 600, 628, 637, 1981; *CC*, 1083, 1085, 1980; *Heterocycl.*, 16, 65, 1981; *JCS Perkin I*, 2282, 1981; DT 2718782; USP 4141986

41214-6226

! NAME: MM-22383 /N-ACETYL-EPI-THIENAMYCIN-D, M-22383/
IDENTICAL: EPITHIENAMYCIN-D, PS-4
REFERENCES:

JA, 32, 961, 1239, 1979; 33, 878, 1980; 34, 600, 628, 637, 1981; *CC*, 1083, 1085, 1980; *Heterocycl.*, 16, 65, 1981; *JCS Perkin I*, 2282, 1981; DT 2718782; USP 4141986

41214-6227

NAME: 17927-A1
IDENTICAL: EPITHIENAMYCIN-A, MM-22380
REFERENCES:

JP 80/29909; *CA*, 93, 130561

41214-6228

NAME: 17927-A2
IDENTICAL: EPITHIENAMYCIN-B, MM-22382
REFERENCES:

JP 78/109997, 78/103401; 80/29909; *CA*, 90, 20820; 93/130561

41214-6265

! NAME: EPITHIENAMYCIN-C, MSD-890-A3
IDENTICAL: MM-22381, PS-3 /N-ACETYL-8-EPI-THIENAMYCIN-C/
PO: /S.CATTLEYA, S.FUNGICIDICUS/
UV: HCL: (301, 375,)
SOL-GOOD: W
SOL-POOR: BUOH, HEX
REFERENCES:

Abst. AAC, 17, 80, 81, 1977; *JA*, 34, 628, 637, 1981; USP 4162324, 4235967

41214-6568

NAME: N-ACETYL-DEHYDROTHIENAMYCIN
MW: 312
UV: W: (228, ,)(308, ,)
REFERENCES:
CC, 1085, 1980

41215-5057

NAME: CLAVULANIC ACID
TRADE NAMES: AUGMENTIN (WITH AMPICILLIN)
PO: S.KATSURAHAESIS
UTILITY: ANTIBACTERIAL DRUG
REFERENCES:
JA, 31, 1162, 1978; *AAC*, 14, 650, 1978; USP 4123540; EP 26044; JP 78/104796; 80/162993; *CA*, 90, 119758; 94, 137803

41215-6610

NAME: FORMYLOXYMETHYLCLAVAM
TO: (FUNGI,)
REFERENCES:

41215-6611

NAME: CLAVAM-2-CARBOXYLIC ACID
OR: (-124, ETOAC)
UV: ETOH-NAOH: (272, 1080,)
UV: NAOH: (258, 415,) /(258, , 415)/
TO: (FUNGI,)
REFERENCES:

41220-1523

NAME: AUREOTHRICIN
PO: STV.SP.
REFERENCES:
Chem. Ph. Bull., 28, 3157, 1980

41220-1524

NAME: HOLOMYCIN
IDENTICAL: MM-21801
PO: S.CLAVULIGERUS
UV: MEOH: (246, , 4665)(301, , 2354)(388, , 7918)
TO: (B.SUBT.,)(S.LUTEA)
REFERENCES:
JA, 32, 549, 1979

41220-1525

NAME: THIOLUTIN
PO: STV.SP.
REFERENCES:
Chem. Ph. Bull., 28, 3157, 1980

41220-1528

NAME: ISOBUTYRROPYRROTHIN
PO: STV.SP.
REFERENCES:
Chem. Ph. Bull., 28, 3157, 1980

41230-1534

NAME: LYDIMYCIN
REFERENCES:
Biochem., 19, 3069, 1980

41310-1540

NAME: PIPERAZINEDIONE
REFERENCES:
JACS, 102, 2122, 1980; *JOC*, 44, 296, 1979; *Canc. Tmt. Rep.*, 63, 939, 1979

41321-1543

NAME: GLIOTOXIN
PO: ASP.DEIQUESCENS
REFERENCES:
JACS, 102, 1885, 1980; *Tetr.*, 37, 2045, 1981; *CA*, 92, 211541

41321-5259

NAME: EPICORAZINE A
CT: NEUTRAL
EA: (N, 6)(S, 16)
UV: MEOH: (215, , 24400)
TO: (S.AUREUS, 20) /ANTIBACTERIAL/
REFERENCES:

41322-1554

NAME: CHAETOCIN
PO: CHAETOMIUM THIELAVIOIDEUM
TV: HELA
REFERENCES:
Can. J. Micr., 25, 170, 1979; 27, 766, 1981

41322-1563

NAME: CHAETOMIN, CHETOMIN
MW: /546/
REFERENCES:
JCS Perkin I, 1248, 1978; *Can. J. Micr.*, 25, 170, 1979; 27, 766, 1981

41323-6495

NAME: SPORIDESMIN-H
UV: MEOH: (216, , 23440)(252, , 11200)(290, , 6360)
STR-NEW:
REFERENCES:
JCS Perkin I, 1476, 1978

41324-1571

NAME: HYALODENDRIN
REFERENCES:
JOC, 45, 2625, 1980; *Tetr.*, 37, 2045, 1981

41330-1584

NAME: NEOASPERGILLIC ACID
MW: 224 /229/
UV: MEOH: (234, , 10557)(330, , 8143)
REFERENCES:
Chem. Ph. Bull., 26, 1320, 1978

42
HOMOPEPTIDE ANTIBIOTICS

421 Oligopeptides
422 Linear Homopeptides
423 Cyclic Homopeptides

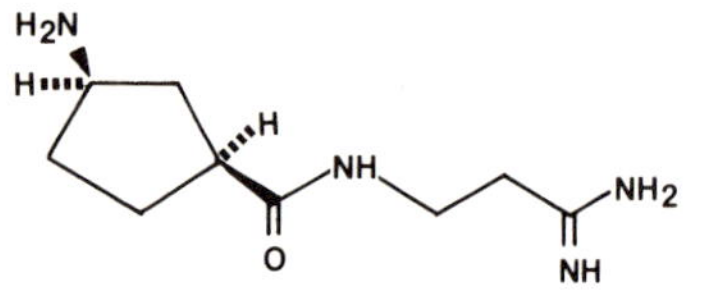

Amidinomycin (Myxoviromycin)

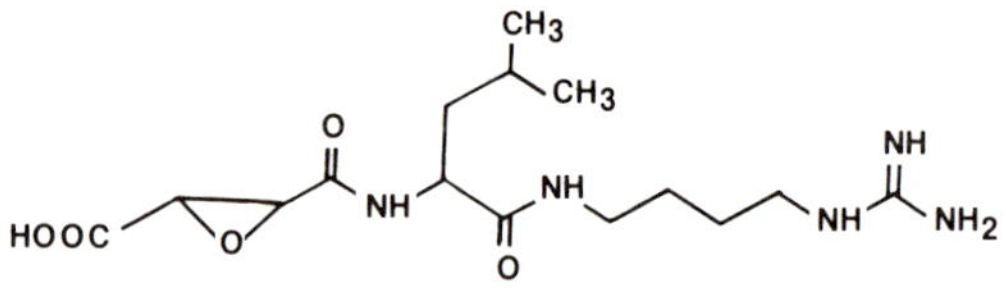

E-64

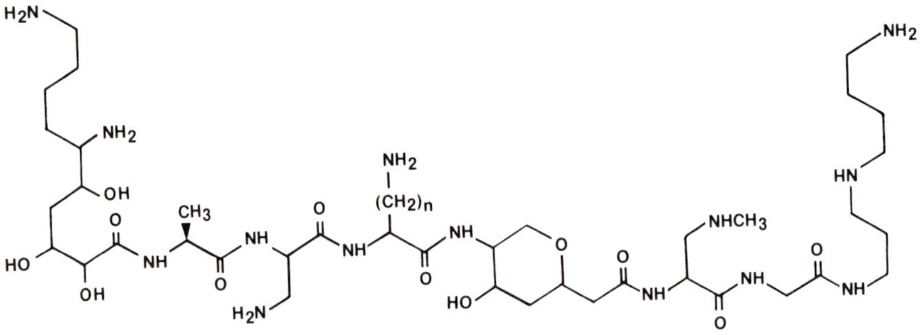

Galantin I n = 3
Galantin II n = 4

Ac-Trp—Ile—Gln—X—Ile—Thr—Aib—Leu—Aib—Hyp—Gln—Aib—Hyp—Aib—Pro—Phol

Emerimycin IIA X: Aib
Emerimycin IIB X: Iva

Ac-Trp—Ile—Glu—Iva—Val—Thr—Aib—Leu—Aib—Hyp—Gln—Aib—Hyp—Aib—Pro—Phol

Zervamicin IA

Val-Orn-Leu-Pro-Tyr
| |
Val-Orn-Leu-Pro-Tyr

Gratizin

L–Ala–Aib–Aib–Aib–X–L–Ala–Aib–Aib–NH–C–$CH_2NCH_2CH_2OH$

Trichopolyn-A X: L-Ile
Trichopolyn-B X: L-Val

42111-1598

NAME: DISTAMYCIN-A
REFERENCES:
Tetr. 34, 2389, 1978; *Bioorg. Khim.*, 4, 1065, 1978; *J. Med. Chem.*, 22, 1296, 1979; *JOC*, 46, 3492, 1981

42112-1601

NAME: MYXOVIROMYCIN, AMIDINOMYCIN
STR-CORR:
REFERENCES:
JA, 33, 778, 1980

* 42113-2630 /52300/

NAME: SF-98
CT: OLIGOPEPTIDE, KIKUMYCIN T. /ADENINE GLYCOSIDE L./
EA: (N, 25) /(N, 18)/
MW: 340 /450/
UV: HCL: (238, 410,)(324, 765,)
LD50: (50, IV)(200|50, IP)
REFERENCES:

42120-1609

NAME: NEGAMYCIN
REFERENCES:
JA, 30, 725, 914, 1978; *Tetr.*, 36, 1763, 1980; *BBRC*, 100, 1497, 1981; Fr. P 2408579

42120-1610

NAME: LEUCYLNEGAMYCIN
REFERENCES:
JA, 32, 531, 1979

42120-1612

NAME: MALONOMYCIN
REFERENCES:
JCS Perkin I, 2017, 1979

42120-1617

NAME: <u>TETAINE</u>
REFERENCES:
Pol. P 95385; *CA*, 90, 101967

42120-1618

NAME: <u>BACILYSIN</u>
IDENTICAL: KM-208
REFERENCES:
CA, 93, 62284

42120-1624

NAME: <u>SF-1293</u>
REFERENCES:
JP 80/21754; *CA*, 93, 43924

42120-4940

NAME: <u>E-64</u>
STR-NEW:
REFERENCES:

42120-5989

NAME: <u>BESTATIN</u>
UTILITY: ANTITUMOR DRUG
REFERENCES:
JA, 33, 653, 1980

42130-1627

NAME: <u>BICYCLOMYCIN</u>
TRADE NAMES: BICOZAMYCIN
PO: S.GRISEOFLAVUS-BICYCLOMYCETICUS, S.ECHINATUS, S.WERRAENSIS
REFERENCES:
JACS, 100, 6786, 1978; *JA*, 32, 689, 1979; 33, 402, 480, 1980; *TL*, 2009, 4155, 4973, 1981; BP 2003727

42210-1632

NAME: <u>GRAMICIDIN-A</u>
REFERENCES:
J. Chrom., 208, 414, 1981

42220-1639

NAME: EDEIN-A1, EDEINE-A
FORMULA: C33H58N10010 /C30H57N10010/
MW: 754 /1600, 753/
REFERENCES:
JA, 34, 28, 1981

42220-1640

NAME: EDEIN-B1, EDEINE-B
MW: /800/
REFERENCES:

42220-1641

NAME: EDEIN-D
FORMULA: C33H58N1009 /C33H57N1009/
MW: 738
REFERENCES:

* 42220-3898 /40000/

NAME: GALANTIN-II
CT: EDEIN T.
FORMULA: C42H84N14012
MW: 976
REFERENCES:
CA, 95, 43646

* 42220-3899 /40000/

NAME: GALANTIN-I
CT: EDEIN T.
FORMULA: C41H82N14012
MW: 962
STR-NEW:
REFERENCES:
CA, 95, 43646

42220-6122

NAME: EDEIN-F
PO: B.BREVIS
FORMULA: C34H60N12O9 /C31H59N12O9/
MW: 880
REFERENCES:
Proc. Antib. Symp., Weimar, 1979, B-11, B-12; Pol P 88098

42250-1697

NAME: ALAMETHICIN-I
REFERENCES:
Proc. Nat. Acad. Sci., 74, 115, 1977; *JCS Perkin I*, 1866, 1980; *Tetr.*, 37, 1263, 1981; *JACS*, 103, 1493, 6127, 6373, 1981

42250-1869

NAME: EMERIMICIN-II
STR-NEW:
REFERENCES:
JACS, 103, 6517, 1981

42250-1872

NAME: ZERVAMICIN-I
STR-NEW:
REFERENCES:
JACS, 103, 6517, 1981

42250-5193

NAME: SAMAROSPORIN
REFERENCES:
JP 79/19475; *CA*, 91, 138859

42250-6071

! NAME: ANTIAMOEBIN-II
REFERENCES:

42260-3905

NAME: CEREXIN-A
REFERENCES:
JA, 32, 313, 1979

42260-3906

NAME: CEREXIN-B
REFERENCES:
JA, 32, 313, 1979

42260-5424

NAME: CEREXIN-C
REFERENCES:
JA, 32, 313, 1979

42260-5425

NAME: CEREXIN-D
REFERENCES:
JA, 32, 313, 1979

* 42270-1877 /40000/

NAME: LEUCINOSTATIN-A
FORMULA: C62H111N11O13
MW: 1217
REFERENCES:
CC, 94, 1982

* 42270-2100 /40000/

NAME: ICI-13595
REFERENCES:

* 42270-5994 /43100/

NAME: TRICHOPOLIN-A, TRICHPOLYN-I
CT: PEPTIDE /LIPOPEPTIDE/
FORMULA: C61H111N11O13
MW: 1205 /2000/
PC: WH, POW.
STR-NEW:
REFERENCES:
CC, 585, 1981; JP 80/72104

* 42270-5995 /43100/

NAME: TRICHOPOLIN-B, TRICHOPOLYN-II
CT: PEPTIDE /LIPOPEPTIDE/
FORMULA: C60H109N11013
MW: 1191 /2000/
PC: WH, POW.
STR-NEW:
REFERENCES:
CC, 585, 1981; JP 80/72104

42310-1669

NAME: GRAMICIDIN-S
REFERENCES:
Top. Enzyme Ferm. Biotech., 5, 187, 1980 (Wiley); *CA*, 90, 166525, 166526; 91, 173311, 173312

42310-1706

NAME: GRATIZIN
FORMULA: C78H110N14014
EA: (N, 13) /(N,)/
MW: 1466
STR-NEW:
REFERENCES:
Vestn. Moscow Univ., Biol. Fac., 28, 123, 1973; *JA*, 34, 1227, 1981

42320-1673

NAME: BACITRACIN-A
REFERENCES:
Ind. J. Biochem. Biophys., 16(1. Suppl.), S-16, 1979; USP 4101539, 4164572

* 42320-4393 /40000/

NAME: GALLERIN, 26A
PO: B.SUBTILUS /B.BREVIS/
CT: CYCLOPEPTIDE, BACITRACIN T. /PEPTIDE/
FORMULA: C77H128N22022S
EA: (N, 17)(S, 2) /(N, 16)/
MW: 1747
EW: 910, 878
UV: W: /(200, ,)/
LD50: (900, IP)(1000, SC)
REFERENCES:
Acta Micr. Pol., 27, 213, 225, 359, 375, 1978; *CA*, 89, 213533, 213534; 90, 184844; 91, 13770

42330-1677

NAME: VIOMYCIN
IDENTICAL: XK-33-F-3
PO: S.OLIVORETICULI-CELLULOPHYLUS
REFERENCES:
J. Gen. Micr., 120, 95, 1980; *CA*, 94, 188287

42330-1680

NAME: CAPREOMYCIN-I-A
MW: 668 /750, 1115/
REFERENCES:
Tetr., 34, 921, 1978; *Bull. Ch. Soc. Jap.*, 52, 1709, 1979

42330-1685

NAME: TUBERACTINOMYCIN-N
REFERENCES:
JA, 31, 792, 1978; 32, 1078, 1979; *Heterocycl.*, 15, 999, 1981

42330-5682

NAME: LL-BM-547-B" /LL-BB-547-B"/
REFERENCES:
Chem. Lett., 589, 1978

42350-4873

NAME: CYCLOSPORIN-A
PO: FUS.SOLANI
OR: (-182, ETOH)
UTILITY: IMMUNOSUPPRESSIVE DRUG
REFERENCES:
Nature, 280, 149, 1979; *Agr. Biol. Ch.*, 45, 1223, 1981; USP 4215199

42350-5060

NAME: CYCLOSPORIN-C
PO: FUS, SOLANI
REFERENCES:
Agr. Biol. Ch., 45, 1223, 1981

42350-5575

NAME: CYCLOSPORIN-D
PO: TOLYPOCLADIUM INFLETUM
REFERENCES:
USP 4117118; DT 2819094; Belg. P, 866810

42300-1689

NAME: CHLAMYDOCIN
REFERENCES:
JACS, 101, 5412, 1979

43
HETEROMER PEPTIDES

431 (Cyclic) Lipopeptide Antibiotics
432 Thiapeptides
433 Chelate-Forming Peptides

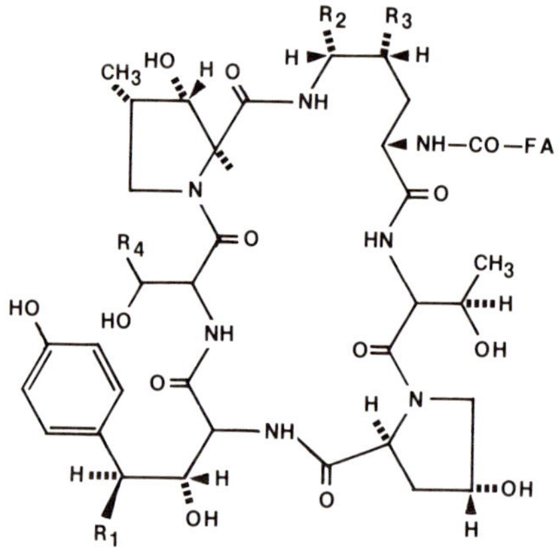

	R_1	R_2	R_3	R_4	FA
Echinocandin A	OH	OH	OH	CH_3	linoleoyl
Echinocandin C (A-30912-B)	H	OH	OH	CH_3	linoleoyl
Echinocandin D (A-30912-C)	H	H	H	CH_3	linoleoyl
A-30912-H	OH	OCH_3	OH	CH_3	linoleoyl
Aculeacin A	OH	OH	OH	CH_3	palmitoyl
S-31794 F/1	OH	OH	OH	CH_2CONH_2	myristoyl

linoleoyl: $-(CH_2)_7CH{=}CH{-}CH_2{-}CH{=}CH{-}(CH_2)_4CH_3$
palmitoyl: $-(CH_2)_{14}CH_3$
myristoyl: $-(CH_2)_{12}CH_3$

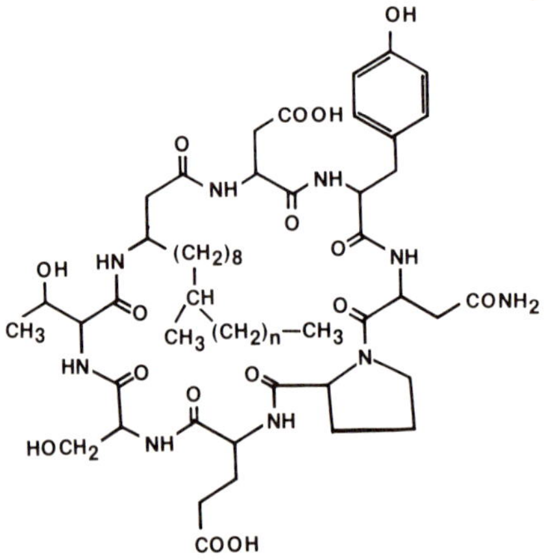

Bacillomycin D:n = 0, 1
("Raubitschek substance")

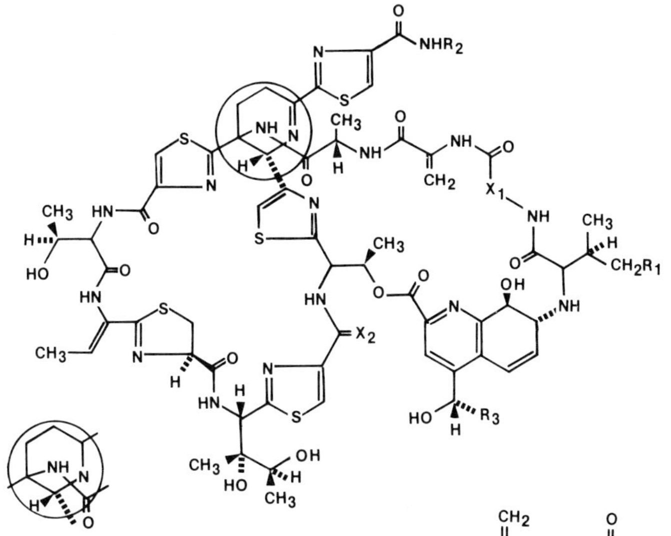

*: in SCH-18640 and thiopeptin "a" series

A: 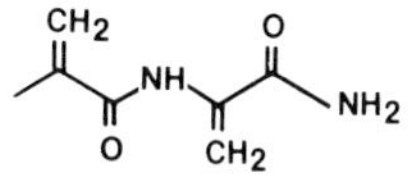 B: 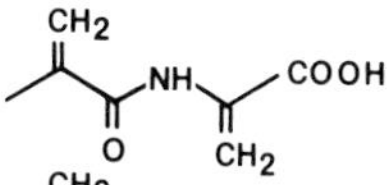 C: 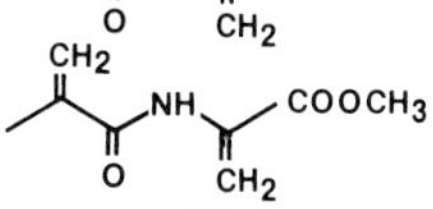 D:

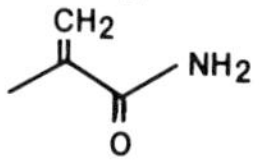

	R_1	R_2	X_1	X_2	R_3
Thiostrepton	CH_3	A	>CH–CH_3	=O	CH_3
Siomycin-A	H	A	>C=CH_2	=O	CH_3
SCH-18640*	CH_3	A	>CH–CH_3	=S	CH_3
Siomycin-D_1	H	A	>C=CH_2	=O	H
Thiopeptin-Bb,-Ba*	H	B	>CH–CH_3	=S	CH_3
Thiopeptin-A_{1b},-A_{1a}*	H	C	>CH–CH_3	=S	CH_3
Siomycin-C	H	C	>CH–CH_3	=O	CH_3
Thiopeptin A_{4b},-A_{4a}*	H	D	>CH–CH_3	=S	CH_3
Thiopeptin A_{3b},-A_{3a}*	H	H	>CH–CH_3	=S	CH_3
Siomycin-B	H	H	>C=CH_2	=O	CH_3
Thiostrepton-B	CH_3	H	>CH–CH_3	=O	CH_3

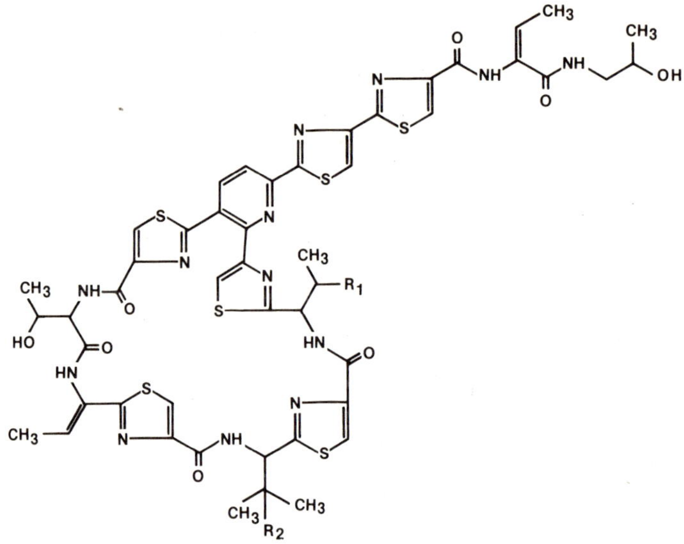

	R_1	R_2
Thiocillin I	OH	OH
Thiocillin II	OCH_3	OH
Thiocillin III	OCH_3	H

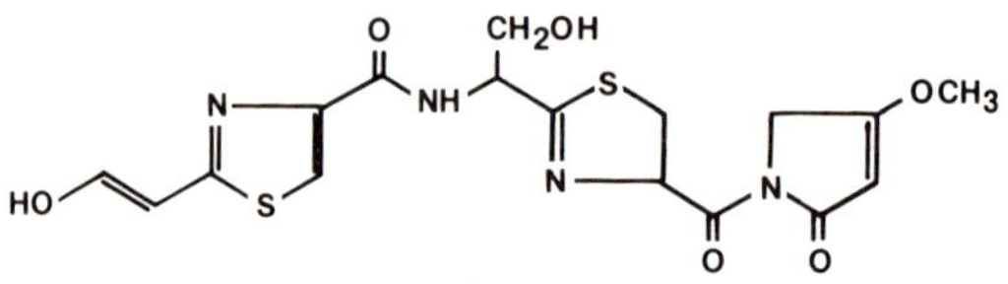

Althiomycin

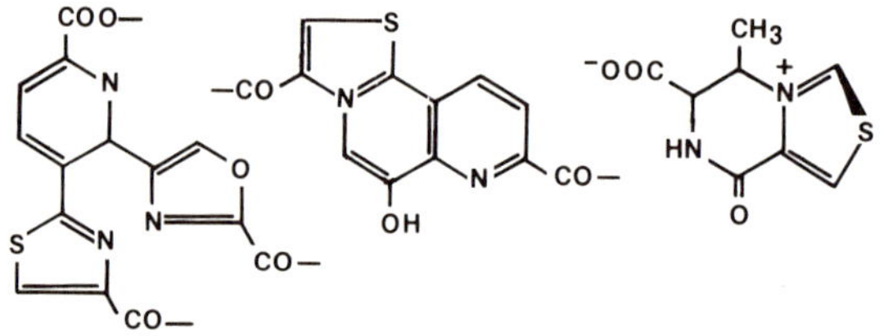

Sulfomycin I (degradation products)

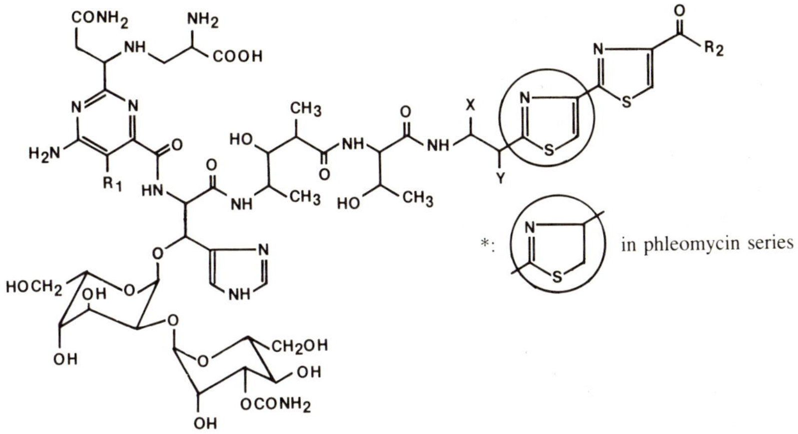

*: in phleomycin series

	R_1	X, Y	R_2
Bleomycins	CH_3	H, H	different amines
Phleomycin G*	CH_3	H, H	$NH[(CH_2CH_2CH_2CH_2NHCHNH)_3]–H$ (with =NH on the CH)
SF-1961-A; SF-1961-B	H	H, sugar	different amines

43110-1714

NAME: AMPHOMYCIN
IDENTICAL: 33B
REFERENCES:

Chemother. (Tokyo), 5, 365, 1975; *JA*, 32, 978, 1979; *BBRC*, 86, 902, 1979; *Biochem.*, 20, 1561, 1981; *Antib.*, 883, 1981; Belg. P 884399

43110-1716

NAME: LASPARTOMYCIN
IDENTICAL: RP-18887
REFERENCES:

* 43110-4052 /00000/

NAME: RP-18887
IDENTICAL: LASPARTOMYCIN
CT: LIPOPEPTIDE, AMPHOMYCIN T.
REFERENCES:

43121-1722

NAME: POLYMIXIN-A
FORMULA: C51H98N16014 /C51H97N16013.4HCL/
MW: 1158 /1293/
REFERENCES:

43121-1723

NAME: POLYMIXIN-M
FORMULA: C51H98N16014 /C51H96N16014.5HCL/
MW: 1158 /1185/
REFERENCES:

43121-1724

NAME: POLYMYXIN-B1
FORMULA: C56H99N16013 /C56H99N16014/
MW: 1203 /1220/
REFERENCES:

J. Chrom., 173, 313, 1979; 218, 653, 1981

43121-1725

NAME: POLYMIXIN-B2
FORMULA: C55H97N16013 /C55H97N16014/
REFERENCES:

43121-1726

NAME: POLYMYXIN-C, POLYMYXIN-C1, POLYMYXIN-C COMPLEX
IDENTICAL: POLYMYXIN-P
REFERENCES:

43121-1727

NAME: POLYMIXIN-D1
FORMULA: C50H94N15015 /C50H94N15014/
MW: 1144 /1150/
REFERENCES:

43121-1728

NAME: POLYMIXIN-E1
IDENTICAL: COLISTIN-A
FORMULA: C54H103N16013 /C53H102N16013.5HCL/
MW: 1183 /1250/
REFERENCES:

43121-5991

NAME: POLYMYXIN-F1
FORMULA: C56H106N15013 /C54H101N15013/
MW: 1196 /1167/
REFERENCES:
DT 2810540

43122-1742

NAME: OCTAPEPTIN-A2
REFERENCES:
JA, 33, 760, 1980; *J. Chrom.*, 173, 313, 1979; *Adv. Appl. Micr.*, 24, 187, 1978

43122-1743

NAME: OCTAPEPTIN-A1
REFERENCES:
JA, 33, 760, 1980; *J. Chrom.*, 173, 313, 1979; *Adv. Appl. Micr.*, 24, 187, 1978

43122-1744

NAME: <u>OCTAPEPTIN-B2</u>
REFERENCES:
JA, 33, 760, 1980; *J. Chrom.*, 173, 313, 1979; *Adv. Appl. Micr.*, 24, 187, 1978

43122-1745

NAME: <u>OCTAPEPTIN-B1</u>
REFERENCES:
JA, 33, 760, 1980; *J. Chrom.*, 173, 313, 1979; *Adv. Appl. Micr.*, 24, 187, 1978

43124-5426

! NAME: <u>TRIDECAPTIN-AA"</u>
REFERENCES:
JA, 31, 646, 652, 1978; 32, 313, 1979; *Adv. Appl. Micr.*, 24, 187, 1978; *CA*, 94, 16054

43124-6229

! NAME: <u>TRIDECAPTIN-BA"</u>
REFERENCES:
JA, 32, 305, 313, 1979

43124-6230

! NAME: <u>TRIDECAPTIN-CA"1</u>
REFERENCES:
JA, 32, 305, 313, 1979

43130-1749

NAME: <u>ECHINOCANDIN-B</u>
IDENTICAL: A-30912-A
MW: 1059 /1100/
REFERENCES:
Exp., 34, 1670, 1978; *Helv.*, 63, 220, 1980; *Eur. J. Bioch.*, 97, 345, 1979

43130-3902

NAME: ACULEACIN COMPLEX
REFERENCES:
Arch. Mikr., 127, 11, 1980; Can. P 1041446; BP 2065130

43130-5064

! NAME: ECHINOCANDIN-C
IDENTICAL: A-30912-B /ECHINOCANDIN-C/
PO: ASP.NIDULANS
FORMULA: C52H81N7015 /C52H81N7016/
MW: 1043 /1018/
UV: ETOH: (226, , 12500)(278, , 1350)
STR-NEW:
REFERENCES:
Helv., 62, 1252, 1979; BP 2065130; Belg. P 886577, 886578

43130-5065

! NAME: ECHINOCANDIN-D
IDENTICAL: A-30912-C /ECHINOCANDIN-D/
PO: ASP.NIDULANS
FORMULA: C52H81N7013
MW: 1011 /1000|200/
UV: ETOH: (224, , 12550)(278, , 1450)
STR-NEW:
REFERENCES:
Helv., 62, 1252, 1979; *Eur. . Bioch.*, 97, 345, 1979; BP 2065663, 2065130

43130-5066

NAME: ACULEACIN-A
FORMULA: C50H83N7016 /C57H94N8020, C60H102N8021/
MW: 1037 /1021/
STR-NEW:
REFERENCES:
Arch. Mikr., 127, 11, 1980; BP 2065130

43130-5429

NAME: S-31794-F-1
PO: ASP.SP.
FORMULA: C49H78N8017
MW: 1038
STR-NEW:
REFERENCES:
USP 4173629; BP 2065130

43130-5685

NAME: A-30912-A
IDENTICAL: ECHINOCANDIN-B, SL-7810-F
FORMULA: C52H81N7O16 /C52H81N7O18/
MW: 1059 /1100/
REFERENCES:
BP 2050385, 2065663, 2065130; Belg. P 883592, 886578, 886577, 886776

43130-5686

NAME: A-30912-B
IDENTICAL: ECHINOCANDIN-C, SL-7810-FII
REFERENCES:
BP 2050385, 2065663, 2065130; Belg. P 883592, 886578, 886577, 886776

43130-5687

NAME: A-30912-C
IDENTICAL: ECHINOCANDIN-D, SL-7810-FIII
REFERENCES:
BP 2050385, 2065663, 2065130; Belg. P 883592, 886578, 886577, 886776

43140-1650

NAME: "RAUBITSCHEK SUBSTANCE"
IDENTICAL: BACILLOMYCIN-D
STR-NEW:
REFERENCES:
JA, 33, 1146, 1980; *Eur. J. Bioch.*, 118, 323, 1981

43140-1663

NAME: MYCOSUBTILIN
REFERENCES:
JA, 32, 828, 1979

43140-1750

NAME: ITURIN-A
REFERENCES:
JA, 32, 828, 1979; *Biochem.*, 19, 3992, 1978

43140-5639

NAME: BACILLOMYCIN-L
REFERENCES:
JA, 29, 1043, 1976; 32, 828, 1979

43211-1752

NAME: THIOSTREPTON
REFERENCES:
JA, 32, 1072, 1978; 34, 123, 1981; *Anal. Profil Drug Subst.*, 7, 423, 1978; *CC*, 577, 1978; *J. Bact.*, 142, 455, 1980; *J. Gen. Micr.*, 124, 291, 1981; *CA*, 94, 4233

43211-1754

NAME: SIOMYCIN-A
REFERENCES:
JA, 32, 1072, 1979; 34, 123, 800, 1981; 33, 1565, 1980; *CA*, 94, 4233

43211-1757

NAME: SIOMYCIN-C
FORMULA: C72H82N18O19S5 /C69H87N14O31S4.5/
MW: 1662 /1700/
STR-NEW:
REFERENCES:
JA, 34, 123, 1981; *Proc. 17th Symp. Peptide Chem.*, 1979, 19—24; *CA*, 94, 4233

43211-1758

NAME: THIOPEPTIN-A1A
STR-NEW:
REFERENCES:
TL, 3648, 1978; *JA*, 32, 1072, 1979; 34, 123, 1981; *Proc. 17th Symp. Peptide Chem.*, 1979, 13; JP 80/122303; *CA*, 93, 112303; 94, 103798

43211-1759

NAME: THIOPEPTIN-A2A
REFERENCES:
TL, 3648, 1978; *JA*, 32, 1072, 1979; 34, 123, 1981; *Proc. 17th Symp. Peptide Chem.*, 1979, 13; JP 80/122303; *CA*, 93, 112303; 94, 103798

43211-1760

NAME: THIOPEPTIN-A3A
STR-NEW:
REFERENCES:
TL, 3648, 1978; *JA*, 32, 1072, 1979; 34, 123, 1981; *Proc. 17th Symp. Peptide Chem.*, 1979, 13; JP 80/122303; *CA*, 93, 112303; 94, 103798

43211-1761

NAME: THIOPEPTIN-A4A
STR-NEW:
REFERENCES:
TL, 3648, 1978; *JA*, 32, 1072, 1979; 34, 123, 1981; *Proc. 17th Symp. Peptide Chem.*, 1979, 13; JP 80/122303; *CA*, 93, 112303; 94, 103798

43211-1762

NAME: THIOPEPTIN-BA
STR-NEW:
REFERENCES:
TL, 3648, 1978; *JA*, 32, 1072, 1979; 34, 123, 1981; *Proc. 17th Symp. Peptide Chem.*, 1979, 13; JP 80/122303; *CA*, 93, 112303; 94, 103798

* 43211-1771 /43212/

NAME: ACTINOTHIOCIN
CT: THIOSTREPTOM T. /ALTHIOMYCIN T./
REFERENCES:

* 43211-1980 /44130/

NAME: TAITOMYCIN COMPLEX
CT: THIAZOLYL-PEPTIDE, THIOSTREPTON T. /PEPTIDE L., TAITOMYCIN T./
REFERENCES:

* 43211-1981 /44130/

NAME: TAITOMYCIN-B
CT: THIAZOLYL-PEPTIDE, THIOSTREPTON T. /PEPTIDE L., TAITOMYCIN T./
REFERENCES:

* 43211-1982 /44130/

NAME: 3354-1
CT: THIAZOLYL-PEPTIDE, THIOSTREPTON T. /PEPTIDE L., TAITOMYCIN T./
REFERENCES:

* 43211-1983 /44130/

NAME: 3354-1
CT: THIAZOLYL-PEPTIDE, THIOSTREPTON T. /PEPTIDE L., TAITOMYCIN T./
REFERENCES:

* 43211-1984 /44130/

NAME: 1542-19
CT: THIAZOLYL-PEPTIDE, THIOSTREPTON T. /PEPTIDE L., TAITOMYCIN T./
REFERENCES:

* 43211-1985 /44130/

NAME: RP-9671
CT: THIAZOLYL-PEPTIDE, THIOSTREPTON T. /PEPTIDE L., TAITOMYCIN T./
REFERENCES:

43211-5534

NAME: NOSIHEPTID
TO: (S.AUREUS, .0009)(S.LUTEA, .0011)(B.SUBT., .03) /(S.AUREUS,)(B.SUBT.,)/
LD50: (3000/500, IP)
IS-EXT: (CH2CL2-I.PROH, , MIC.)
IS-CHR: (SILG, CHL-ETOH-W)
REFERENCES:
Exp., 36, 414, 1980; *Rev. Agricult. (Brusseles)*, 33, 1069, 1980; *J. Gen. Micr.*, 126, 185, 1981; *CA*, 94, 101869

43211-5572

NAME: SCH-18640
FORMULA: C72H87N19017S6
MW: 1683
STR-NEW:
REFERENCES:
JACS, 103, 5231, 1981

43211-6235

NAME: THIOPEPTIN-BB
STR-NEW:
REFERENCES:
JA, 32, 1072, 1979; *CA*, 94, 103798

43211-6236

NAME: THIOPEPTIN-A1B
STR-NEW:
REFERENCES:
JA, 32, 1072, 1979; *CA*, 94, 103798

43211-6237

NAME: THIOPEPTIN-A3B
STR-NEW:
REFERENCES:
JA, 32, 1072, 1979; *CA*, 94, 103798

43211-6238

NAME: THIOPEPTIN-A4B
STR-NEW:
REFERENCES:
JA, 32, 1072, 1979; *CA*, 94, 103798

43212-1767

NAME: ALTHIOMYCIN
STR-CORR:
REFERENCES:
Heterocycl., 13(Suppl.), 175, 1979

43214-5194

NAME: THIOCILLIN-I
FORMULA: C48H49N13O10S6 /C50H59N13O12S6/
MW: 1159 /1292/
STR-NEW:
REFERENCES:
JA, 34, 1126, 1981

43214-5195

NAME: THIOCILLIN-II
FORMULA: C49H51N13O10S6
MW: 1173 /1392/
STR-NEW:
REFERENCES:
JA, 34, 1126, 1981

43214-5196

NAME: THIOCILLIN-III
FORMULA: C49H51N13O9S6
MW: 1157
STR-NEW:
REFERENCES:
JA, 34, 1126, 1981

43220-1773

NAME: BOTTROMYCIN-A
REFERENCES:
JA, 32, 1046, 1979; *Bull. Ch. Soc. Jap.*, 51, 878, 1978; *CA*, 94, 47727, 84483

43220-1774

NAME: BOTTROMYCIN-A2
REFERENCES:
JA, 32, 1046, 1979; *Bull. Ch. Soc. Jap.*, 51, 878, 1978; *CA*, 94, 47727, 84483

43220-1775

NAME: BOTTROMYCIN-B1
REFERENCES:
JA, 32, 1046, 1979; *Bull. Ch. Soc. Jap.*, 51, 878, 1978; *CA*, 94, 47727, 84483

43220-1777

NAME: BOTTROMYCIN-C2
REFERENCES:
JA, 32, 1046, 1979; *Bull. Ch. Soc. Jap.*, 51, 878, 1978; *CA*, 94, 47727, 84483

43230-1782

NAME: SULFOMYCIN-I
STR-NEW:
REFERENCES:
TL, 2791, 1978

43230-1785

NAME: BERNINAMYCIN-A
REFERENCES:
JACS, 101, 5069, 1979

43200-5990

NAME: MC-902-I
PO: S.PLATENSIS /S.SP./
MW: 900
UV: PROH-NH4OH: (218, 370,)
TO: (S.AUREUS,)(PHYT.FUNGI,) /(PHYT.BACT.,)/
IS-EXT: (BUOH, 9, FILT.)
REFERENCES:
CA, 88, 20528

43200-6127

NAME: MC-902-I′
MW: 900
OR: (-59, PROH-NH4OH) /(-59,)/
UV: PROH-NH4OH: (218, 408,)
SOL-GOOD: DMSO, DMFA, ACOH
SOL-POOR: W, MEOH, HEX
REFERENCES:
JP 77/102201; *CA*, 88, 20528

43320-1826

NAME: BLEOMYCIN COMPLEX
STR-CORR:
REFERENCES:

JA, 31, 801, 1070, 1073, 1316, 1978; 32, 453, 756, 1979; 33, 435, 1980; *Jap. J. Ant.*, 32, 720, 1979; *BBRC*, 89, 534, 1979; 91, 721, 1979; *J. Chrom.*, 170, 443, 1979; *Canc. Tmt. Rep.*, 62, 1227, 1978; *J. Heterocycl. Chem.*, 17, 1799, 1980; *Antib.* 675, 887, 1980; *JACS*, 102, 6630, 1980; *Acta Micr. Sinica*, 20, 76, 1980; *Biken J.*, 23, 143, 1980; *BBA*, 517, 526, 1978; *JOC*, 46, 1413, 1981; *Tox. Appl. Pharm.*, 57, 355, 1981; *TL*, 671, 1981; *J. Pharm. Sci.*, 70, 878, 1981; *CA*, 92, 213487; 94, 41962

43320-1828

NAME: BLEOMYCIN-A2
IDENTICAL: ZHENGGUANGMYCIN-A2
REFERENCES:

TL, 671, 1981; *Acta Micr. Sinica*, 20, 76, 1980; *CA*, 94, 96926; *TL*, 521, 1982

43320-1835

NAME: BLEOMYCIN-A5
IDENTICAL: BLEOMYCETIN
PO: STV.GRISEOCARNEUM-BLEOMYCINI
REFERENCES:

Antib., 887, 1980

43320-1838

NAME: BLEOMYCIN-B2
IDENTICAL: ZHENGGUANGMYCIN-B2
REFERENCES:

Acta Micr. Sinica, 20, 76, 1980

43320-1848

NAME: PHLEOMYCIN-G
STR-NEW:
REFERENCES:

Heterocycl., 13(Suppl.), 271, 1979

43320-5431

NAME: TALLYSOMYCIN-A
UTILITY: ON CLINICAL TRIAL
REFERENCES:

JA, 31, 497, 667, 1978; 34, 659, 1981; *Proc. Am. Ass. Cancer Res.*, 18, 35, 1977; *Cancer Res.*, 38, 3322, 1978; *Canc. Tmt. Rep.*, 63, 1821, 1979; 64, 659, 1980; DT 3026425; Belg. P 884291

43320-5432

NAME: TALLYSOMYCIN-B
UTILITY: ON CLINICAL TRIAL
REFERENCES:

JA, 31, 497, 667, 1978; 34, 659, 1981; *Proc. Am. Ass. Cancer Res.*, 18, 35, 1977; *Cancer Res.*, 38, 3322, 1978; *Canc. Tmt. Rep.*, 63, 1821, 1979; 64, 659, 1980; DT 3026425; Belg. P 884291

43320-5689

NAME: SF-1771, SF-1771-A
FORMULA: C58H94N17O33S2CU.3HCL
REFERENCES:

JA, 33, 1236, 1980; JP 77/133902

43320-5690

NAME: SF-1771-B, CU-FREE, SF-1771
FORMULA: C58H94N17O33S2
REFERENCES:

JA, 33, 1236, 1980

43320-6353

NAME: PEP-BLEOMYCIN
REFERENCES:

JA, 32, 36, 1979; *Canc. Tmt. Rep.*, 64, 659, 1981; JP 79/41387; 91/138847; 94/114027; *CA*, 91, 9408

43320-6618

! NAME: SF-1961-B /SF-1961-A/
STR-NEW:
REFERENCES:

JA, 33, 1243, 1980

43320-6619

! NAME: SF-1961-A /SF-1961-B/
MW: 2000
OR: (-15, W)
SOL-GOOD: MEOH
SOL-POOR: ACET, CHL, BENZ
QUAL: (NINH., -)
TO: (S.AUREUS, 25)(E.COLI, .78)(P.VULG., 6.25)
(S.LUTEA, 3.13)(K.PNEUM., .78)(SHYG., .78)
/(E.COLI,)(K.PNEUM.,)(P.VULG, .)/
STR-NEW:
REFERENCES:
JA, 33, 1243, 1980; *220th Sci. Meet. JARA*, 1980

43330-1963

NAME: VIRIDOMYCIN-A
REFERENCES:
Proc. Antib. Symp., Weimar, 1979, C-11; *AAC*, 20, 558, 1981

40
OTHER LESS KNOWN PEPTIDE ANTIBIOTICS

Gly—Phe—Glu—Cys—[X]—Taurine

Monoketoorganomycin

* 40000-4096 /00000/

NAME: AQUINOMYCIN
CT: PEPTIDE
REFERENCES:

40000-5059

NAME: GARDIMYCIN
REFERENCES:
CA, 93, 155888

40000-5197

NAME: FR-3383
PO: S.ODAINENSIS /S.ADANIENSIS/
QUAL: (EHRL.,-)(SAKA.,-)
TO: (E.COLI,33)(P.VULG.,63)(PS.AER.,63)(SHYG,125)
/(E.COLI,)(P.VULG.,)(PS.AER.,)(SHYG.,)/
REFERENCES:
207th Sci. Meet. JARA, 1977

40000-5198

NAME: NRC-501
REFERENCES:
Zbl. Bakt. Parasit., 133, 169, 1978; *CA*, 89, 127693

40000-5455

NAME: S-19
PO: S.YOKOSUKAENSIS
SOL-FAIR: MEOH
REFERENCES:
JP 78/18597

40000-5536

NAME: SC-4
REFERENCES:
Planta Med., 39, 202, 1980

40000-5574

NAME: MONOKETOORGANOMYCIN
STR-NEW:
REFERENCES:
JA, 32, 935, 943, 1979

40000-5993

NAME: AM-2504
REFERENCES:
JP 78/73502; *CA*, 89, 161519

40000-6125

NAME: BN-192
MW: 1500
PC: WH.,POW.
IS-EXT: (BUOH,,FILT.)
REFERENCES:

40000-6219

NAME: SYRINGOTOXIN
TO: (G.POS.,)
REFERENCES:
CA, 92, 16306

44
PEPTOLIDES

441 Chromopeptolides
442 Lipopeptolides
443 Heteropeptolides
444 Simple Peptolides
445 Depsipeptide Antibiotics

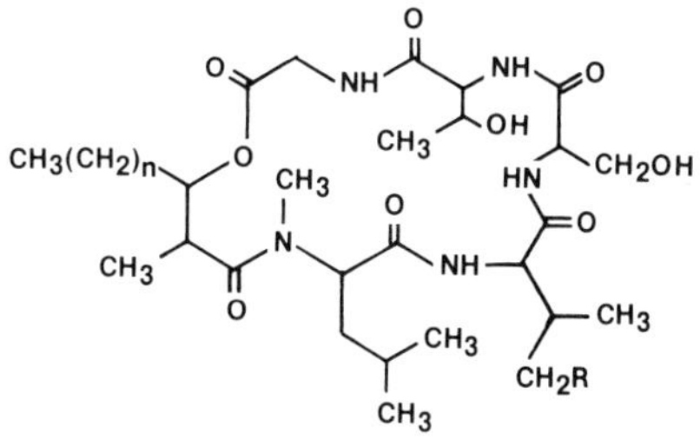

	R	n
Globomycin (SF-1902-A_1)	CH_3	5
SF-1902-A_5	CH_3	7

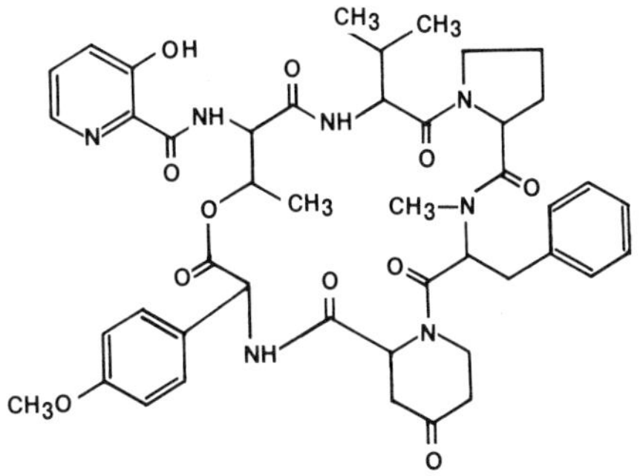

Plauracin 37277

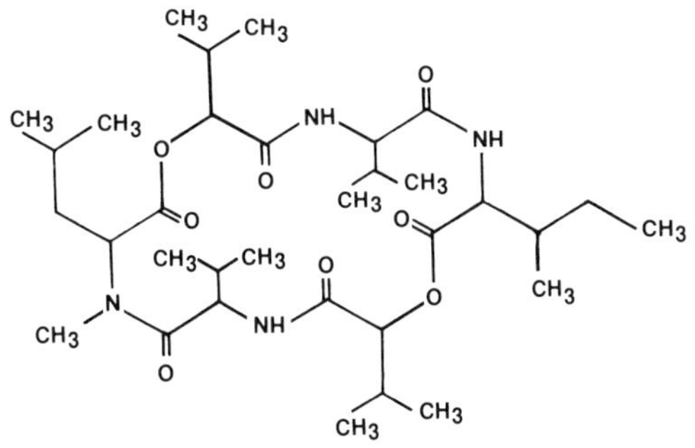

Sporidesmolide I

* 44110-1818 /43314/

NAME: GLUCONIMYCIN
CT: CHROMOPEPTOLIDE, ACTINOMYCIN T. /PSEUDOSIDEROMYCIN/
UV: MEOH: (442, 100,)
REFERENCES:
JA, 33, 1208, 1980

44110-1895

NAME: ACTINOMYCIN-IV
REFERENCES:
Meiji Yakka Daig. Kenk., 9, 1, 1979; JP 80/68297; Hung. P 16432; *CA*, 91, 122155; 93, 112308, 130495

44110-1904

NAME: ACTINOMYCIN-C COMPLEX
IDENTICAL: KENGSHENGMYCIN
REFERENCES:
Antib., 7, 1979; *Proc. Egypt Acad. Sci.*, 27, 107, 1977; *CA*, 90, 4382

44110-1912

! NAME: ACTINOLEVALINE
PO: S.OLIVOLEVALINE + L-LEUCINE
REFERENCES:
Neoplasma, 15, 623, 1968; *Antib.*, 409, 1972

44110-1913

! NAME: ACTINOLEUCINE
PO: S.OLIVOBRUNEUS
OR: (-343, MEOH)
REFERENCES:
Antib., 409, 1972

44110-1925

NAME: ACTINOMYCIN-PIP-1B"
TV: P-388, MELANOMA
REFERENCES:
Bioch. J., 73, 458, 1959

44110-1940

NAME: ETABETACIN
PO: S.PADANUS /S.SP./
REFERENCES:
Giorn. Micr., 16, 9, 1968

* 44110-4149 /00000/

NAME: KENGSHENGMYCIN
IDENTICAL: ACTINOMYCIN-C COMPLEX
CT: CHROMOPEPTOLIDE, ACTINOMYCIN T., NEUTRAL
TV: ANTITUMOR
REFERENCES:

44110-6281

NAME: TOXIFERTILIN
CT: CHROMOPEPTOLIDE, ACTINOMYCIN T., NEUTRAL
EA: (N, 12)
MW: 1451
UV: HCL: (240, ,)(442, ,)
UV: MEOH: (237, ,)(420, ,)(442, ,)
UV: NAOH: (222, ,)(283, ,)(342, ,)
SOL-GOOD: MEOH, ET2O, DIOXIN
SOL-POOR: W, HEX
QUAL: (NINH., -)(FEHL., -)(FECL3, -)(SAKA., -)
(BIURET,)
STAB: (BASE, -)(ACID, -)
TO: (S.AUREUS, .097)
REFERENCES:
Nippon Daigaku, 37, 20, 1980; *CA*, 93, 19371; *Bull. Coll. Agr. & Veter. Med., Nihon Univ.*, 35, 1, 1978

44120-1963

NAME: TRIOSTIN-A
REFERENCES:
JCS Perkin I, 1313, 1979

44120-1972

NAME: QUINONMYCIN-B
REFERENCES:
Anal. Biochem., 89, 213, 1979; *Egypt. J. Micr.*, 15, 99, 1980—81; *CA*, 95, 40765

44120-1973

NAME: QUINONMYCIN-C
REFERENCES:
Anal. Biochem., 89, 213, 1979; *Egypt. J. Micr.*, 15, 99, 1980—81; *CA*, 95, 40765

44120-1974

NAME: QUINONMYCIN-A
REFERENCES:
Anal. Biochem., 89, 213, 1979; *Egypt. J. Micr.*, 15, 99, 1980—81; *CA*, 95, 40765

44230-1960

NAME: SURFACTIN
REFERENCES:
J. Ph. Soc. Jap., 98, 1432, 1978

44230-1988

NAME: ISARIIN
REFERENCES:
JA, 32, 569, 1979; 34, 1261, 1266, 1981

44250-5941

! NAME: SF-1902-A1
REFERENCES:

44250-6081

NAME: GLOBOMYCIN
IDENTICAL: SF-1902-A1 /SF-1902/
REFERENCES:
JP 78/141202; *CA*, 90, 101947

44250-6624

NAME: SF-1902-A5
STR-NEW:
REFERENCES:
JP 80/47644; *JA*, 34, 1416, 1981

44311-2001

NAME: STAPHYLOMYCIN-S
IDENTICAL: 16-5B
PO: S.ALBORECTUS
REFERENCES:
JA, 31, 1313, 1978

44311-5882

! NAME: PLAURACIN-37277
PO: ACTINOPLANES AZUREUS
FORMULA: C45H53N7O11
MW: 867
STR-NEW:
REFERENCES:
Abst. AAC, 17, 242, 243, 1977; *Ann. Rev. Microb.*, 33, 389, 1979

44311-5883

NAME: PLAURACIN-37932
PO: ACTINOPLANES AZUREUS
REFERENCES:
Abst. AAC, 17, 242, 243, 1977; *Ann. Rev. Microb.*, 33, 389, 1979

44311-5884

NAME: PLAURACIN-40042
PO: ACTINOPLANES AZUREUS
REFERENCES:
Abst. AAC, 17, 242, 243, 1977; *Ann. Rev. Microb.*, 33, 389, 1979

44312-2010

NAME: ETAMYCIN
PO: S.GRISEOVIRIDUS
REFERENCES:
Zh. Obsch. Khim., 39, 891, 1969; *JA*, 32, 392, 1979; *CA*, 91, 173307

44312-6082

NAME: NEOVIRIDOGRISEIN-I
PO: S.GRISEOVIRIDUS + L-A"-AMINOBUTYRIC ACID /S.SP./
OR: (+13, MEOH)
REFERENCES:
JA, 32, 575, 584, 1002, 1130, 1979

44312-6083

NAME: NEOVIRIDOGRISEIN-II
IDENTICAL: ETAMYCIN-B
PO: S.GRISEOVIRIDUS + L-PROLINE /S.SP./
OR: (-39.3, MEOH)
REFERENCES:
JA, 32, 575, 584, 1002, 1130, 1979

44312-6084

NAME: NEOVIRIDOGRISEIN-III
PO: S.GRISEOVIRIDUS + L-A"-AMINOBUTYRIC ACID /S.SP./
OR: (+73.7, MEOH)
REFERENCES:
JA, 32, 575, 584, 1002, 1130, 1979

44320-2015

NAME: MONAMYCIN COMPLEX
REFERENCES:
JCS Perkin I, 1451, 1979

44430-3900

NAME: TL-119
IDENTICAL: A-3302-B /A-3309-B/
REFERENCES:
Adv. Appl. Micr., 24, 187, 1978; *Proc. 14th Symp. Peptide Chem.*, 1976, 123

44510-2040

NAME: VALINOMYCIN, ISOLEUCINOMYCIN
REFERENCES:
Biopolymers, 19, 1517, 1980

44510-2043

NAME: ENNIATIN-A
REFERENCES:
Aust. J. Chem., 31, 1397, 1978; *Agr. Biol. Ch.*, 43, 1079, 1979; *Mycopath.*, 70, 103, 1980

44510-2044

NAME: ENNIATIN-B
REFERENCES:
Aust. J. Chem., 31, 1397, 1978; *Agr. Biol. Ch.*, 43, 1079, 1979; *Mycopath.*, 70, 103, 1980

44510-2045

NAME: ENNIATIN-C
REFERENCES:
Aust. J. Chem., 31, 1397, 1978; *Agr. Biol. Ch.*, 43, 1079, 1979; *Mycopath.*, 70, 103, 1980

44510-2046

NAME: ENNIATIN-A1
REFERENCES:
Aust. J. Chem., 31, 1397, 1978; *Agr. Biol. Ch.*, 43, 1079, 1979; *Mycopath.*, 70, 103, 1980

44510-2052

NAME: BEAUVERICIN
TO: (INSECTICID,)
REFERENCES:
Mycopath., 70, 103, 1979

44510-5447

NAME: BASSIANOLIDE
REFERENCES:
Agr. Biol. Ch., 43, 1079, 1979

44510-6240

NAME: ENNIATIN-B1
REFERENCES:
Aust. J. Chem., 31, 1397, 1978; *Agr. Biol. Ch.*, 43, 1079, 1979; *Mycopath.*, 70, 103, 1980

44520-2056

NAME: SPORIDESMOLIDE-I
STR-NEW:
REFERENCES:

44530-1629

NAME: GRISEOVIRIDIN
REFERENCES:
JACS, 102, 870, 1980

44530-2057

NAME: OSTREOGRICIN-A
IDENTICAL: 16-5A
PO: S.ALBORECTUS
REFERENCES:
JA, 31, 1313, 1978; *JACS*, 102, 5964, 1980; *Microb. Rev.*, 43, 145, 1979

44530-2059

NAME: PRISTINAMYCIN-IIA
REFERENCES:
JA, 30, 665, 1977

44530-2065

NAME: OSTREOGRICIN-G
IDENTICAL: 16-5D
PO: S.ALBORECTUS
REFERENCES:
JA, 31, 1313, 1978

44530-2066

NAME: A-2315-A
IDENTICAL: PLAURACIN-35763, A-17002
REFERENCES:
JA, 32, 108, 1979; *Ann. Rev. Microb*, 33, 389, 1979; USP 4001397

44530-2067

NAME: A-2315-B
IDENTICAL: PLAURACIN-36926, MADUMYCIN-I
REFERENCES:
JA, 32, 108, 1979; *Ann. Rev. Microb*, 33, 389, 1979; USP 4001397

44530-2069

NAME: MADUMYCIN-I
IDENTICAL: PLAURACIN-36295, A-2315-B
REFERENCES:

44530-2070

NAME: MADUMYCIN-II
IDENTICAL: PLAURACIN-35763, A-17002-F, A-15104-V
REFERENCES:

44530-2071

NAME: YAKUSIMYCIN-A
PO: S.ANTIBIOTICUS /S.SP./
MW: 475 /552/
REFERENCES:
JP 73/10294

44530-5085

NAME: 1745-Z3-BW
PC: WH., CRYST.
IS-EXT: (ETOAC, , FILT.)
REFERENCES:
Rep. Yamanochi Centr. Res. Lab., 2, 173, 1974

* 44530-5680 /40000/

NAME: GEMINIMYCINS
CT: DEPSIPEPTIDE /PEPTIDE/
REFERENCES:

44530-5885

! NAME: PLAURACIN-36926
IDENTICAL: A-2315-B, MADUMYCIN-I
PO: ACTINOPLANES AZUREUS
REFERENCES:
Abst. AAC, 17, 242,243, 1977; *Ann. Rev. Microb.*, 33, 389, 1979

44530-5886

NAME: PLAURACIN-35763
IDENTICAL: A-2315-A, MADUMYCIN-II, A-17002-F
PO: ACTINOPLANES AZUREUS
REFERENCES:
Abst. AAC, 17, 242,243, 1977; *Ann. Rev. Microb.*, 33, 389, 1979

44540-2075

NAME: DETOXIN-D1
REFERENCES:
Exp., 37, 365, 926, 1981; *Heterocycl.*, 13, 1622, 1979

45
MACROMOLECULAR (PEPTIDE) ANTIBIOTICS

451 Polypeptide Antibiotics
452 Protein Antibiotics
453 Proteide Antibiotics

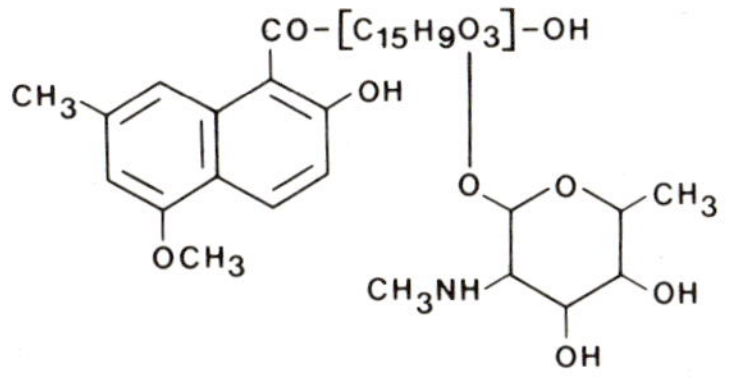

Partial structure
of the chromophore
part of Neocarzinostatin

H-ala-asn-cys-ser-cys-ser-thr-ala-ser-asp-tyr-cys-pro-ile-leu-

-thr-phe-cys-thr-thr-gly-thr-ala-cys-ser-tyr-thr-pro-thr-gly-

-cys-gly-thr-gly-trp-val-tyr-cys-ala-cys-asn-gly-asn-phe-tyr-OH

Primary Structure of Fulvocin C

```
                                   10                                      20                                     30
Leu-Pro-Asn-Ile-Thr-Ile-Leu-Ala-Thr-Gly-Gly-Thr-Ile-Ala-Gly-Gly-Gly-Asp-Ser-Ala-Thr-Lys-Ser-Asn-Tyr-Thr-Ala-Gly-Lys-Val-

                                   40                                      50                                     60
Gly-Val-Glu-Asn-Leu-Val-Asn-Ala-Val-Pro-Gln-Leu-Lys-Asp-Ile-Ala-Asn-Val-Lys-Gly-Glu-Gln-Val-Val-Asn-Ile-Gly-Ser-Gln-Asp-

                                   70                                      80                                     90
Met-Asn-Asp-Asp-Val-Trp-Leu-Thr-Leu-Ala-Lys-Lys-Ile-Asn-Thr-Asp-Cys-Asp-Lys-Thr-Asp-Gly-Phe-Val-Ile-Thr-His-Gly-Thr-Asp-

                                  100                                     110                                    120
Thr-Met-Glu-Glu-Thr-Ala-Tyr-Phe-Leu-Asp-Leu-Thr-Val-Lys-Cys-Asp-Lys-Pro-Val-Met-Val-Gly-Ala-Met-Arg-Pro-Ser-Thr-Ser-Met-

                                  130                                     140                                    150
Ser-Ala-Asp-Gly-Pro-Phe-Asn-Leu-Tyr-Asn-Ala-Val-Thr-Ala-Ala-Asp-Lys-Ala-Ser-Ala-Asn-Arg-Gly-Val-Leu-Val-Met-Asn-Asp-Thr-

                                  160                                     170                                    180
Val-Leu-Asp-Gly-Arg-Asp-Val-Thr-Lys-Thr-Asn-Thr-Thr-Asp-Val-Ala-Thr-Phe-Lys-Ser-Val-Asn-Tyr-Gly-Pro-Leu-Gly-Tyr-Ile-His-

                                  190                                     200                                    210
Asp-Gly-Lys-Ile-Asp-Tyr-Gln-Arg-Thr-Pro-Ala-Arg-Lys-His-Thr-Ser-Asp-Thr-Pro-Phe-Asp-Val-Ser-Lys-Leu-Asn-Glu-Leu-Pro-Lys-

                                  220                                     230                                    240
Val-Gly-Ile-Val-Tyr-Asn-Tyr-Ala-Asn-Ala-Ser-Asp-Leu-Pro-Ala-Lys-Ala-Leu-Val-Asp-Ala-Gly-Tyr-Asp-Gly-Ile-Val-Ser-Ala-Gly-

                                  250                                     260                                    270
Val-Gly-Asp-Gly-Asn-Leu-Tyr-Lys-Thr-Val-Phe-Asp-Thr-Leu-Ala-Thr-Ala-Ala-Lys-Asp-Gly-Thr-Ala-Val-Arg-Ser-Ser-Arg-Val-Pro-

                                  280                                     290                                    300
Thr-Gly-Ala-Thr-Thr-Gln-Asp-Ala-Glu-Val-Asp-Asp-Ala-Lys-Tyr-Gly-Phe-Val-Ala-Ser-Gly-Thr-Leu-Asn-Pro-Gln-Lys-Ala-Arg-Val-

                                  310                                     320
Leu-Leu-Gln-Ala-Leu-Thr-Gln-Thr-Lys-Asp-Pro-Gln-Gln-Ile-Gln-Gln-Ile-Phe-Asn-Gln-Tyr
```

The primary structure of L-asparaginase from *Escherichia coli* A-1-3.

L-asparaginase

45110-2078

NAME: PHYTOACTIN
REFERENCES:
CA, 94, 170744

45110-6621

NAME: BACILEUCINE-A
UV: W: (273, 10,)
QUAL: (BIURET, +)(NINH., -)
IS-CRY: (PREC., HCL, FILT.)
REFERENCES:

45110-6622

NAME: BACILEUCINE-B
UV: W: (260, 10,)
QUAL: (BIURET, +)(NINH., -)
REFERENCES:
JP 78/113088

45130-2105

NAME: SYRINGOMYCIN /SYRINGOTOXIN/
REFERENCES:
CA, 92, 16306

45140-2120

NAME: NISIN
REFERENCES:
Adv. Appl. Micr., 27, 85, 1981; *Antib.*, 872, 1978; *Prikl. Biokh. Mikr.*, 15, 712, 1979; DT 2000818; *CA*, 92, 99496

45140-2200

NAME: DIPLOCOCCIN
PO: STREPTOCOCCUS CREMORIS
CT: BACTERIOCIN
MW: 5300
TO: (STREPTOCOCCUS LACTIS,)
REFERENCES:
Appl. Micr., 41, 84, 1981

45211-2137

NAME: NEOCARZINOSTATIN
MW: 10700
UV: HCL: (270, , 23000)(340, , 8000)
STR-CORR:
REFERENCES:

JA, 31, 468, 1978; 33, 110, 342, 347, 744, 1980; 33, 1545, 1586, 1590, 1980; *BBRC*, 89, 635, 1979; 94, 255, 1980; 95, 1351, 1980; *Gann*, 69, 407, 1978; 70, 545, 1979; *Agr. Biol. Ch.*, 43, 371, 1979; *Biochem.*, 20, 4155, 1981; *Canc. Tmt. Rep.*, 65, 699, 1981; DT 2813017

45211-2138

NAME: ACTINOXANTHIN
REFERENCES:

Proc. Antib. Symp., Weimar, 1979, A-28; *Bioorg. Khim.*, 5, 1605, 1979; 7, 835, 1981

45211-2140

NAME: LYMPHOMYCIN COMPLEX
PO: S.SP.
REFERENCES:

JA, 27, 346, 1974; *Progr. AAC*, 93, 1970; JP 79/3957; *CA*, 90, 184905

45211-2143

NAME: MACROMOMYCIN
REFERENCES:

JA, 31, 875, 1978; 32, 340, 1979; *J. Biol. Chem.*, 253, 3259, 1978; *Cancer Res.*, 39, 1180, 1979; *BBRC*, 86, 1133, 1979; 89, 1281, 1979; 94, 769, 1980; BP 2003135

45211-2145

NAME: MITOMALCIN
REFERENCES:

Abst. 157th ACS Meet., 1969, 1188; JP 73/23917

45212-5343

NAME: SPORAMYCIN
UTILITY: ON CLINICAL TRIAL
REFERENCES:

JA, 32, 386, 1201, 1979; *Abst. AAC*, 19, 852, 1979; JP 78/7601; *CA*, 88, 150596

45213-2183

! NAME: PICIBANIL /PACIBILIN/
TO: (C.ALB.,)
REFERENCES:
Gann, 67, 115, 1976; 69, 699, 1978; *Chem. Ph. Bull.*, 27, 166, 1979; *Microbiol. & Immunol.*, 23, 549, 1979; DT 2043971

45220-2207

! NAME: SILLUCIN
PO: MUCOR MIEHEI
EA: (S,)
MW: 3400
UV: NAOH: (282, ,)(288, ,)
REFERENCES:
J. Dairy Sci., 56, 639, 1973; *FEBS Lett.*, 97, 81, 1979; *Dev. Ind. Micr.*, 20, 661, 1979

* 45320-6294 /00000/

NAME: NO NAME
REFERENCES:

45340-2274

NAME: ASPARAGINASE, L-ASPARAGINASE
MW: 34080 /480000, 800000/
STR-NEW:
REFERENCES:
Hoppe Seyler, 361, 105, 1980; *Chem. Ph. Bull.*, 25, 571, 1979; *J. Bact.*, 122, 1017, 1975; 125, 999, 1978; Belg. P 775286; SU P 649746; JP 80/19018

45340-2286

NAME: PTERIDIN DEAMINASE, PTERIN DEAMINASE
REFERENCES:
Agr. Biol. Ch., 38, 1753, 1974; 43, 1983, 1979; *J. Biol. Chem.*, 234, 955, 1959; 238, 1116, 1963

45340-6288

NAME: CARBOXYPEPTIDASE G1, CPD-G1
MW: 92000
UV: W: (280, ,)
IS-CHR: (SEPHADEX G-150, W)
REFERENCES:

Agr. Biol. Ch., 44, 1661, 1980

45350-4802

! NAME: VIRIDIN-B
MW: 87000
REFERENCES:

AAC, 13, 473, 1978; 15, 436, 1979; *Proc. Soc. Exp. B.* 155, 456, 1977; *Inf. Immun.*, 14, 776, 1976

45350-4804

NAME: HAEMOCIN
REFERENCES:

AAC, 19, 668, 1981

45350-5719

NAME: CAROTOVORICIN
REFERENCES:

J. Gen. Appl. Micr., 26, 51, 1980; *Agr. Biol. Ch.*, 44, 1135, 1980

45350-6094

NAME: VIRIDICINE, VIRIDICIN
MW: 115000|5000
REFERENCES:

AAC, 17, 784, 1980

45350-6241

NAME: CLOSTOCIN-O
REFERENCES:

J. Gen. Appl. Micr., 24, 223, 1978; *CA*, 89, 176029

* 45350-6354 /45330/

NAME: GLAUCESCIN
REFERENCES:
J. Gen. Micr., 113, 243, 1979

45350-6496

NAME: FULVOCIN-C
STR-NEW:
REFERENCES:
BBA, 667, 213, 1981

45350-6628

NAME: COLICIN-E-3
REFERENCES:

45360-2330

NAME: AVF, "ANTI VIRAL FACTOR"
REFERENCES:
Curr. Trends Biochem., 31(2), 1981; *CA*, 88, 149098; *Prosp. Virol.*, 11, 129, 1981; *CA*, 95, 78253

45000-4104

NAME: NA-699
CT: MACROMOLECULAR
REFERENCES:

Section II
New Compounds

SECTION II

NEW MICROBIAL METABOLITES

In recent years new microbial antibiotics were described at an ever increasing rate. Each year almost 400 new antibiotic compounds were described in the scientific literature including journals, patents, congress reports, etc. An especially great number of new aminoglycosides, mainly gentamicin and fortimicin derivatives (112 compounds), new β-lactam type compounds (33 antibiotics), anthracyclines (66 compounds), anasamycins and ansa-macrolactams (34 compounds), thiostrepton type antibiotics (20 compounds), and polyether antibiotics (23 compounds) were described and listed in this section. In recent years the increasing application of modern biosynthetic techniques in antibiotic research led to an especially great number of new biosynthetically derived aminoglycosides, anthracyclines, and bleomycin type glycopeptides. The main source of new compounds besides the various modified screening techniques was the extensive research of rare and until now neglected microorganism species (rare Actinomycetales, *Myxococcus* species etc.), the biosynthetically first mutasynthetically modified compounds, and the isolation of the minor components of the fermentation broths. On the other hand, the number of recently discovered new macrolides polyenes, streptothricin type and aliphatic compounds are definitely declining.

These results indicate the main trends of the recent antibiotic research, namely the increasing interest in aminoglycosides and new types of β-lactams serving mainly or exclusively as starting compounds for synthetic or biosynthetic modifications, as well as the great efforts to find new useful antitumor agents and cheap compounds potentially utilizable in agriculture as feed additive or veterinary drugs.

Numerous recently developed specific screening methods led to the isolation of several narrow-spectrum compounds with unique chemical structures (tetrocarcins, CC-1065, kirrothricin, cytovaricin etc.) and to a great number of antimicrobially inactive derivatives having other bioactivities (enzyme inhibiting activity, ion-transport effects etc.). The number of these physiologically active compounds derived from microbes was already near 500.

Among the entirely new structures representing the new antibiotic types the monobactam type β-lactams, tetrocarcin and nargenicin type complicated macrocyclic compounds, indolyloligopeptide CC-1065, combimicin type and the 1-deamino-1-hydroxy-desoxystreptamine containing aminoglycosides, ophiocordin containing azepine ring, glysperins, vinenomycins, clazamycins, neplanocins, izumenolide, setomimycin, and fredericamycin are worth mentioning. The discovery of the full chemical structures of the first members of several long known antibiotic types such as moenomycin, concanamycin (humidin type), medemycin (luteomycin type) toromycin (chartreusin-like), and the recognition of the structures of monazomycin, elaiophyllin, galantine etc., all were the results of recent work. The hundreds of antibiotics with entirely unknown chemical structures and several surprising structural corrections of long known compounds (validamycins, chromomycin, bleomycins, neocarzinostatin, thermorubin, etc.), however, indicate that our knowledge about the chemical nature of antibiotic compounds is far from complete.

LISTING OF NEW COMPOUNDS

An interesting feature of the research of natural antibiotic compounds in the past years, is the greatly increasing rate of isolation of active derivatives, especially antitumor compounds from higher organisms (algae, plants, animals). In the last 3 to 4 years almost 1000 new active compounds were isolated from these organisms, mainly derived from the sea. The ratio of the number of newly discovered microbial and other metabolites among antibiotic compounds has shifted from 3:1 to 3:2 in these years. The fundamental results of research on the compounds isolated from organisms other than microbes will be summarized in Volume XII.

The selection and covering of the new microbial metabolites in this supplement was made according to the unchanged rules applied in the Main Volumes. In the listing of compounds, only the titles of Antibiotic Families (CARBOHYDRATE ANTIBIOTICS), Subfamilies (**Pure Saccharides**), and the Groups (Polysaccharides) are indicated. No new introduced material are inserted to any group of compounds. The chemical structures belonging to compounds in every subfamily are inserted before the listing of these compounds. For general properties of a given group or type of compounds one should see the Main Volumes. In some cases, to enhance the recognition of relationships between similar structures given in summarized forms as derivatives of a basic skeleton, the structures (substituents) of a few previously known compounds were also given.

Referencing of new compounds includes Patents and Chemical Abstract references also. This section contains about 1200 new compounds. The editing work was finished at the end of 1981.

Indexes for compounds are attached to this Supplementary Volume including Indexes of Name, Producing Organism, and Numbers and Names. The index of newly published names and identities as well as producing species of old compounds, appearing in Section I of this volume, are merged with the indexes of new compounds. In this cumulative index therefore all compounds (referred by their Compound Number also) having sequence number (second part of the 9-digit Compound Number) less than 6700 are found in Section I of this Volume. Further elaboration of the organization of the indexes is to be found in the explanatory material which precedes the Index of Names of Antibiotics.

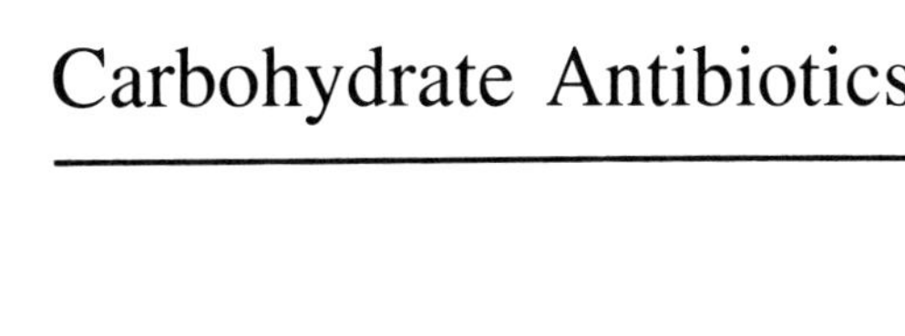

Carbohydrate Antibiotics

11

PURE SACCHARIDES

111 Mono- and Oligosaccharide Types

112 Polysaccharides

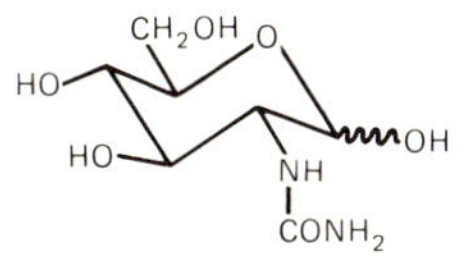

SF-1993
(N-Carbamoyl-β-D-Glucosaminide)

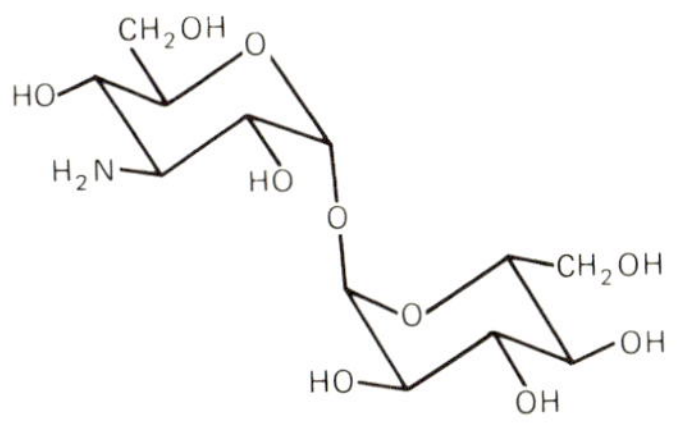

3-Trehalosamine

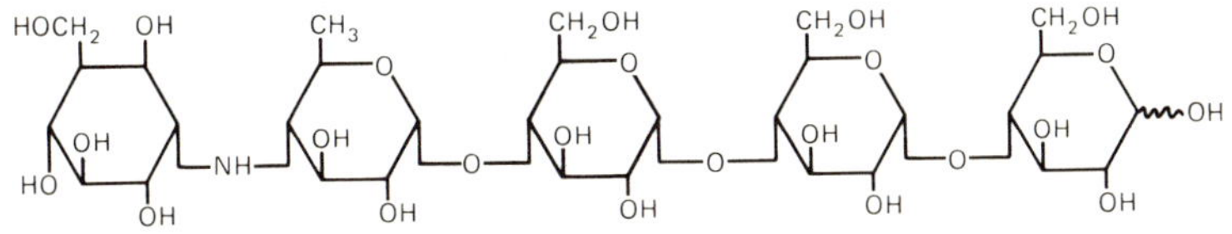

Oligostatin C (SF-1130-X_3)

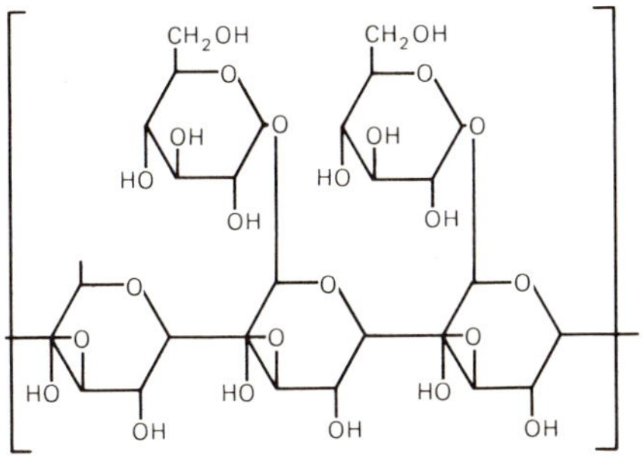

Water-soluble D-glucan

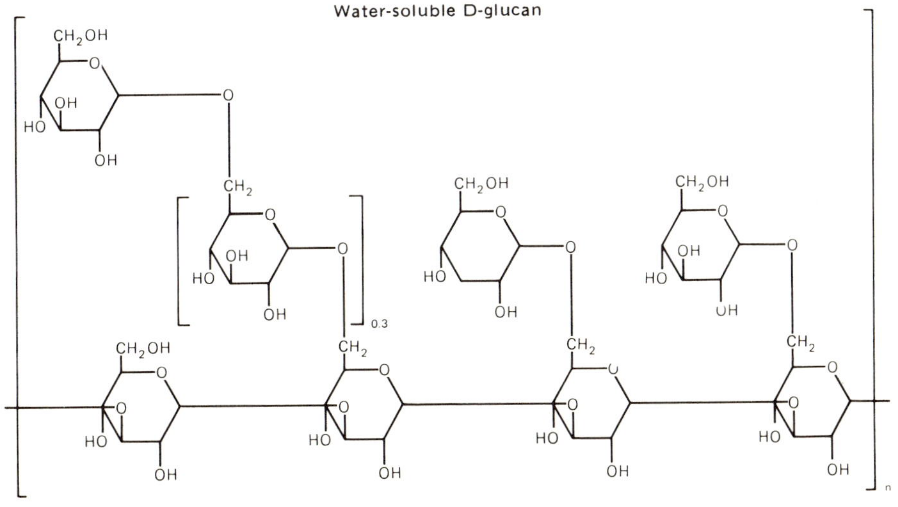

Alkali-insoluble D-glucan

Glucan-I

11121-6731

NAME: SF-1993, N-CARBAMOYL-D-GLUCOSAMINE
IDENTICAL: N-CARBAMOYL-B"-D-GLUCOSAMINIDE
PO: S.HALSTEDII
CT: NEUTRAL, AMINO SUGAR
FORMULA: C7H14N2O6
EA: (N, 12)
MW: 222
PC: WH., CRYST.
OR: (+73, W)(+58, W)
UV: W: (200, ,)
SOL-GOOD: W
SOL-POOR: MEOH, HEX
QUAL: (EHRL., +)(NINH., -)
TO: (E.COLI,)(SHYG.,)(P.VULG.,)(FUNGI,)
LD50: NONTOXIC
IS-ABS: (CARBON, W)
IS-CHR: (SEPHADEX LH-20, MEOH)
IS-CRY: (CRYST., ETOH-W)
REFERENCES:

JA, 32, 427, 436, 1979; JP 79/103801; *CA*, 92, 56810

11121-7553

NAME: N-CARBAMOYL-B"-D-GLUCOSAMINIDE, IA-II
IDENTICAL: SF-1993
PO: S.ABURAVIENSIS-RUFUS
CT: AMINO SUGAR, NEUTRAL
FORMULA: C7H14N2O6
EA: (N, 12)
MW: 222
PC: WH.
OR: (+58, W)
TO: (P.VULG., 12.5)
LD50: (4000|1000, IV)
IS-FIL: ORIG.
IS-ABS: (CARBON, MEOH-W)
IS-CHR: (DEAE-SEPHADEX A-25,)
REFERENCES:

JP 79/154713; *CA*, 92, 196384

11122-7601

NAME: 3-TREHALOSAMINE, U-59834
PO: NOCARDIOPSIS TREHALOSEI
CT: BASIC, AMINODISACCHARIDE
FORMULA: C12H23NO10
EA: (N, 4)
MW: 341
PC: WH., CRYST.
OR: (+161, W)
UV: W: (200, ,)
SOL-GOOD: W
SOL-POOR: MEOH, HEX
QUAL: (NINH., +)
TO: (B.SUBT.,)(S.AUREUS,)
LD50: (1200|400, IP)
IS-ION: (DX-50-4, AMMONIUM SULPHATE)(XE-348, ACET-W)
IS-CHR: (SILG, CHL-MEOH-NH4OH-BUOH)
IS-CRY: (LIOF.,)(CRYST., ETOH-W-HCL)
REFERENCES:

JA, 33, 690, 1980, USP 4276412

11132-7602

NAME: TAI-A
PO: S.CULBUS
CT: BASIC, AMINOOLIGOSACCHARIDE
EA: (C, 41)(H, 6)(N, 1.3)
MW: 1000|50
PC: WH., POW.
OR: (+157.3, W)
UV: W: (200, ,)
SOL-GOOD: W, DMSO
SOL-POOR: MEOH, HEX, PYR
TO: (E.COLI,)(SHYG.,)(PHYT.FUNGI,)
REFERENCES:

JP 80/64509; *CA*, 93, 166097

11132-7603

NAME: TAI-B
PO: S.CULBUS
CT: BASIC, AMINOOLIGOSACCHARIDE
EA: (C, 42)(H, 6)(N, 2)
MW: 650
PC: WH., POW.
OR: (+142, W)
UV: W: (200, ,)
SOL-GOOD: W, DMSO
SOL-FAIR: MEOH, PYR
SOL-POOR: ETOH, HEX
TO: (B.SUBT.,)(E.COLI,)(SHYG.,)(PHYT.FUNGI,)
IS-ION: (IR-120-H,)(SP-SEPHADEX C-25-H,)
REFERENCES:

JP 80/64524; *CA*, 94, 45580

11132-8465

NAME: OLIGOSTATIN-C, SF-1130-X3
PO: S.MYXOGENES
CT: AMINOOLIGOSACCHARIDE, BASIC
FORMULA: C31H55NO22
EA: (N, 2)
MW: 830, 793
PC: WH., POW.
OR: (+154, W)
UV: W: (200, ,)
SOL-GOOD: W, DMSO
SOL-FAIR: MEOH, ETOH
SOL-POOR: ACET, HEX
TO: (E.COLI,)
LD50: NONTOXIC
IS-FIL: 2
IS-ION: (IR-120-H, NH4OH)(DX-50-PYR, PH3.1 BUFF.)
IS-ABS: (CARBON, ACET-HCL)
IS-CHR: (BIOGEL P-2, W)
IS-CRY: LIOF.
REFERENCES:

DT 3035193; *JA*, 34, 1424, 1429, 1981

11210-6732

NAME: G-I-2A"B"
PO: POLYPORACEAE SP., GANODERMA APPLANATUM, GONODERMA APPLANATUM
CT: POLYSACCHARIDE, GLUCAN
MW: 312000
PC: YELLOW, BROWN, POW.
OR: (+23, W)
SOL-GOOD: W
SOL-POOR: ETOH, MEOH, HEX
TV: EHRLICH, YOSHIDA, S-180
IS-EXT: (MEOH-W, , MIC.)
REFERENCES:
Acta Med. Univ. Kagoshima, 20, 209, 1978; *Agr. Biol. Ch.*, 45, 323, 1981; *CA*, 90, 66712

11210-6733

PO: PRODISCULUS SP.
CT: POLYSACCHARIDE
MW: 1000|5000
PC: BROWN, POW.
QUAL: (NINH., +)
TV: ANTITUMOR
IS-EXT: (NAOH, , MIC.)
REFERENCES:
JP 78/94019; *CA*, 90, 4433

11210-6734

NAME: CORIOLAN, "GLUCAN", D-II
PO: CORIOLUS VERSICOLOR
CT: POLYSACCHARIDE, GLUCAN
EA: (C, 38)(H, 6)(O, 56)
MW: 2000000
PC: WH., POW.
OR: (+22, W)
SOL-GOOD: W, BASE, DMSO
SOL-POOR: ETOH, HEX
STAB: (HEAT, +)(ACID, +)
LD50: NONTOXIC
TV: S-180, EHRLICH
IS-EXT: (W, 100 DEGREE, MIC.)
IS-CRY: (PREC., W, ETOH)
REFERENCES:
Jap. J. Pharm., 30, 503, 1980; *CA*, 93, 230733; DT 2845765

11210-6735

NAME: B"-1.3-GLUCAN
PO: CORIOLUS HIRSUTIS
CT: NEUTRAL, POLYSACCHARIDE, GLUCAN
TV: S-180
REFERENCES:
JP 78/39519; *CA*, 90, 53133

11210-7062

NAME: HELVELLAN
PO: HELVELLA LACUNOSA
CT: POLYSACCHARIDE, GLUCAN
EA: (C, 44)(H, 6)(O, 50)
MW: 10000
PC: WH., POW.
OR: (+16, NAOH)
SOL-GOOD: W
SOL-POOR: ACET
LD50: NONTOXIC
TV: S-180
IS-CRY: (PREC., W, ACET)
REFERENCES:
JP 79/107512; *CA*, 92, 179069

11210-7063

NAME: A-5 MANNAN
PO: SACCHAROMYCES CEREVISIAE
CT: POLYSACCHARIDE, GLUCAN, NEUTRAL
EA: (C,)(H,)(O,)
MW: 24000
PC: WH., POW.
OR: (+62.5, W)
SOL-GOOD: W
SOL-POOR: ETOH
TV: S-180
IS-EXT: (W, , MIC.)
IS-CRY: (PREC., W, ETOH)
REFERENCES:
JP 79/97692; *CA*, 91, 191364

11210-7509

NAME: "YEAST GLUCAN", "YEAST-B"-GLUCAN"
PO: SACCHAROMYCES SP.+ACHROMOBACTER ENZYME
CT: POLYSACCHARIDE, GLUCAN
EA: (C, 39)(H, 6)(O, 55)
MW: 32000|20000, 85000|35000
PC: WH., POW.
OR: (-17|3, DMSO)
UV: W: (200, ,)
SOL-GOOD: BASE, DMSO
SOL-FAIR: W
SOL-POOR: MEOH, HEX
QUAL: (NINH., -)(BIURET, -)
TV: ANTITUMOR
REFERENCES:
JP 80/709; 79/138115

11210-7604

NAME: NEOSCHYZOPHYLLAN, TSH
PO: SCHYZOPHYLLUM COMMUNE
CT: POLYSACCHARIDE, GLUCAN, NEUTRAL
EA: (C,)(H,)(O,)
MW: 368000|262000
PC: WH., POW.
SOL-GOOD: W
SOL-POOR: MEOH, HEX
LD50: (30, IV)
TV: S-180
REFERENCES:
DT 2648834; Belg. P 879273

11210-7605

NAME: "FRUCTAN", "GLUCAN"
PO: B.LAEVOLACTICUS
CT: POLYSACCHARIDE, GLUCAN, NEUTRAL
EA: (C,)(H,)(O,)
MW: 44000, 640000, 2000000
PC: WH.
SOL-GOOD: W
TV: ANTITUMOR
REFERENCES:
JP 80/37188; 80/37189; *CA*, 93, 68672, 68673

11210-7606

NAME: "MANNAN"
PO: CANDIDA TROPICALIS
CT: POLYSACCHARIDE, NEUTRAL, GLUCAN
EA: (C,)(H,)(N,)
MW: 14000
PC: WH.
TV: TMV, ANTIVIRAL
IS-EXT: (W, , MIC.)
REFERENCES:

SU P 687120

11210-8093

NAME: "HYDROLYSED CORTICANE"
PO: CORTICIUM VAGUM
CT: POLYSACCHARIDE, GLUCAN, NEUTRAL
FORMULA: C6H10O5.N
EA: (C, 40)(H, 7)(O, 54)
MW: 155000|5000
PC: WH., POW.
OR: (-15, DMSO)
SOL-GOOD: W, DMSO, DMFA
SOL-POOR: ETOH, ET2O
STAB: (HEAT, +)
LD50: NONTOXIC
TV: S-180
IS-EXT: (W, ,)
IS-CRY: (PREC., ETOH, W)(LIOF.,)
REFERENCES:

Belg. P 883444

11210-8162

NAME: "GLUCAN-I"
PO: AURICULARIA AURICULA-JUDAE, AURICULARIA POLYTCHA, AURICULARIA MESENTERICA
CT: POLYSACCHARIDE, GLUCAN, NEUTRAL
EA: (C, 40)(H, 8)(O, 52)
MW: 140000
PC: WH., POW.
OR: (-10.1, NAOH)
SOL-GOOD: W
TV: S-180
REFERENCES:

JP 80/25409; 79/63012; *CA*, 93, 155834; *Agr. Biol. Ch.*, 42, 417, 1978; *Carb. Res.*, 92, 115, 1981

11210-8466

NAME: B"-1.3-GLUCAN-POLYOL
PO: PESTALOTIA SP.+H104
CT: POLYSACCHARIDE, GLUCAN, NEUTRAL
EA: (C, 40)(H, 6)(O, 54)
PC: WH., POW.
TV: ANTITUMOR
REFERENCES:
DT 3032636

11210-8467

PO: GRIFOLA FRONDOSA-TOKACHIANA
CT: POLYSACCHARIDE, GLUCAN, NEUTRAL
PC: WH., POW.
SOL-GOOD: W
TV: ANTITUMOR
REFERENCES:
JP 80/150892

11210-8857

NAME: "GLUCAN"
PO: POLYPORUS UMBELLATUS
CT: POLYSACCHARIDE, GLUCAN, NEUTRAL
PC: WH., POW.
SOL-GOOD: W
SOL-POOR: ETOH
TV: S-180
IS-EXT: (W, , MIC.)
REFERENCES:
JP 81/76401; *CA*, 95, 185554

11220-7065

NAME: C-45
PO: CLOSTRIDIUM SP.
CT: POLYSACCHARIDE-PROTEIN COMPLEX, AMPHOTER
EA: (C, 44)(H, 7)(N, 10)
PC: WH., POW.
SOL-GOOD: DMSO, W
SOL-POOR: ETOH
TV: ANTITUMOR
IS-EXT: (DMSO, , MIC.)
IS-CRY: (PREC., DMSO, ETOH)
REFERENCES:
JP 79/76896; *CA*, 91, 191355

11220-7066

PO: CORIOLUS VERSICOLOR, PLEUROTUS OSTREATUS, LENTINUS EDODES
CT: POLYSACCHARIDE-PROTEIN COMPLEX
EA: (N,)
TV: S-180
IS-EXT: (W, , MIC.)
REFERENCES:

Lloydia 42, 684, 1979

11220-7413

NAME: KS-SUBSTANE
PO: DAEDALEA DICKINSII, LENTINUS EDODES
CT: POLYSACCHARIDE-PROTEIN COMPLEX, NEUTRAL
EA: (C, 19)(H, 3)(N, 1.5)(ASH, 11)
MW: 70000
PC: WH., POW.
SOL-GOOD: W
SOL-POOR: ETOH, HEX
TV: ANTITUMOR
IS-EXT: (NACL, , MIC.)
REFERENCES:

JP 79/140788; *CA*, 92, 213532

11220-7414

NAME: KS-2B
PO: DAEDALEA DICKINSII, LENTINUS EDODES
CT: POLYSACCHARIDE-PROTEIN COMPLEX, ACIDIC, AMPHOTER
EA: (C, 44)(H, 7)(N, 1.5)
MW: 75000|15000
PC: WH., POW.
OR: (+61.8, W)
UV: W: (200, ,)
SOL-GOOD: W
SOL-POOR: ETOH, HEX, PHENOL
QUAL: (NINH., +)(BIURET, +)
TV: ANTITUMOR, ANTIVIRAL
IS-EXT: (ETOH-W, , MIC.)
UTILITY: ON CLINICAL TRIAL
REFERENCES:

JA, 31, 1079, 1978; *Agr. Biol. Ch.*, 44, 1863, 1980; DT 2813353, 2919132; JP 79/140711; 80/27101; *CA*, 93, 236923

11220-7607

PO: ACHROMOBACTER ALCOHOLYTICUS
CT: POLYSACCHARIDE-PROTEIN COMPLEX
EA: (C,)(H,)(N,)
PC: WH., YELLOW, POW.
OR: (+49, W)
SOL-GOOD: W
QUAL: (NINH., +)(BIURET, +)
TV: S-180
REFERENCES:
JP 80/34024; *CA*, 93, 68679

11220-7608

PO: ACHROMOBACTER SLIMOGEN
CT: POLYSACCHARIDE-PROTEIN COMPLEX
EA: (C, 33)(H, 5)(N, 1)(O,)
PC: WH., YELLOW, POW.
OR: (+55, NAOH)
SOL-GOOD: W, BASE
QUAL: (NINH., +)(BIURET, +)
TV: S-180
REFERENCES:
JP 80/29974; *CA*, 93, 68675

11220-7609

NAME: DH-6665
PO: MICROELLOBOSPORIA GRISEA
CT: POLYSACCHARIDE-PROTEIN COMPLEX
EA: (C, 39)(H, 5)(N, 2)
MW: 150000|140000
SOL-GOOD: W, FA
SOL-FAIR: DMSO
SOL-POOR: MEOH, HEX, PYR
LD50: (1000, IV)
TV: ANTITUMOR
IS-EXT: (W, , MIC.)
IS-CHR: (SEPHADEX G-100, NACL)
IS-CRY: (PREC., W, ETOH)
REFERENCES:
JP 80/51096; *CA*, 93, 148114; *Mikrob.*, 36, 309, 1967

11220-7610

PO: MICROPORUS AFFINIS
CT: POLYSACCHARIDE-PROTEIN COMPLEX
EA: (N,)
TV: S-180
REFERENCES:
CA, 93, 66147

11220-7612

NAME: SP-MT-1, T1-B1, T0-B1, T2-B1
PO: CORYNEBACTERIUM PARVUM, MYCOBACTERIUM BOVIS, NOC.RUBRA
CT: POLYSACCHARIDE-PROTEIN COMPLEX, NEUTRAL
EA: (N,)
PC: WH., POW.
LD50: (1000, IP.)
TV: S-180, MELANOMA
IS-CHR: (CM-CEL,)
REFERENCES:
JP 80/50894; 80/50895; *CA*, 93, 148119

11220-8094

NAME: "PROTEOGLYCAN"
PO: AGARICUS HETEROSITES
CT: POLYSACCHARIDE, GLUCAN
PC: BROWN, POW.
SOL-GOOD: W
SOL-POOR: ETOH, ET2O
TV: S-180
IS-EXT: (W, ,)
IS-CRY: (PREC., ETOH, W)
REFERENCES:
JP 80/108292; *CA*, 93, 236941

11220-8095

NAME: JA-AS-15-712, D-MANNAN
PO: SACCHAROMYCES CEREVISIAE
CT: POLYSACCHARIDE-PROTEIN COMPLEX, AMPHOTER
EA: (C, 41)(H, 6)(N, 2)(O, 51)
MW: 180000|40000
PC: WH., POW.
OR: (+65|10, W)
UV: W: (200, ,)
SOL-GOOD: W
SOL-POOR: MEOH, ET2O
QUAL: (NINH., +)(BIURET, +)
TV: S-180
REFERENCES:

DT 3017372

11220-8468

NAME: KS-2D
PO: LENTINUS EDODES
CT: POLYSACCHARIDE-PROTEIN COMPLEX, NEUTRAL
EA: (C, 42)(H, 6)(N, 1)(ASH, 7)
MW: 70000|40000
PC: WH., POW.
OR: (81.8|15.7, W)
UV: W: (200, ,)
SOL-GOOD: W
SOL-POOR: ETOH, HEX
QUAL: (NINH., +)(BIURET, +)
TV: ANTITUMOR
REFERENCES:

JP 81/53101; *CA*, 96, 18643

11220-8469

PO: GONODERMA LUCIDUM
CT: POLYSACCHARIDE-PROTEIN COMPLEX
EA: (N,)
PC: BROWN, POW.
TV: S-180
IS-EXT: (NAOH, , MIC.)
REFERENCES:

Honguk Saenghwa Hokoku Chi, 14, 101, 1981; *CA*, 95, 93841, 144002

11220-8783

NAME: IES-1638
PO: BIFIDOBACTERIUM INFANTIS, BIFIDOBACTERIUM ADOLESCENSIS
CT: POLYSACCHARIDE, GLUCAN, NEUTRAL
EA: (C, 42)(H, 6)(N, .5)
MW: 10000|1000000
PC: WH., BROWN, POW.
OR: (+155.6, W)
UV: W: (200, ,)
SOL-GOOD: W
SOL-POOR: MEOH, BENZ
QUAL: (NINH., +)
TV: S-180
IS-CRY: LIOF.
REFERENCES:
JP 81/58491; *CA*, 95, 130973

11230-8784

NAME: "ARABINOMANNAN"
PO: MYCOBACTERIUM TUBERCULOSIS
CT: LIPOPOLYSACCHARIDE
EA: (C,)(H,)(O,)
MW: 12000|5000
PC: WH., POW.
SOL-GOOD: W
TV: ANTITUMOR
IS-EXT: (W, ,)
REFERENCES:
J. Ph. Soc. Jap., 101, 713, 1981; JP 81/8320

11240-6736

NAME: PS-9-5H
PO: PS.HYDROGENOVORA
CT: NEUTRAL, POLYSACCHARIDE
EA: (H, 6)(N, .2)(C, 39)
MW: 200000
PC: WH., POW.
UV: W: (200, ,)
SOL-GOOD: W, ACOH
SOL-POOR: MEOH, HEX
QUAL: (NINH., -)
TV: TMV, ANTITUMOR, S-180, EHRLICH
REFERENCES:
JP 78/113091; 78/113092; 78/121926; *Agr. Biol. Ch.*, 44, 2925, 1980; *CA*, 90, 53136, 53137, 70590

11240-8163

NAME: PH-2
PO: PS.HYDROGENOTHERMOPHILA
CT: POLYSACCHARIDE, NEUTRAL
EA: (C, 34)(H, 6)(O, 54)(N, .5)(ASH, 6)
MW: 200000
PC: WH., POW.
SOL-GOOD: W
SOL-POOR: MEOH, HEX
TO: (E.COLI,)
REFERENCES:
JP 80/147502; *CA*, 94, 154991

11240-8164

NAME: AP
PO: ISARIA ATYPICOLA
CT: POLYSACCHARIDE, NEUTRAL
EA: (C, 36)(H, 6)(N, .5)
MW: 20000
PC: WH., BROWN, POW.
SOL-GOOD: W, BASE
SOL-POOR: MEOH, HEX
QUAL: (NINH., +)
LD50: (3000|1000, IP)
TV: S-180, EHRLICH
IS-EXT: (W, ,)
REFERENCES:
DT 3030107

11240-8858

PO: PSEUDOMONAS SP.+METHANOL
CT: POLYSACCHARIDE, ACIDIC
EA: (C, 37)(H, 5)(O,)(N, 1)
PC: WH., POW.
SOL-GOOD: W
SOL-POOR: MEOH, ACET
TV: ANTIVIRAL, ANTITUMOR
REFERENCES:
JP 81/48891; 81/99795; *CA*, 95, 185563

11200-7064

CT: POLYSACCHARIDE
LD50: NONTOXIC
TV: S-180, CYTOTOXIC
REFERENCES:

Hsueh Hiu Tso Chi, 78, 549, 1979; *CA*, 92, 33820

11200-7524

NAME: LEVAN
PO: AEROBACTER LEVANICUM
CT: POLYSACCHARIDE, NEUTRAL
MW: 2000000
PC: WH., POW.
TV: ANTITUMOR, LEWIS
REFERENCES:

J. Nat. Cancer Inst., 65, 391, 1980; *Cancer Res.*, 35, 1921, 1975; 36, 1593, 1976; *Isr. J. Med. Sci.*, 13, 859, 1977; *Brit. J. Exp. Path.*, 123, 157, 1977; *Exp.*, 34, 1362, 1978

11200-7554

PO: SACCHAROMYCES UVARUM+SEO2
CT: POLYSACCHARIDE
EA: (SE,)
TV: ANTITUMOR
REFERENCES:

JP 80/3710

11200-7611

PO: POLYPORUS TUBERASTER
CT: POLYSACCHARIDE, NEUTRAL
MW: 8000
PC: WH., YELLOW, POW.
SOL-GOOD: W
SOL-POOR: MEOH, HEX
QUAL: (BIURET, -)
TV: S-180, EHRLICH
REFERENCES:

DT 2938039

11200-8161

PO: LENTINUS EDODES
CT: POLYSACCHARIDE
PC: BROWN, POW.
LD50: (18.6|1, IP)
TV: ANTIVIRAL
IS-EXT: (W, ,)
IS-CHR: (SEPHADEX LH-20, MEOH)
REFERENCES:
JP 80/157517; *CA*, 94, 114718

11200-8859

NAME: PK-1
PO: PESTALOTIA SP.
CT: POLYSACCHARIDE
EA: (C, 41)(H, 6)(N, 3)
SOL-GOOD: W
SOL-POOR: MEOH, ETOH
TV: S-180, LEWIS
REFERENCES:
JP 81/49395; *CA*, 95, 185547

12
AMINOGLYCOSIDE ANTIBIOTICS

121 Streptamine Derivatives
122 2-Deoxystreptamine Derivatives
123 Inositol-1 Inoseamine Derivatives
124 Other Aminocyclitols
125 Aminohexitols

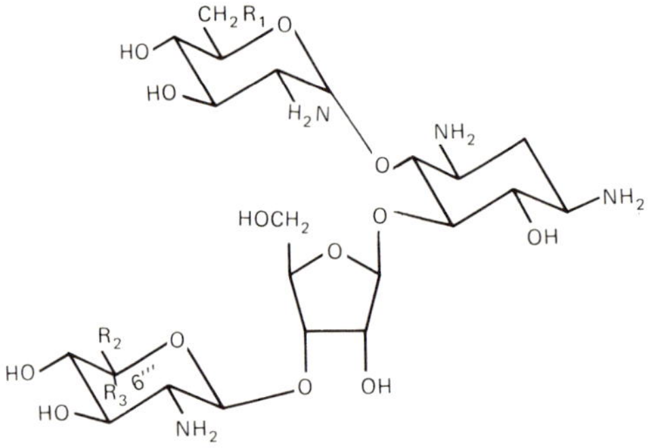

	R_1	R_2	R_3
6‴-Deamino-6‴-hydroxyparomomycin-I	OH	H	CH_2OH
6‴-Deamino-6‴-hydroxyparomomycin-II	OH	CH_2OH	H
6‴-Deamino-6‴-hydroxyneomycin-C	NH_2	H	CH_2OH

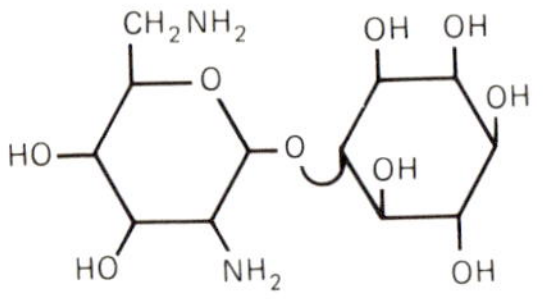

Neomycin G

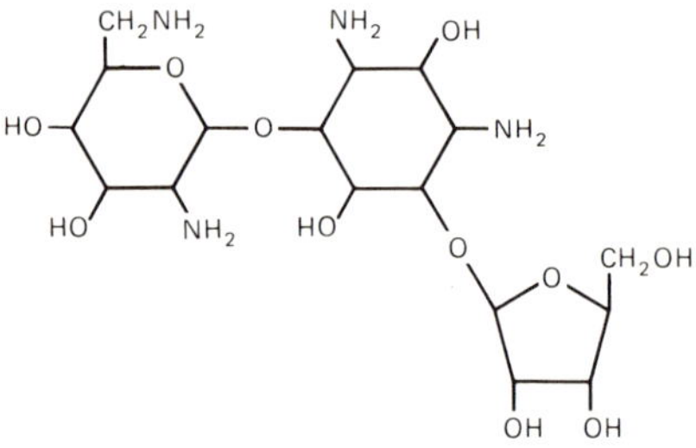

Neomycin K

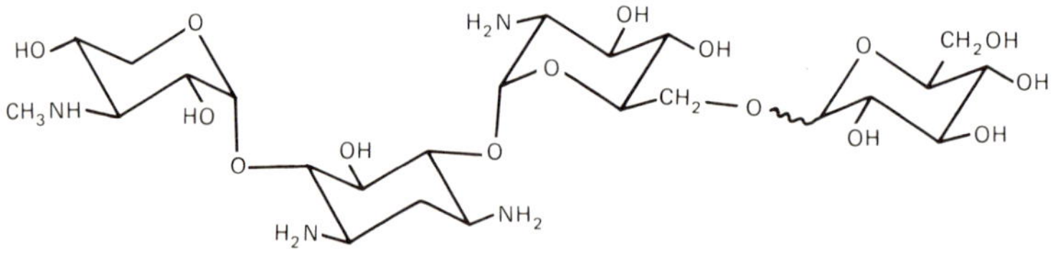

477-2h

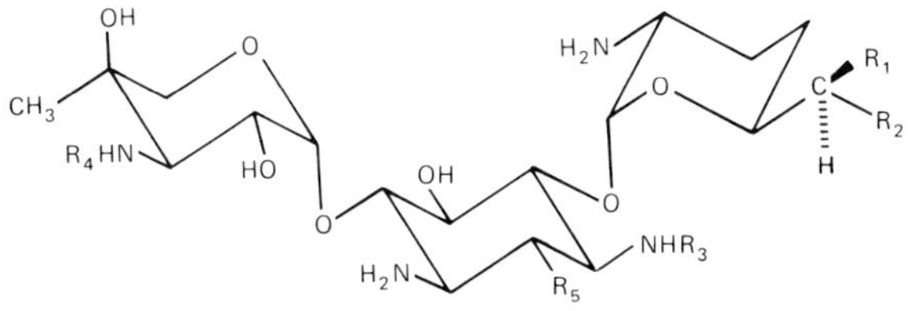

	R_1	R_2	R_3	R_4	R_5
Sagamicin (XK-62-2)⁻	H	$NHCH_3$	H	CH_3	H
XK-62-3	H	NH_2	CH_3	CH_3	H
XK-62-4	H	$N(CH_3)_2$	H	CH_3	H
XK-62-5	H	$NHCH_3$	H	H	H
XK-62-6	H	$NHCH_3$	CH_3	CH_3	H
XK-62-7	H	$N(CH_3)_2$	CH_3	CH_3	H
XK-62-8	CH_3	$NHCH_3$	CH_3	CH_3	H
2-Hydroxysagamicin	H	$NHCH_3$	H	CH_3	OH
Gentoxymicin A	H	NH_2	H	CH_3	H
Gentoxymicin B	CH_3	NH_2	H	CH_3	H
3″-N-Demethylgentamicin C_{1a}	H	NH_2	H	H	H

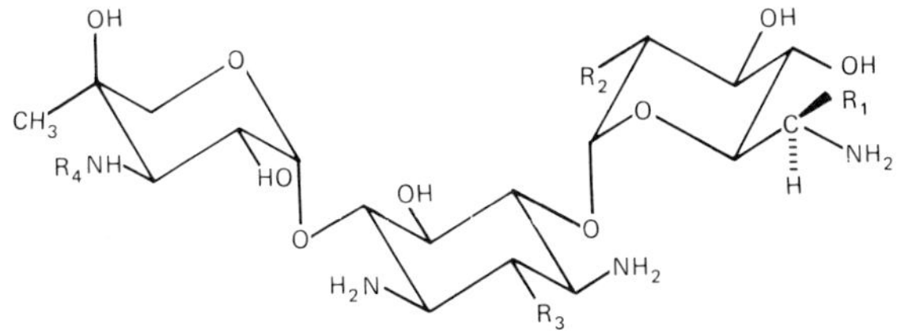

	R_1	R_2	R_3	R_4
TPJ-B	CH_3	NH_2	H	H
2-Hydroxygentamicin B	H	OH	OH	CH_3
2-Hydroxygentamicin B_1	CH_3	OH	OH	CH_3
2-Hydroxy-JI-20A	H	NH_2	OH	CH_3
2-Hydroxy-JI-20B	CH_3	NH_2	OH	CH_3

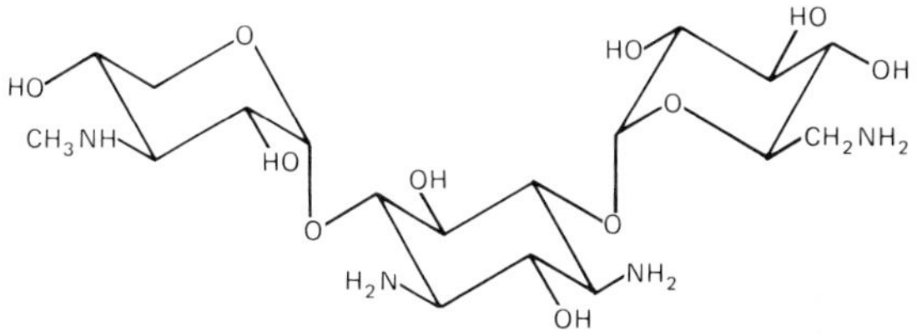

2-Hydroxygentamicin A_3

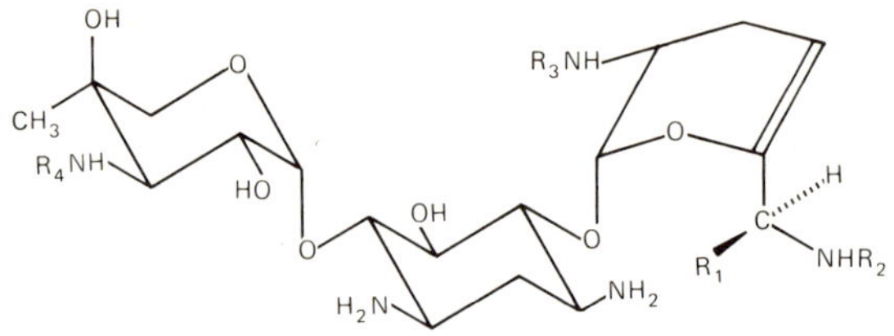

	R_1	R_2	R_3	R_4
3″-N-Demethylsisomicin	H	H	H	H
D-53 (3″-N-Demethyl-G-52)	H	CH_3	H	H
G-367-1	H	H	CHO	CH_3
K-26-2 (6′-N-Methylverdamicin)	CH_3	CH_3	H	CH_3

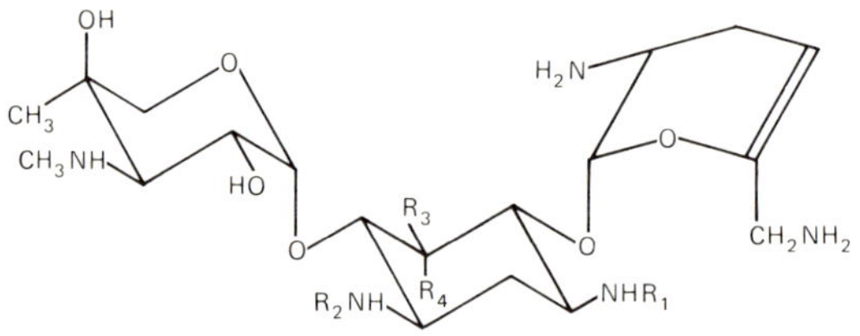

	R_1	R_2	R_3	R_4
Mutamicin-7 (Mu-7)	CH_3	H	OH	H
Mutamicin-8 (Mu-8)	H	CH_3	H	H
5-Epifluoro-5-deoxysisomicin	H	H	H	F

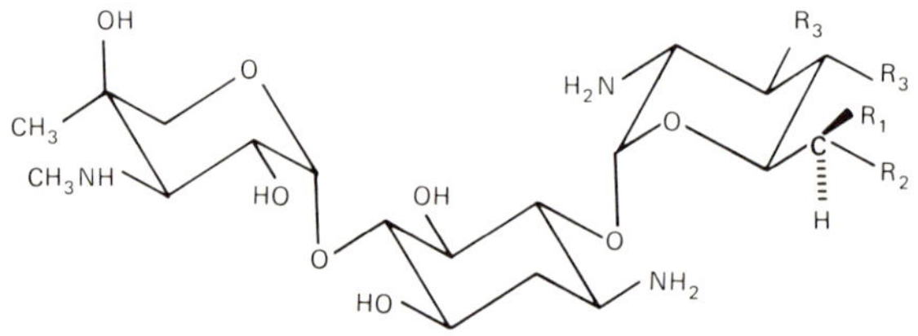

	R_1	R_2	R_3
SU-1	H	NH_2	H
SU-2	CH_3	NH_2	H
SUM-3	H	$NHCH_3$	H
SUM-4	CH_3	NH_2	H,OH
K-114-E	H	OH	OH
K-114-G	CH_3	OH	OH

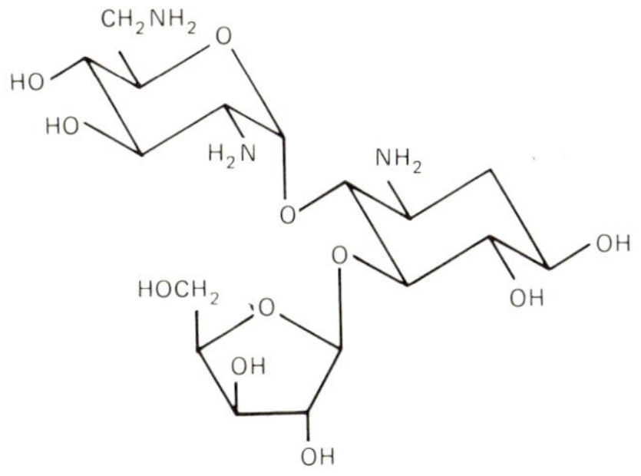

S-11-A

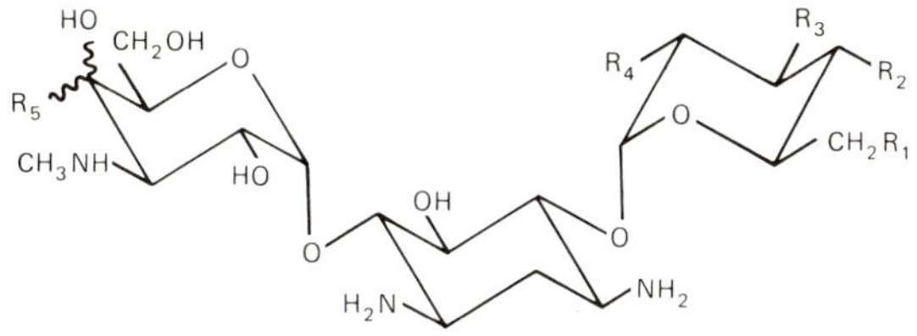

	R_1	R_2	R_3	R_4	R_5
Combimicin A_2 (I-A_2)	NH_2	H	H	OH	CH_3
Combimicin B_1 (I-B_1)	$NHCH_3$	H	H	NH_2	CH_3
Combimicin B_2 (I-B_2)	NH_2	H	H	NH_2	CH_3
Combimicin A_3 (3″-N-Methyl-4″-C-Methyl-kanamycin A)	NH_2	OH	OH	OH	CH_3
Combimicin A_4 (3″-N-Methylkanamycin A)	NH_2	OH	OH	OH	H
Combimicin B_3 (I-B_3)	$NHCH_3$	H	H	NH_2	H
Combimicin B_4 (3″-N-Methyldibekacin)	NH_2	H	H	NH_2	H
Combimicin B_5 (3″-N-Methylkanamycin B)	NH_2	OH	OH	NH_2	H
I-T_{1a}	NH_2	OH	H	OH	CH_3
I-T_2 (3″-N-Methyltobramycin)	NH_2	OH	H	NH_2	H
I-T_{1b} (3″-N-Methyl-4″-C-methyltobramycin)	NH_2	OH	H	NH_2	CH_3
3″-N-Methyl-4″-C-methylkanamycin B	NH_2	OH	OH	NH_2	CH_3
3″-N-Methyltobramycin	NH_2	OH	H	NH_2	H

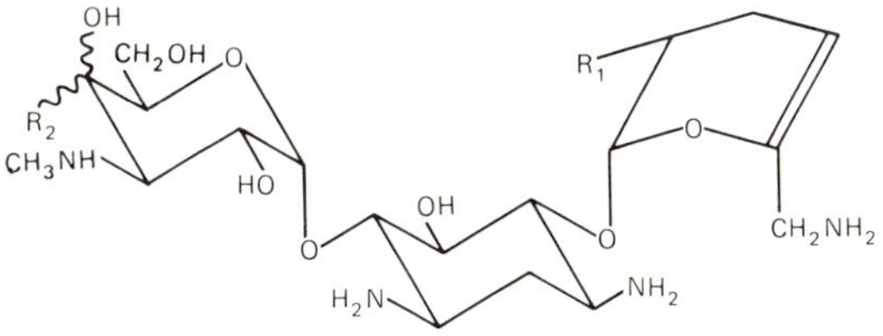

	R_1	R_2
I-SK-A_1	OH	CH_3
I-SK-B_1	NH_2	CH_3
I-SK-B_2	NH_2	H

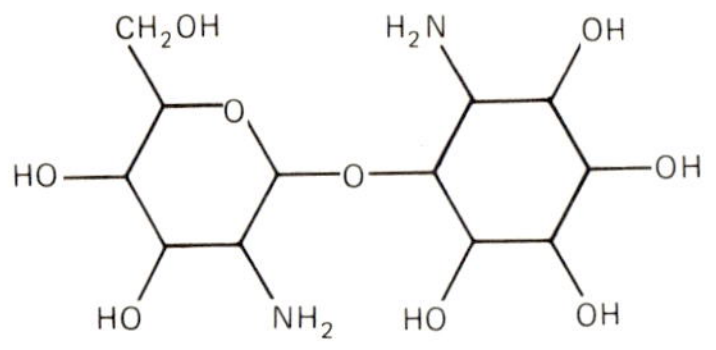

FU-10

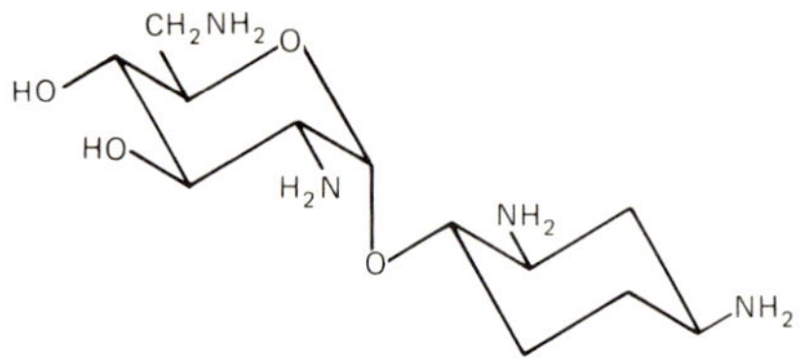

5.6-Dideoxyneamine

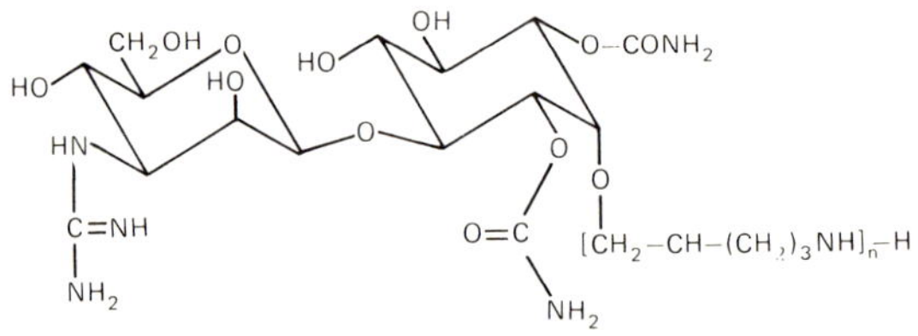

	n
LL-BM-782α_2	3
LL-BM-782α_1	4
LL-BM-782α_{1a}	5
Myomycin-B	2
DC-5-4	0 (tentative)

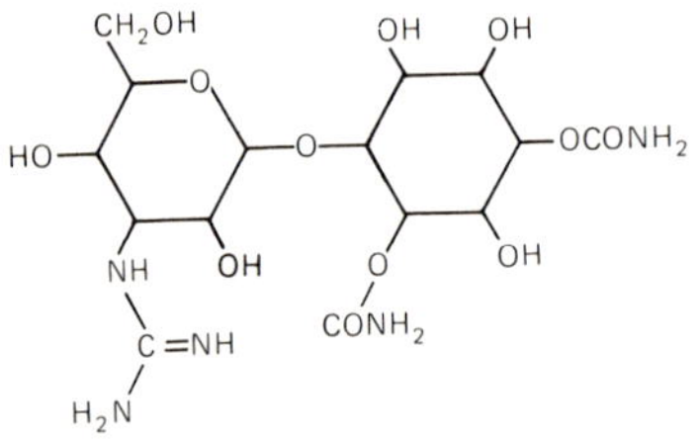

DC-5-4

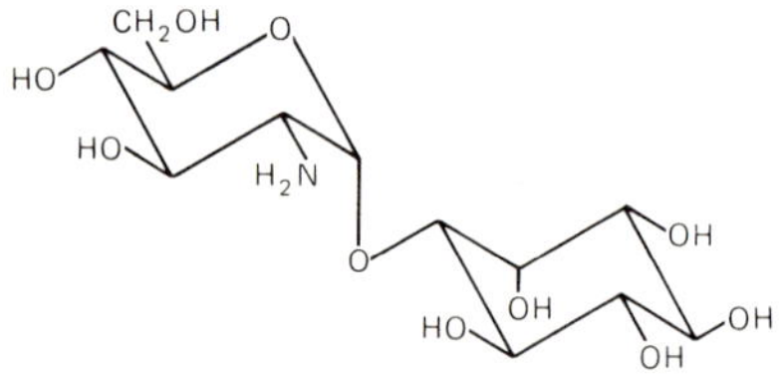

X-14847

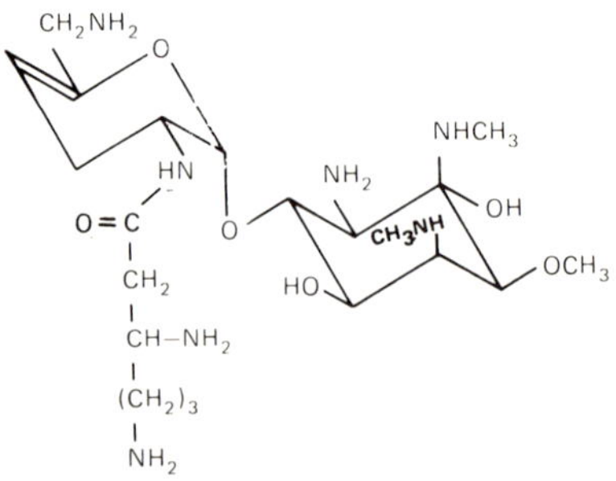

Lysinomycin
(AX-127-B)

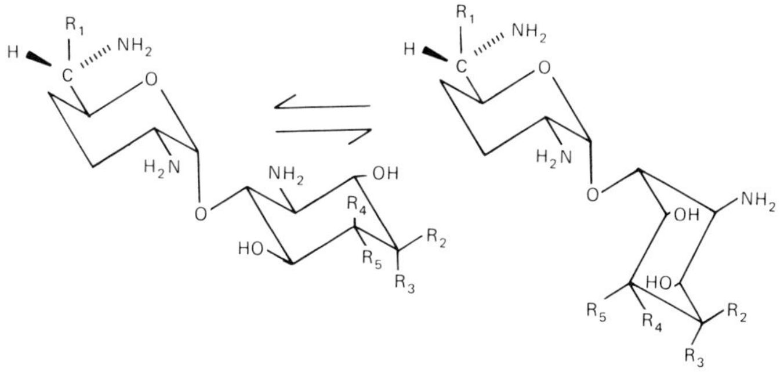

A B

	R_1	R_2	R_3	R_4	R_5	Conf.*
Fortimicin E	CH_3	H	OCH_3	$NHCH_3$	H	A
Fortimicin KO_1	CH_3	OCH_3	H	$NHCH_3$	H	A
Fortimicin AE	CH_3	OCH_3	H	H	$NHCH_3$	A
Fortimicin AM	CH_3	H	OH	H	$NHCH_3$	A
Fortimicin AP	CH_3	OH	H	H	$NHCH_3$	A
Fortimicin AQ	CH_3	H	OCH_3	$N(CH_3)_2$	H	B
Fortimicin AS	CH_3	H	OCH_3	$N(CH_3)CH_2CH_2OH$	H	B
3-O-Demethyl-fortimicin A	CH_3	H	OH	$N(CH_3)COCH_2NH_2$	H	B
Fortimicin KR_1	H	H,OH		H,$NHCH_3$		A
3-O-Demethyl-fortimicin-KE	H	H	OH	$NHCH_3$	H	A
Fortimicin A	CH_3	H	OCH_3	$N(CH_3)$ $COCH_2NH_2$	H	B
Fortimicin B	CH_3	H	OCH_3	$NHCH_3$	H	A
Fortimicin KE	H	H	OCH_3	H	$NHCH_3$	A

* Conf.: favorable conformation state of the molecule

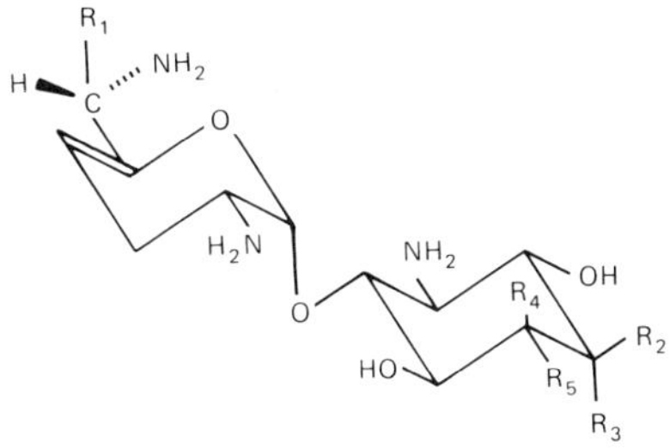

	R_1	R_2	R_3	R_4	R_5
Fortimicin KG	CH_3	H	OCH_3	H	$NHCH_3$
Fortimicin KF	H	H	OCH_3	H	$NHCH_3$
Fortimicin KG_1	CH_3	OCH_3	H	H	$NHCH_3$
Fortimicin KG_2	CH_3	H	OCH_3	$NHCH_3$	H
Fortimicin KG_3	CH_3	H	OCH_3	$N(CH_3)COCH_2NH_2$	H
Fortimicin AL	CH_3	OH	H	H	$NHCH_3$
Fortimicin KQ	CH_3	H,OH		H,$NHCH_3$	

	R_1	R_2	R_3	R_4	R_5	R_6	R_7	R_8	Conf.*
Sannamycin A (Istamycin A)	H	CH_3	NH_2	H	H	OCH_3	N–$COCH_2NH_2$ ╲CH_3	H	B
Sannamycin B (Istamycin A_0)	H	CH_3	NH_2	H	H	OCH_3	$NHCH_3$	H	A
Istamycin B	H	CH_3	H	NH_2	H	OCH_3	N–$COCH_2NH_2$ ╲CH_3	H	B
Istamycin B_0	H	CH_3	H	NH_2	H	OCH_3	$NHCH_3$	H	A
Sannamycin E (KA-7038 II)	H	H	NH_2	H	OH	H	H	$NHCH_3$	A
Sannamycin D (KA-7038 IV)	H	H	NH_2	H	H,OH		H,$NHCH_3$		A
Sannamycin G (KA-7038 V)	H	H	NH_2	H	H	OCH_3	H	$NHCH_3$	A
Sannamycin C (KA-7038 VI)	H	CH_3	NH_2	H	OCH_3	H	$NHCH_3$	H	A
Sannamycin H (KA-7038 VII)	H	H	NH_2	H	OCH_3	H	H	$NHCH_3$	A
KA-6606 V (3-Epi-Sporaricin-B)	CH_3	H	NH_2	H	H	OCH_3	H	$NHCH_3$	A
Sporaricin E (K-6606-VI)	CH_3	H	NH_2	H	H	OCH_3	$NHCH_3$	H	A
Dactimicin (SF-2052-A)	CH_3	H	NH_2	H	H	OCH_3	N–$COCH_2NHCH{=}NH$ ╲CH_3	H	B
SF-2052-B (SF-1854)	CH_3	H	NH_2	H	H	OCH_3	N–$COCH_2NHCHO$ ╲CH_3	H	B

* Conf.: favorable conformation state of the molecule

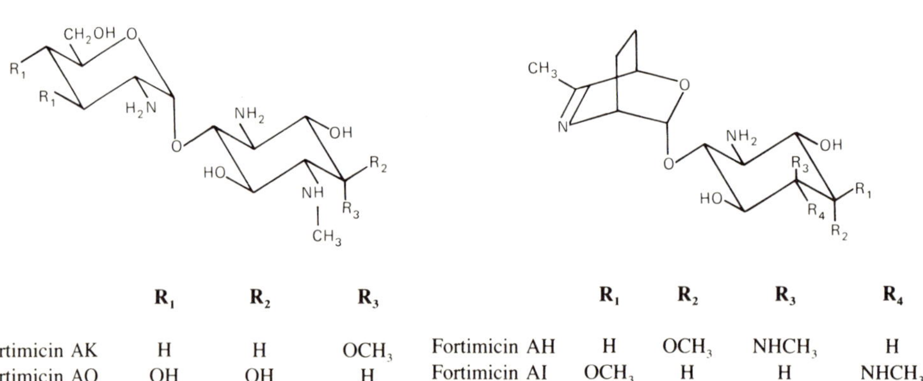

	R_1	R_2	R_3
Fortimicin AK	H	H	OCH_3
Fortimicin AO	OH	OH	H

	R_1	R_2	R_3	R_4
Fortimicin AH	H	OCH_3	$NHCH_3$	H
Fortimicin AI	OCH_3	H	H	$NHCH_3$

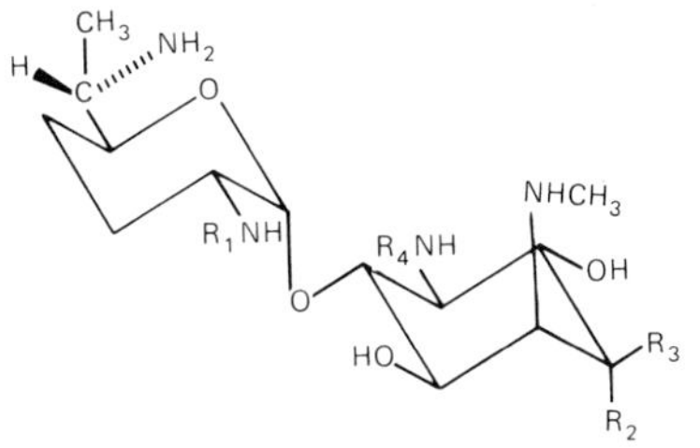

	R_1	R_2	R_3	R_4
Isofortimicin A	$COCH_2NH_2$	OCH_3	H	H
Fortimicin AN	H	OH	H	$COCH_2NH_2$
Fortimicin KO	$COCH_2NH_2$	H	OCH_3	H
3-O-Demethyl-2′-N-glycyl-fortimicin B	$COCH_2NH_2$	OH	H	H
Demethylisofortimicin C	H	OH	H	$COCH_2NHCONH_2$

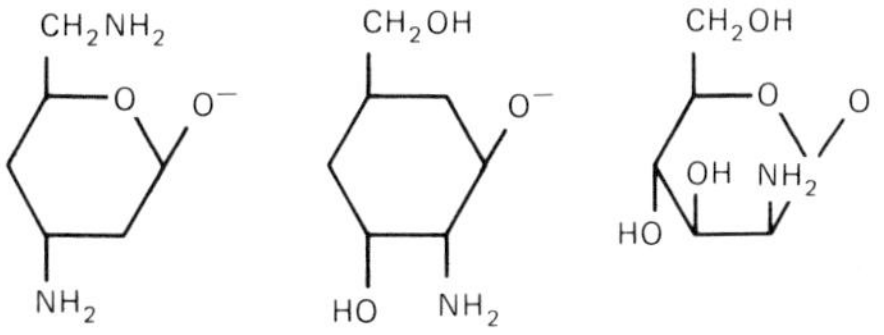

Desomycin (degradation products)

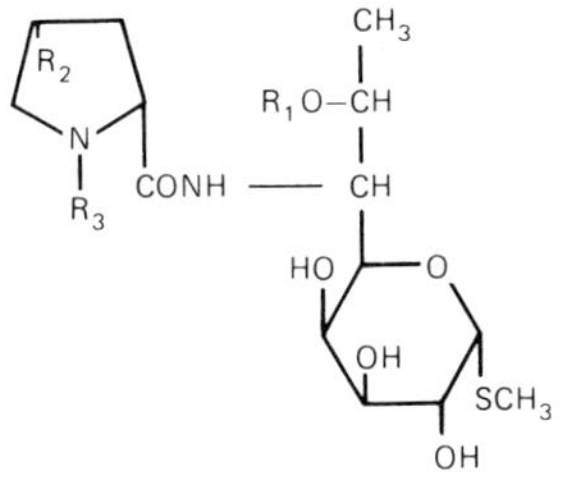

	R_1	R_2	R_3
BU-2545	CH_3	H	CH_3
U-24166	H	C_3H_7	C_2H_5

12110-6740

NAME: <u>741</u>
PO: ACT.SP.
CT: BASIC, AMINOGLYCOSIDE, STREPTOMYCIN T.
EA: (N,)
SOL-GOOD: W
STAB: (BASE, -)
TO: (G.POS.,)(G.NEG.,)(MYCOB.SP.,)(PHYT.BACT.,)
REFERENCES:
Farmatsiya (Sofia), 28, 43, 1978; *CA*, 91, 3842

12211-8470

NAME: <u>6'''-DEAMINO-6'''-HYDROXYPAROMOMYCIN-I</u>, COMP.X1
PO: S.RIMOSUS-PAROMOMYCINUS
CT: AMINOGLYCOSIDE, NEOMYCIN T., BASIC
FORMULA: C23H44N4O15
EA: (N, 9)
MW: 616
PC: WH., POW.
UV: W: (200, ,)
SOL-GOOD: W
QUAL: (NINH., +)(FEHL., -)
STAB: (BASE, +)
TO: (S.AUREUS, 5)(B.SUBT., 2)(K.PNEUM., 5)(E.COLI, 40)
IS-ION: (IRC-50-NH4, NH4OH)(XE-64-NH4, NH4OH)
IS-CHR: (SILG, MEOH-NH4OH)
REFERENCES:
JA, 34, 536, 544, 1981

12211-8471

NAME: <u>6'''-DEAMINO-6'''-HYDROXYPAROMOMYCIN-II</u>, COMP.X2
PO: S.RIMOSUS-PAROMOMYCINUS
CT: AMINOGLYCOSIDE, NEOMYCIN T., BASIC
FORMULA: C23H44N4O15
EA: (N, 9)
MW: 616
PC: WH., POW.
OR: (+54.1, H2SO4)
UV: W: (200, ,)
SOL-GOOD: W
QUAL: (NINH., +)(FEHL., -)
STAB: (BASE, +)
TO: (S.AUREUS, 40)(B.SUBT., 10)(K.PNEUM., 40)
REFERENCES:
JA, 34, 536, 544, 1981

12211-8472

NAME: 6'''-DEAMINO-6'''-HYDROXYNEOMYCIN-C, COMP.Y2
PO: S.FRADIAE
CT: AMINOGLYCOSIDE, NEOMYCIN T., BASIC
FORMULA: C23H45N5O14
EA: (N, 11)
MW: 615
PC: WH., POW.
OR: (+111.4, H2SO4)
UV: W: (200, ,)
SOL-GOOD: W
QUAL: (NINH., +)(FEHL., -)
TO: (S.AUREUS, 5)(S.LUTEA,)(K.PNEUM., 2)(E.COLI, 10)
IS-ION: (IRC-50-NH4, NH4OH)(XE-64-NH4, NH4OH)
IS-CHR: (SILG, MEOH-NH4OH)
REFERENCES:
JA, 34, 536, 544, 1981

12211-8473

NAME: NEOMYCIN-G
PO: S.FRADIAE
CT: AMINOGLYCOSIDE, BASIC, NEOMYCIN T.
FORMULA: C12H24N2O9
EA: (N, 8)
MW: 340
PC: WH., POW.
UV: W: (200, ,)
SOL-GOOD: W
QUAL: (NINH., +)
TO: (G.POS.,)
IS-CHR: (CG-50-NH4, NH4OH)(DX-1X2-OH, W)
REFERENCES:
JA, 27, 931, 1974

12211-8474

NAME:	NEOMYCIN-K
PO:	S.FRADIAE
CT:	AMINOGLYCOSIDE, NEOMYCIN T., BASIC
FORMULA:	C17H34N4O11
EA:	(N, 12)
MW:	470
PC:	WH., POW.
UV:	W: (200, ,)
SOL-GOOD:	W
QUAL:	(NINH., +)
TO:	(G.POS.,)

REFERENCES:

JA, 27, 931, 1974

12212-6739

NAME:	H-60
IDENTICAL:	RIBOSTAMYCIN
PO:	S.SP.
CT:	BASIC, AMINOGLYCOSIDE, RIBOSTAMYCIN T.
FORMULA:	C17H34N4O10
EA:	(N, 12)
MW:	470, 454
PC:	WH., POW.
OR:	(+43, W)
UV:	W: (200, ,)
SOL-GOOD:	W
SOL-FAIR:	MEOH
SOL-POOR:	BUOH, BENZ, HEX
QUAL:	(NINH., +)(FEHL., -)(BIURET, -)
TO:	(G.POS.,)(G.NEG.,)
IS-ION:	(IRC-50-NH4, NH4OH)
IS-ABS:	(CARBON, HCL)
IS-CHR:	(CG-50-NH4, NH4OH)(DX-1X2-OH, W)

REFERENCES:

JP 79/3958; *CA*, 91, 3915

12222-7071

NAME: 477-2H, 6'-O-GLUCOSYLGENTAMICIN-A
PO: MIC.SAGAMIENSIS
CT: AMINOGLYCOSIDE, BASIC, GENTAMICIN T.
FORMULA: C24H46N4015
EA: (N, 9)
MW: 630
PC: WH., POW.
SOL-GOOD: W
QUAL: (NINH., +)
TO: (G.POS.,)(G.NEG.,)
REFERENCES:

JP 79/132550; *CA*, 92, 109104

12222-7555

NAME: MUTAMICIN-7, MU-7, 3"-N-METHYLSISOMICIN
PO: MIC.INYOENSIS+2-DEOXY-N-METHYLSTREPTAMINE
CT: AMINOGLYCOSIDE, GENTAMICIN T., BASIC
FORMULA: C20H39N507
EA: (N, 15)
MW: 461
PC: WH.
UV: W: (200, ,)
SOL-GOOD: W
QUAL: (NINH., +)
TO: (G.POS., 50)(G.NEG., 100)
REFERENCES:

Microbiology 1979/1980; 314; USP 4053591

12222-7556

NAME: MUTAMICIN-8, MU-8, 1-N-METHYL-5-DEOXYSISOMICIN
PO: MIC.INYOENSIS+2.5-DIDEOXY-N-METHYLSTREPTAMINE
CT: GENTAMICIN T., BASIC, AMINOGLYCOSIDE
FORMULA: C20H39N506
EA: (N, 16)
MW: 445
PC: WH.
UV: W: (200, ,)
SOL-GOOD: W
QUAL: (NINH., +)
TO: (G.POS., .1)(G.NEG., 1)
REFERENCES:

Microbiology 1979/1980, 314; USP 4053591

12222-7557

NAME: 5-EPIFLUORO-5-DEOXYSISOMICIN, MU-X
PO: MIC.INYOENSIS+5-FLUORODEOXYSTREPTAMINE
CT: AMINOGLYCOSIDE, GENTAMICIN T., BASIC
FORMULA: C19H37N506F
EA: (N,)(F,)
PC: WH.
UV: W: (200, ,)
SOL-GOOD: W
QUAL: (NINH., +)
TO: (G.POS., .1)(G.NEG., 1)
REFERENCES:

Microbiology 1979/1980, 314

12222-7614

NAME: XK-62-5, 3"-N-DEMETHYLSAGAMICIN
PO: MIC.SAGAMIENSIS-NONREDUCTANS
CT: AMINOGLYCOSIDE, GENTAMICIN T., BASIC
FORMULA: C19H39N507
EA: (N, 15)
MW: 449
PC: WH., POW.
UV: W: (200, ,)
SOL-GOOD: W, MEOH
SOL-FAIR: ETOH, ACET
SOL-POOR: BUOH, HEX
QUAL: (NINH., +)(BIURET, -)
TO: (S.AUREUS, .0021)(B.SUBT., .0011)(E.COLI, .033)
(K.PNEUM., .016)(P.VULG., .26)(SHYG., .13)
(PS.AER., .13)
IS-FIL: 2
IS-ION: (IRC-50-NH4, NH4OH)
IS-CHR: (CG-50-NH4, NH4OH)(DX-1X2-OH, W)
IS-CRY: (LIOF.,)(PREC., ACET-W,)
REFERENCES:

JP 80/50896; *CA*, 93, 184281; Fr. P 2317939

12222-7615

NAME: XK-62-6
PO: MIC.SAGAMIENSIS-NONREDUCTANS
CT: AMINOGLYCOSIDE, GENTAMICIN T., BASIC
FORMULA: C21H43N5O7
EA: (N, 14)
MW: 477
PC: WH., POW.
OR: (+165, W)
UV: W: (200, ,)
SOL-GOOD: W
QUAL: (NINH., +)
TO: (S.AUREUS, .016)(B.SUBT., .0082)(E.COLI, .52)
(K.PNEUM., .26)(P.VULG., .52)(SHYG., .52)
REFERENCES:
JP 80/51098; *CA*, 93, 130563

12222-7616

NAME: XK-62-7
PO: MIC.SAGAMIENSIS-NONREDUCTANS
CT: AMINOGLYCOSIDE, GENTAMICIN T., BASIC
FORMULA: C22H45N5O7
EA: (N, 14)
MW: 491
PC: WH., POW.
UV: W: (200, ,)
SOL-GOOD: W
QUAL: (NINH., +)
TO: (S.AUREUS, .066)(B.SUBT., .033)(E.COLI, 1)
(K.PNEUM., .52)(P.VULG., 1)(SHYG., 1)
REFERENCES:
JP 80/51098; *CA*, 93, 130563

12222-7617

NAME: XK-62-8
PO: MIC.SAGAMIENSIS-NONREDUCTANS
CT: AMINOGLYCOSIDE, GENTAMICIN T., BASIC
FORMULA: C22H45N5O7
EA: (N, 14)
MW: 491
PC: WH., POW.
UV: W: (200, ,)
SOL-GOOD: W
QUAL: (NINH., +)
TO: (S.AUREUS, .033)(B.SUBT., .0082)(E.COLI, .52)
(K.PNEUM., 1)(P.VULG., 1)(SHYG., .52)
REFERENCES:
JP 80/51098; *CA*, 93, 130563

12222-7618

NAME: UAA-3
PO: MIC.SAGAMIENSIS
CT: AMINOGLYCOSIDE, GENTAMICIN T., BASIC
FORMULA: C19H38N4O8
EA: (N, 12)
MW: 450
PC: WH., POW.
UV: W: (200, ,)
SOL-GOOD: W
QUAL: (NINH., +)(BIURET, -)
TO: (S.AUREUS, .3.12)(B.SUBT., 3.12)(E.COLI, 12.5)
(K.PNEUM., 25)
IS-FIL: 2
IS-ION: (DIAION HPK-NH4, NH4OH)(IRC-50-NH4, NH4OH)
IS-CHR: (CG-50-NH4, NH4OH)
IS-CRY: (LIOF.,)
REFERENCES:
JP 80/23909; *CA*, 93, 68659

12222-8064

NAME: <u>TPJ-B</u>
PO: MIC.SAGAMINENSIS+JI-20B
CT: AMINOGLYCOSIDE, GENTAMICIN T., BASIC
FORMULA: C20H41N5O9
EA: (N, 13)
MW: 495
PC: WH., POW.
UV: W: (200, ,)
SOL-GOOD: W
SOL-FAIR: ACET, ETOH
SOL-POOR: HEX, BUOH
QUAL: (NINH., +)(BIURET, -)
TO: (S.AUREUS, .4)(B.SUBT., .78)(E.COLI, .78)
(K.PNEUM., .4)(P.VULG., 3.12)(PS.AER., 25)
IS-ION: (IRC-50-NH4, NH4OH)(DIAION HPK-25-NH4, NH4OH)
IS-CHR: (SILG, CHL-MEOH-NH4OH)(CG-50-NH4, NH4OH)
REFERENCES:
JP 80/100395; *CA*, 94, 45583

12222-8065

NAME: <u>K-26-2</u>, 6′-N-METHYLVERDAMICIN
PO: MIC.SAGAMIENSIS, MIC.SP.+VERDAMICIN, MIC.SP.+G-418, MIC.ZIONENSIS
CT: AMINOGLYCOSIDE, GENTAMICIN T., BASIC
FORMULA: C21H41N5O7
EA: (N, 15)
MW: 475
PC: WH., POW.
UV: W: (200, ,)
SOL-GOOD: W
QUAL: (NINH., +)
TO: (B.SUBT.,)(S.AUREUS,)(E.COLI,)(K.PNEUM.,)
(S.AUREUS, 1.56)(B.SUBT., 3.12)(E.COLI, 6.25)
(P.VULG., 12.5)(K.PNEUM., 3.12)(PS.AER., 25)
IS-ION: (IRC-50-NH4, NH4OH)
IS-CHR: (SILG, CHL-MEOH-NH4OH)(CG-50-NH4, NH4OH)
REFERENCES:
JP 80/108891; *CA*, 94, 14022

12222-8083

NAME: 2-HYDROXYSAGAMICIN
PO: MIC.SAGAMIENSIS+STREPTAMINE
CT: AMINOGLYCOSIDE, GENTAMICIN T., BASIC
FORMULA: C20H41N5O8
EA: (N, 15)
MW: 479
PC: WH., POW.
UV: W: (200, ,)
SOL-GOOD: W
QUAL: (NINH., +)
TO: (S.AUREUS, .4)(E.COLI, .78)(PS.AER., .78)
LD50: (250, IV)
IS-ION: (DIAION HPK, NH4OH)
REFERENCES:

JP 80/151597; *CA*, 94, 190307

12222-8096

NAME: G-367-1, 2′-N-FORMYLSISOMICIN
PO: DACTYLOSPORANGIUM THAILANDENSE
CT: AMINOGLYCOSIDE, GENTAMICIN T., BASIC
FORMULA: C20H37N5O8
EA: (N, 14)
MW: 475
PC: WH., POW.
OR: (+188.9, W)
UV: W: (200, ,)
SOL-GOOD: W, MEOH
SOL-POOR: ACET, BENZ
QUAL: (NINH., +)(BIURET, -)
STAB: (BASE, +)
TO: (S.AUREUS, 6.3)(B.SUBT., .8)(E.COLI, 1.6)
(K.PNEUM., 1.6)(SHYG., 1.6)(PS.AER., 25)
IS-FIL: 2
IS-ION: (IRC-50-NH4, NH4OH)
IS-CHR: (CM-SEPHADEX C-25-NH4, NH4OH)
IS-CRY: (LIOF.,)
REFERENCES:

DT 3013210; BP 2053895; Fr. P. 2452932

12222-8097

NAME: G-367-2, EPI-SISOMICIN
PO: DACTYLOSPORANGIUM THAILANDENSE
CT: AMINOGLYCOSIDE, GENTAMICIN T., BASIC
FORMULA: C19H37N5O7
EA: (N, 15)
MW: 447
PC: WH., POW.
OR: (+159.8, W)
UV: W: (200, ,)
SOL-GOOD: W, MEOH
SOL-POOR: ACET, BENZ
QUAL: (NINH., +)(BIURET, -)
STAB: (BASE, +)
TO: (S.AUREUS, 12.5)(B.SUBT., 1.6)(E.COLI, 1.6)
(K.PNEUM., 1.6)(SHYG., 3.1)(PS.AER., 25)
IS-FIL: 2
IS-ION: (IRC-50-NH4, NH4OH)
IS-CHR: (CM-SEPHADEX C-25-NH4, NH4OH)
IS-CRY: (LIOF.,)
REFERENCES:
DT 3013210; BP 2053895; Fr. P 2452932

12222-8098

NAME: D-53, 3"-N-DEMETHYL-G-52
PO: MIC.SAGAMIENSIS
CT: AMINOGLYCOSIDE, GENTAMICIN T., BASIC
FORMULA: C19H39N5O7
EA: (N, 15)
MW: 449
PC: WH., POW.
UV: W: (200, ,)
SOL-GOOD: W
QUAL: (NINH., +)
TO: (G.POS.,)(G.NEG.,)
IS-ION: (DIAION HPK-25,)(IRC-50-NH4, NH4OH)
IS-CHR: (BIOREX-70, NH4OH)
REFERENCES:
JP 80/69594; *CA*, 93, 219314

12222-8475

NAME: GENTOXIMICIN-A, 3′.4′-DIDEOXYGENTAMICIN-B, 2′-DEAMINO-2′-HYDROXYGENTAMICIN-C1A
PO: MIC.PURPUREA-NIGRESCENS
CT: AMINOGLYCOSIDE, GENTAMICIN T., BASIC
FORMULA: C19H38N4O8
EA: (N, 12)
MW: 450
PC: WH., POW.
OR: (+163, W)
UV: W: (200, ,)
SOL-GOOD: W
SOL-FAIR: MEOH
SOL-POOR: BUOH, HEX
QUAL: (NINH., +)(FEHL., -)
STAB: (BASE, +)(HEAT, +)
TO: (B.SUBT., 1)(S.AUREUS, 4)(E.COLI, 4)(K.PNEUM, 4)(P.VULG., 32)(PS.AER., 2)(SHYG., 8)(MYCOB.TUB., .4)
LD50: (200, IP)
IS-FIL: 2
IS-EXT: (IRC-50-NH4, NH4OH)
IS-CHR: (CG-50-NH4, NH4OH)(CM-SEPHADEX C-25-NH4, NH4OH)
IS-CRY: LIOF.
REFERENCES:

CA, 95, 78387; Unpublished results; USP 4212859; *JA*, 30, 945, 1977

12222-8476

NAME: GENTOXIMICIN-B, 3′.4′-DIDEOXYGENTAMICIN-B1, 2′-DEAMINO-2′-HYDROXYGENTAMICIN-C2
PO: MIC.PURPUREA-NIGRESCENS
CT: AMINOGLYCOSIDE, GENTAMICIN T., BASIC
FORMULA: C20H40N4O8
EA: (N, 12)
MW: 464
PC: WH., POW.
OR: (+158, W)
UV: W: (200, ,)
SOL-GOOD: W
SOL-FAIR: MEOH
SOL-POOR: BUOH, HEX
QUAL: (NINH., +)(FEHL., -)
STAB: (BASE, +)(HEAT, +)
TO: (S.AUREUS, 8)(B.SUBT., .5)(E.COLI, 2)(K.PNEUM., 16)(P.VULG., 32)(PS.AER., 4)(SHYG., 8)(MYCOB.TUB., .8)
LD50: (250, IP)
REFERENCES:

CA, 95, 78387; Unpublished results; USP 4212859; *JA*, 30, 945, 1977

12222-8477

NAME: 3"-N-DEMETHYLGENTAMICIN-C1A
PO: MIC.SAGAMIENSIS
CT: AMINOGLYCOSIDE, GENTAMICIN T., BASIC
FORMULA: C18H37N5O7
EA: (N, 16)
MW: 435
PC: WH., POW.
UV: W: (200, ,)
SOL-GOOD: W
QUAL: (NINH., +)
TO: (G.POS.,)(G.NEG.,)
REFERENCES:
JP 81/1892; *CA*, 95, 40840; DT 2601490

12222-8478

NAME: 3"-N-DEMETHYLSISOMICIN
IDENTICAL: 66-40-G
PO: MIC.SAGAMIENSIS
CT: AMINOGLYCOSIDE, GENTAMICIN T., BASIC
FORMULA: C18H35N5O7
EA: (N, 16)
MW: 433
PC: WH., POW.
UV: W: (200, ,)
SOL-GOOD: W
QUAL: (NINH., +)
TO: (G.POS.,)(G.NEG.,)
IS-FIL: 2
IS-ION: (IRC-50-NH4, NH4OH)
IS-CHR: (CG-50-NH4, NH4OH)
IS-CRY: LIOF.
REFERENCES:
JP 81/1893

12222-8479

NAME:	2-HYDROXYGENTAMICIN-A3
PO:	MIC.PURPUREA+STREPTAMINE
CT:	AMINOGLYCOSIDE, GENTAMICIN T., BASIC
FORMULA:	C18H36N4O11
EA:	(N, 10)
MW:	484
PC:	WH., POW.
OR:	(+109, W)
UV:	W: (200, ,)
SOL-GOOD:	W
QUAL:	(NINH., +)
TO:	(S.AUREUS, .78)(B.SUBT., .2)(E.COLI, .39) (K.PNEUM., 1.56)(PS.AER., 25)(P.VULG., .39)
IS-FIL:	2
IS-ION:	(IRC-50-NH4, NH4OH)
IS-CHR:	(CG-50-NH4, NH4OH)
IS-CRY:	LIOF.

REFERENCES:

DT 3042075

12222-8480

NAME:	2-HYDROXYGENTAMICIN-B
PO:	MIC.PURPUREA+STREPTAMINE
CT:	AMINOGLYCOSIDE, GENTAMICIN T., BASIC
FORMULA:	C19H38N4O11
EA:	(N, 10)
MW:	498
PC:	WH., POW.
OR:	(+156, W)
UV:	W: (200, ,)
SOL-GOOD:	W
QUAL:	(NINH., +)
TO:	(S.AUREUS, 3.12)(B.SUBT., .39)(E.COLI, .78) (K.PNEUM., 3.12)(P.VULG., .78)(PS.AER., 7.8)
IS-FIL:	2
IS-ION:	(IRC-50-NH4, NH4OH)
IS-CHR:	(CG-50-NH4, NH4OH)
IS-CRY:	LIOF.

REFERENCES:

DT 3042075

12222-8481

NAME: 2-HYDROXYGENTAMICIN-B1
PO: MIC.PURPUREA+STREPTAMINE
CT: AMINOGLYCOSIDE, GENTAMICIN T., BASIC
FORMULA: C20H40N4O11
EA: (N, 10)
MW: 512
PC: WH., POW.
OR: (+143, W)
UV: W: (200, ,)
SOL-GOOD: W
QUAL: (NINH., +)
TO: (S.AUREUS, 3.12)(B.SUBT., .78)(E.COLI, .78)
(K.PNEUM., 3.12)(PS.AER., 6.25)(P.VULG., 1.56)
IS-FIL: 2
IS-ION: (IRC-50-NH4, NH4OH)
IS-CHR: (CG-50-NH4, NH4OH)
IS-CRY: LIOF.
REFERENCES:
DT 3042075

12222-8482

NAME: 2-HYDROXY-JI-20A
PO: MIC.PURPUREA+STREPTAMINE
CT: AMINOGLYCOSIDE, GENTAMICIN T., BASIC
FORMULA: C19H39N5O10
EA: (N, 12)
MW: 497
PC: WH., POW.
OR: (+146, W)
UV: W: (200, ,)
SOL-GOOD: W
QUAL: (NINH., +)
TO: (S.AUREUS, 1.56)(B.SUBT., .39)(E.COLI, .78)
(K.PNEUM., 1.56)(PS.AER., 12.5)(P.VULG., .39)
IS-FIL: 2
IS-ION: (IRC-50-NH4, NH4OH)
IS-CHR: (CG-50-NH4, NH4OH)
IS-CRY: LIOF.
REFERENCES:
DT 3042075

12222-8483

NAME: 2-HYDROXY-JI-20B
PO: MIC.PURPUREA+STREPTAMINE
CT: AMINOGLYCOSIDE, GENTAMIAN T., BASIC
FORMULA: C20H41N5O10
EA: (N, 12)
MW: 511
PC: WH., POW.
OR: (+150, W)
UV: W: (200, ,)
SOL-GOOD: W
QUAL: (NINH., +)
TO: (S.AUREUS, 12)(B.SUBT., 3.12)(E.COLI, 6.25)
(K.PNEUM., 12.5)(P.VULG., 3.12)
IS-FIL: 2
IS-ION: (IRC-50-NH4, NH4OH)
IS-CHR: (CG-50-NH4, NH4OH)
IS-CRY: LIOF.
REFERENCES:

DT 3042075

12224-6738

NAME: SU-2, 1-DEAMINO-1-HYDROXYGENTAMICIN-C1A
PO: MIC.SAGAMIENSIS
CT: AMINOGLYCOSIDE, BASIC
FORMULA: C19H38N4O8
EA: (N, 12)
MW: 450
PC: WH., POW.
OR: (+172, W)
UV: W: (200, ,)
SOL-GOOD: W, MEOH
SOL-FAIR: ETOH, ACET
SOL-POOR: ETOH, HEX
QUAL: (NINH., +)(BIURET, -)
TO: (S.AUREUS, .008)(B.SUBT., .00025)(E.COLI, .008)
(P.VULG., .0167)(K.PNEUM., .004)
LD50: NONTOXIC
IS-ION: (IRC-50-NH4, NH4OH)
REFERENCES:

JP 79/59202; *CA*, 91, 191354

12224-7068

NAME: SU-1, 1-DEAMINO-1-HYDROXYGENTAMICIN-C2
PO: MIC.SAGAMIENSIS
CT: AMINOGLYCOSIDE, BASIC
FORMULA: C20H40N408
EA: (N, 12)
MW: 464
PC: WH., POW.
UV: W: (200, ,)
SOL-GOOD: W
SOL-POOR: BUOH, CHL
QUAL: (NINH., +)(BIURET, -)
TO: (G.POS.,)(G.NEG.,)(S.AUREUS,)(K.PNEUM.,)
IS-FIL: 2
IS-ION: (DIAION HPK-25-NH4, NH4OH)
IS-CHR: (CG-50-NH4, NH4OH)
REFERENCES:
JP 79/86788; 79/135705; *CA*, 92, 109109

12224-7069

NAME: SU-3, 1-DEAMINO-1-HYDROXYSAGAMICIN, SUM-3
PO: MIC.SAGAMIENSIS
CT: AMINOGLYCOSIDE, BASIC
FORMULA: C20H40N408
EA: (N, 12)
MW: 464
PC: WH., POW.
OR: (+146, W)
UV: W: (200, ,)
SOL-GOOD: W, MEOH
SOL-FAIR: ETOH, ACET
SOL-POOR: BUOH, CHL
QUAL: (NINH., +)(BIURET, -)
TO: (S.AUREUS, .78)(B.SUBT., .2)(E.COLI, .78)
(P.VULG., 31)(K.PNEUM., .39)(PS.AER., 6.2)
IS-FIL: 2
IS-ION: (CG-50-NH4, NH4OH)(IRC-50-NH4, NH4OH)
IS-CHR: (DIAION HPK-25-NH4, NH4OH)(CEL,)(BIOREX-70-NH4, NH4OH-NH4OAC)
IS-CRY: (LIOF.,)
REFERENCES:
JP 79/117477; *CA*, 92, 92700; *Abst. Papers, Kanto Area Symp. Agr. Ch. Soc. Jap.*, No. 4, 1980

12224-7070

NAME: SU-4, HYDROXY-SU-2, SUM-4
PO: MIC.SAGAMIENSIS
CT: AMINOGLYCOSIDE, BASIC
FORMULA: C19H38N4O9
EA: (N, 12)
MW: 466
PC: WH., POW.
UV: W: (200, ,)
SOL-GOOD: W
SOL-POOR: BUOH, CHL
QUAL: (NINH., +)(BIURET, -)
TO: (G.POS.,)(G.NEG.,)(PS.AER.,)
IS-FIL: 2
IS-ION: (DIAION HPK-25-NH4, NH4OH)
IS-CHR: (CG-50-NH4, NH4OH)
REFERENCES:

JP 79/135704; *CA*, 92, 109110

12224-7889

NAME: S-11-A, 1-DEAMINO-1-HYDROXYXYLOSTATIN
PO: B.CIRCULANS-S-11
CT: AMINOGLYCOSIDE, BASIC
FORMULA: C17H33N3O11
EA: (N, 9)
MW: 455
PC: WH., POW.
OR: (+38.2, W)
UV: W: (200, ,)
SOL-GOOD: W
QUAL: (NINH., +)(FECL3, -)
TO: (S.AUREUS, 50)(E.COLI, 25)(P.VULG., 25)
(K.PNEUM., 3.13)
IS-ION: (IRC-50-NH4, NH4OH)
IS-ABS: (CARBON, MEOH-HCL)
IS-CHR: (CG-50-NH4, NH4OH)(DX-1X2-OH, W)(CM-SEPHADEX C-25-NH4, NH4OH)
IS-CRY: (LIOF.,)
REFERENCES:

JA, 33, 836, 1980; JP 81/68699; *CA*, 95, 185552

12224-8165

NAME: K-114-E
PO: MIC.SAGAMIENSIS
CT: AMINOGLYCOSIDE, BASIC
FORMULA: C19H37N3O11
EA: (N, 8)
MW: 483
PC: WH., POW.
UV: W: (200, ,)
SOL-GOOD: W
QUAL: (NINH., +)
TO: (G.POS.,)(G.NEG.,)
IS-ION: (DIAION HPK-25-NH4, NH4OH)(IRC-50-NH4, NH4OH)
IS-CHR: (CG-50-NH4, NH4OH)(SILG, CHL-MEOH-NH4OH)
IS-CRY: (LIOF.,)
REFERENCES:

JP 80/99196; *CA*, 94, 119467

12224-8166

NAME: K-114-G
PO: MIC.SAGAMIENSIS
CT: AMINOGLYCOSIDE, BASIC
FORMULA: C20H39N3O11
EA: (N, 8)
MW: 497
PC: WH., POW.
UV: W: (200, ,)
SOL-GOOD: W
QUAL: (NINH., +)
TO: (G.POS.,)(G.NEG.,)
REFERENCES:

JP 80/99196; *CA*, 94, 119467

12225-7415

NAME:	<u>COMBIMICIN-A2</u>, 3"-N-METHYL-4"-C-METHYL-3'.4'-DIDEOXYKANAMYCIN-A, I-A2
PO:	MIC.SP.+KANAMYCIN-A, MIC.ECHINOSPORA+KANAMYCIN-A
CT:	AMINOGLYCOSIDE, BASIC, COMBIMICIN T.
FORMULA:	C20H40N4O9
EA:	(N, 11)
MW:	480
PC:	WH., POW.
OR:	(+147.5, W)
UV:	W: (200, ,)
SOL-GOOD:	W, MEOH
SOL-FAIR:	ETOH, ACET
SOL-POOR:	ETOAC, HEX
QUAL:	(NINH., +)
TO:	(S.AUREUS, 3.13)(S.LUTEA, .78)(B.SUBT.,) (E.COLI, 1.56)(SHYG., 3.13)(P.VULG., 3.13) (PS.AER., 6.25)(MYCOB.SP., 1.56)
IS-FIL:	2
IS-ION:	(IRC-50-NH4, NH4OH)
IS-CHR:	(CG-50-NH4, NH4OH)
IS-CRY:	(LIOF.,)

REFERENCES:

JA, 34, 777, 1981; DT 2900315; JP 79/98741; *CA*, 92, 56807, 56808

12225-7416

NAME: COMBIMICIN-B1, 3"-N-METHYL-4"-C-METHYL-3'.4'-DIDEOXY-6-N'-METHYLKANAMICIN-B, I-B1, COMBIMICIN-B1
PO: MIC.SP.+KANAMYCIN-B, MIC.ECHINOSPORA+KANAMYCIN-B
CT: AMINOGLYCOSIDE, BASIC, COMBIMICIN T.
FORMULA: C21H43N5O8
EA: (N, 14)
MW: 493
PC: WH., POW.
OR: (+128, W)
UV: W: (200, ,)
SOL-GOOD: W, MEOH
SOL-FAIR: ETOH, ACET
SOL-POOR: ETOAC, HEX
QUAL: (NINH., +)
TO: (S.AUREUS, .19)(S.LUTEA, .19)(B.SUBT.,)
(E.COLI, .78)(SHYG., .78)(P.VULG., .78)
(PS.AER., 1.56)(MYCOB.SP., .78)
IS-FIL: 2
IS-ION: (IRC-50-NH4OH,)
IS-CHR: (CG-50-NH4, NH4OH)
IS-CRY: (LIOF.,)
REFERENCES:
JA, 34, 777, 1981; DT 2900315; JP 79/98741

12225-7417

NAME: COMBIMICIN-B2, 3"-N-METHYL-4"-C-METHYL-3'.4'-DIDEOXYKANAMYCIN-B
PO: MIC.SP.+KANAMYCIN-B, MIC.ECHINOSPORA+KANAMYCIN-B
CT: AMINOGLYCOSIDE, BASIC, COMBIMICIN T.
FORMULA: C20H41N5O8
EA: (N, 14)
MW: 479
PC: WH., POW.
OR: (+142, W)
UV: W: (200, ,)
SOL-GOOD: W, MEOH
SOL-FAIR: ETOH, ACET
SOL-POOR: ETOAC, HEX
QUAL: (NINH., +)
TO: (S.AUREUS, .39)(S.LUTEA, .39)(B.SUBT.,)
(E.COLI, .78)(SHYG., .39)(P.VULG., .78)
(PS.AER., 1.56)(MYCOB.SP., .39)
REFERENCES:
JA, 34, 777, 1981; DT 2900315; JP 79/98741

12225-8099

NAME: I-SK-A1, I-A1, 2′-DEAMINO-2′-HYDROXY-5″-HYDROXYMETHYLSISOICIN
PO: MIC.INYOENSIS+KANAMYCIN-A
CT: AMINOGLYCOSIDE, BASIC, COMBIMICIN T.
FORMULA: C20H38N4O9
EA: (N, 11)
MW: 478
PC: WH., POW.
OR: (+151.5, W)
UV: W: (200, ,)
SOL-GOOD: W
QUAL: (NINH., +)
TO: (S.AUREUS, .78)(S.LUTEA, .78)(E.COLI, 1.56)(SHYG., 3.13)(P.VULG., 3.13)(PS.AER., 3.13)(MYCOB.PHLEI, 1.56)
IS-ION: (IRC-50-NH4, NH4OH)
IS-CHR: (CA-30-NH4, NH4OH)
REFERENCES:

JP 80/115896; *CA*, 94, 64774

12225-8100

NAME: I-B1, I-SK-B1, 5″-HYDROXYMETHYLSISOMICIN
PO: MIC.INYOENSIS+KANAMYCIN-B
CT: AMINOGLYCOSIDE, BASIC, COMBIMICIN T.
FORMULA: C20H39N5O8
EA: (N, 13)
MW: 477
PC: WH., POW.
OR: (+162.5, W)
UV: W: (200, ,)
SOL-GOOD: W
QUAL: (NINH., +)
TO: (S.AUREUS, .39)(S.LUTEA, .39)(E.COLI, 1.50)(SHYG., 3.13)(P.VULG., 3.13)
REFERENCES:

JP 80/115896; *CA*, 94, 63774

12225-8101

NAME: I-SK-B2, I-B2, 5"-HYDROXYMETHYL-4"-DEMETHYLSISOMICIN
PO: MIC.INYOENSIS+KANAMYCIN-B
CT: AMINOGLYCOSIDE, BASIC, COMBIMICIN T.
FORMULA: C19H37N5O8
EA: (N, 15)
MW: 463
PC: WH., POW.
OR: (+141.9, W)
UV: W: (200, ,)
SOL-GOOD: W
QUAL: (NINH., +)
TO: (S.AUREUS, .78)(MYCOB.PHLEI,)
REFERENCES:
JP 80/115896; *CA*, 94, 63774

12225-8102

NAME: 3"-N-METHYLKANAMYCIN-A
IDENTICAL: COMBIMICIN-A4
PO: MIC.INYOENSIS+KANAMYCIN-A, MIC.PURPUREA+KANAMYCIN-A
CT: AMINOGLYCOSIDE, BASIC, COMBIMICIN T.
FORMULA: C19H38N4O11
EA: (N, 11)
MW: 498
PC: WH., POW.
UV: W: (200, ,)
SOL-GOOD: W
QUAL: (NINH.,)
TO: (S.AUREUS,)(B.SUBT.,)(E.COLI,)(PS.AER.,)
IS-FIL: 3.5
IS-ION: (IRC-50-NH4, NH4OH)
IS-CHR: (SILG, CHL-MEOH-NH4OH)
IS-CRY: (LIOF.,)
REFERENCES:
USP 4234685

12225-8103

NAME: 3"-N-METHYL-4"-C-METHYLKANAMYCIN-A
IDENTICAL: COMBIMICIN-A3
PO: MIC.INYOENSIS+KANAMYCIN-A,
MIC.PURPUREA+KANAMYCIN-A
CT: AMINOGLYCOSIDE, BASIC, COMBIMICIN T.
FORMULA: C20H40N4O11
EA: (N, 11)
MW: 512
PC: WH., POW.
UV: W: (200, ,)
SOL-GOOD: W
QUAL: (NINH., +)
TO: (S.AUREUS,)(B.SUBT.,)(E.COLI,)(PS.AER.,)
REFERENCES:
USP 4234685

12225-8104

NAME: 3"-N-METHYLKANAMYCIN-B
IDENTICAL: COMBIMICIN-B5
PO: MIC.INYOENSIS+KANAMYCIN-B,
MIC.PURPUREA+KANAMYCIN-B
CT: AMINOGLYCOSIDE, BASIC, COMBIMICIN T.
FORMULA: C19H39N5O10
EA: (N, 14)
MW: 497
PC: WH., POW.
UV: W: (200, ,)
SOL-GOOD: W
QUAL: (NINH., +)
TO: (S.AUREUS,)(B.SUBT.,)(E.COLI,)(PS.AER.,)
REFERENCES:
USP 4234685

12225-8105

NAME:	3"-N-METHYL-4"-C-METHYLKANAMYCIN-B
PO:	MIC.INYOENSIS+KANAMYCIN-B, MIC.PURPUREA+KANAMYCIN-B
CT:	AMINOGLYCOSIDE, BASIC, COMBIMICIN T.
FORMULA:	C20H47N5O10
EA:	(N, 14)
MW:	511
PC:	WH., POW.
UV:	W: (200, ,)
SOL-GOOD:	W
QUAL:	(NINH., +)
TO:	(S.AUREUS,)(B.SUBT.,)(E.COLI,)(PS.AER.,)
REFERENCES:	

USP 4234685

12225-8106

NAME:	3"-N-METHYLTOBRAMYCIN
IDENTICAL:	COMBIMICIN-T2
PO:	MIC.INYOENSIS+TOBRAMYCIN, MIC.PURPUREA+TOBRAMYCIN
CT:	AMINOGLYCOSIDE, KANAMYCIN T., BASIC, COMBIMICIN T.
FORMULA:	C19H39N5O9
MW:	481
PC:	WH.
UV:	W: (200, ,)
SOL-GOOD:	W
QUAL:	(NINH., +)
TO:	(S.AUREUS,)(B.SUBT.,)(E.COLI,)
REFERENCES:	

USP 4234685

12225-8107

NAME:	3"-N-METHYLDIBEKACIN
IDENTICAL:	COMBIMICIN-B4
PO:	MIC.INYOENSIS+DIBEKACIN, MIC.PURPUREA+DIBEKACIN
CT:	AMINOGLYCOSIDE, KANAMYCIN T., BASIC, COMBIMICIN T.
FORMULA:	C19H39N5O8
EA:	(N, 15)
MW:	465
PC:	WH.
SOL-GOOD:	W
TO:	(B.SUBT.,)(S.AUREUS,)(E.COLI,)
REFERENCES:	

USP 4234685

12225-8108

NAME: 3"-N-METHYL-4"-C-METHYLTOBRAMYCIN
IDENTICAL: I-T1B, COMBIMICIN-T1B
PO: MIC.INYOENSIS+TOBRAMYCIN,
MIC.PURPUREA+TOBRAMYCIN
CT: AMINOGLYCOSIDE, BASIC, COMBIMICIN T.
FORMULA: C20H41N509
EA: (N, 13)
MW: 495
PC: WH.
UV: W: (200, ,)
SOL-GOOD: W
QUAL: (NINH., +)
TO: (S.AUREUS,)(B.SUBT.,)(E.COLI,)(PS.AER.,)
REFERENCES:

USP 4234685

12225-8484

NAME: COMBIMICIN-A3, I-A3
IDENTICAL: 3"-N-METHYL-4"-C-METHYLKANAMYCIN-A
PO: MIC.ECHINOSPORA+KANAMYCIN-A,
MIC.PURPUREA+KANAMYCIN-A, MIC.SP.+KANAMYCIN-A
CT: AMINOGLYCOSIDE, COMBIMICIN T., BASIC
FORMULA: C20H40N4011
EA: (N, 10)
MW: 512
PC: WH., POW.
OR: (+163.6, W)
UV: W: (200, ,)
SOL-GOOD: W
QUAL: (NINH., +)
TO: (S.AUREUS, .78)(P.VULG., .78)(PS.AER., 12.5)
IS-FIL: 2
IS-ION: (IRC-50-NH4, NH4OH)(DX-50-NH4, NH4OH)
IS-CHR: (CG-50-NH4, NH4OH)(SILG, CHL-MEOH-NH4OH)
REFERENCES:

JP 81/20598; *CA*, 95, 78462

12225-8485

NAME: COMBIMICIN-A4, I-A4
IDENTICAL: 3"-N-METHYLKANAMYCIN-A
PO: MIC.ECHINOSPORA+KANAMYCIN-A, MIC.PURPUREA+KANAMYCIN-A, MIC.SP.+KANAMYCIN-A
CT: AMINOGLYCOSIDE, COMBIMICIN T., BASIC
FORMULA: C19H38N4O11
EA: (N, 11)
MW: 498
PC: WH., POW.
OR: (+,)
UV: W: (200, ,)
SOL-GOOD: W
QUAL: (NINH., +)
TO: (S.AUREUS, .78)(P.VULG., .78)(PS.AER., 50)
IS-FIL: 2
IS-ION: (IRC-50-NH4, NH4OH)(DX-50-NH4, NH4OH)
IS-CHR: (CG-50-NH4, NH4OH)(SILG, CHL-MEOH-NH4OH)
REFERENCES:
JP 81/20598; *CA*, 95, 78462

12225-8486

NAME: COMBIMICIN-B3, I-B3, 6′-N-METHYL-3′.4′-DIDEOXY-3"-N-METHYLKANAMICIN-B
PO: MIC.ECHINOSPORA+KANAMYCIN-B, MIC.SP.+KANAMYCIN-B
CT: AMINOGLYCOSIDE, COMBIMICIN T., BASIC
FORMULA: C20H41N5O8
EA: (N, 15)
MW: 479
PC: WH., POW.
OR: (+134.8, W)
UV: W: (200, ,)
SOL-GOOD: W
QUAL: (NINH., +)
TO: (S.AUREUS, .39)(E.COLI, 25)(P.VULG., 3.13)(PS.AER., 3.13)
IS-FIL: 2
IS-ION: (IRC-50-NH4, NH4OH)(DX-50-NH4, NH4OH)
IS-CHR: (CG-50-NH4, NH4OH)(SILG, CHL-MEOH-NH4OH)
REFERENCES:
JP 81/20598; *CA*, 95, 78462

12225-8487

NAME: COMBIMICIN-B4, I-B4, 3"-N-METHYL-3'.4'-DIDEOXYKANAMYCIN-B
IDENTICAL: 3"-N-METHYLDIBEKACIN
PO: MIC.ECHINOSPORA+KANAMYCIN-B, MIC.SP.+KANAMYCIN-B
CT: AMINOGLYCOSIDE, COMBIMIAN T., BASIC
FORMULA: C19H39N5O8
EA: (N, 15)
MW: 465
PC: WH., POW.
OR: (+131.1, W)
UV: W: (200, ,)
SOL-GOOD: W
QUAL: (NINH., +)
TO: (S.AUREUS, .2)(E.COLI, 3.13)(P.VULG., 1.56)(PS.AER., 25)
IS-FIL: 2
IS-ION: (IRC-50-NH4, NH4OH)(DX-50-NH4, NH4OH)
IS-CHR: (CG-50-NH4, NH4OH)(SILG, CHL-MEOH-NH4OH)
REFERENCES:
JP 81/20598; *CA*, 95, 78462

12225-8488

NAME: COMBIMICIN-B5, I-B5
IDENTICAL: 3"-N-METHYLKANAMYCIN-B
PO: MIC.ECHINOSPORA+KANAMYCIN-B, MIC.SP.+KANAMYCIN-B
CT: AMINOGLYCOSIDE, COMBIMICIN T., BASIC
FORMULA: C19H39N5O10
EA: (N, 14)
MW: 497
PC: WH., POW.
UV: W: (200, ,)
SOL-GOOD: W
QUAL: (NINH., +)
TO: (S.AUREUS, .2)(P.VULG., 1.56)(PS.AER., 12.5)
IS-FIL: 2
IS-ION: (IRC-50-NH4, NH4OH)(DX-50-NH4, NH4OH)
IS-CHR: (CG-50-NH4, NH4OH)(SILG, CHL-MEOH-NH4OH)
REFERENCES:
JP 81/20598; *CA*, 95, 78462

12225-8489

NAME: COMBIMICIN-T1A, 3′-DEOXY-4″-C-METHYL-3″-N-METHYLKANAMYCIN-A, I-T1A
PO: MIC.ECHINOSPORA+TOBRAMYCIN, MIC.SP.+TOBRAMYCIN
CT: AMINOGLYCOSIDE, COMBIMICIN T., BASIC
FORMULA: C20H40N4O10
EA: (N, 11)
MW: 496
PC: WH., POW.
OR: (+152.8, W)
UV: W: (200, ,)
SOL-GOOD: W
QUAL: (NINH., +)
TO: (E.COLI, .78)(K.PNEUM., .2)(P.VULG., 3.13)
IS-FIL: 2
IS-ION: (IRC-50-NH4, NH4OH)(DX-50-NH4, NH4OH)
IS-CHR: (CG-50-NH4, NH4OH)(SILG, CHL-MEOH-NH4OH)
REFERENCES:

JP 81/20599; *CA*, 95, 59910

12225-8490

NAME: COMBIMICIN-T2, 3′-DEOXY-3″-N-METHYLKANAMYCIN-B, I-T2
IDENTICAL: 3″-N-METHYLTOBRAMYCIN
PO: MIC.ECHINOSPORA+TOBRAMYCIN, MIC.SP.+TOBRAMYCIN
CT: AMINOGLYCOSIDE, COMBIMICIN T., BASIC
FORMULA: C19H39N5O9
EA: (N, 14)
MW: 481
PC: WH., POW.
OR: (+135.3, W)
UV: W: (200, ,)
SOL-GOOD: W
QUAL: (NINH., +)
TO: (E.COLI, .78)(P.VULG., 50)
IS-FIL: 2
IS-ION: (IRC-50-NH4, NH4OH)(DX-50-NH4, NH4OH)
IS-CHR: (CG-50-NH4, NH4OH)(SILG, CHL-MEOH-NH4OH)
REFERENCES:

JP 81/20599; *CA*, 95, 59910

12225-8491

NAME: COMBIMICIN-T1B, 3′-DEOXY-4″-C-METHYL-3″-N-METHYLKANAMYCIN-B, I-T1B
IDENTICAL: 3″-N-METHYL-4″-C-METHYLTOBRAMYCIN
PO: MIC.ECHINOSPORA+TOBRAMYCIN, MIC.SP.+TOBRAMYCIN
CT: AMINOGLYCOSIDE, COMBIMICIN T., BASIC
FORMULA: C20H41N5O9
EA: (N, 13)
MW: 495
PC: WH., POW.
OR: (+102, W)
UV: W: (200, ,)
SOL-GOOD: W
QUAL: (NINH., +)
TO: (E.COLI, .78)(K.PNEUM., .2)
REFERENCES:
JP 81/20599; *CA*, 95, 59910

12230-8785

NAME: A-9594
PO: S.HYGROSCOPICUS
CT: AMINOGLYCOSIDE, HYGROMYCIN-B T.
FORMULA: C20H37N3O13
EA: (N, 8)
MW: 527
PC: WH., POW.
UV: W: (200, ,)
SOL-GOOD: W
SOL-FAIR: MEOH, ETOH, ACET
SOL-POOR: ETOAC, HEX
QUAL: (NINH., +)(FEHL., -)(FECL3, -)(SAKA., -)
REFERENCES:
JP 81/82094

12250-7072

NAME: FU-10
PO: MIC.OLIVOASTEROSPORA
CT: AMINOGLYCOSIDE, NEAMINE T., BASIC
FORMULA: C12H24N2O9
EA: (N, 8)
MW: 340
PC: WH., POW.
OR: (+89, W)
UV: W: (200, ,)
SOL-GOOD: W
QUAL: (NINH., +)
TO: (G.POS.,)
IS-FIL: 2.5
IS-ION: (IRC-50-NH4, NH4OH)
IS-CHR: (CG-50-NH4, NH4OH)(SILG, CHL-MEOH-NH4OH)
REFERENCES:
JP 79/128547; *CA*, 92, 92701

12250-7558

NAME: 5.6-DIDEOXYNEAMINE
PO: S.FRADIAE+2.5.6-TRIDEOXYSTREPTAMINE
CT: AMINOGLYCOSIDE, NEAMINE T., BASIC
FORMULA: C12H26N4O4.2H2SO4
EA: (N, 19)
MW: 290
PC: WH., POW.
OR: (+40, W)
UV: W: (200, ,)
SOL-GOOD: W
QUAL: (NINH., +)
TO: (B.SUBT., .25)(E.COLI, 4)
REFERENCES:
JACS, 102, 857, 1980; *Carb. Res.*, 53, 239, 1977

12330-7073

NAME: DC-5-4
PO: CORYNEBACTERIUM SP.
CT: AMINOGLYCOSIDE L., BASIC
FORMULA: C15H27N5O12
EA: (N, 15)
MW: 469
PC: WH., POW.
OR: (-1.6, W)
UV: W: (200, ,)
SOL-GOOD: W, MEOH, ETOH, ACET
SOL-POOR: CHL, HEX, BUOH
QUAL: (NINH., -)(SAKA., +)
TO: (S.AUREUS, 50)(E.COLI, 25)(K.PNEUM., 12)
(PHYT.BACT.,)(B.SUBT., 12)
LD50: NONTOXIC
IS-ABS: (CARBON, MEOH-HCL)
IS-CHR: (SEPHADEX LH-20, MEOH)
IS-CRY: (PREC., W, MEOH)
REFERENCES:

JP 79/117094; 80/31008; *CA*, 92, 109106; 93, 39504

12330-7074

NAME: LL-BM-782-A"1, BM-782-A"1
PO: NOC.SP.
CT: AMINOGLYCOSIDE L., BASIC
FORMULA: C39H75N13O16
EA: (N, 18)
MW: 981
PC: WH., POW.
OR: (0, W)
UV: W: (200, ,)
SOL-GOOD: W
SOL-POOR: HEX, ACET
QUAL: (NINH., +)(SAKA., +)
STAB: (ACID, +)(BASE, -)
TO: (S.AUREUS, .25)(E.COLI, 4)(K.PNEUM., 1)
(P.VULG., 4)(SHYG., .12)(PS.AER., 32)
IS-FIL: 2
IS-ION: (IRC-50-NA, H2SO4)
IS-CHR: (CARBON, ACET-W)(CM-SEPHADEX)
IS-CRY: (LIOF.,)
REFERENCES:

JOC, 46, 792, 1981; *CC*, 1134, 1981; DT 2912054; USP 4234717; *Abst. AAC*, 20, 71, 1980; EP 4768

12330-7075

NAME: LL-BM-782-A"1A, BM-782-A"1A
PO: NOC.SP.
CT: AMINOGLYCOSIDE L., BASIC
FORMULA: C45H87N15017
EA: (N, 19)
MW: 1110
PC: WH., POW.
OR: (+4, W)
UV: W: (200, ,)
SOL-GOOD: W
SOL-POOR: HEX, ACET
QUAL: (NINH., +)(SAKA., +)
STAB: (ACID, +)(BASE, -)
TO: (S.AUREUS, 1)(E.COLI, 64)(K.PNEUM., 8)(P.VULG., 128)(SHYG., 1)(PS.AER., 128)
IS-FIL: 2
IS-ION: (IRC-50-NA, H2SO4)
IS-CHR: (CARBON, ACET-W)
IS-CRY: (LIOF.,)
REFERENCES:

JOC, 46, 792, 1981; *CC*, 1134, 1981; DT 2912054; USP 4234717; *Abst. AAC*, 20, 71, 1980; EP 4768

12330-7076

NAME: LL-BM-782-A"2, BM-782-A"2
PO: NOC.SP.
CT: AMINOGLYCOSIDE L., BASIC
FORMULA: C33H63N11015
EA: (N, 18)
MW: 851
PC: WH., POW.
OR: (0, W)
UV: W: (200, ,)
SOL-GOOD: W
SOL-POOR: HEX, ACET
QUAL: (NINH., +)(SAKA., +)
STAB: (ACID, +)(BASE, -)
TO: (S.AUREUS, .25)(E.COLI, 4)(K.PNEUM., 2)(P.VULG., 2)(SHYG., .25)(PS.AER., 64)
IS-FIL: 2
IS-ION: (IRC-50-NA, H2SO4)
IS-CHR: (CARBON, ACET-W)
IS-CRY: (LIOF.,)
REFERENCES:

JOC, 46, 792, 1981; *CC*, 1134, 1981; DT 2912054; USP 4234717; *Abst. AAC*, 20, 71, 1980; EP 4768

12330-8492

NAME:	MYOMYCIN-C
PO:	NOC.SP.
CT:	AMINOGLYCOSIDE L., BASIC
EA:	(N,)
SOL-GOOD:	W
TO:	(G.POS.,)(MYCOB.SP.,)

REFERENCES:

JA, 26, 272, 1973; USP 3795668

12350-8109

NAME:	X-14847, 2-AMINO-2-DEOXY-A"-D-GLUCOPYRANOSYL-1-O-D-MYO-INOSITOL
PO:	MIC.ECHINOSPORA
CT:	AMINOGLYCOSIDE L., BASIC
FORMULA:	C12H23NO10
EA:	(N, 4)
MW:	341
PC:	WH., POW.
OR:	(+88.5, W)
UV:	W: (200, ,)
SOL-GOOD:	W
SOL-POOR:	BUOH, HEX
QUAL:	(NINH., +)(FEHL., -)
TO:	(B.SUBT., 200)
IS-FIL:	2
IS-ION:	(IRC-50-NH4, NH4OH)
IS-CHR:	(DX-1X4-OH, W)

REFERENCES:

JA, 33, 1431, 1980; *JOC*, 46, 378, 1981

12410-6741

NAME: FORTIMICIN-KG
PO: MIC.OLIVOASTEROSPORA
CT: BASIC, AMINOGLYCOSIDE L., FORTIMICIN T.
FORMULA: C15H30N4O5
EA: (N, 16)
MW: 346
PC: WH., POW.
OR: (+90, W)
UV: W: (200, ,)
SOL-GOOD: W, MEOH
SOL-FAIR: ETOH, ACET
SOL-POOR: HEX, BUOH, ETOAC
QUAL: (NINH., +)(BIURET, -)
TO: (S.AUREUS, 1.1)(B.SUBT., 4.5)(K.PNEUM., 4.5)
(E.COLI, 4.5)(P.VULG., 4.5)(SHYG., 9)
IS-FIL: 2.5
IS-ION: (IRC-50-NH4, NH4OH)
IS-ABS: (CARBON, H2SO4)
IS-CHR: (SILG, CHL-ETOH-ACET, NH4OH)(CG-50-NH4, NH4OH)
REFERENCES:

USP 4209612; JP 79/66679; *CA*, 91, 156031; *Biotech. Bioeng.*, 22, Suppl. 1, 65, 1980

12410-6742

NAME: FORTIMICIN-KF
PO: MIC.OLIVOASTEROSPORA
CT: BASIC, AMINOGLYCOSIDE L., FORTIMICIN T.
FORMULA: C14H28N4O5
EA: (N, 16)
MW: 332
PC: WH., POW.
OR: (+127, W)
UV: W: (200, ,)
SOL-GOOD: W, MEOH
SOL-FAIR: ETOH, ACET
SOL-POOR: HEX, BUOH, ETOAC
QUAL: (NINH., +)(BIURET, -)
TO: (S.AUREUS, 2.1)(B.SUBT., 4.2)(K.PNEUM., 4.2)
(E.COLI, 4.5)(P.VULG., 4.2)(SHYG., 8.4)
IS-FIL: 2.5
IS-ION: (IRC-50-NH4, NH4OH)
IS-ABS: (CARBON, H2SO4)
IS-CHR: (SILG, CHL-ETOH-ACET, NH4OH)(CG-50-NH4, NH4OH)
REFERENCES:

USP 4209612; JP 79/66679. *CA*, 91, 156031; *Biotech. Bioeng;* 22, Suppl. 1, 65, 1981

12410-6743

NAME: FORTIMICIN-E, 3.4-DIEPIFORTIMICIN-B, FORTIMICIN-KH
PO: MIC.OLIVOASTEROSPORA+BASE
CT: BASIC, AMINOGLYCOSIDE L., FORTIMICIN T.
FORMULA: C15H32N4O5
EA: (N, 16)
MW: 348
PC: WH., POW.
OR: (+56.9, MEOH)
UV: W: (200, ,)
SOL-GOOD: W
QUAL: (NINH., +)
TO: (G.POS.,)(G.NEG.,)
IS-FIL: 7
IS-ION: (IRC-50-NH4, NH4OH)
IS-CHR: (IRC-50-NH4, NH4OH)
IS-CRY: (LIOF.,)
REFERENCES:

Belg. P 872916; 872913; *JA*, 32, 884, 1979

12410-6744

NAME: ISOFORTIMICIN-A, ISOFORTIMICIN
PO: MIC.OLIVOASTEROSPORA+BASE
CT: BASIC, AMINOGLYCOSIDE L., FORTIMICIN T.
FORMULA: C17H35N5O6
EA: (N, 14)
MW: 405
PC: WH., POW.
OR: (+41.6, MEOH)
UV: W: (200, ,)
SOL-GOOD: W
TO: (G.POS.,)(G.NEG.,)
IS-ION: (IRC-50-NH4, NH4OH)
IS-CHR: (IRC-50-NH4, NH4OH)
REFERENCES:

Carb. Res., 85, 61, 1980; USP 4207314; Belg. P 872915; JP 79/92941

12410-7077

NAME: SANNAMYCIN-A, KA-7038-I
IDENTICAL: ISTAMYCIN-A
PO: S.SANNANENSIS
CT: AMINOGLYCOSIDE L., FORTIMICIN T., BASIC
FORMULA: C17H35N5O5
EA: (N, 17)
MW: 389
PC: WH., POW.
OR: (+120.5, W)
UV: W: (200, ,)
SOL-GOOD: W, MEOH
SOL-FAIR: ETOH, ACET
SOL-POOR: ETOAC, HEX
QUAL: (NINH., +)(SAKA., -)(FECL3, -)(FEHL., -)
STAB: (ACID, +)(BASE, -)
TO: (S.AUREUS, 9.9)(B.SUBT., .20)(E.COLI, 1.56)
(P.VULG., 3.13)(K.PNEUM., 1.56)(PS.AER., 6.25)
LD50: (150|50, IV)
IS-FIL: 2
IS-ION: (IRC-50-NH4, NH4OH)
IS-CHR: (CM-SEPHADEX C-25-NH4, NH4OH)(CEL, CHL-MEOH-NH4OH)
IS-CRY: (LIOF.,)
UTILITY: ON CLINICAL TRIAL
REFERENCES:

JA, 32, 1061, 1066, 1979; DT 2928373; JP 79/141701

12410-7078

NAME:	SANNAMYCIN-B, KA-7038-III
IDENTICAL:	ISTAMYCIN-AO
PO:	S.SANNANENSIS
CT:	AMINOGLYCOSIDE L., FORTIMICIN T., BASIC
FORMULA:	C15H32N4O4
EA:	(N, 17)
MW:	332
PC:	WH., POW.
OR:	(+78, W)
UV:	W: (200, ,)
SOL-GOOD:	W, MEOH, ETOH
SOL-FAIR:	ACET
SOL-POOR:	ETOAC, HEX
QUAL:	(NINH., +)(SAKA., -)(FECL3, -)(FEHL., -)
STAB:	(ACID, +)(BASE, -)
TO:	(B.SUBT., 12)(S.AUREUS, 25)(E.COLI, 100) (P.VULG., 100)
LD50:	(600|200, IV)
IS-FIL:	2
IS-ION:	(IRC-50-NH4, NH4OH)
IS-CHR:	(CM-SEPHADEX C-25-NH4, NH4OH)(CEL, CHL-MEOH-W)
IS-CRY:	(LIOF.,)

REFERENCES:

JA, 32, 1061, 1066, 1979; DT 2928373

12410-7079

NAME:	ISTAMYCIN-A
IDENTICAL:	SANNAMYCIN-A
PO:	S.TENJIMARIENSIS
CT:	AMINOGLYCOSIDE L., FORTIMICIN T., BASIC
FORMULA:	C17H35N5O5
EA:	(N, 17)
MW:	389
PC:	WH., POW.
OR:	(+155, W)
UV:	W: (200, ,)
SOL-GOOD:	W, MEOH
SOL-POOR:	ETOH, HEX
STAB:	(BASE, +)
TO:	(S.AUREUS, .39)(S.LUTEA, 1.56)(B.SUBT., .39) (E.COLI, 3.13)(K.PNEUM., 6.25)(P.VULG., 1.56)
LD50:	(150|50, IV)
IS-FIL:	2
IS-ION:	(IRC-50-NH4, NH4OH)
IS-CHR:	(CG-50-NH4, NH4OH)(SILG, CHL-MEOH-NH4OH)(DX- 1X4, W)
IS-CRY:	(LIOF.,)

REFERENCES:

JA, 32, 964, 1365, 1979; *Jap. J. Ant.*, 32, S-228, 1979; *214th Meet. JARA*, 12, 1979; *JA*, 33, 1281, 1502, 1510, 1515, 1980; DT 3012014; *JA*, 34, 824, 1981

12410-7080

NAME: ISTAMYCIN-B, 1-EPI-ISTAMYCIN-A
PO: S.TENJIMARIENSIS
CT: AMINOGLYCOSIDE L., FORTIMICIN T., BASIC
FORMULA: C17H35N5O5
EA: (N, 17)
MW: 389
PC: WH., POW.
OR: (+165, W)
UV: W: (200, ,)
SOL-GOOD: W, MEOH
SOL-POOR: ETOH, HEX
QUAL: (NINH., +)
STAB: (BASE, +)
TO: (S.AUREUS, .1)(S.LUTEA, 3.13)(B.SUBT., .05)
(E.COLI, 1.56)(SHYG., 6.25)(P.VULG., .78)
(K.PNEUM., 3.15)(PS.AER., 12.5)
LD50: (120|40, IV)
IS-FIL: 2
IS-ION: (IRC-50-NH4, NH4OH)
IS-CHR: (CG-50-NH4, NH4OH)(SILG, CHL-MEOH-NH4OH)(DX-1X4, W)
IS-CRY: (LIOF.,)
REFERENCES:

JA, 32, 964, 1365, 1979; *Jap. J. Ant.*, 32, S-288, 1979; *214th Meet. JARA*, 12, 1979; *JA*, 33, 1281, 1502, 1510, 1515, 1980; DT 3012014; *JA*, 34, 824, 1981

12410-7081

NAME: FORTIMICIN-KG1
PO: MIC.OLIVOASTEROSPORA
CT: AMINOGLYCOSIDE L., FORTIMICIN T., BASIC
FORMULA: C15H30N4O5
EA: (N, 16)
MW: 346
PC: WH., POW.
OR: (+58.3, W)
UV: W: (200, ,)
SOL-GOOD: W, MEOH
SOL-FAIR: ETOH, ACET
SOL-POOR: BUOH, HEX
QUAL: (NINH., +)(BIURET, -)
TO: (B.SUBT.,)(S.AUREUS, 13.1)(E.COLI, 26.1)
(P.VULG., 26.1)
IS-FIL: 2.5
IS-ION: (IRC-50-NH4, NH4OH)
IS-CHR: (SILG, CHL-MEOH-NH4OH)(BIOREX-70, NH4OH-NH4OAC)
IS-CRY: (LIOF.,)
REFERENCES:

Abst. ACS/CSJ Symp. Ser., 125, 309, 1980; DT 2908150; BP 2018286; USP 4241182; JP 79/117478

12410-7082

NAME: FORTIMICIN-KG2
PO: MIC.OLIVOASTEROSPORA
CT: AMINOGLYCOSIDE L., FORTIMICIN T., BASIC
FORMULA: C15H30N4O5
EA: (N, 16)
MW: 346
PC: WH., POW.
OR: (+30, W)
UV: W: (200, ,)
SOL-GOOD: W, MEOH
SOL-FAIR: ETOH, ACET
SOL-POOR: BUOH, HEX
QUAL: (NINH., +)(BIURET, -)
TO: (B.SUBT.,)(S.AUREUS, 10.5)(E.COLI, 20.9)
(P.VULG., 83.3)
IS-FIL: 2.5
IS-ION: (IRC-50-NH4OH, NH4OH)
IS-CRY: (LIOF.,)
REFERENCES:

Abst. ACS/CJS Symp. Ser., 125, 309, 1980; DT 2908150; BP 2018286; USP 4241182; JP 79/117478

12410-7083

NAME: FORTIMICIN-KG3
PO: MIC.OLIVOASTEROSPORA
CT: AMINOGLYCOSIDE L., FORTIMICIN T., BASIC
FORMULA: C17H33N5O6
EA: (N, 17)
MW: 403
PC: WH., POW.
OR: (+185, W)
UV: W: (200, ,)
SOL-GOOD: W, MEOH
SOL-FAIR: ETOH, ACET
SOL-POOR: BUOH, HEX
QUAL: (NINH., +)(BIURET, -)
TO: (B.SUBT., .0045)(S.AUREUS, .083)(E.COLI, .33)
(K.PNEUM., .18)(SHYG., .7)(P.VULG., .66)
(PS.AER., .5.2)
LD50: (225, IV)
IS-FIL: 2.5
IS-ION: (IRC-50-NH4OH, NH4OH)
IS-CHR: (SILG, CHL-MEOH-NH4OH)
IS-CRY: (LIOF.,)
REFERENCES:

Abst. ACS/CSJ Symp. Ser., 125, 309, 1980; DT 2908150; BP 2018286; USP 4241182; JP 79/117478

12410-7084

NAME: FORTIMICIN-KO1, 3-EPI-FORTIMICIN-B
PO: MIC.OLIVOASTEROSPORA
CT: AMINOGLYCOSIDE L., FORTIMICIN T., BASIC
FORMULA: C15H32N4O5
EA: (N, 16)
MW: 348
PC: WH., POW.
OR: (+79, W)
UV: W: (200, ,)
SOL-GOOD: W
QUAL: (NINH., +)
TO: (B.SUBT., .04)(E.COLI, .1)(P.VULG., .35)
(K.PNEUM., .3)(PS.AER., 5.3)(S.AUREUS, .4)
REFERENCES:

Abst. Pap. Ann. Meet. Agr. Ch. Soc. Jap., 1980, 3P-14; *Abst. Pap. 100th Ann. Meet. Pharm. Soc. Jap.*, 1980, 5G11-3; JP 79/109948; *CA*, 92, 162176

12410-7418

NAME: KA-7038-II
PO: S.SANNANENSIS
CT: AMINOGLYCOSIDE L., FORTIMICIN T., BASIC
FORMULA: C13H28N4O4
EA: (N, 18)
MW: 304
PC: WH., POW.
OR: (+61, W)
UV: W: (200, ,)
SOL-GOOD: W, MEOH
SOL-FAIR: ETOH
SOL-POOR: ACET, HEX
QUAL: (NINH., +)(SAKA., -)(FECL3, -)(FEHL., -)
STAB: (ACID, +)(BASE, -)
TO: (S.AUREUS, 25)(B.SUBT., 6)(E.COLI, 100)
(K.PNEUM., 100)(PS.AER., 100)
REFERENCES:
DT 2928373, 2942194; JP 79/141701

12410-7419

NAME: DACTIMICIN, SF-2052
PO: DACTYLOSPORANGIUM MATSUZUKIENSE, MIC.SP.
CT: AMINOGLYCOSIDE L., FORTIMICIN T., BASIC
FORMULA: C18H36N6O6
EA: (N, 19)
MW: 432
PC: WH., POW.
OR: (+81, W)(+87, HCL)
UV: W: (200, ,)
SOL-GOOD: W
SOL-FAIR: MEOH
SOL-POOR: ACET, ETOAC
QUAL: (NINH., +)(SAKA., -)(EHRL., -)(DNPH, -)(FEHL., -)(BIURET, -)
STAB: (BASE, -)
TO: (S.AUREUS, .39)(B.SUBT., .78)(E.COLI, .78)
(SHYG., 3.13)(K.PNEUM., 3.13)(P.VULG., 6.25)
(PS.AER., .78)(S.LUTEA, .78)
LD50: (300, IV)
IS-FIL: 2
IS-ION: (IRC-50-NA, HCL)
IS-ABS: (CARBON, MEOH-W)
IS-CHR: (CM-SEPHADEX C-25-NA, NACL)
IS-CRY: (PREC., W, ACET)
REFERENCES:
JA, 32, 1354, 1979; 33, 510, 924, 1980; 34, 1090, 1981; DT 2944143;
JP 81/40697

12410-7510

NAME: KA-7038-IV
PO: S.SANNANENSIS
CT: AMINOGLYCOSIDE L., FORTIMICIN T., BASIC
FORMULA: C14H30N4O5
EA: (N, 16)
MW: 334
PC: WH., POW.
OR: (+115, W)
UV: W: (200, ,)
SOL-GOOD: W, MEOH
SOL-FAIR: ETOH, ACET
SOL-POOR: ETOAC, HEX
QUAL: (NINH., +)(SAKA., -)(FECL3, -)(FEHL., -)
STAB: (ACID, +)(BASE, -)
TO: (S.AUREUS, 12)(B.SUBT., 3)(E.COLI, 50)(P.VULG., 50)(K.PNEUM., 100)(PS.AER., 100)
LD50: (500|200, IV)
REFERENCES:
DT 2928373; 2942194

12410-7511

NAME: KA-7038-V
PO: S.SANNANENSIS
CT: AMINOGLYCOSIDE L., FORTIMICIN T., BASIC
FORMULA: C14H30N4O4
EA: (N, 17)
MW: 318
PC: WH., POW.
OR: (+98, W)
UV: W: (200, ,)
SOL-GOOD: W, MEOH
SOL-FAIR: ETOH, ACET
SOL-POOR: ETOAC, HEX
QUAL: (NINH., +)(SAKA., -)(FECL3, -)(FEHL., -)
STAB: (ACID, +)(BASE, -)
TO: (S.AUREUS, 25)(B.SUBT., 3)(E.COLI, 50)(P.VULG., 50)
LD50: (500|200, IV)
REFERENCES:
DT 2928373; 2942194

12410-7512

NAME: SANNAMYCIN-C, KA-7038-VI
PO: S.SANNANENSIS
CT: AMINOGLYCOSIDE L., FORTIMICIN T., BASIC
FORMULA: C15H32N4O4
EA: (N, 17)
MW: 332
PC: WH., POW.
OR: (+58, W)(+59, W)
UV: W: (200, ,)
SOL-GOOD: W, MEOH
SOL-FAIR: ETOH, ACET
SOL-POOR: ETOAC, HEX
QUAL: (NINH., +)(SAKA., -)(FECL3, -)(FEHL., -)
STAB: (ACID, +)(BASE, -)
TO: (S.AUREUS, 6)(B.SUBT., 12)(E.COLI, 100)
(P.VULG., 100)(K.PNEUM., 100)(PS.AER., 100)
LD50: (500|200, IV)
REFERENCES:
JA, 33, 1274, 1980; *Abst. Pap. 214th Meet. JARA*, 1979, 16; DT 2928373; 2942194

12410-7513

NAME: KA-7038-VII
PO: S.SANNANENSIS
CT: AMINOGLYCOSIDE L., FORTIMICIN T., BASIC
FORMULA: C14H30N4O4
EA: (N, 17)
MW: 318
PC: WH., POW.
OR: (+59, W)
UV: W: (200, ,)
SOL-GOOD: W, MEOH
SOL-FAIR: ETOH, ACET
SOL-POOR: ETOAC, HEX
QUAL: (NINH., +)(SAKA., -)(FECL3, -)(FEHL., -)
STAB: (ACID, +)(BASE, +)
TO: (S.AUREUS, 25)(B.SUBT., 6)(E.COLI, 50)(P.VULG., 100)(K.PNEUM., 100)
LD50: (500|200, IV)
REFERENCES:
DT 2928373; 2942194

12410-7514

NAME: FORTIMICIN-AE
PO: MIC.OLIVOASTEROSPORA
CT: AMINOGLYCOSIDE L., FORTIMICIN T., BASIC
FORMULA: C15H32N4O5
EA: (N, 16)
MW: 348
PC: WH., POW.
UV: W: (200, ,)
SOL-GOOD: W
QUAL: (NINH., +)
TO: (G.POS.,)(G.NEG.,)
IS-ION: (IRC-50-NH4, NH4OH)
REFERENCES:
JP 79/95548; *CA*, 92, 109099

12410-7619

NAME: ISTAMYCIN-A0
IDENTICAL: SANNAMYCIN-B
PO: S.TENJIMARIENSIS
CT: AMINOGLYCOSIDE L., FORTIMICIN T., BASIC
FORMULA: C15H32N4O4
EA: (N, 16)
MW: 332
PC: WH., CRYST.
OR: (+76, W)
UV: W: (200, ,)
SOL-GOOD: W, MEOH
SOL-POOR: HEX, ETOH
QUAL: (NINH., +)
STAB: (BASE, +)
TO: (G.NEG.,)(S.AUREUS, 80)(B.SUBT., 80)(S.LUTEA, 300)(E.COLI, 600)
IS-ION: (IRC-50-NH4, NH4OH)
IS-CHR: (SILG, CHL-MEOH-NH4OH)(DX-1X4, W)
REFERENCES:
Abst. Pap. 214th Meet, JARA, 1979, 12; DT 3012014

12410-7620

NAME: ISTAMYCIN-BO, 1-EPI-SANNAMYCIN-B
PO: S.TENJIMARIENSIS
CT: AMINOGLYCOSIDE L., FORTIMICIN T., BASIC
FORMULA: C15H32N4O4
EA: (N, 16)
MW: 332
PC: WH., CRYST.
OR: (+160, W)
UV: W: (200, ,)
SOL-GOOD: W, MEOH
SOL-POOR: HEX, ETOH
QUAL: (NINH., +)
STAB: (BASE, +)
TO: (G.NEG.,)(P.VULG., 160)(S.AUREUS, 20)(B.SUBT., 10)(S.LUTEA, 300)(E.COLI, 300)
IS-ION: (IRC-50-NH4, NH4OH)
IS-CHR: (SILG, CHL-MEOH-NH4OH)(DX-1X4, W)
REFERENCES:

Abst. Pap. 214th Meet. JARA, 1979, 12; DT 3012014

12410-7621

NAME: KA-6606-V, 3-EPI-SPORARICIN-B
PO: SACCHAROPOLYSPORA HIRSUTA-KOBENSIS
CT: AMINOGLYCOSIDE L., FORTIMICIN T., BASIC
FORMULA: C15H32N4O4
EA: (N, 16)
MW: 332
PC: WH., POW.
OR: (+103, W)
UV: W: (200, ,)
SOL-GOOD: W
QUAL: (NINH., +)
TO: (S.AUREUS, 10)(B.SUBT., 3)(E.COLI, 12)(P.VULG., 6)(K.PNEUM., 100)
IS-ION: (IRC-50-NH4, NH4OH)
IS-CHR: (SILG, CHL-MEOH-NH4OH)(CG-50-NH4, NH4OH)
REFERENCES:

DT 2813021; 2942194; JP 80/111497; *CA*, 94, 14023

12410-7622

NAME: KA-6606-VI, 1-EPI-SPORARICIN-B
PO: SACCHAROPOLYSPORA HIRSUTA-KOBENSIS
CT: AMINOGLYCOSIDE L., FORTIMICIN T., BASIC
FORMULA: C15H32N4O4
EA: (N, 16)
MW: 332
PC: WH., POW.
OR: (+54, W)
UV: W: (200, ,)
SOL-GOOD: W
QUAL: (NINH., +)
TO: (S.AUREUS, 50)(B.SUBT., 12)(E.COLI, 50)
(K.PNEUM., 200)(P.VULG., 50)
IS-ION: (IRC-50-NH4, NH4OH)
IS-CHR: (SILG, CHL-MEOH-NH4OH)(CG-50-NH4, NH4OH)
REFERENCES:
DT 2813021, 2942194; JP 80/111497; *CA*, 94, 14023

12410-7623

NAME: FORTIMICIN-AH
PO: MIC.OLIVOASTEROSPORA
CT: AMINOGLYCOSIDE L., FORTIMICIN T., BASIC
FORMULA: C15H26N3O5
EA: (N, 13)
MW: 328
PC: WH., POW.
UV: W: (200, ,)
SOL-GOOD: W
REFERENCES:
Abst. ACS/CSJ Symp. Ser., 125, 295, 1980; USP 4213972; 4219644

12410-7624

NAME: FORTIMICIN-AI
PO: MIC.OLIVOASTEROSPORA
CT: AMINOGLYCOSIDE L., FORTIMICIN T., BASIC
FORMULA: C15H26N3O5
EA: (N, 13)
MW: 328
PC: WH., POW.
UV: W: (200, ,)
SOL-GOOD: W
QUAL: (NINH., +)
REFERENCES:
Abst. ACS/CSJ Symp. Ser., 125, 295, 1980; USP 4213972, 4219644

12410-7625

NAME: FORTIMICIN-AK
PO: MIC.OLIVOASTEROSPORA
CT: AMINOGLYCOSIDE L., FORTIMICIN T., BASIC
FORMULA: C14H29N3O6
EA: (N, 12)
MW: 335
PC: WH., POW.
UV: W: (200, ,)
SOL-GOOD: W
TO: (G.POS., 100)
REFERENCES:

Abst. ACS/CSJ Symp. Ser., 125, 295, 1980; USP 4213971 4226979

12410-7626

NAME: FORTIMICIN-AL, 3-O-DEMETHYLFORTIMICIN-KG1
PO: MIC.OLIVOASTEROSPORA
CT: AMINOGLYCOSIDE L., FORTIMICIN T., BASIC
FORMULA: C14H28N4O5
EA: (N, 16)
MW: 332
PC: WH., POW.
UV: W: (200, ,)
SOL-GOOD: W
QUAL: (NINH., +)
REFERENCES:

Abst. ACS/CSJ Symp. Ser., 125, 295, 1980; USP 4124079, 4124078

12410-7627

NAME: FORTIMICIN-AM, 3-O-DEMETHYL-4-EPI-FORTIMICIN-B
PO: MIC.OLIVOASTEROSPORA
CT: AMINOGLYCOSIDE L., FORTIMICIN T., BASIC
FORMULA: C14H30N4O5
EA: (N, 16)
MW: 334
PC: WH., POW.
UV: W: (200, ,)
SOL-GOOD: W
REFERENCES:

Abst. ACS/CSJ Symp. Ser., 125, 295, 1980; USP 4124080

12410-7628

NAME: FORTIMICIN-AN
PO: MIC.OLIVOASTEROSPORA
CT: AMINOGLYCOSIDE L., FORTIMICIN T., BASIC
FORMULA: C16H33N5O6
EA: (N, 17)
MW: 391
PC: WH., POW.
UV: W: (200, ,)
SOL-GOOD: W
REFERENCES:
Abst. ACS/CSJ Symp. Ser., 125, 295, 1980; USP 4219643; Belg. P. 882509; *JA*, 33, 1071, 1980

12410-7629

NAME: FORTIMICIN-AO
PO: MIC.OLIVOASTEROSPORA
CT: AMINOGLYCOSIDE L., FORTIMICIN T., BASIC
FORMULA: C13H27N3O8
EA: (N, 12)
MW: 353
PC: WH.
REFERENCES:
Abst. ACS/CSJ Symp. Ser., 125, 295, 1980; USP 4213974; 4219642

12410-7630

NAME: FORTIMICIN-AP, 3-O-DEMETHYLFORTIMICIN-E
PO: MIC.OLIVOASTEROSPORA
CT: AMINOGLYCOSIDE L., FORTIMICIN T., BASIC
FORMULA: C14H30N4O5
EA: (N, 16)
MW: 334
PC: WH., POW.
UV: W: (200, ,)
SOL-GOOD: W
REFERENCES:
Abst. ACS/CSJ Symp. Ser., 125, 295, 1980; USP 4214080

12410-7631

NAME: FORTIMICIN-AQ
PO: MIC.OLIVOASTEROSPORA
CT: AMINOGLYCOSIDE L., FORTIMICIN T., BASIC
FORMULA: C16H34N4O5
EA: (N, 15)
MW: 362
PC: WH., POW.
UV: W: (200, ,)
SOL-GOOD: W
REFERENCES:
Abst. ACS/CSJ Symp. Ser., 125, 295, 1980

12410-7632

NAME: FORTIMICIN-AS
PO: MIC.OLIVOASTEROSPORA
CT: AMINOGLYCOSIDE L., FORTIMICIN T., BASIC
FORMULA: C17H36N4O5
EA: (N, 15)
MW: 376
PC: WH., POW.
UV: W: (200, ,)
SOL-GOOD: W
REFERENCES:
Abst. ACS/CSJ Symp. Ser., 125, 295, 1980

12410-7633

NAME: FORTIMICIN-KO
PO: MIC.OLIVOASTEROSPORA
CT: AMINOGLYCOSIDE L., FORTIMICIN T., BASIC
FORMULA: C17H35N5O6
EA: (N, 17)
MW: 405
PC: WH., POW.
UV: W: (200, ,)
SOL-GOOD: W
TO: (G.NEG.,)
REFERENCES:
JP 79/46747; *CA*, 91, 106595; *Abst. Pap. Ann. Meet. Agr. Ch. Soc. Jap.*, 1980, 3P-14; *Abst. Pap. 100th Ann. Meet. Pharm. Soc. Jap.*, 1980, 5G11- 3

12410-7634

NAME: 3-O-DEMETHYLFORTIMICIN-A, A-49759
PO: MIC.OLIVOASTEROSPORA
CT: AMINOGLYCOSIDE L., FORTIMICIN T., BASIC
FORMULA: C16H33N5O6
EA: (N, 17)
MW: 391
PC: WH., POW.
OR: (+82, W)(+7., MEOH)
UV: W: (200, ,)
SOL-GOOD: W
SOL-FAIR: MEOH
SOL-POOR: BUOH, HEX
QUAL: (NINH., +)
TO: (B.SUBT.,)(S.AUREUS, .25)(K.PNEUM., 1)
(P.VULG., .25)(PS.AER., 16)
IS-FIL: 2.5
IS-ION: (IRC-50-NH4, NH4OH)
IS-CHR: (CG-50-NH4, NH4OH)(SILG, CHL-MEOH-NH4OH)
IS-CRY: (LIOF.,)
REFERENCES:

AAC, 18, 761, 766, 773, 1980; USP 4242503; DT 3012131; JP 80/54898; *CA*, 93, 202700

12410-7635

NAME: 3-O-DEMETHYL-2′-N-GLYCYLFORTIMICIN-B
PO: MIC.OLIVOASTEROSPORA
CT: AMINOGLYCOSIDE L., FORTIMICIN T., BASIC
FORMULA: C16H33N5O6
EA: (N, 18)
MW: 391
PC: WH., POW.
OR: (+57.4, W)
UV: W: (200, ,)
SOL-GOOD: W, MEOH
SOL-FAIR: ETOH, ACET
SOL-POOR: ETOAC, HEX, BUOH
QUAL: (NINH., +)(BIURET, -)
TO: (B.SUBT., 10.5)(S.AUREUS, 5.3)(K.PNEUM., 41.7)
(P.VULG., 41.7)(SHYG., 41.7)
IS-EXT: (IRC-50, NH4OH)
IS-CHR: (CG-50-NH4, NH4OH)
REFERENCES:

JP 80/45631; *CA*, 93, 184262

12410-7888

NAME: LYSINOMYCIN, AX-127-B
PO: MIC.PILOSOSPORA
CT: BASIC, AMINOGLYCOSIDE L., FORTIMICIN T.
FORMULA: C20H40N6O6
EA: (N, 18)
MW: 460
PC: WH., POW.
OR: (+49, W)
UV: W: (200, ,)
SOL-GOOD: W
SOL-FAIR: MEOH
SOL-POOR: BUOH, HEX
QUAL: (NINH., +)
TO: (S.AUREUS, .05)(E.COLI, .1)(K.PNEUM., .05)
(P.VULG., .39)(PS.AER., 1.6)(SHYG., .4)
IS-FIL: 2
IS-ION: (IRC-84-NH4, NH4OH)
IS-CHR: (REXYN-102-NH4, NH4OH)(DX-1-OH, W)
IS-CRY: (LIOF.,)
REFERENCES:

DT 3003497; USP 4283529; Belg. P 881476

12410-7890

NAME: FORTIMICIN-KR1, DEMETHYLFORTIMICIN-B
PO: MIC.OLIVOASTEROSPORA
CT: AMINOGLYCOSIDE L., FORTIMICIN T., BASIC
FORMULA: C13H28N4O5
EA: (N, 18)
MW: 320
PC: WH., POW.
OR: (+69.5, W)
UV: W: (200, ,)
SOL-GOOD: W, MEOH
SOL-FAIR: ETOH, ACET
SOL-POOR: BUOH, ETOAC, HEX
QUAL: (NINH., +)(BIURET, -)
TO: (B.SUBT., .7)(S.AUREUS, 2.7)(E.COLI, 20.9)
(K.PNEUM., 10.5)(P.VULG., 20.9)(SHYG., 41.7)
IS-ION: (IRC-50-NH4, NH4OH)
IS-CHR: (CG-50-NH4, NH4OH-NH4CL)
REFERENCES:

JP 80/45631; *CA*, 93, 184262

12410-7891

NAME:	FORTIMICIN-KQ
PO:	MIC.OLIVOASTEROSPORA
CT:	AMINOGLYCOSIDE L., FORTIMICIN T., BASIC
FORMULA:	C14H28N4O5
EA:	(N, 17)
MW:	332
PC:	WH., POW.
OR:	(+15.5, W)
UV:	W: (200, ,)
SOL-GOOD:	W, MEOH
SOL-FAIR:	ETOH, ACET
SOL-POOR:	BUOH, ETOAC, HEX
QUAL:	(NINH., +)(BIURET, -)
TO:	(S.AUREUS, .33)(E.COLI, 5.2)(P.VULG., 26)
IS-ION:	(IRC-50-NH4, NH4OH)
IS-CHR:	(CG-50-NH4, NH4OH-NH4CL)

REFERENCES:

JP 80/45631; *CA*, 93, 184262

12410-7892

NAME:	3-O-DEMETHYLFORTIMICIN-KE
PO:	MIC.OLIVOASTEROSPORA
CT:	AMINOGLYCOSIDE L., FORTIMICIN T., BASIC
FORMULA:	C13H28N4O5
EA:	(N, 18)
MW:	320
PC:	WH., POW.
OR:	(+33.1, W)
UV:	W: (200, ,)
SOL-GOOD:	W, MEOH
SOL-FAIR:	ETOH, ACET
SOL-POOR:	BUOH, ETOAC, HEX
QUAL:	(NINH., +)(BIURET, -)
TO:	(B.SUBT., 10.5)(S.AUREUS, 20.9)(E.COLI, 83.3) (K.PNEUM., 41.7)
IS-ION:	(IRC-50-NH4, NH4OH)
IS-CHR:	(CG-50-NH4, NH4OH-NH4CL)

REFERENCES:

JP 80/45631; *CA*, 93, 184262

12410-7893

NAME: DEMETHYL-ISOFORTIMICIN-C, 3-O-DEMETHYL-1-N-HYDANTHOYLFORTIMICIN-B
PO: MIC.OLIVOASTEROSPORA
CT: AMINOGLYCOSIDE L., FORTIMICIN T., BASIC
FORMULA: C17H34N6O7
EA: (N, 19)
MW: 434
PC: WH., POW.
OR: (+15.3, W)
UV: W: (200, ,)
SOL-GOOD: W, MEOH
SOL-FAIR: ETOH, ACET
SOL-POOR: BUOH, ETOAC, HEX
QUAL: (NINH., +)(BIURET, -)
TO: (S.AUREUS, 4.17)
IS-ION: (IRC-50-NH4, NH4OH)
IS-CHR: (CG-50-NH4, NH4OH-NH4CL)
REFERENCES:

JP 80/45631; *CA*, 93, 184262

12410-8786

NAME: DACTIMICIN-B, SF-2052-B
IDENTICAL: SF-1854
PO: DACTYLOSPORANGIUM MATSUZAKIENSE
CT: AMINOGLYCOSIDE L., FORTIMICIN T., BASIC
FORMULA: C18H35N5O7
EA: (N, 16)
MW: 433
PC: WH., POW.
OR: (+89.2, W)(+90.3, MEOH)
UV: W: (200, ,)
SOL-GOOD: W
QUAL: (NINH., +)
TO: (G.POS.,)(G.NEG.,)
IS-ION: (IRC-50-NA, HCL)
IS-CHR: (CM-SEPHADEX C-25-NH4, NH4OH)
REFERENCES:

JA, 34, 1090, 1981; JP 81/40697; CA, 95, 167125

12510-7085

NAME: GL-A1
IDENTICAL: SORBISTIN-A1, LL-AM-31-B"
PO: STV.NETROPSIS
CT: AMINOGLYCOSIDE L., AMINOHEXITOL DERIV., BASIC
FORMULA: C15H31N3O9
EA: (N, 11)
MW: 397
PC: WH., POW.
OR: (+87.1, W)
UV: W: (200, ,)
SOL-GOOD: W
TO: (G.POS.,)(G.NEG.,)
LD50: (3000|1000, IP)
IS-ION: (IRC-50-NA, NH4OH)
IS-CHR: (SILG,)
REFERENCES:
JP 77/128313; 78/116312

12510-7086

NAME: GL-A2
IDENTICAL: SORBISTIN-B, LL-AM-31-G"
PO: STV.NETROPSIS
CT: AMINOGLYCOSIDE L., AMINOHEXITOL DERIV., BASIC
FORMULA: C14H29N3O9
EA: (N, 11)
MW: 383
PC: WH., POW.
UV: W: (200, ,)
TO: (G.POS.,)(G.NEG.,)
REFERENCES:
JP 77/128313, 78/116312

12000-7067

NAME: DESOMYCIN
PO: S.SP.
CT: AMINOGLYCOSIDE, BASIC
EA: (N,)
TO: (G.POS., .1)(G.NEG., .1)
REFERENCES:
FEMS Symp., Basel, 1977, Acad. Press 1978, p.13

12000-7613

NAME:	SF-1999
PO:	S.PLATENSIS
CT:	AMINOGLYCOSIDE, BASIC
EA:	(C, 44)(H, 7)(N, 13)
MW:	700
PC:	WH., POW.
OR:	(-65, W)
UV:	(263, 63,)
UV:	1: (261, 80,)
UV:	HCL: (261, 80,)
UV:	MEOH: (260, 63,)
UV:	W: (261, 80,)
SOL-GOOD:	W
SOL-FAIR:	MEOH
SOL-POOR:	ACET, HEX
QUAL:	(SAKA., +)(NINH., +)(FECL3, -)
STAB:	(HEAT, +)(ACID, +)(BASE, -)
TO:	(E.COLI, 6.25)(SHYG., 6.25)(K.PNEUM., 25) (P.VULG., 400)(PS.AER., 400)(MYCOB.SP., 3.13)
LD50:	(200|100, IV)
IS-ION:	(XAD-2,)(IRC-50-H,)
IS-ABS:	(CARBON, ACET-W)
IS-CHR:	(CM-SEPHADEX C-25-H,)

REFERENCES:

JP 81/99495; *CA*, 96, 18645

13
OTHER GLYCOSIDES

131 Streptothricin Group
132 Glycopeptides, C-Glycosides

13110-7894

NAME: Y-U17W-C2
PO: S.LAVENDULAE
CT: STREPTOTHRICIN T., BASIC
EA: (N,)
PC: WH., POW.
UV: W: (200, ,)
SOL-GOOD: W
TO: (G.POS.,)(G.NEG.,)
LD50: (25, IV)
IS-ION: (IRC-50-NA, HCL)
REFERENCES:
Rep. Yamanochi Cent. Res. Lab., 2, 164, 1974; *CA*, 84, 15641

13110-8167

NAME: 402
PO: S.LAVENDULORECTUS
CT: STREPTOTHRICIN T., BASIC
FORMULA: C20H44N8O11
EA: (N, 18)
PC: WH., POW.
OR: (-23.2, W)
UV: W: (200, ,)
SOL-GOOD: W, MEOH
SOL-FAIR: ETOH
SOL-POOR: BUOH, HEX
STAB: (ACID, +)(BASE, -)
TO: (G.POS.,)(G.NEG.,)
IS-ION: (IRC-50,)
IS-CHR: (AL,)
REFERENCES:
Wei Sheng Hsueh T'ung Pao, 7, 145, 1980; *CA*, 94, 137730

13110-8168

NAME: S-15-1-C1
PO: S.PURPEOFUSCUS
CT: STREPTOTHRICIN T., BASIC
TO: (B.SUBT.,)(E.COLI,)
IS-CHR: (CM-SEPHADEX C-25, ACOH-W-HCL)
REFERENCES:
JA, 34, 292, 1981

13110-8493

NAME: AY-24488
PO: S.SP.+B"-LYSINE
CT: STREPTOTHRICIN T., BASIC
FORMULA: C18H33N7O7
EA: (N, 21)
MW: 459
PC: WH., POW.
UV: W: (200, ,)
SOL-GOOD: W
QUAL: (NINH., +)
TO: (G.POS.,)(G.NEG.,)
REFERENCES:

Microbiology 1976/1977, 528

13200-6771

NAME: AZURENOMYCIN-A, AM-3696
PO: PSEUDONOCARDIA AZUREA
CT: GLYCOPEPTIDE, BASIC
FORMULA: C32|2H40|4N3O22.5|1.5
EA: (C,44)(H,4)(N,5)
MW: 850|40
PC: WH., POW.
OR: (-124,W)
UV: HCL: (282,54,)
UV: NAOH: (301,81,)
SOL-GOOD: W
SOL-FAIR: DMSO, ME-CELLOSOLV
SOL-POOR: MEOH, HEX
QUAL: (NINH.,-)(SAKA.,-)
TO: (S.AUREUS, 6.25)(B.SUBT., 1.56)(S.LUTEA, .78)
LD50: (1000|500, IP)
IS-ION: (IRC-50-H, HCL)
IS-ABS: (CARBON, ACET-H2SO4)
IS-CHR: (AVICEL, BUOH-ACOH-W)
REFERENCES:

JA, 32, 978, 987, 995, 1979; *J. Bioch. (Tokyo)*, 88, 565, 1980; *Proc. Ann. Meet. ASM*, 1979, A (4)11; JP 80/115894; *CA*, 94, 45584

13200-7087

NAME: AZURENOMYCIN-B, AM-3696-B
PO: PSEUDONOCARDIA AZUREA
CT: GLYCOPEPTIDE, BASIC
FORMULA: C35H46N3O19, C36H48N3O21
EA: (N, 5)(CL, 4)
MW: 870|20
PC: WH., POW.
OR: (-116, W)
UV: HCL: (279, 50,)
UV: NAOH: (286, 77,)
SOL-GOOD: W
SOL-FAIR: DMSO, MEOH-HCL
SOL-POOR: MEOH, HEX
QUAL: (NINH., -)(SAKA., -)
TO: (S.AUREUS, 12.5)(B.SUBT., 1.56)(S.LUTEA, .78)
REFERENCES:

JA, 32, 987, 1979; JP 80/14798; *CA*, 94, 137792

13200-7768

NAME: OA-7653
PO: S.HYGROSCOPICUS-HIWASAENSIS
CT: BASIC, AMPHOTER, GLYCOPEPTIDE
EA: (C, 46)(H, 5)(N, 7)
PC: WH., CRYST.
OR: (+60.3, W)
UV: HCL: (278, 56,)
UV: NAOH: (298, 100.8,)
UV: W: (278, 56)
SOL-GOOD: W, ACID, BASE
SOL-FAIR: ACID
SOL-POOR: MEOH, HEX
TO: (B.SUBT., 6.25)(S.LUTEA, 3.13)(S.AUREUS, 3.13)
(MYCOB.SP., 1.56)
LD50: (2000|500, IV)
IS-ION: (OX-50X4,)(DX-50X4-NA, PH10 BUFF-NACL)
IS-ABS: (CARBON, MEOH-NH4OH)
IS-CHR: (ECTEOLA-CEL., NACL)(SILG, ETOH-W)
REFERENCES:

JP 80/19246; *CA*, 93, 6127

13200-8110

NAME: A-16686
PO: ACTINOPLANES SP.
CT: GLYCOPEPTIDE, BASIC
EA: (C, 52)(H, 6)(N, 10)(CL, 6)
MW: 2500|500
PC: WH., CRYST.
OR: (+49.7, DMFA)
UV: MEOH: (232, 178,)(265, 107,)
UV: MEOH-HCL: (231, 167,)(270, 96,)
UV: MEOH-NAOH: (250, 232,)
SOL-GOOD: W, DMFA, MEOH, BUOH
SOL-POOR: ET2O, HEX
QUAL: (NINH., +)(BIURET, +)(FEHL., -)
TO: (S.AUREUS, 1.6)(S.LUTEA, .02)(B.SUBT., .062)
LD50: (625|125, IP)
IS-FIL: 3.5
IS-EXT: (MEOH, 5.5, MIC.)(BUOH, 7, W)(BUOH, 3.5, FILT.)
IS-CHR: (SILG, ACCN-HCL)(SEPHADEX LH-20, MEOH-W)
REFERENCES:
BP 2045231; Belg. P 882632

13220-6745

NAME: AVOPARCIN-A'', LL-AV-290-A''
PO: S.CANDIDUS
CT: GLYCOPEPTIDE, VANCOMYCIN T., BASIC
FORMULA: C89H101N9036CL
EA: (N, 6)(CL, 2)
MW: 1907
PC: WH., POW.
TO: (S.AUREUS,)(S.LUTEA,)(B.SUBT.,)
IS-CHR: (HPLC, W-ACCN-ACOH)
REFERENCES:
JACS, 101, 2237, 1979; 102, 1671, 1980; 103, 6522, 1981

13230-7895

NAME: SR-1768-F
PO: S.KATSUNUMAENESIS
CT: CHROMOMYCIN T., ACIDIC
PC: YELLOW, POW.
TO: (G.POS.,)(XANTHOMONAS SP.,)
LD50: (4, IP)
TV: L-1210
REFERENCES:
JA, 26, 701, 1973; JP 75/13593

13230-7896

NAME: 144-3-I
PO: S.ABURAVIENSIS-VERRUCOSUS
CT: CHROMOMYCIN T., ACIDIC
PC: YELLOW, POW.
OR: (+33, MEOH)
UV: MEOH: (270,438)(414,51,)
SOL-GOOD: MEOH, BENZ
SOL-POOR: W
TO: (G.POS.,)
TV: ANTITUMOR
IS-EXT: (ETOAC, ,FILT.)
REFERENCES:
Antib., 882, 1972; *Mikrob.*, 43, 857, 1974

13230-7897

NAME: 144-3-II
PO: S.ABURAVIENSIS-VERRUCOSUS
CT: CHROMOMYCIN T., ACIDIC
PC: YELLOW, POW.
OR: (+75, MEOH)
UV: MEOH: (280, 415,)(420, 93)
SOL-GOOD: MEOH, BENZ
SOL-POOR: W
TO: (G.POS.,)
TV: ANTITUMOR
IS-EXT: (ETOAC, , FILT.)
REFERENCES:
Antib., 882, 1972; *Mikrob.*, 43, 857, 1974

13230-7898

NAME: 144-3-III
PO: S.ABURAVIENSIS-VERRUCOSUS
CT: CHROMOMYCIN T., ACIDIC
PC: YELLOW, POW.
OR: (+68, MEOH)
UV: MEOH: (280, 425,)(418, 87,)
SOL-GOOD: MEOH, BENZ
SOL-POOR: W
TO: (G.POS.,)
TV: ANTITUMOR, S-180
IS-EXT: (ETOAC, , FILT.)
REFERENCES:
Antib., 882, 1972; *Mikrob.*, 43, 857, 1974

14
SUGAR DERIVATIVES

141 Sugar Esters, Sugar Amides
142 Sugar Lipids

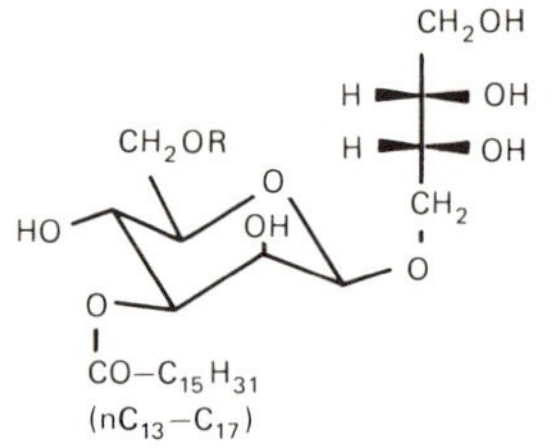

	R
Schizonellin	A H
Schizonellin	B $COCH_3$

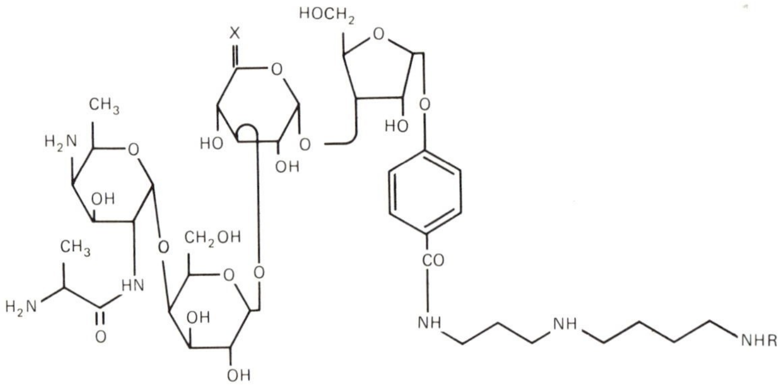

	X	R
Glysperin A	$=CH_2$	$(CH_2)_4NH_2$
Glysperin B	$=CH_2$	H
Glysperin C	CH_2OH,H	$(CH_2)_4NH_2$

14110-7088

NAME:	SF-2033
PO:	STREPTOSPORANGIUM VULGARE
CT:	EVERNINOMYCIN T., ACIDIC
FORMULA:	C54H115029CL4
EA:	(CL, 10)
MW:	1200|100
PC:	WH., POW.
OR:	(-29.2, MEOH)
UV:	HCL: (288, 53,)
UV:	NAOH: (296, 110,)
SOL-GOOD:	DMFA, MEOH, ETOH, ACET, ETOAC, CHL, BASE
SOL-POOR:	ET2O, HEX, W
STAB:	(ACID, -)
TO:	(S.AUREUS, 1.50)(B.SUBT., .78)(S.LUTEA, .39)
IS-ION:	(IONEX HP-20,)(DIAION HP-20, MEOH)
IS-CHR:	(SILG, ACET)(SEPHADEX LH-20, MEOH)
IS-CRY:	(DRY, MEOH)

REFERENCES:

JP 79/122202; *CA*, 92, 179071

14110-7636

NAME:	SE-73
PO:	ACTINOPLANES SP.
CT:	EVERNINOMYCIN T., NEUTRAL
FORMULA:	C73H110037
MW:	1578
PC:	YELLOW, POW.
UV:	MEOH: (277.5|2.5, 60,)(365, 406,)
SOL-GOOD:	MEOH, CHL, ACCN, DMFA, DMSO
SOL-FAIR:	ET2O
SOL-POOR:	W, HEX
TO:	(S.AUREUS,)(B.SUBT.,)(E.COLI,)(K.PNEUM.,) (P.VULG.,)
IS-EXT:	(ACET, , MIC.)(ETOAC, 6.5, FILT.)
IS-CHR:	(SILG, CHL-MEOH-NH4OH)
IS-CRY:	(PREC., ETOAC, C.HEX)

REFERENCES:

DT 2848793; DT 2510161; USP 4031208

14120-7637

NAME: BU-2545
PO: S.SP.
CT: LINCOMYCIN T., BASIC
FORMULA: C16H30N2O6S
EA: (N, 7)(S, 8)
MW: 378
PC: WH., POW.
OR: (+140, CHL)
UV: MEOH: (200, ,)
SOL-GOOD: MEOH, CH2CL2, ACID
SOL-POOR: BASE, W, HEX
TO: (S.AUREUS, 6.3)(BACTEROIDES SP., 1.6)
LD50: NONTOXIC
IS-EXT: (BUOH, 10, W)(W, 2, BUOH)(CH2CL2, 10, W)
IS-ION: (DIAION HP-20, ACET-W)
IS-CHR: (SILG, BENZ-MEOH)
IS-CRY: (DRY, MEOH)
REFERENCES:

JA, 33, 751, 1980; 34, 596, 1981

14120-7899

NAME: U-24166, N-ETILLINCOMYCIN
PO: S.LINCOLNENSIS-LINCOLNENSIS
CT: LINCOMYCIN T., BASIC
FORMULA: C19H36N2O6S
EA: (N, 7)(S, 8)
MW: 384
PC: WH.
TO: (G.POS.,)
REFERENCES:

Antib., 554, 1967

14220-7420

NAME: SCHIZONELLIN-A
PO: SCHIZONELLA MELANOGRAMA
CT: GLYCOLIPID, NEUTRAL
FORMULA: C28H52O11
MW: 564
PC: WH., GUM
OR: (-44.3, CHL)
UV: MEOH: (200, ,)
SOL-GOOD: MEOH, CHL
SOL-FAIR: W
TO: (G.POS., 2)(P.VULG., 100)(FUNGI, 10)(MYCOB.SP., 20)
TV: EHRLICH
IS-EXT: (MEOH, , MIC.)
IS-CHR: (SEPHADEX LH-20, MEOH)(AL, BUOH-ETOH-W)(SILG, CHL-MEOH)
REFERENCES:

Phytoch., 19, 83, 1980

14220-7421

NAME: SCHIZONELLIN-B
PO: SCHIZONELLA MELANOGRAMA
CT: GLYCOLIPID, NEUTRAL
FORMULA: C30H54O12
MW: 594
PC: WH., OIL
OR: (-42.8, CHL)
UV: MEOH: (200, ,)
SOL-GOOD: MEOH, CHL
SOL-FAIR: W
TO: (B.SUBT., 2)(P.VULG., 100)(MYCOB.SP., 20)(FUNGI, 10)
TV: EHRLICH
REFERENCES:

Phytoch., 19, 83, 1980

14230-8084

NAME:	GLYSPERIN-C, BU-2349-C
PO:	B.CEREUS
CT:	SUGAR DERIV., BASIC
FORMULA:	C44H77N7O19
EA:	(N, 9)
MW:	1007
PC:	WH., POW.
OR:	(+157, W)
UV:	HCL: (247, ,)
UV:	NAOH: (247, ,)
UV:	W: (247, 133,)
SOL-GOOD:	W
SOL-FAIR:	MEOH, ETOH, DMFA, DMSO
SOL-POOR:	BUOH, HEX
QUAL:	(NINH., +)(FECL3, -)(SAKA., -)
TO:	(S.AUREUS, 3.1)(B.SUBT., 6.3)(E.COLI, 3.1) (K.PNEUM., 12.5)(P.VULG., 6.3)(MYCOB.SP., 1.6)

REFERENCES:

JA, 34, 381, 390, 1980

14230-8111

NAME:	GLYSPERIN-A, BU-2349-A
PO:	B.CEREUS
CT:	SUGAR DERIV., BASIC
FORMULA:	C44H75N7O18
EA:	(N, 9)
MW:	989
PC:	WH., POW.
OR:	(+109, W)(+113, W)
UV:	HCL: (247, ,)
UV:	NAOH: (247, ,)
UV:	W: (247, 131,)
SOL-GOOD:	W
SOL-FAIR:	MEOH, ETOH, DMSO, DMFA
SOL-POOR:	ACET, HEX
QUAL:	(NINH., +)(FEHL., +)(FECL3, -)(SAKA., -)
STAB:	(BASE, +)
TO:	(S.AUREUS, 1.6)(B.SUBT., 1.6)(E.COLI, .8) (K.PNEUM., 3.1)(P.VULG., 3.1)(MYCOB.SP., 1.6)
LD50:	(35, IV)(285, IM)
IS-FIL:	6.5
IS-ION:	(IRC-50-NH4, NH4OH)
IS-CHR:	(CG-50-NH4, NH4OH)(HP-20, PH7BUFF.)
IS-CRY:	(LIOF.,)

REFERENCES:

JA, 34, 381, 390, 1980; *Abst. AAC*, 20, 60, 61, 1980; USP 4250170; JP 80/102398

14230-8112

NAME:	GLYSPERIN-B, BU-2349-B
PO:	B.SP.
CT:	SUGAR DERIV., BASIC
FORMULA:	C40H66N6018
EA:	(N, 8)
MW:	918
PC:	WH., POW.
OR:	(+115, W)(+132, W)
UV:	HCL: (247, ,)
UV:	NAOH: (247, ,)
UV:	W: (247, 148,)
SOL-GOOD:	W
SOL-FAIR:	MEOH, ETOH
SOL-POOR:	ACET, HEX
QUAL:	(NINH., +)(FEHL., +)(FECL3, -)(SAKA., -)
TO:	(B.SUBT., 25)(S.AUREUS, 6.3)(E.COLI, .8) (K.PNEUM., 12.5)(P.VULG., 15)(MYCOB.SP., 3.1)

REFERENCES:

JA, 34, 381, 390, 1980; *Abst. AAC*, 20, 60, 61, 1980; USP 4250170; JP 80/102398

10
OTHER PROBABLY CARBOHYDRATE ANTIBIOTICS

10000-7885

NAME: 460-B
PO: MIC.CHALCEA-FLAVIDA
CT: BASIC
EA: (C, 38)(H, 7)(N,)
PC: WH., POW.
UV: W: (200, ,)
SOL-GOOD: W
TO: (G.POS.,)(G.NEG.,)
IS-ION: (IRC-50-NA, H2SO4)
REFERENCES:
USP 3454696

10000-7886

NAME: OS-469
PO: S.SP.
CT: BASIC
SOL-GOOD: W
TO: (G.POS.,)
IS-ION: (IRC-50,)
REFERENCES:
Abst. 91st Ann. Meet. Pharm. Soc. Jap., 1971, 7C-10-3

10000-7887

NAME: K-7
PO: S.SP.
CT: BASIC
PC: GRAY, WH., POW.
SOL-GOOD: W
TO: (G.POS.,)(G.NEG.,)(MYCOB.SP.,)
IS-ABS: (CARBON, MEOH-HCL)
REFERENCES:
Anais Biol. Pernambuco, 11, 11, 15, 1953

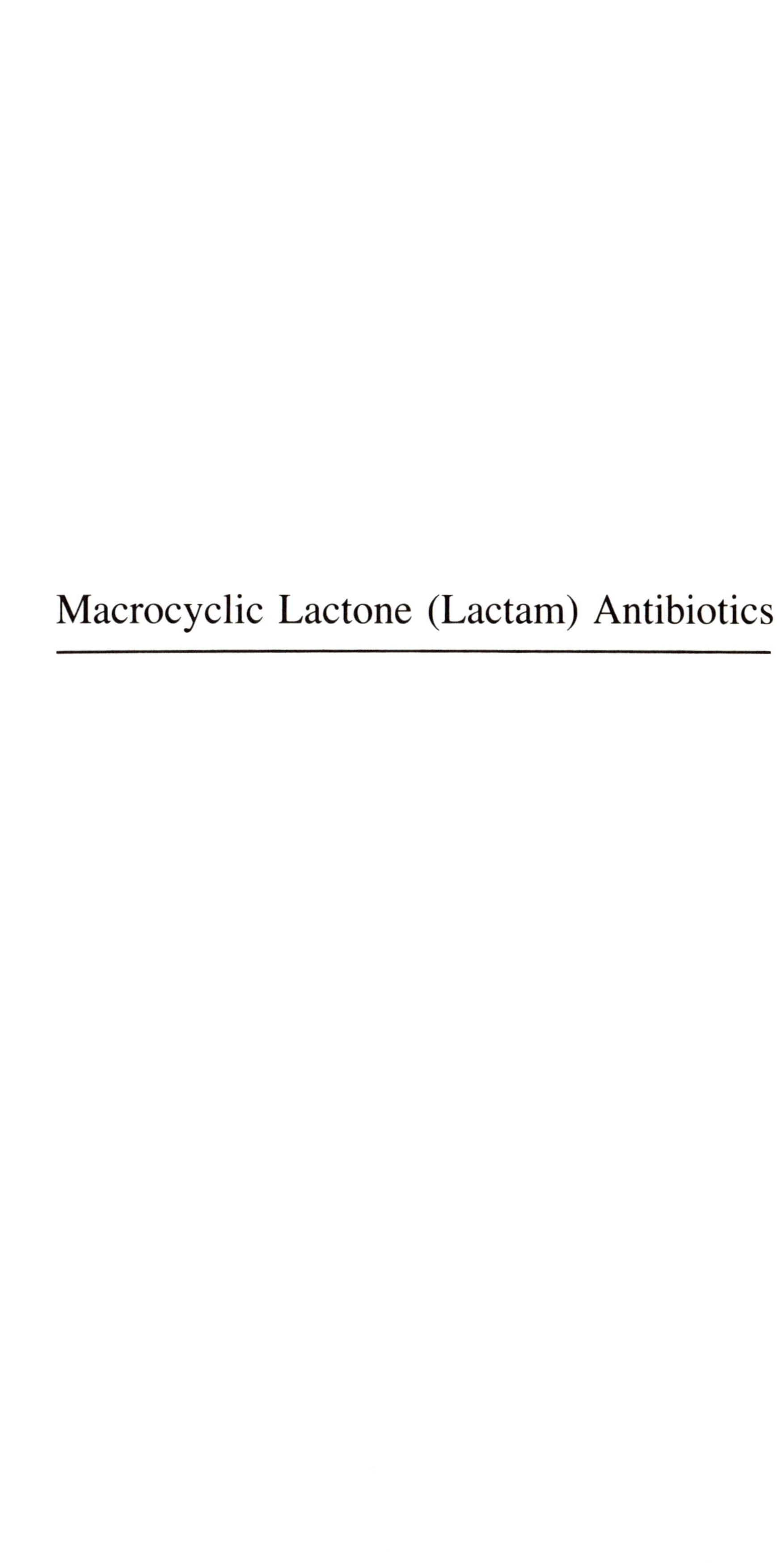

Macrocyclic Lactone (Lactam) Antibiotics

21
MACROLIDE ANTIBIOTICS

211 Small Macrolides
212 16-Membered Macrolides
213 Other Macrolides

	R_1	R_2
Leucomycin A_{12}	$COCH_3$	$CO(CH_2)_4CH_3$
Leucomycin A_{13}	H	$CO(CH_2)_4CH_3$
Leucomycin A_{14}	$COCH_3$	$COCH_2CH_2CH(CH_3)_2$
Leucomycin A_{15}	H	$COCH_2CH_2CH(CH_3)_2$
SF-837-M_1	COC_2H_5	H

Deepoxy-carbomycin-A

Staphococcomycin

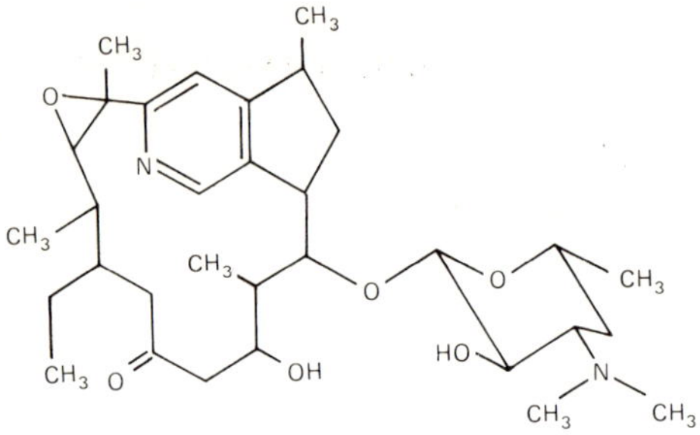

SCH-23831

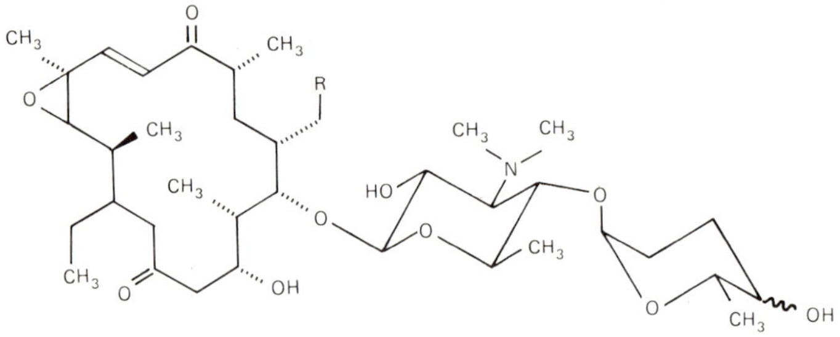

	R
A-6888	C CHO
A-6888	X CH_2OH

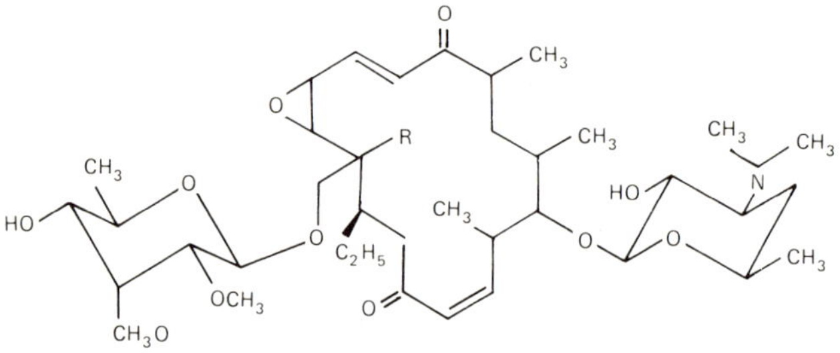

	R
Mycinamycin I	H
Mycinamycin II	OH

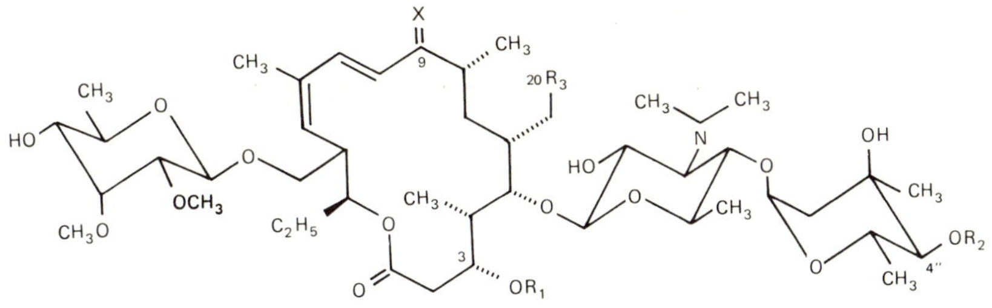

	R_1	R_2	R_3	X
9.20-Tetrahydrotylosin	H	H	CH_2OH	H,OH
3-Acetyl-4″-butyryltylosin	$COCH_3$	$CO(CH_2)_2CH_3$	CHO	=0
3-Acetyl-4″-isovaleryltylosin	$COCH_3$	$COCH_2CH(CH_3)_2$	CHO	=0
3-Propionyl-4″-butyryltylosin	COC_2H_5	$CO(CH_2)_2CH_3$	CHO	=0
3-Propionyl-4″-isovaleryltylosin	COC_2H_5	$COCH_2CH(CH_3)_2$	CHO	=0

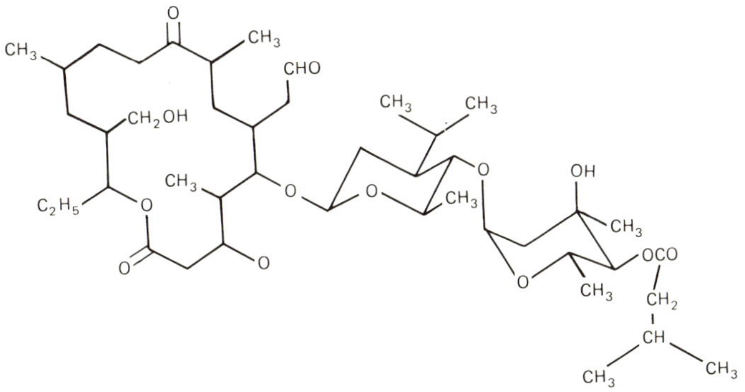

N-1

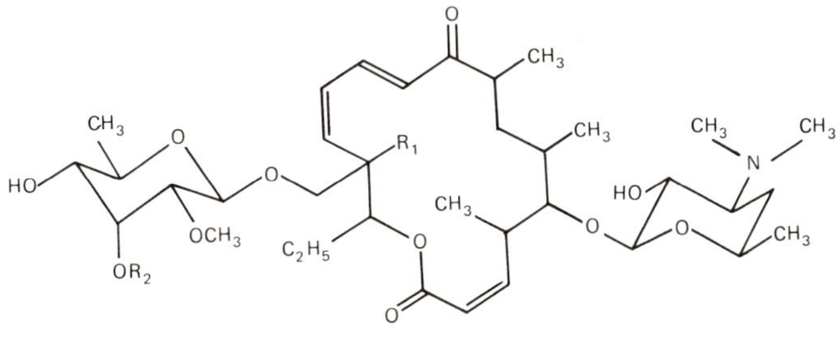

	R_1	R_2
Mycinamycin III	H	H
Mycinamycin IV	H	CH_3
Mycinamycin V	OH	CH_3

21211-6746

NAME: LEUCOMYCIN-A12, KITASAMYCIN-A12
PO: S.KITASATOENSIS
CT: BASIC, MACROLIDE, LEUCOMYCIN T.
FORMULA: C43H71NO15
EA: (N, 1.5)
MW: 841
PC: WH.
TO: (B.SUBT.,)
IS-EXT: (ETOAC, 8, FILT.)
REFERENCES:
AAC, 15, 738, 1979; Can. P 1046438

21211-6747

NAME: LEUCOMYCIN-A13, KITASAMYCIN-A13
CT: BASIC, MACROLIDE, LEUCOMYCIN T.
FORMULA: C41H69NO14
EA: (N, 1.5)
MW: 799
PC: WH.
TO: (B.SUBT.,)
REFERENCES:
AAC, 15, 738, 1979; Can. P 1046438

21211-6748

NAME: LEUCOMYCIN-A14, KITASAMYCIN-A14
PO: S.KITASATOENSIS
CT: BASIC, MACROLIDE, LEUCOMYCIN T.
FORMULA: C43H71NO15
EA: (N, 1.5)
MW: 841
PC: WH.
TO: (B.SUBT.,)
REFERENCES:
AAC, 15, 738, 1979; Can. P 1046438

21211-6749

NAME: LEUCOMYCIN-A15, KITASAMYCIN-A15
PO: S.KITASATOENSIS
CT: BASIC, MACROLIDE, LEUCOMYCIN T.
FORMULA: C41H69NO14
EA: (N, 1.5)
MW: 799
PC: WH.
TO: (B.SUBT.,)
REFERENCES:

AAC, 15, 738, 1979; Can. P 1046438

21211-6750

NAME: SF-837-MI, 4′-DEACYL-MYDECAMYCIN, 4′-DEACYL-SF-837-A1
PO: S.MYCAROFACIENS
CT: BASIC, MACROLIDE, LEUCOMYCIN T.
FORMULA: C38H62NO14
EA: (N, 2)
MW: 756
PC: WH., YELLOW, POW.
OR: (-61.8, CHL)(-61.8,)
UV: ETOH: (230, ,)
UV: MEOH: (230, ,)
TO: (S.AUREUS, 1.56)(S.LUTEA, .72)(B.SUBT., 6.25)
IS-EXT: (BUOAC, 9.5, FILT.)(W, 2.2, CHL)
REFERENCES:

JP 79/20197; *CA*, 202223

21221-7090

NAME: DEEPOXY-CARBOMYCIN-A
PO: B.SUBTILIS+CARBOMYCIN-A
CT: MACROLIDE, CARBOMYCIN T., BASIC
FORMULA: C42H69NO16
EA: (N, 2)
MW: 812
PC: WH., CRYST.
TO: (G.POS.,)
IS-EXT: (ETOAC, 8.5, W)
IS-CHR: (SILG, BENZ-ACET)
REFERENCES:

JP 79/81288; *CA*, 91, 191362

21222-7422

NAME: STAPHCOCCOMYCIN, DESMYCAROSYL-ANGOLAMYCIN
PO: S.SP.
CT: MACROLIDE, ANGOLAMYCIN T., BASIC
FORMULA: C39H65NO14
EA: (N, 2)
MW: 771
PC: WH., POW.
UV: ETOH: (240, 165,)
SOL-GOOD: MEOH, BENZ
SOL-FAIR: W
SOL-POOR: HEX
TO: (S.AUREUS, 1.56)(B.SUBT., 12.5)(S.LUTEA, .39)
LD50: (500|200, IV)
IS-EXT: (ETOAC-CHL, 7, FILT.)
IS-CRY: (PREC., CHL, HEX)
REFERENCES:

JA, 32, 1248, 1979

21223-6751

NAME: SCH-23831
PO: MIC.ROSARIA
CT: BASIC, MACROLIDE, CIRRAMYCIN T.
FORMULA: C31H48N2O7
EA: (N, 5)
MW: 560
PC: WH., POW.
OR: (-60.2, ETOH)
UV: MEOH: (229, , 8800)(269, , 2350)(278, , 2100)
SOL-GOOD: MEOH, BENZ
SOL-POOR: W
TO: (G.POS., 100)
REFERENCES:

TL, 2767, 1979

21223-7091

NAME: A-6888-C
PO: S.FLOCCULUS
CT: MACROLIDE, CIRRAMYCIN T., BASIC
FORMULA: C37H61NO12
EA: (N, 2)
MW: 711
PC: WH., CRYST.
UV: MEOH: (237, , 13400)(272, , 2700)
UV: MEOH-HCL: (237, , 13400)(272, , 3200)
UV: MEOH-NAOH: (239, , 16050)(283, , 2700)
SOL-GOOD: MEOH, BENZ
SOL-POOR: HEX, W
TO: (SHYG.,)(S.AUREUS, 1)(E.COLI,)
IS-EXT: (ETOAC, 7, FILT.)
IS-CHR: (SILG, CHL-MEOH)
REFERENCES:

Abst. AAC, 19, 1026, 1979; USP 4252898; JP 80/154994; *CA*, 93, 166021; 94, 137798

21223-7092

NAME: A-6888-X
PO: S.FLOCCULUS
CT: MACROLIDE, CIRRAMYCIN T., BASIC
FORMULA: C37H63NO12
EA: (N, 2)
MW: 713
PC: WH., CRYST.
UV: MEOH: (239, , 14900)
UV: MEOH-HCL: (239, , 14560)
UV: MEOH-NAOH: (239, , 17900)
SOL-GOOD: MEOH, BENZ
SOL-POOR: HEX, W
TO: (E.COLI,)(K.PNEUM.,)(SHYG.,)(S.AUREUS, 32)
IS-EXT: (ETOAC, 7, FILT.)
IS-CHR: (SILG, CHL-MEOH)
REFERENCES:

Abst. AAC, 19, 1026, 1979; USP 4252898; JP 80/154994; *CA*, 93, 166021; 94, 137798

21224-7093

NAME:	MYCINAMYCIN-I, A-11725-I
IDENTICAL:	AR5-1
PO:	MIC.GRISEORUBIDA, MIC.POLYTROTA
CT:	MACROLIDE, BASIC
FORMULA:	C37H61NO12
EA:	(N, 2)
MW:	711
PC:	WH., CRYST.
OR:	(-40, MEOH)
UV:	ETOH: (217, 329, 23400)(245, 170, 12070)
UV:	MEOH: (218, 340,)(240, 180,)(218, , 23700) (240, , 12900)
SOL-GOOD:	MEOH, BENZ, ACID
SOL-FAIR:	W, BASE
SOL-POOR:	HEX
QUAL:	(NINH., -)(SAKA., -)(EHRL., -)(BIURET, -)
TO:	(S.AUREUS, .05)(S.LUTEA, .05)(B.SUBT., .4) (E.COLI, 100)(SHYG., 100)(P.VULG., 100) (K.PNEUM., 25)(C.ALB., 256)(MYCOPLASMA SP., .006)
LD50:	NONTOXIC(140, IV)(310, IP)
IS-EXT:	(ETOAC, 7, FILT.)(W, 2.5, ETOAC)(CHL, 8.5, W)
IS-CHR:	(SILG, CHL-MEOH-NH4OH)(SILG, CHL-MEOH)
IS-CRY:	(CRYST., ACET-HEX)

REFERENCES:

JA, 33, 364, 1980; 34, 346, 619, 1981; *CC*, 119, 1980; DT 2918711; BP 2020647; EP 33433

21224-7094

NAME: MYCINAMYCIN-II, A-11725-II
IDENTICAL: AR5-2
PO: MIC.GRISEORUBIDA, MIC.POLYTROTA
CT: MACROLIDE, BASIC
FORMULA: C37H61NO13
EA: (N, 2)
MW: 727
PC: WH., CRYST.
OR: (-31, MEOH)(+22.9, ETOH)
UV: ETOH: (216, 324, 23390)(246, 138, 10000)
UV: MEOH: (217, 337,)(218, , 23200)(240, 180, 12200)
SOL-GOOD: MEOH, BENZ, ACID
SOL-FAIR: W, BASE
SOL-POOR: HEX
QUAL: (NINH., -)(SAKA., -)(EHRL., -)(BIURET, -)
TO: (S.AUREUS, .1)(S.LUTEA, .05)(B.SUBT., .4) (E.COLI, 25)(P.VULG., 50)(K.PNEUM., 25)(C.ALB., 256)(MYCOPLASMA SP., .031)
LD50: (310, IV)(610, IP)
IS-EXT: (ETOAC, , FILT.)(W, 2.5, ETOAC)(CHL, 8.5, W)
IS-CHR: (SILG, CHL-MEOH-NH4OH)(SILG, CHL-MEOH)
IS-CRY: (CRYST., ACET-HEX)
REFERENCES:

JA, 33, 364, 1980; 34, 346, 619, 1981; *CC*, 119, 1980; DT 2918711; BP 2020647; EP 33433

21231-7900

NAME: PLATENOMYCIN-W3, YL-704-W3
PO: S.PLATENSIS-MALVINUS
CT: MACROLIDE, CARBOMYCIN-B T., BASIC
FORMULA: C43H71NO15
EA: (N, 2)
MW: 841
PC: WH., POW.
UV: ETOH: (287, 220,)
SOL-GOOD: MEOH, BENZ
SOL-FAIR: W
TO: (G.POS.,)
IS-EXT: (ETOAC, 8, FILT.)(W, 2, ETOAC)(ETOAC, 8, W)
IS-CHR: (SILG, BENZ-MEOH)
REFERENCES:

JA, 27, 95, 1974; JP 72/20394; 76/22076

21232-7901

NAME:	9.20-TETRAHYDROTYLOSIN, S-2
PO:	S.AMBOFACIENS+TYLOSIN+CERULENIN
CT:	MACROLIDE, TYLOSIN T., BASIC
FORMULA:	C46H81NO17
EA:	(N, 1.5)
MW:	919
PC:	WH., CRYST.
UV:	MEOH: (231.5, , 26200)
TO:	(S.LUTEA, 5)(B.SUBT., 6)(S.AUREUS, 10)
IS-EXT:	(BENZ, , FILT.)
IS-CHR:	(SILG, CHL-MEOH-NH4OH)

REFERENCES:

JA, 33, 911, 1980

21232-8113

NAME:	3-ACETYL-4"-BUTYRYLTYLOSIN
PO:	S.THERMOTOLERANS+TYLOSIN
CT:	MACROLIDE, TYLOSIN T., BASIC
FORMULA:	C52H85NO19
EA:	(N, 1.5)
MW:	1027
PC:	WH., CRYST.
OR:	(-31, MEOH)
UV:	ETOH: (282, 240,)
SOL-GOOD:	ACID, MEOH, BENZ
SOL-POOR:	W, HEX
QUAL:	(NINH., -)(BIURET, -)(EHRL., -)
STAB:	(HEAT, +)
TO:	(S.AUREUS, .39)(B.SUBT., .78)(S.LUTEA, .19) (E.COLI, 100)

REFERENCES:

JA, 32, 542, 1979; 33, 1300, 1309, 1980; USP 4092473; EP 1841

21232-8114

NAME:	3-ACETYL-4"-ISOVALERYLTYLOSIN
PO:	S.THERMOTOLERANS+TYLOSIN
CT:	MACROLIDE, TYLOSIN T., BASIC
FORMULA:	C53H87NO19
EA:	(N, 1.5)
MW:	1041
PC:	WH., CRYST.
OR:	(-34.3, MEOH)
UV:	ETOH: (282, 222,)
SOL-GOOD:	ACID, MEOH, BENZ
SOL-POOR:	W, HEX
QUAL:	(NINH., -)(BIURET, -)(EHRL., -)
STAB:	(HEAT, +)
TO:	(S.AUREUS, .39)(B.SUBT., .39)(S.LUTEA, .19) (E.COLI, 100)

REFERENCES:

JA, 32, 542, 1979; 33, 1300, 1309, 1980; USP 4092473; EP 1841

21232-8115

NAME:	3-PROPIONYL-4"-BUTYRYLTYLOSIN
PO:	S.THERMOTOLERANS+TYLOSIN
CT:	MACROLIDE, TYLOSIN T., BASIC
FORMULA:	C53H87NO19
EA:	(N, 1.5)
MW:	1041
PC:	WH., CRYST.
OR:	(-35.2, MEOH)
UV:	ETOH: (282, 210,)
SOL-GOOD:	ACID, MEOH, BENZ
SOL-POOR:	W, HEX
QUAL:	(NINH., -)(BIURET, -)(EHRL., -)
STAB:	(HEAT, +)
TO:	(S.AUREUS, .39)(B.SUBT., .78)(S.LUTEA, .19) (E.COLI, 200)

REFERENCES:

JA, 32, 542, 1979; 33, 1300, 1309, 1980; USP 4092473; EP 1841

21232-8116

NAME:	3-PROPIONYL-4"-ISOVALERYLTYLOSIN
PO:	S.THERMOTOLERANS+TYLOSIN
CT:	MACROLIDE, TYLOSIN T., BASIC
FORMULA:	C54H89NO19
EA:	(N, 1.5)
MW:	1055
PC:	WH., CRYST.
OR:	(-33.1, MEOH)
UV:	ETOH: (282, 227,)
SOL-GOOD:	W, MEOH, BENZ
SOL-POOR:	W, HEX
QUAL:	(NINH., -)(BIURET, -)(EHRL., -)
STAB:	(HEAT, +)
TO:	(S.AUREUS, .39)(B.SUBT., .39)(S.LUTEA, .19) (E.COLI, 200)

REFERENCES:

JA, 32, 542, 1979; 33, 1300, 1309, 1980; USP 4092473; EP 1841

21232-8169

NAME:	N-1
PO:	S.THERMOTOLERAUS+MYCAVNINOSYL-TYLONOLIDE
CT:	MACROLIDE, TYLOSIN T., BASIC
FORMULA:	C45H73NO15
EA:	(N, 2)
MW:	867
PC:	WH., POW.
UV:	MEOH: (280, ,)
SOL-GOOD:	MEOH, BENZ, TOL
SOL-POOR:	W
TO:	(G.POS.,)
IS-FIL:	4
IS-EXT:	(TOL, 7, FILT.)(W, 4, TOL)(TOL, 7, W)
IS-CHR:	(SILG, BENZ-ACET-MEOH)

REFERENCES:

JP 80/43013; *CA*, 93, 68663

21234-7095

NAME: MYCINAMYCIN-IV, A-11725-IV
IDENTICAL: 12.13-DEEPOXY-AR5-1
PO: MIC.GRISEORUBIDA, MIC.POLYTROTA
CT: BASIC, MACROLIDE
FORMULA: C37H61NO11
EA: (N, 2)
MW: 695
PC: WH., CRYST.
OR: (+2.7, MEOH)(+7.2, ETOH)
UV: ETOH: (214, 290, 20160)(280, 295, 20500)
UV: MEOH: (215, 326, 20700)(281, , 21500)(283, 333,)
SOL-GOOD: MEOH, BENZ, ACID
SOL-FAIR: W, BASE
SOL-POOR: HEX
QUAL: (NINH., -)(FECL3, -)(SAKA., -)(BIURET, -)(EHRL., -)
TO: (S.AUREUS, .05)(S.LUTEA, .025)(B.SUBT., .4)(C.ALB., 256)(E.COLI, 75)(SHYG., 128)(MYCOPLASMA SP., .031)
LD50: (160, IV)(275, IP)
IS-EXT: (ETOAC, , FILT.)(W, 2.5, ETOAC)(CHL, 8.5, W)
IS-CHR: (SILG, CHL-MEOH-NH4OH)(SILG, CHL-MEOH)
IS-CRY: (CRYST., ACET-HEX)
REFERENCES:

JA, 33, 364, 1980; 34, 346, 619, 1981; *CC*, 119, 1980; DT 2918711; BP 2020647; EP 33433

21234-7096

NAME: MYCINAMYCIN-V, A-11725-V
IDENTICAL: 12.13-DEEPOXY-AR5-2
PO: MIC.GRISEORUBIDA, MIC.POLYTROTA
CT: MACROLIDE, BASIC
FORMULA: C37H61NO12
EA: (N, 2)
MW: 711
PC: WH., CRYST.
OR: (+18.7, MEOH)(+68.1, CHL)
UV: ETOH: (214, , 20067)(278, , 19702)
UV: MEOH: (215, 323, 20800)(280, 323, 21400)
SOL-GOOD: MEOH, BENZ, ACID
SOL-FAIR: BASE, W
SOL-POOR: HEX
QUAL: (NINH., -)(FECL3, -)(SAKA., -)(BIURET, -)
(EHRL., -)
TO: (C.ALB., 256)(S.AUREUS, .2)(S.LUTEA, .05)
(B.SUBT., 1.6)(E.COLI, 16)(MYCOPLASMA, .031)
IS-EXT: (ETOAC, , FILT.)(W, 2.5, ETOAC)(CHL, 2.5, W)
IS-CHR: (SILG, CHL-MEOH-NH4OH)(SILG, CHL-MEOH)
IS-CRY: (CRYST., ACET-HEX)
REFERENCES:

JA, 33, 364, 1980; 34, 346, 619, 1981; *CC*, 119, 1980; DT 2918711; BP 2020647; EP 33433

21234-7423

NAME: MYCINAMYCIN-III, A-11725-III
PO: MIC.GRISEORUBIDA
CT: MACROLIDE, BASIC
FORMULA: C36H59NO11
EA: (N, 2)
MW: 681
PC: WH., POW.
OR: (-2.3, MEOH)
UV: MEOH: (216, 310,)(283, 310,)
SOL-GOOD: MEOH, BENZ, ACID
SOL-FAIR: W, BASE
SOL-POOR: HEX
QUAL: (NINH., -)(SAKA., -)(FECL3, -)(BIURET, -)
(EHRL., -)
TO: (S.AUREUS, .2)(S.LUTEA, .05)(B.SUBT., 1.6)
(K.PNEUM., 100)
IS-EXT: (ETOAC, 7, FILT.)(W, 2.5, ETOAC)(CHL, 8.5, W)
IS-CHR: (SILG, CHL-MEOH-NH4OH)(SILG, CHL-MEOH)
IS-CRY: (CRYST., ACET-HEX)
REFERENCES:

JA, 33, 364, 1980; 34, 276, 1981; DT 2918711, BP 2020647

22
POLYENE ANTIBIOTICS

221	**Trienes**
222	**Tetraenes**
223	**Pentaenes**
224	**Hexaenes**
225	**Heptaenes**
226, 227, 228	**Octaenes and Other Polyenes**

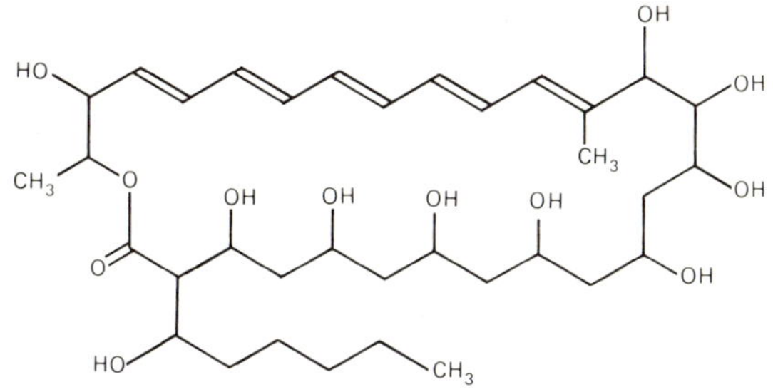

Elizabethin
(isomeric with lagosine)

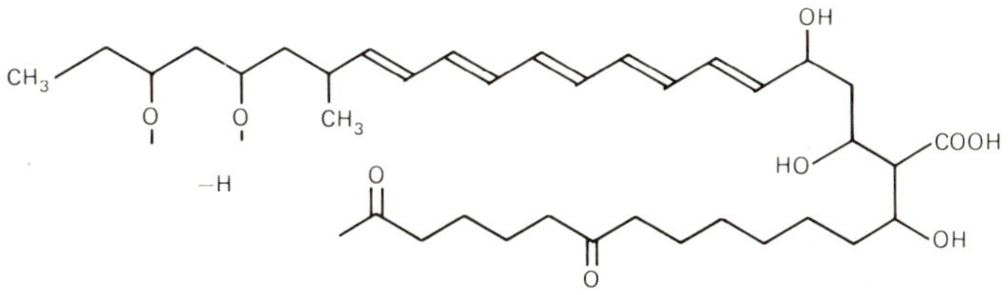

Gannibamycin (aglycone)

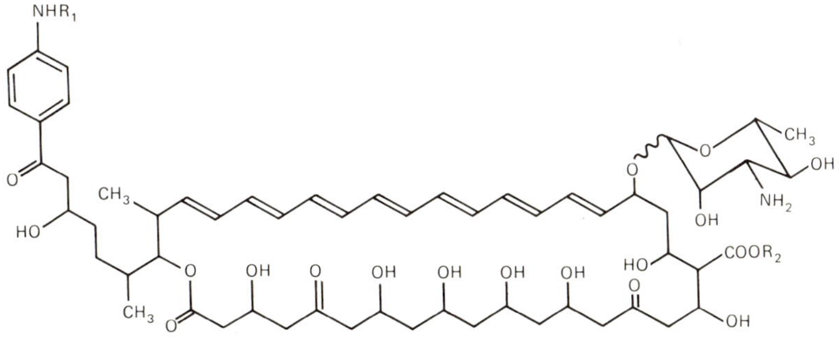

	R_1	R_2
Partricin A	CH_3	H
Particin B	H	H
Mepartricin	CH_3,H	CH_3

22110-8494

NAME:	M-5070
PO:	WESTERDYKEDIA DISPERSA
CT:	TRIENE
FORMULA:	C17H27NO4
EA:	(N, 4)
MW:	309
PC:	WH., YELLOW, POW.
OR:	(-82, MEOH)
UV:	MEOH: (265, 715,)(274, 900,)(285, 718,)
UV:	MEOH-HCL: (265, ,)(274, ,)(285, ,)
SOL-GOOD:	MEOH, ET2O
SOL-POOR:	W, HEX
QUAL:	(NINH., +)(FECL3, -)(FEHL., -)
TO:	(E.COLI, 500)(FUNGI,)
LD50:	(400, IP)
IS-EXT:	(CHL, 7.8, FILT.)
IS-CHR:	(SILG, CHL-MEOH)
IS-CRY:	(DRY, CHL)

REFERENCES:

JP 81/18592

22210-7638

NAME:	ALBICIDIN
PO:	S.SP.
CT:	TETRAENE, PIMARICIN T., AMPHOTER
FORMULA:	C32H51NO14
EA:	(N, 2)
MW:	673
PC:	YELLOW, POW.
OR:	(+65.6, PYR)
TO:	(C.ALB.,)(S.CEREV.,)
IS-CHR:	(BIOGEL P-2, W)

REFERENCES:

Abst. Pap. 7th Int. Symp. Med. Chem., 1980, 68

22200-7904

NAME:	TETRAENE BG-6, TETRAENE OS-1
PO:	S.SP.
CT:	TETRAENE
PC:	YELLOW
TO:	(PHYT.FUNGI,)

REFERENCES:

Ind. J. Exp. Biol., 18, 79, 1980; *CA*, 93, 180167

22200-7906

NAME: HA-236
PO: CHAINIA CINNAMOMEA
CT: TETRAENE, AMPHOTER
PC: YELLOW, POW.
UV: MEOH: (290, ,)(303, ,)(318, ,)
SOL-GOOD: MEOH
SOL-POOR: W
TO: (G.POS.,)(G.NEG.,)(FUNGI,)(C.ALB.,)
(PHYT.FUNGI,)
IS-EXT: (BUOH, , WB.)
REFERENCES:
HAB, 15, 1, 1972

22300-7905

NAME: PENTAENE EG-4, PENTAENE DG-15
PO: S.SP.
CT: PENTAENE
PC: YELLOW
TO: (PHYT.FUNGI,)
REFERENCES:
Ind. J. Exp. Biol., 18, 78, 1980; *CA*, 93, 180167

22310-8495

NAME: ELIZABETHIN
PO: S.ELIZABETHII
CT: PENTAENE, METHYLPENTAENE, NEUTRAL
FORMULA: C36H63O11
MW: 668
PC: YELLOW, POW.
TO: (FUNGI,)(C.ALB.,)
REFERENCES:
J. Chem. Tech. Biotech., 31, 167, 368, 1981; *CA*, 95, 78388, 201992

22320-7907

NAME: LIA-0371
PO: STV.SP.
CT: PENTAENE, AMPHOTER
EA: (N,)
PC: YELLOW, POW.
TO: (C.ALB.,)
LD50: (15, IP)
IS-EXT: (MEOH, , MIC.)
REFERENCES:
CA, 81, 176073

22320-8496

NAME: <u>GANNIBAMYCIN</u>
PO: S.SP.
CT: PENTAENE, EUSOCIDIN T.
EA: (C,)(H,)(N,)(O,)
PC: YELLOW, POW.
TO: (C.ALB.,)(FUNGI,)
REFERENCES:

Antib., 19, 1981

22511-7908

NAME: <u>PARTRICIN-B</u>
IDENTICAL: VACIDIN-A, AUREOFACIN, AYFACTIN
PO: S.AUREOFACIENS
CT: AROMATIC HEPTAENE, CANDICIDIN T., AMPHOTER
FORMULA: C58H84N2O19
EA: (N, 2.5)
MW: 1112
PC: YELLOW, POW.
TO: (FUNGI,)(PROTOZOA,)
REFERENCES:

JA, 33, 904, 1980; *ASM Sci. Meet.*, 1977

22511-7909

NAME: <u>HEPTAFUNGIN-B</u>
IDENTICAL: TRICHOMYCIN-B
PO: S.ACHROMO-LAVENDULAE, S.LONGISPORO-LAVENDULAE
CT: AROMATIC HEPTAENE, CANDICIDIN T., AMPHOTER
PC: YELLOW, POW.
UV: MEOH: (360, ,)(380, ,)(402, ,)
SOL-GOOD: DMFA, MEOH
SOL-POOR: W
TO: (C.ALB.,)(S.CEREV.,)
IS-EXT: (ACET, 7.8, MIC.)
REFERENCES:

Acta Micr. Hung., 19, 111, 1972; Hung. P 155813

22511-8497

NAME: "ANTIBIOTIC A"
PO: S.SP.
CT: AROMATIC HEPTAENE, CANDICIDIN T.
EA: (N,)
MW: 1100
PC: YELLOW, POW.
TO: (C.ALB.,)
REFERENCES:
J. Chrom., 208, 365, 1981

22511-8498

NAME: 67-121-D
PO: ACTINOPLANES CAERULEUS
CT: AROMATIC HEPTAENE, CANDICIDIN T., AMPHOTER
EA: (N,)
PC: YELLOW, POW.
UV: MEOH: (342, ,)(363, ,)(382, ,)(403, ,)
TO: (C.ALB.,)
IS-EXT: (MEOH, 2.5, MIC.)
REFERENCES:
USP 4223130

22520-7639

NAME: MYCOHEPTIN-A1
PO: S.NETROPSIS, STV.MYCOHEPTINICUM
CT: NONAROMATIC HEPTAENE, AMPHOTERICIN-B T., AMPHOTER
EA: (C, 59)(H, 8)(N, 2)
PC: YELLOW, POW.
OR: (+260, DMSO)
UV: MEOH: (347, 460,)(363, 860,)(383, 1400,) (406, 1580,)
SOL-GOOD: DMFA, DMSO
SOL-POOR: W, BUOH, HEX
TO: (C.ALB., .19)(FUNGI, .1)
LD50: (59.5, IP)
REFERENCES:
Antib., 499, 1980

22520-8499

NAME: LIA-0735
PO: S.COERULATUS
CT: NONAROMATIC HEPTAENE
EA: (C, 61)(H, 8)(N, 1.5)
PC: YELLOW, POW.
UV: MEOH: (364, 880,)(383, 1350,)(405, 1390,)
SOL-GOOD: DMSO, PYR
SOL-FAIR: MEOH
SOL-POOR: W, BUOH, HEX
TO: (C.ALB., 1.3)(S.CEREV., 1.3)(FUNGI,)
IS-EXT: (MEOH, , MIC.)
REFERENCES:
Antib., 8, 1981

22610-6752

NAME: OCTAMYCIN, 4041-II
PO: ACTINOPLANES IANTHIOGENES-OCTAMYCINI
CT: OCTAENE
FORMULA: C55H99NO26
EA: (N, 1)
PC: YELLOW, CRYST.
OR: (+4.4, PYR)
UV: ETOH: (360, 272,)(375, 392,)(398, 512,)
(420, 617,)
SOL-GOOD: PYR, ACOH-DMFA
SOL-FAIR: MEOH, ETOH
SOL-POOR: W, ACET, HEX
QUAL: (NINH., -)
TO: (G.POS.,)
TV: NK-LY
IS-EXT: (ACET, , MIC.)(BUOH, 7, W)
IS-CHR: (SILG, MEOH)
IS-CRY: (PREC., BUOH, HEX)(CRYST., PROH-W)
REFERENCES:
Antib., 582, 1979

22710-7097

NAME: NIGROFUNGIN, LIA-0167
PO: S.ALBOCYANEUS-NIGER
CT: OXO-PENTAENE, FLAVOFUNGIN T., NEUTRAL
EA: (C,)(H,)(O,)
PC: YELLOW, POW.
TO: (C.ALB.,)
IS-EXT: (BUOH-ETOAC, , WB.)
IS-CHR: (AL, CHL)
REFERENCES:
Antib., 736, 1979; *Nauch. Dokl. Vys. Skol., Biol. Nauk,* (4), 71, 1979

23
MACROCYCLIC LACTONE ANTIBIOTICS

231 Macrolide-Like Antibiotics
232 Simple Lactones
233 Dilactones
234 Cyclopolylactones
235 Condensed Macrolactones

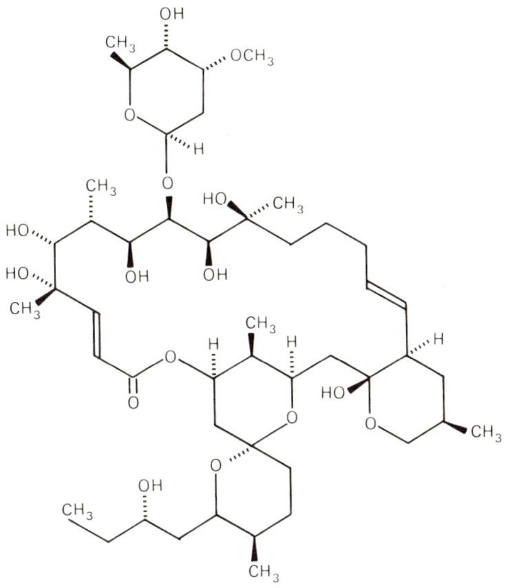

Cytovaricin

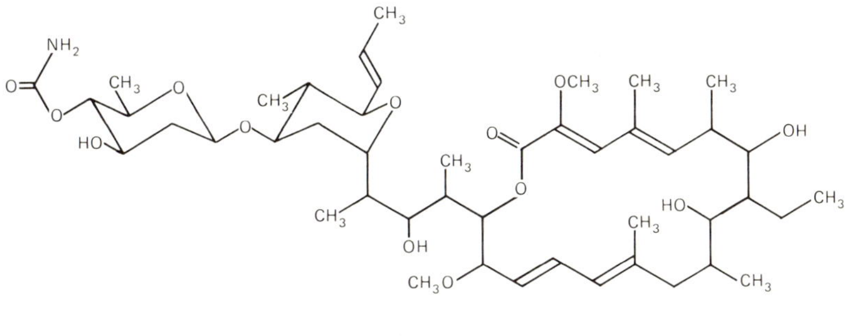

Concanamycin A

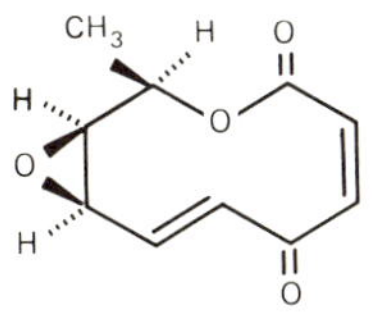

Pyrenolide A

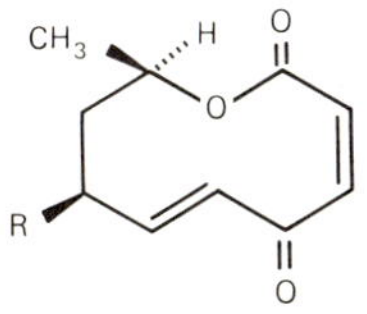

	R
Pyrenolide B	H
Pyrenolide C	OH

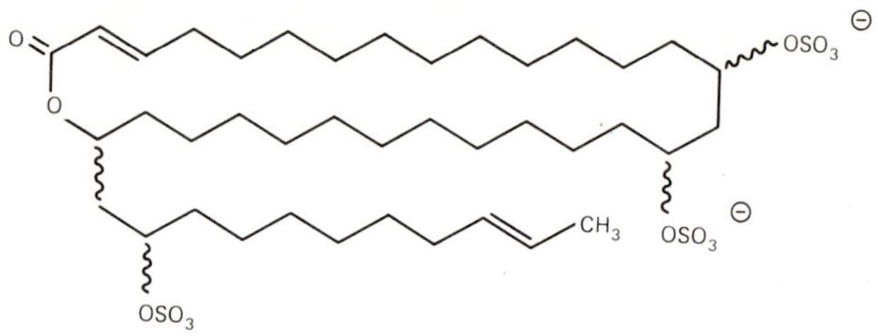

Izumenolide

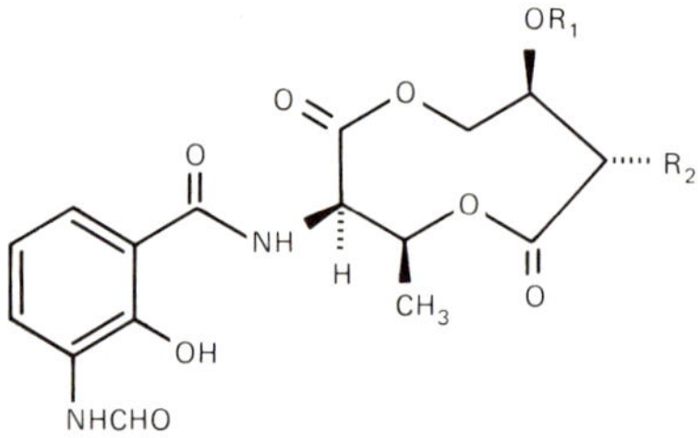

	R_1	**R_2**
Antimycin A_{oa}	$CO(CH_2)_4CH_3$	nC_6H_{13}
Antimycin A_{ob}	$CO(CH_2)_5CH_3$	nC_4H_9
Antimycin A_{oc}	$CO(CH_2)_2CH_3$	nC_8H_{17}
Antimycin A_{od}	$COCH_2CH(CH_3)_2$	nC_7H_{15}

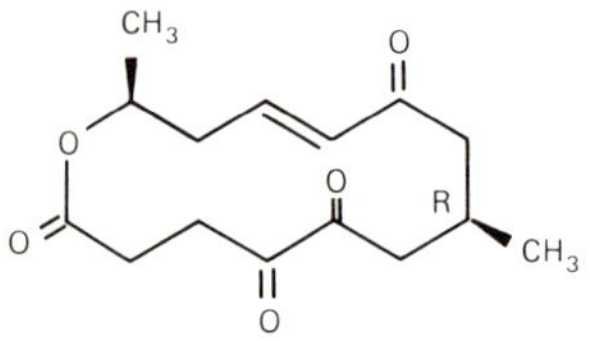

Grahamimycin A_1

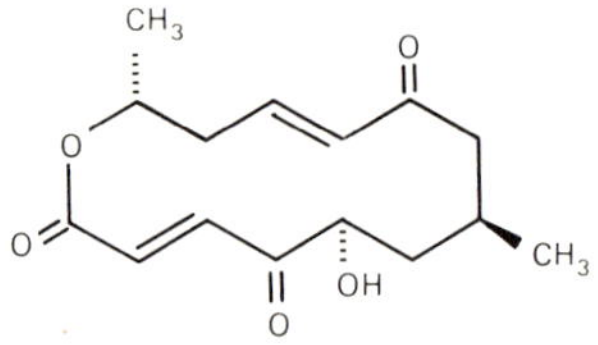

Grahamimycin A

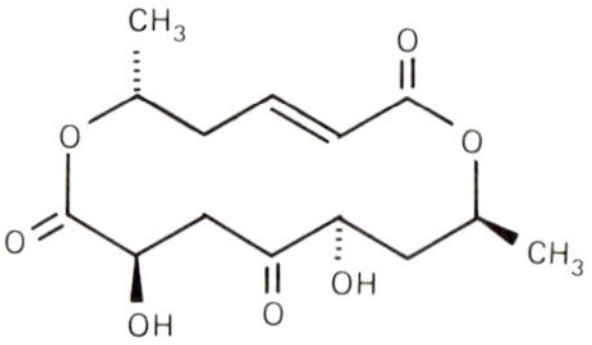

Grahamimycin B

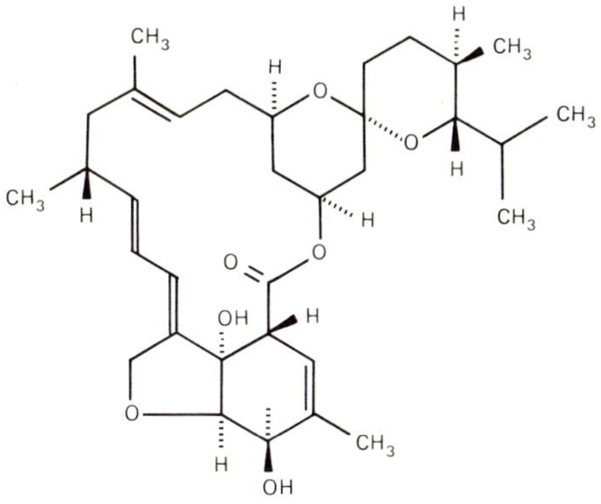

Milbemycin D (B-41-D)

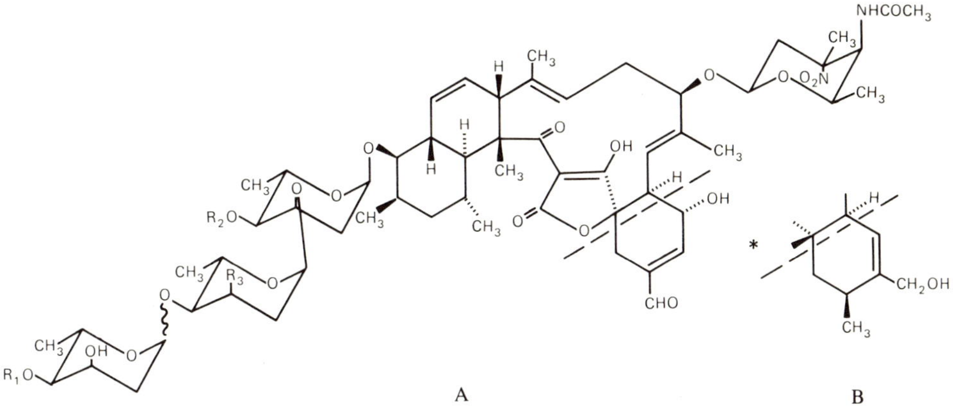

A B

	R_1	R_2	R_3	Str.
Tetrocarcin B	H	CH_3CO	H	A
Kijanimycin A*	CH_3	H	X	B

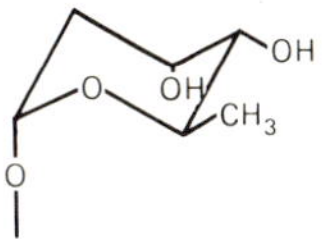

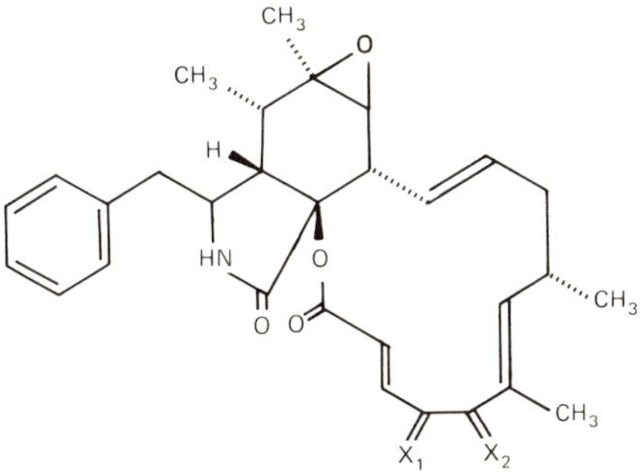

	X_1	X_2
Cytochalasin L	=O	||||H,▶OCOCH$_3$
Cytochalasin M	||||H,▶OH	=O

Zearalenone

Zeanol

Hypothemycin

Dihydrohypothemycin

Trans-Resorcylide

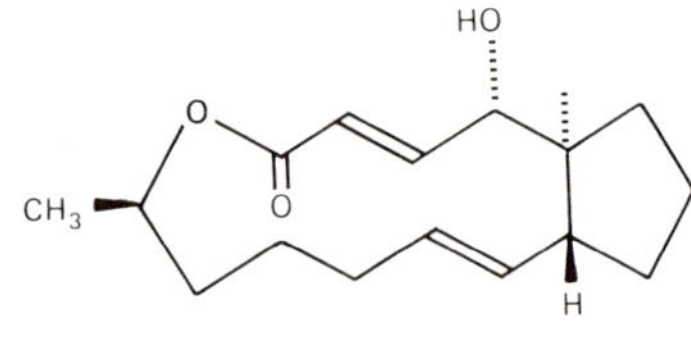

Brefeldin C

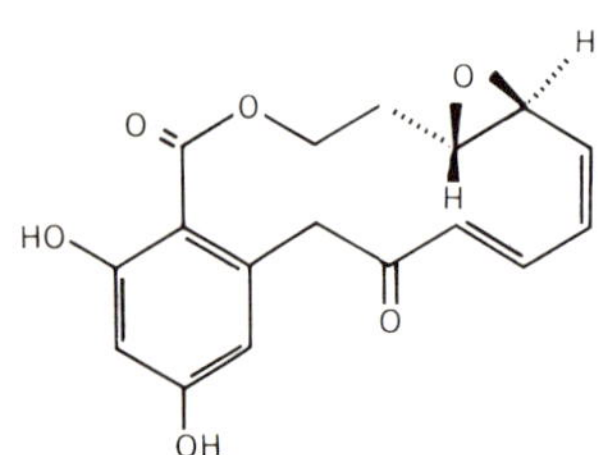

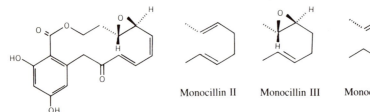

Monocillin I

Monocillin II

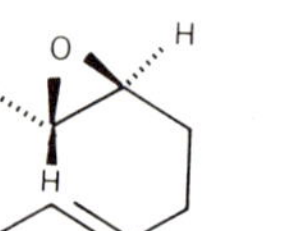

Monocillin III

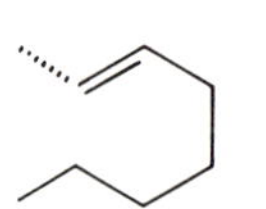

Monocillin IV

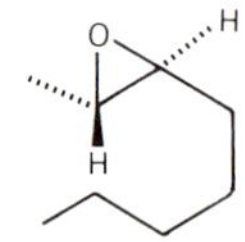

Monocillin V

Aspochalasin A

Aspochalasins B

R = H

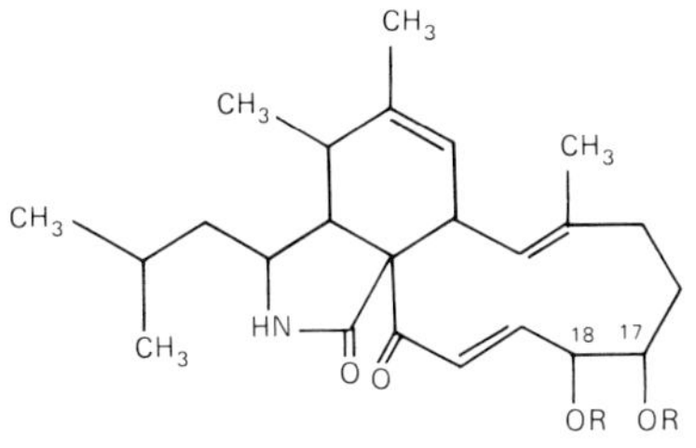

Aspochalasins C (29) and D (1)
1:R = H
29:epimer to 1 at C(17) or C(18)

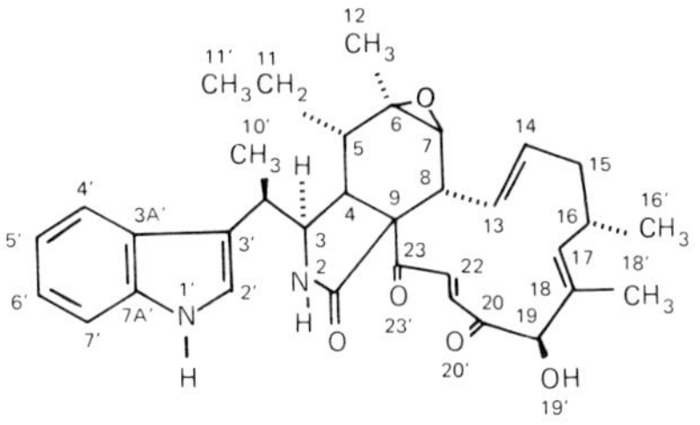

Structure and numbering system of chaetoglobosin K

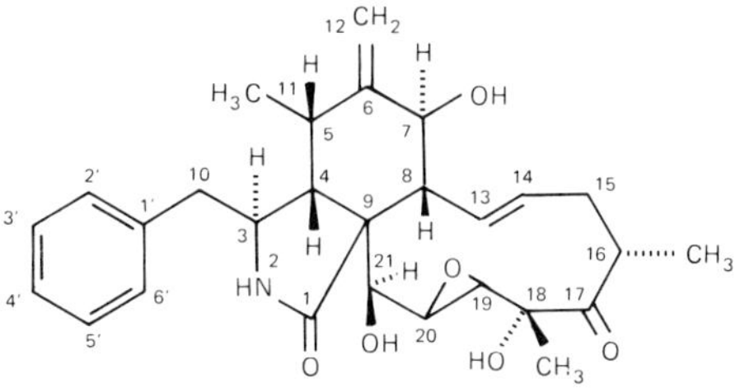

Engleromycin

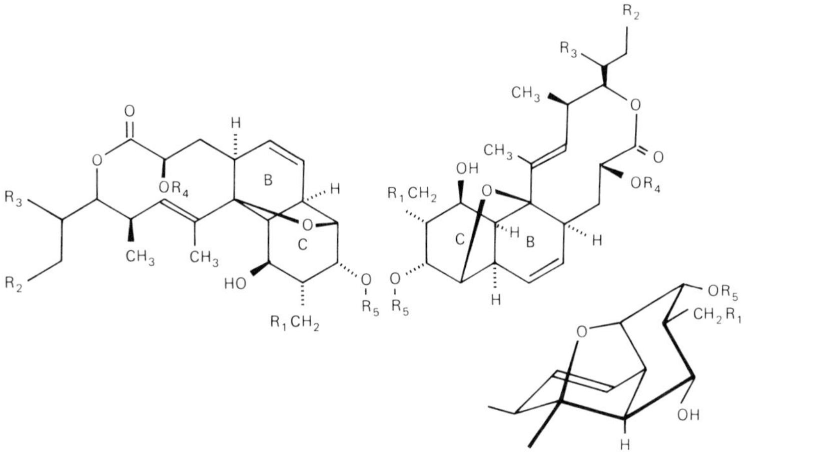

	R_1	R_2	R_3	R_4	R_5
Nargericin A_1(CP-47444)	H	H	OH	OCH_3	X
Nargericin B_1(CP-51467)	OCH_3	OCH_3	OH	OH	X
Nargericin B_2(CP-52726)	OCH_3	OH	H	OH	X
Nargericin B_3(CP-52748)	H	OCH_3	OH	OH	X
Nodusmycin	H	H	OH	OCH_3	H

X:

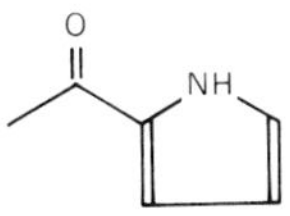

23110-8787

NAME:	CYTOVARICIN
IDENTICAL:	H-230
PO:	S.SP.
CT:	MACROLIDE L., OLIGOMYCIN T., NEUTRAL
FORMULA:	C47H80O16
MW:	900
PC:	WH., CRYST.
OR:	(+8.1, CHL)
UV:	MEOH: (200, ,)(215, , 11460)
SOL-GOOD:	MEOH, CHL, ACCN
SOL-FAIR:	ET2O, W
SOL-POOR:	HEX
QUAL:	(NINH., -)(FEHL., -)
TO:	(PHYT.FUNGI,)
LD50:	(2.3|.8, IP)
TV:	YOSHIDA
IS-EXT:	(ETOAC, 7.6, FILT.)
IS-CHR:	(SILG, CHL-MEOH)
IS-CRY:	(CRYST., ACCN)

REFERENCES:

JA, 33, 1073, 1981

23150-7098

NAME:	TENDOMYCIN, AB-99, TENDAMYCIN
PO:	S.HYGROSCOPICUS
CT:	MACROLIDE L., MELANOSPORIN T., BASIC
FORMULA:	C51H91N3O19
EA:	(N, 4)
MW:	1061
PC:	WH., POW.
OR:	(+21.6, DMSO)
UV:	MEOH: (218, 100,)
SOL-GOOD:	CHL-MEOH, DMSO
SOL-FAIR:	MEOH
SOL-POOR:	ETOH, HEX-W
QUAL:	(NINH., -)(SAKA., -)(DNPH, -)(EHRL., -)(FECL3, -)
TO:	(MYCOB.SP., 10)(C.ALB., 3)(S.CEREV., 10) (S.AUREUS, 6.25)(B.SUBT., 6.25)(S.LUTEA, 6.25)
LD50:	(50, IP)
TV:	ANTITUMOR, HELA, KB, EHRLICH
IS-EXT:	(ACET-W, , MIC.)(BUOH, , W)
IS-CHR:	(SEPHADEX LH-20, MEOH)
IS-CRY:	(PREC., CHL-MEOH, HEX)

REFERENCES:

Jap. J. Ant., 33, 227, 1980; JP 79/106401; *CA*, 92, 39810; 93, 68586

23150-8078

NAME:	SF-2077
PO:	S.HYGROSCOPICUS
CT:	NEUTRAL, MACROLIDE L., MELANOSPORIN T.
EA:	(C, 59)(H, 9)(N, 4)
MW:	1100
PC:	WH., CRYST.
OR:	(+6.2, MEOH)
UV:	MEOH: (200, ,)
SOL-GOOD:	MEOH, BUOH, PYR
SOL-FAIR:	ACET
SOL-POOR:	ETOAC, HEX, W
QUAL:	(NINH., -)(FECL3, -)
STAB:	(ACID, +)(BASE, +)(HEAT, +)
TO:	(B.SUBT., 100)(S.AUREUS, 25)(C.ALB., 3.13) (PHYT.FUNGI, .78)(FUNGI, 3.13)
LD50:	(7.5|2.5, IP)
IS-EXT:	(BUOH, , W)
IS-ION:	(DIAION HP-20, ACET-W)

REFERENCES:

JP 80/100396; *CA*, 94, 28882

23150-8171

NAME:	ENDOMYCIN COMPLEX
PO:	S.ANTIMYCOTICUS
CT:	BASIC
UV:	MEOH: (292.5, ,)(306, ,)(318, ,)(335, ,) (355, ,)(382, ,)
SOL-GOOD:	W, DMSO, MEOH, ETOH
SOL-POOR:	BUOH, HEX
STAB:	(HEAT, -)
TO:	(B.SUBT.,)(S.LUTEA,)(S.AUREUS,)(S.CEREV.,)
IS-ABS:	(CARBON,)

REFERENCES:

Phytoprotection, 61, 79, 1980; *CA*, 94, 99371

23150-8836

NAME:	<u>NIPHITHRICIN-A</u>
PO:	S.VIOLACEONIGER
CT:	MACROLIDE L., MELANOSPORIN T.
FORMULA:	C54H95N3O17
EA:	(N, 4)
MW:	1057
EW:	1128
PC:	WH., POW.
OR:	(+14, MEOH)
UV:	ETOH: (215, , 15800)
SOL-GOOD:	MEOH, PYR, DMFA, DMSO, BUOH
SOL-POOR:	W, ETOAC, HEX
TO:	(B.SUBT., 30)(S.AUREUS, 30)(S.CEREV., 10) (C.ALB., 10)(FUNGI, 3)(PHYT.FUNGI, 3)
IS-EXT:	(MEOH, , MIC.)(BUOH, , W)
IS-CHR:	(SILG, CHL-MEOH)(SEPHADEX LH-20, MEOH)
IS-CRY:	(PREC., MEOH, W)

REFERENCES:

JA, 34, 1107, 1981

23150-8837

NAME:	<u>NIPHITHRICIN-B</u>
PO:	S.VIOLACEONIGER
CT:	MACROLIDE L., MELANOSPORIN T.
FORMULA:	C54H95N3O17
EA:	(N, 4)
MW:	1057
EW:	1155
PC:	WH., POW.
OR:	(+15, MEOH)
UV:	ETOH: (215, , 15800)
SOL-GOOD:	MEOH, BUOH, PYR, DMFA, DMSO
SOL-POOR:	W, ETOAC, HEX
TO:	(B.SUBT., 100)(S.AUREUS, 100)(C.ALB., 100) (S.CEREV., 30)(FUNGI, 10)(PHYT.FUNGI, 10)

REFERENCES:

JA, 34, 1107, 1981

23160-7910

NAME: SCOPATHRICIN-II
PO: S.SP.
CT: MACROLIDE L., HUMIDIN T., NEUTRAL
FORMULA: C45H75NO15
EA: (N, 1.5)
MW: 809
PC: WH., CRYST.
UV: MEOH: (245, , 37200)(286, , 17880)
TO: (FUNGI,)
REFERENCES:
Abst. AAC, 13, 141, 1973

23160-8500

NAME: CONCANAMYCIN-A
PO: S.DIASTATOCHROMOGENES
CT: MACROLIDE-L., HUMIDIN T.
FORMULA: C46H75NO14
EA: (N, 2)
MW: 865
PC: WH., CRYST.
OR: (-21.7, MEOH)
UV: MEOH: (245, , 40500)(284, , 19300)
REFERENCES:
TL, 3857, 3861, 1981

23210-7100

NAME: PYRENOLIDE-A
PO: PYRENOPHORA TERES
CT: SIMPLE LACTONE, NEUTRAL
FORMULA: C10H10O4
MW: 194
PC: WH., CRYST.
OR: (-262, CHL)
UV: MEOH: (222, , 6700)(245, , 7200)
TO: (FUNGI, 1)
REFERENCES:
TL, 301, 1980; *Proc. 22nd Symp. Chem. Nat. Prod.*, 362, 1979; *CA*, 92, 177095

23210-8117

NAME: PYRENOLIDE-B
PO: PYRENOPHORA TERES
CT: SIMPLE LACTONE, NEUTRAL
FORMULA: C10H12O3
MW: 180
PC: WH., CRYST.
OR: (-49, CHL)
UV: MEOH: (222, , 6500)(239, , 6500)
SOL-GOOD: MEOH, BENZ
SOL-POOR: W
TO: (PHYT.FUNGI,)
IS-EXT: (ETOAC, , FILT.)
IS-CHR: (SILG, ETOAC-HEX)(SEPHADEX LH-20, ACET)
IS-CRY: (CRYST., ETOAC-HEX)
REFERENCES:

Agr. Biol. Ch., 44, 2761, 1980

23210-8118

NAME: PYRENOLIDE-C
PO: PYRENOPHORA TERES
CT: SIMPLE LACTONE, NEUTRAL
FORMULA: C10H12O4
MW: 196
PC: WH., OIL
OR: (+24, CHL)
UV: MEOH: (218, , 700)(238, , 6300)
SOL-GOOD: MEOH, BENZ
SOL-POOR: W
TO: (PHYT.FUNGI,)
IS-EXT: (ETOAC, , FILT.)
IS-CHR: (SILG, ETOAC-HEX)
IS-CRY: (DRY, ETOAC)
REFERENCES:

Agr. Biol. Ch., 44, 2761, 1980

23240-7426

NAME: IZUMENOLIDE, EM-4615-A
PO: MIC.CHALCEA-IZUMENSIS
CT: ACIDIC, MACROLACTONE
FORMULA: C40H74014S3
EA: (S, 11)
MW: 1100|200, 1294
PC: WH., POW.
OR: (-8.5, W)
UV: W: (214, 95,)
SOL-GOOD: W, MEOH
SOL-POOR: ACET, CHL
TO: (E.COLI, 62)(K.PNEUM., 250)(P.VULG., 250)
(PS.AER., 125)
LD50: (2, SC)
IS-EXT: (BUOH, 2, FILT.)(W, 7, BUOH)
IS-CHR: (SEPHADEX G-15, W)(SEPHADEX LH-20, MEOH)
REFERENCES:
JA, 33, 1256, 1262, 1560, 1980; *Tetr.*, 37(Suppl.), 275, 1981; DT 2920293; BP 2021096

23310-7911

NAME: ANTIMYCIN-AOA
PO: S.KITAZAWAENSIS
CT: ANTIMYCIN T., ACIDIC
FORMULA: C29H42N209
EA: (N, 5)
MW: 562
PC: WH.
TO: (S.CEREV.,)(C.ALB.,)
IS-EXT: (CH2CL2, , CL)
REFERENCES:
JA, 23, 75, 81, 1970; 26, 215, 1973; 29, 248, 265, 1976

23310-7912

NAME: ANTIMYCIN-AOC
PO: S.KITAZAWAENSIS
CT: ANTIMYCIN T., ACIDIC
FORMULA: C29H42N209
EA: (N, 5)
MW: 562
PC: WH.
TO: (S.CEREV.,)(C.ALB.,)
IS-EXT: (CH2CL, , MIC.)
REFERENCES:
JA, 23, 75, 81, 1970; 26, 215, 1973; 29, 248, 265, 1976

23310-7913

NAME: ANTIMYCIN-AOD
PO: S.KITAZAWAENSIS
CT: ANTIMYCIN T., ACIDIC
FORMULA: C29H42N2O9
EA: (N, 5)
MW: 562
PC: WH.
TO: (S.CEREV.,)(C.ALB.,)
IS-EXT: (CH2CL2, , MIC.)
REFERENCES:
JA, 23, 75, 81, 1970; 26, 215, 1973; 29, 248, 265, 1976

23310-7914

NAME: ANTIMYCIN-AOB
PO: S.KITAZAWAENSIS
CT: ANTIMYCIN T., ACIDIC
FORMULA: C28H40N2O9
EA: (N, 5)
MW: 548
PC: WH.
TO: (C.ALB.,)(S.CEREV.,)
REFERENCES:
JA, 23, 75, 81, 1970; 26, 215, 1973; 29, 248, 265, 1976

23320-7099

NAME: GRAHAMIMYCIN-A1
PO: CYTOSPORA SP.
CT: NEUTRAL, DILACTONE
FORMULA: C14H18O6
MW: 282
PC: YELLOW, CRYST.
OR: (-14.7, CHL)
UV: ETOH: (200, ,)(426, , 16)
SOL-GOOD: MEOH, HEX
SOL-POOR: W
TO: (B.SUBT., 1)(S.AUREUS, 1)
IS-EXT: (CHL, , FILT.)(CHL, , MIC.)
IS-CHR: (SILG, CH2CL2-ET2O)
IS-CRY: (CRYST., ET2O)
REFERENCES:
TL, 681, 1980; *AAC*, 19, 153, 1981; USP 4220718; *TL*, 159, 1982

23320-8119

NAME: GRAHAMIMYCIN-A
PO: CYTOSPORA SP.
CT: DILACTONE, NEUTRAL
FORMULA: C14H1806
MW: 282
PC: WH., CRYST.
OR: (-34, CHL)
UV: MEOH: (353, , 36)
SOL-GOOD: MEOH, ET2O
SOL-FAIR: W, CCL4
SOL-POOR: HEX
QUAL: (NINH., -)
STAB: (HEAT, +)
TO: (B.SUBT., 8)(S.LUTEA,)(SHYG.,)(K.PNEUM.,)
(PS.AER, .125)(P.VULG., 125)(S.CEREV., 15)
(S.AUREUS, 5)(E.COLI, 100)
LD50: (200|100, IP)
IS-EXT: (CHL, , FILT.)(CHL, , MIC.)
IS-CHR: (SILG, CH2CL2-ET2O)
IS-CRY: (PREC., CHL, HEX)(CRYST., CH2CL2-HEX)
REFERENCES:
AAC, 19, 153, 1981; USP 4220718; 4239690

23320-8120

NAME: GRAHAMIMYCIN-B
PO: CYTOSPORA SP.
CT: DILACTONE, NEUTRAL
FORMULA: C14H2007
MW: 300
PC: WH., CRYST.
OR: (-82.5, CHL)
UV: MEOH: (200, ,)
SOL-GOOD: MEOH, ET2O
SOL-FAIR: W, CCL4
SOL-POOR: HEX
QUAL: (NINH., -)
STAB: (HEAT, +)
TO: (B.SUBT., 80)(S.LUTEA,)(SHYG.,)(K.PNEUM.,)
(S.AUREUS, 50)(S.CEREV.,)(SHYG., 20)(PS.AER.,
50)(S.CEREV., 150)
IS-CHR: (SILG, CH2CL2-ET2O-MEOH)
IS-CRY: (CRYST., ET2O-CH2CL2)
REFERENCES:
AAC, 19, 153, 1981; USP 4220718; 4239690

23410-7915

NAME: SP-351-D
PO: S.PHACEOCHROMOGENES
CT: NONACTIN T., NEUTRAL, CYCLOPOLYLACTONE
MW: 778
PC: WH., CRYST.
UV: MEOH: (200, ,)
SOL-POOR: W
TO: (G.POS.,)(G.NEG.,)(C.ALB.,)
IS-CHR: (SILG, BENZ-ETOAC)
REFERENCES:
Agr. Biol. Ch., 37, 661, 1973

23512-8170

NAME: MILBEMYCIN-D, B-41-D
PO: S.CHATTANOOGAENSIS
CT: MACROLACTONE, MILBEMYCIN T., NEUTRAL
FORMULA: C33H48O7
MW: 556
PC: WH., POW.
UV: ETOH: (243, , 30500)
SOL-GOOD: MEOH, ETOAC
SOL-POOR: W
TO: INSECTICID, ACARICID
REFERENCES:
Holl. P 80/4791; JP 81/49308; *CA*, 95, 75483

23513-7193

NAME:	TETROCARCIN-A, DC-11
IDENTICAL:	ANTLERMICIN-A
PO:	MIC.CHALCEA
CT:	ACIDIC, MACROLACTONE L.
FORMULA:	C67H96N2O24
EA:	(N, 2)
MW:	1340|136, 1312
PC:	WH., POW.
OR:	(-74.3, ACET)
UV:	MEOH: (268, , 10400)(278, , 8960)
SOL-GOOD:	MEOH, CHL
SOL-FAIR:	BENZ, W
SOL-POOR:	ET2O, HEX
TO:	(S.AUREUS, 100)(B.SUBT., 1.6)
LD50:	(54, IP)
TV:	P-388, S-180
IS-ION:	(HP-10-ACET,)
IS-ABS:	(CARBON, ACET)
IS-CHR:	(SILG, CHL-MEOH)
IS-CRY:	(DRY,)

REFERENCES:

JA, 33, 668, 940, 946, 1980; *TL*, 2559, 1980; DT 2916155; BP 2019388; JP 79/138501; 80/79322; *CA*, 94, 2928

23513-7523

NAME:	ANTLERMICIN-A
IDENTICAL:	TETROCARCIN-A
PO:	MIC.CHALCEA-KAZUNOENSIS
CT:	ACIDIC, MACROLACTONE L.
FORMULA:	C67H96N2O24
EA:	(N, 2)
MW:	1306, 1312
EW:	1365
PC:	WH., POW.
OR:	(-67.6, MEOH)
UV:	MEOH: (232, 154,)(265, 95,)
UV:	MEOH-HCL: (257|4, 63,)
UV:	MEOH-NAOH: (235, 132,)(265, 96,)
SOL-GOOD:	MEOH, ETOAC
SOL-FAIR:	CHL, ET2O
SOL-POOR:	W, HEX
QUAL:	(DNPH, +)(FECL3, +)(FEHL., -)
TO:	(B.SUBT., .015)
LD50:	(37.5|12.5, IP)
TV:	YOSHIDA
IS-FIL:	7
IS-EXT:	(ETOAC, 7, FILT.)(MEOH, , MIC.)
IS-CHR:	(SILG, BENZ-MEOH)(SILG, ETOAC-ACET)(SILG, CHL-MEOH)
IS-CRY:	(PREC., ETOAC, HEX)

REFERENCES:

JA, 33, 244, 1980; JP81/29595; *CA*, 95, 40856

23513-7647

NAME: TETROCARCIN-B, DC-11-A2
PO: MIC.CHALCEA
CT: MACROLACTONE L.
FORMULA: C61H86N2O22
EA: (N, 2)
MW: 1198
PC: WH., POW.
OR: (-55.8, ACET)
UV: MEOH: (268, , 10700)
SOL-GOOD: MEOH, CHL
SOL-FAIR: BENZ-W
SOL-POOR: ET2O, HEX
QUAL: (DNPH, +)(FECL3, +)(NINH., -)
TO: (B.SUBT., .1)(S.AUREUS, 20)
LD50: (70|10, IP)
TV: S-180, P-388
IS-EXT: (ETOAC, , W)
IS-ION: (DIAION HP-20, ACET-W)
IS-ABS: (CARBON, ACET)
IS-CHR: (SILG, BENZ-ACET)
REFERENCES:
JA, 33, 668, 940, 946, 1980; *TL*, 2559, 1980; *Proc. 23rd Symp. Chem. Nat. Prod., Nagoya*, 76, 1980; *Proc. Ann. Meet. Agr. Ch. Soc. Jap.*, 1981, IJ-18; DT 2916155

23513-7648

NAME: TETROCARCIN-C, DC-11-A3
PO: MIC.CHALCEA
CT: MACROLACTONE L.
FORMULA: C69H98N2O25
EA: (N, 2)
MW: 1354
PC: WH., POW.
OR: (-62.5, CHL)
UV: MEOH: (268, , 10500)
SOL-GOOD: MEOH, CHL
SOL-FAIR: BENZ-W
SOL-POOR: ET2O, HEX
QUAL: (DNPH, +)(FECL3, +)(NINH., -)
TO: (B.SUBT., .1)(S.AUREUS, 30)
LD50: (70|10, IP)
TV: S-180, P-388
IS-CHR: (SILG, BENZ-ACET)
REFERENCES:
JA, 33, 668, 940, 946, 1980; DT 2916155

23513-7649

NAME: ANTLERMICIN-B
PO: MIC.CHALCEA-KAZUNOENSIS
CT: ACIDIC, MACROLACTONE L.
FORMULA: C63H92N2O23
EA: (N, 2)
MW: 1260
PC: WH., POW.
OR: (-67.1, MEOH)
UV: MEOH: (233, 154,)(265, 96,)
UV: MEOH-HCL: (257|5, 70,)
UV: MEOH-NAOH: (233, 148,)(265, 100,)(276, 93,)
SOL-GOOD: MEOH, ACET
SOL-FAIR: CHL, ET2O
SOL-POOR: HEX, W
QUAL: (FEHL., -)
TO: (B.SUBT., .05)
LD50: (100|50, IP)
TV: YOSHIDA, MYELO
IS-FIL: 7
IS-EXT: (ETOAC, 7, FILT.)
IS-CHR: (SILG, BENZ-MEOH)(SILG, ETOAC-ACET)
IS-CRY: (PREC., ETOAC, HEX)
REFERENCES:
JA, 33, 772, 1980; JP 81/108719; *CA*, 96, 4945

23513-7650

NAME: ANTLERMICIN-C
PO: MIC.CHALCEA-KAZUNOENSIS
CT: ACIDIC, MACROLACTONE L.
FORMULA: C48H68N2O18
EA: (N, 3)
MW: 930
PC: WH., POW.
OR: (-71.3, MEOH)
UV: MEOH: (240, 132,)(265, 102,)
UV: MEOH-HCL: (257, 86,)
UV: MEOH-NAOH: (239, 134,)(266, 106,)(275, 105,)
SOL-GOOD: MEOH, ETOAC
SOL-FAIR: CHL, ET2O
SOL-POOR: W, HEX
QUAL: (DNPH, +)(FEHL., -)
TO: (B.SUBT., .78)
LD50: (150|50, IP)
TV: YOSHIDA, MYELO
REFERENCES:
JA, 33, 772, 1980; JP 81/108719; *CA*, 96, 4945

23513-8517

NAME: KIJANMYCIN, KIJANIMICIN, SCH-25663
PO: ACTINOMADURA KIJANIATA
CT: ACIDIC, MACROLACTONE L.
FORMULA: C67H100N2O24
EA: (N, 2)
MW: 1316
PC: WH., POW.
OR: (-124, MEOH)(-116.4, MEOH)
UV: ETOH: (199, 328,)(242, 56.2,)(264, 63.1,)
(274, 60.7,)
UV: MEOH: (200, , 42832)(241, , 8946)(274, , 9446)
UV: MEOH-HCL: (205, , 38313)(258, , 9881)
UV: MEOH-NAOH: (236, , 14677)(276, , 12002)
SOL-GOOD: MEOH, ACET
SOL-POOR: W, HEX, ET2O
STAB: (ACID, -)(BASE, +)(HEAT, +)
TO: (B.SUBT., .13)(S.AUREUS, .64)(FUNGI, 17.5)
(S.LUTEA,)
LD50: (1000, SC)
TV: ANTITUMOR, P-388
IS-EXT: (ETOAC, , FILT.)
IS-CHR: (SILG, CH2CL2-MEOH)
IS-CRY: (DRY, CH2CL2)(PREC., ACET, ET2O-HEX)
REFERENCES:
JACS, 103, 3938, 3940, 1981; *JA*, 34, 1101, 1981; EP 3384

23513-8788

NAME: TETROCARCIN-C, DC-11-A3
PO: MIC.CHALCEA
CT: MACROLACTONE, ACIDIC
EA: (N, 2)
PC: WH., POW.
OR: (-62.5, CHL)
UV: MEOH: (268, ,)
TO: (B.SUBT.,)(S.AUREUS,)
LD50: (80|20, IP)
TV: ANTITUMOR
IS-ION: (HP-10, ACET)
IS-ABS: (CARBON, ACET-W)
IS-CHR: (SILG, CHL-MEOH)
REFERENCES:
JP 81/75500; *CA*, 95, 202061

23520-8502

NAME: CYTOCHALASIN-L
PO: CHALARIA MICROSPORA
CT: CYTOCHALASIN T., NEUTRAL
FORMULA: C32H37NO7
EA: (N, 2.5)
MW: 547
PC: WH., POW.
OR: (-165, ETOH)
UV: ETOH: (200, ,)
TV: CYTOTOXIC
REFERENCES:
TL, 2703, 1981

23520-8503

NAME: CYTOCHALASIN-M
PO: CHALARIA MICROSPORA
CT: CYTOCHALASIN T., NEUTRAL
FORMULA: C30H37NO6
EA: (N, 3)
MW: 507
PC: WH., CRYST.
OR: (+18.7, ETOH)
UV: ETOH: (235, , 11300)
TV: CYTOTOXIC
REFERENCES:
TL, 2703, 1981

23530-6718

NAME:	ZEARALENONE, F-2 TOXIN, FES
PO:	FUS.GRAMINEARUM, GIBBERELLA ZEAE, FUS.MONILIFORME
CT:	ACIDIC, BREFELDIN T.
FORMULA:	C18H22O5
MW:	318
PC:	WH., CRYST.
OR:	(-170.5, MEOH)
UV:	MEOH: (236, , 29700)(274, , 13909)(316, , 6020)
SOL-GOOD:	MEOH, ET2O, ACCN
SOL-FAIR:	HEX
SOL-POOR:	W
TO:	(B.SUBT.,)
LD50:	TOXIC

REFERENCES:

Adv. Appl. Micr., 14, 495, 1967; 22, 59, 1977; *Nature*, 196, 1328, 1962; *TL*, 3109, 1966; *Tetr.*, 24, 2443, 1968; *JOC*, 33, 4176, 1969; *Microb. Toxins*, 7, 107, 1971; *CR Ser. D*, 285, 201, 1977; *Can. J. Micr.*, 25, 421, 1979; *Mycopath.*, 67, 45, 1979; *IRCS Med. Sci.*, 8, 527, 850, 1980; 6, 491, 1978; 7, 204, 1979

23530-6719

NAME:	ZEAENOL
PO:	COCHLIOBOLUS LUNATA
CT:	ACIDIC
FORMULA:	C19H24O7
MW:	364
PC:	YELLOW, WH., POW.
OR:	(-92, MEOH)
UV:	ETOH: (237, , 31500)(274, , 13400)(316, , 6300)
TO:	(COCHLIOBOLUS SP.,)

REFERENCES:

CA, 90, 99737

23530-7559

NAME:	<u>HYPOTHEMYCIN</u>
PO:	HYPOMYCES TRICHOTHECOIDES
CT:	BREFELDIN T., ACIDIC
FORMULA:	C19H22O8
MW:	378
PC:	WH., CRYST.
OR:	(+109, MEOH)
UV:	MEOH: (220, , 38000)(267, , 14000)(307, , 7000)
TO:	(FUNGI, 30)(PROTOZOA, 30)(PHYT.FUNGI, 60)
IS-EXT:	(ETOAC, 7, FILT.)

REFERENCES:

TL, 2011, 1980; *Tetr.*, 37, 2445, 1981

23530-7560

NAME:	<u>TRANS-RESORCYLIDE</u>
PO:	P.SP.
CT:	BREFELDIN T., ACIDIC
FORMULA:	C19H18O5
MW:	290
PC:	WH., CRYST.
OR:	(+78, CHL)
UV:	MEOH: (218, , 27000)(296, , 2800)
QUAL:	(FECL3, +)
TO:	(FUNGI,)(PIRICULARIA ORYZAE, 12.5)

REFERENCES:

Proc. 22nd Symp. Chem. Nat. Prod., Fukuoka 1979, 362; *CA*, 92, 177095

23530-7641

NAME:	<u>BREFELDIN-C</u>, 7-DEOXYBREFELDIN-A
PO:	EUPENICILLIUM BREFELDIANUM
CT:	BREFELDIN T., NEUTRAL
FORMULA:	C16H24O3
MW:	264
PC:	WH., CRYST.
OR:	(+130.6, CHL)
IS-EXT:	(ETOAC, , FILT.)(CHL-MEOH, , MIC.)
IS-CHR:	(SILG, HEX-ETOAC)
IS-CRY:	(CRYST., MEOH)

REFERENCES:

Heterocycl., 13(Spec.), 267, 1979

23530-7642

NAME: MONOCILLIN-I, DECHLOROMONORDEN
PO: MONOCILLIUM NORDINII
CT: BREFELDIN T., ACIDIC
FORMULA: C18H18O6
MW: 330
PC: WH., OIL
UV: MEOH: (252, , 20000)(273, , 24200)(319, , 13000)
SOL-GOOD: CHL, ET2O
SOL-POOR: W
TO: (PHYT.FUNGI, 10)
IS-EXT: (ET2O, , FILT.)
IS-CHR: (SILG, CHL)
REFERENCES:
Can. J. Micr., 26, 766, 1980

23530-7643

NAME: MONOCILLIN-III
PO: MONOCILLIUM NORDINII
CT: BREFELDIN T., ACIDIC
FORMULA: C18H20O6
MW: 332
PC: WH., CRYST.
TO: (PHYT.FUNGI, 250)
REFERENCES:
Can. J. Micr., 26, 766, 1980

23530-7644

NAME: MONOCILLIN-II
PO: MONOCILLIUM NORDINII
CT: BREFELDIN T., ACIDIC
FORMULA: C18H20O5
MW: 316
PC: WH., CRYST.
TO: (PHYT.FUNGI,)
REFERENCES:
Can. J. Micr., 26, 766, 1980

23530-7645

NAME: MONOCILLIN-IV
PO: MONOCILLIUM NORDINII
CT: BREFELDIN T., ACIDIC
FORMULA: C18H22O5
MW: 318
PC: WH.
TO: (PHYT.FUNGI,)
REFERENCES:
Can. J. Micr., 26, 766, 1980

23530-7646

NAME: MONOCILLIN-V
PO: MONOCILLIUM NORDINII
CT: BREFELDIN T., ACIDIC
FORMULA: C18H22O6
MW: 334
PC: WH., OIL
TO: (PHYT.FUNGI,)
REFERENCES:
Can. J. Micr., 26, 766, 1980

23530-8504

NAME: +-DIHYDROHYPOTHEMYCIN
PO: HYPOMYCES TRICHOTECHOIDES
CT: BREFELDIN T., ACIDIC
FORMULA: C19H24O8
MW: 380
PC: WH., CRYST.
OR: (+,)
UV: MEOH: (220, , 31000)(265, , 14900)(305, , 7000)
SOL-GOOD: MEOH, ETOAC
SOL-POOR: W
STAB: (HEAT, -)(LIGHT, -)
TO: (FUNGI,)(PROTOZOA, 1)
IS-EXT: (ETOAC, , FILT.)
REFERENCES:
Tetr., 37, 2445, 1981

23540-6753

NAME: ASPOCHALASIN-A
PO: ASP.MIROCYSTICUS
CT: NEUTRAL, ZYGOSPORIN T.
FORMULA: C24H33NO4
EA: (N, 3)
MW: 399
PC: YELLOW, POW.
OR: (-20, CHL)
UV: ETOH: (265, ,)(410, ,)
TV: CYTOTOXIC
REFERENCES:
Helv., 62, 1501, 1979

23540-6754

NAME: ASPOCHALASIN-B
IDENTICAL: ASPOSTEROL
PO: ASP.MIROCYSTICUS
CT: NEUTRAL, ZYGOSPORIN T.
FORMULA: C24H33NO4
EA: (N, 3)
MW: 399
PC: WH., YELLOW, POW.
OR: (-118, CHL)
UV: ETOH: (225, , 9900)
TO: (G.POS.,)(G.NEG.,)
TV: CYTOTOXIC
IS-CHR: (SILG, BENZ-ETOAC)
REFERENCES:
Helv., 62, 1501, 1979

23540-6755

NAME: ASPOCHALASIN-C
PO: ASP.MIROCYSTICUS
CT: NEUTRAL, ZYGOSPORIN T.
FORMULA: C24H35NO4
EA: (N, 3)
MW: 401
PC: WH., POW.
OR: (-86, CHL)
UV: ETOH: (240, ,)
TV: CYTOTOXIC
REFERENCES:
Helv., 62, 1501, 1979

23540-6756

NAME: ASPOCHALASIN-D
PO: ASP.MIROCYSTICUS
CT: NEUTRAL, ZYGOSPORIN T.
FORMULA: C24H35NO4
EA: (N, 3)
MW: 401
PC: WH., CRYST.
OR: (-81, ETOH)
UV: ETOH: (248, , 9300)
TV: CYTOTOXIC
REFERENCES:
Helv., 62, 1501, 1979

23540-7561

NAME: CHAETOGLOBOSIN-K
PO: DIPLODIA MACROSPORA
CT: ZYGOSPORIN T., NEUTRAL
FORMULA: C34H40N2O5
EA: (N, 5)
MW: 556
PC: WH., CRYST.
UV: ETOH: (219, , 44800)
TV: ANTITUMOR
IS-EXT: (ETOAC, , FILT.)
IS-CHR: (SILG, BENZ-ACET)
REFERENCES:
J. Agr. Food Chem., 28, 139, 1980; *TL*, 1905, 1980

23540-8066

NAME: ENGLEROMYCIN
PO: ENGLEROMYCES GOETZEI
CT: ZYGOSPORIN T., NEUTRAL
FORMULA: C28H35NO6
EA: (N, 3)
MW: 481
PC: WH., CRYST.
OR: (+64, ETOH)
TV: CYTOTOXIC
IS-CRY: (CRYST., ETOH-W)
REFERENCES:
TL, 5079, 1980

23540-8505

NAME: CYTOCHALASIN-K, ACETYLCHAETOGLOBOSIN-A
PO: CHALARIA MICROSPORA
CT: ZYGOSPORIN T.
FORMULA: C32H37NO6
EA: (N, 3)
MW: 531
PC: WH., POW.
OR: (-127, ETOH)
UV: ETOH: (232, , 11300)
TV: CYTOTOXIC
IS-EXT: (ETOAC, , FILT.)
REFERENCES:
TL, 2109, 1973; 2703, 1981

23550-6893

NAME: NARGENICIN-A1, CP-47444, 47444
PO: NOC.ARGENTINENSIS
CT: MACROLACTONE, NARGERICIN T.
FORMULA: C28H37NO8
EA: (N, 2.5)
MW: 515
PC: WH., POW.
OR: (+49, MEOH)
UV: MEOH: (265, 320,)
SOL-GOOD: MEOH, CHL
SOL-POOR: W, HEX, HEPTAN
TO: (S.AUREUS, .1)(B.SUBT., 1.56)
IS-EXT: (ME.I.BU.KETON, , WB.)(ETOAC, , W)
IS-CHR: (SILG, CHL-ETOAC)
IS-CRY: (DRY, ETOAC)
REFERENCES:
JACS, 102, 4203, 1980; *Abst. AAC*, 20, 62, 1980; USP 4148883; 4224314; E 991

23550-7640

NAME: NODUSMICIN
PO: SACCHAROPOLYSPORA HIRSUTA
CT: MACROLACTONE, NEUTRAL, NARGERICIN T.
FORMULA: C23H34O7
MW: 422
PC: WH., CRYST.
OR: (+121, MEOH)
UV: MEOH: (200, ,)
SOL-GOOD: MEOH, CHL
SOL-POOR: W, HEX
TO: (S.AUREUS,)(S.LUTEA,)
IS-EXT: (CH2CL2, , FILT.)
IS-CRY: (CRYST., CHL)(CRYST., ET2O)
REFERENCES:
TL, 3659, 1980

23550-8121

NAME: NARGENICIN-B1, CP-51467
PO: NOC.ARGENTINENSIS
CT: MACROLIDE L., NARGERICIN T.
FORMULA: C29H39NO10
EA: (N, 2)
MW: 561
PC: WH., POW.
OR: (+80.9, MEOH)
SOL-GOOD: MEOH, CHL
SOL-POOR: W, HEX
TO: (S.AUREUS, .78)(S.LUTEA,)(E.COLI, 12.5)
IS-EXT: (ETOAC, 8, WB.)
IS-CHR: (SILG, HEPTAN-CHL)(SILG, CHL)(SEPHADEX LH-20, MEOH)
IS-CRY: (PREC., ETOAC, HEX)
REFERENCES:
Abst. AAC, 20, 62, 1980; USP 4224314

23550-8122

NAME: NARGENICIN-B2, CP-52726
PO: NOC.ARGENTINENSIS
CT: MACROLIDE L., NARGERICIN T.
FORMULA: C28H37NO9
EA: (N, 2)
MW: 531
PC: WH., YELLOW, POW.
UV: MEOH: (265, ,)
SOL-GOOD: MEOH, CHL
SOL-POOR: W, HEX
TO: (S.AUREUS,)(S.LUTEA,)
REFERENCES:
Abst. AAC, 20, 62, 1980; USP 4224314

23550-8123

NAME: NARGENICIN-B3, CP-52748
PO: NOC.ARGENTINENSIS
CT: MACROLIDE L., NARGERICIN T.
FORMULA: C28H37NO9
EA: (N, 2)
MW: 531
PC: WH., YELLOW, POW.
UV: MEOH: (235, ,)
TO: (S.AUREUS,)(S.LUTEA,)
REFERENCES:
Abst. AAC, 20, 62, 1980; USP 4224314

23000-7441

NAME: FA-2713-I
PO: S.HYGROSCOPICUS
CT: NEUTRAL, AZALOMYCIN-F T.
EA: (C, 60)(H, 8)(N, 4)
PC: WH., POW.
UV: MEOH: (240, 311,)(268, 185,)
SOL-GOOD: MEOH
SOL-POOR: W, ET2O, ACET, ETOAC
QUAL: (NINH., -)(FECL3, -)
STAB: (HEAT, -)
TO: (PHYT.FUNGI,)(S.AUREUS, 6.25)(B.SUBT., 6.25)
(S.CEREV., 3.1)(FUNGI, 12.5)(C.ALB., 3.1)
IS-EXT: (MEOH, , WB.)
IS-CRY: (PREC., HCL, FILT.)(PREC., MEOH, ETOAC)(PREC.
BUOH, W)
REFERENCES:
JP 79/151901; *CA*, 92, 196382

23000-7442

NAME:	FA-2713-II
PO:	S.HYGROSCOPICUS
CT:	NEUTRAL, AZALOMYCIN-F T.
EA:	(C, 62)(H, 8)(N, 5)
PC:	WH., POW.
UV:	MEOH: (240, 186,)(268, 126,)
SOL-GOOD:	MEOH
SOL-POOR:	W, ET2O, ACET, ETOAC
QUAL:	(NINH., -)(FECL3, -)
STAB:	(HEAT, -)
TO:	(FUNGI,)(PHYT.FUNGI,)(S.AUREUS, 12.5) (B.SUBT., 6.25)(S.CEREV., 6.25)(C.ALB., 6.25)
IS-EXT:	(MEOH, , WB.)
IS-CRY:	(PREC., HCL, FILT.)(PREC., MEOH, ETOAC)(PREC., BUOH, W)

REFERENCES:

JP 79/151901; *CA*, 92, 196382

23000-8501

NAME:	L-681110
PO:	S.SP.
CT:	MACROLIDE-L.
TO:	(FUNGI,)

REFERENCES:

Abst. AAC, 20, 87, 1980

24
MACROLACTAM ANTIBIOTICS

241 Ansamycins
242 Ansa-Lactams (Maytanosides)
243 Lactone-Lactams

Proansamycin B-M_1

	R_1	R_2	R_3	Str.*
3-Hydroxyrifamycin S	OH	OH	H	A
3,31-Dihydroxyrifamycin S	OH	OH	OH	A
1-Deoxy-1-oxarifamycin S	H	OH	H	B
8-Deoxyrifamycin S	H	H	H	A
8-Deoxy-3-hydroxyrifamycin S	OH	H	H	A
8-Deoxyrifamycin B	H	H	H	C

* Str.: structural fragment existing in the compound

Streptovaricin U

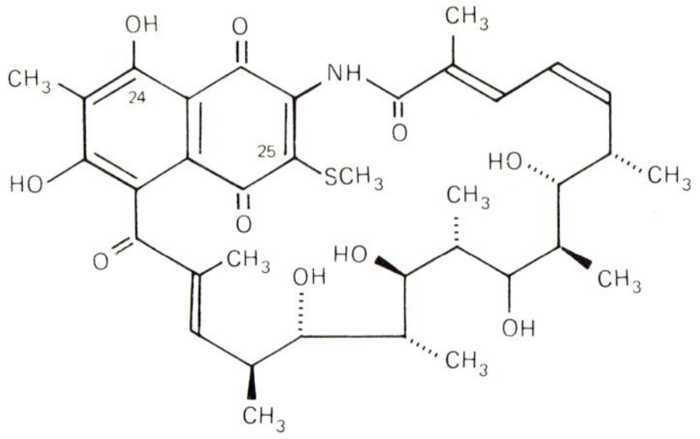

CP-50833

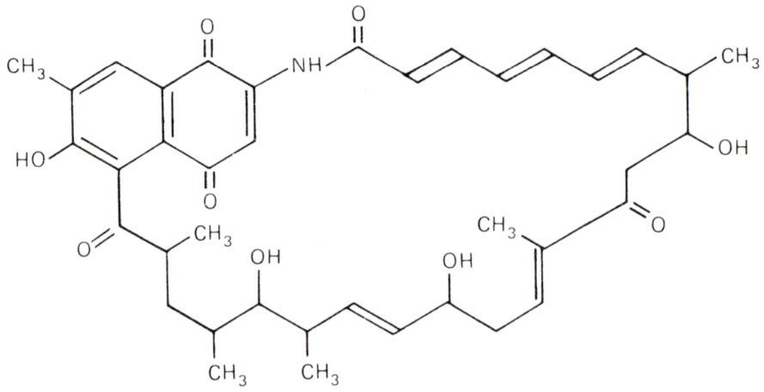

Actamycin

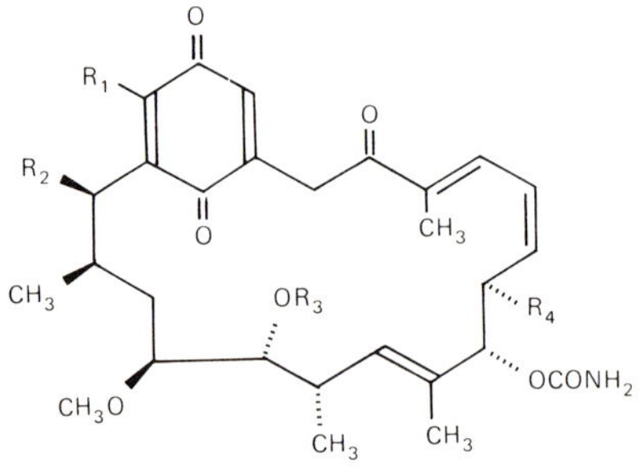

	R_1	R_2	R_3	R_4
Herbimycin A	H	OCH_3	CH_3	OCH_3
Herbimycin B	H	H	H	OCH_3
17-O-Demethylgeldanamycin	OH	H	H	OCH_3
Geldanamycin	OCH_3	H	H	OCH_3
Macbecin I	H	OCH_3	CH_3	CH_3

Ansatrienin B (Mycotrienin II)

Ansatrienin A*
(Mycotrienin I)

Structures of ansamitocin analogs.

C-15003	R_1	R_2	R_3	R_4
PND-4	$COCH_2CH(CH_3)_2$	H	H	H
PND-3	$COCH(CH_3)_2$	H	H	H
PND-2	$COCH_2CH_3$	H	H	H
PND-1	$COCH_3$	H	H	H
PND-0	H	H	H	H
PHM-4	$COCH_2CH(CH_3)_2$	OH	H	CH_3
PHM-3	$COCH(CH_3)_2$	OH	H	CH_3
PHM-2	$COCH_2CH_3$	OH	H	CH_3
PHM-1	$COCH_3$	OH	H	CH_3
P-4-βHY	$COCH_2C(OH)(CH_3)_2$	H	H	CH_3
P-4-γHY	$COCH_2CH(CH_2OH)CH_3$	H	H	CH_3
PND-4-βHY	$COCH_2C(OH)(CH_3)_2$	H	H	H
PHO-3	$COCH(CH_3)_2$	H	OH	CH_3
Ansamitocin P-4	$COCH_2CH(CH_3)_2$	H	H	CH_3
Ansamitocin P-3′	$COCH_2CH_2CH_3$	H	H	CH_3
Ansamitocin P-3	$COCH(CH_3)_2$	H	H	CH_3
Propionyl maytansinol (P-2)	$COCH_2CH_3$	H	H	CH_3
Maytanacine (P-1)	$COCH_3$	H	H	CH_3
Maytansinol (P-0)	H	H	H	CH_3

C-15003	X	Y
deClQND-0	H	H
QND-0	Cl	H
deClQ-0	H	CH_3

24110-7425

NAME: PROANSAMYCIN B-M1
PO: NOC.MEDITERRANEI
CT: ANSAMYCIN, RIFAMYCIN T., ACIDIC
FORMULA: C35H47NO10
EA: (N, 2.5)
MW: 641
PC: YELLOW, CRYST.
OR: (+18, MEOH)
UV: 8: (315, , 14000)(550, , 1300)
UV: ETOH: (212, , 59000)(278, , 50000)(312, , 26400)
UV: NAOH: (298, , 17500)(385, , 9600)
SOL-GOOD: MEOH, CHL
SOL-POOR: W
TO: (NEISSERIA MENIGITIS, 8)
IS-EXT: (ETOAC, 2.2, FILT.)
IS-CHR: (SILG, CHL-MEOH)
IS-CRY: (PREC., ETOAC, HEX)
REFERENCES:
JA, 32, 1267, 1979

24110-7916

NAME: 3-HYDROXYRIFAMYCIN-S
PO: NOC.MEDITERRANEI-CROSSING MUTANT
CT: ANSAMYCIN, RIFAMYCIN T., ACIDIC
FORMULA: C37H45NO13
EA: (N, 2)
MW: 711
PC: RED, POW.
OR: (+700, CHL)
UV: ETOH: (230, , 29600)(602, , 240)
UV: HCL: (267, , 14900)(342, , 4600)
UV: NAOH: (308, , 24500)(440, , 5200)
SOL-GOOD: MEOH, ET2O, DMSO, DMFA, THF, DIOXAN
SOL-POOR: W, HEX
TO: (G.POS.,)(S.AUREUS., .1)(E.COLI, 32)(K.PNEUM., 64)(PS.AER., 8)(S.LUTEA,)
IS-EXT: (CHL, 2.5, FILT.)
IS-CHR: (SILG, CHL-MEOH)
REFERENCES:
JA, 34, 965, 971, 1981; EP 14181

24110-7917

NAME: <u>3.31-DIHYDROXYRIFAMYCIN-S</u>
PO: NOC.MEDITERRANEI-CROSSING MUTANT
CT: ANSAMYCIN, RIFAMYCIN T., ACIDIC
FORMULA: C37H45NO14
EA: (N, 2)
MW: 727
PC: RED, POW.
SOL-GOOD: MEOH, ET2O, DIOXAN, DMSO, THF, DMFA
SOL-POOR: W, HEX
TO: (G.POS.,)(S.AUREUS, .2)(S.LUTEA,)(PS.AER., 64)
IS-EXT: (CHL, 2.5, FILT.)
IS-CHR: (SILG, CHL-MEOH)
REFERENCES:
JA, 34, 965, 971, 1981; EP 14181

24110-7918

NAME: <u>1-DEOXY-1-OXARIFAMYCIN-S</u>, 16.17-DEHYDRORIFAMYCIN-G
PO: NOC.MEDITERRANEI-CROSSING MUTANT
CT: ANSAMYCIN, RIFAMYCIN T., ACIDIC
FORMULA: C36H45NO12
EA: (N, 2)
MW: 683
PC: YELLOW, POW.
SOL-GOOD: MEOH, ET2O, DMSO, DIOXAN, THF, DMFA
SOL-POOR: W, HEX
TO: (G.POS.,)(S.AUREUS, 1)(S.LUTEA,)(PS.AER., 64)
IS-EXT: (CHL, 2.5, FILT.)
IS-CHR: (SILG, CHL-MEOH)
REFERENCES:
JA, 34, 965, 971, 1981; EP 14181

24110-8124

NAME: <u>8-DEOXYRIFAMYCIN-S</u>
PO: NOC.MEDITERRANEI
CT: ANSAMYCIN, RIFAMYCIN T., ACIDIC
FORMULA: C37H45NO11
EA: (N, 2)
MW: 679
TO: (G.POS., 500)(G.NEG.,)(C.ALB.,)
IS-EXT: (ETOAC, 2, FILT.)
IS-CHR: (SILG, CHL-MEOH)
REFERENCES:
JA, 33, 847, 1980; 34, 965, 1981

24110-8125

NAME: <u>8-DEOXY-3-HYDROXYRIFAMYCIN-S</u>
PO: NOC.MEDITERRANEI
CT: ANSAMYCIN, RIFAMYCIN T., ACIDIC
FORMULA: C37H45NO12
EA: (N, 2)
MW: 695
TO: (G.POS., 500)(G.NEG.,)
IS-EXT: (ETOAC, 2, FILT.)
IS-CHR: (SILG, TOL-ETOAC)
REFERENCES:
JA, 33, 847, 1980; 34, 965, 1981

24110-8126

NAME: <u>8-DEOXYRIFAMYCIN-B</u>
PO: NOC.MEDITERRANEI
CT: ANSAMYCIN, RIFAMYCIN T., ACIDIC
FORMULA: C39H49NO13
EA: (N, 2)
MW: 721
PC: ORANGE, POW.
TO: (G.POS., 500)(G.NEG.,)(C.ALB.,)
IS-EXT: (ETOAC, 2, FILT.)
IS-CHR: (SILG, CHL-MEOH)
REFERENCES:
JA, 33, 847, 1980; 34, 965, 1981

24120-7515

NAME: <u>STREPTOVARICIN-U</u>
PO: S.SPECTABILIS
CT: ANSAMYCIN, STREPTOVARICIN T., ACIDIC
FORMULA: C36H49NO10
EA: (N, 2)
MW: 655
PC: ORANGE, POW.
OR: (-110.2, MEOH)
TV: ANTIVIRAL
IS-CHR: (SILG, CHL-MEOH-W)(SILG, ETOAC-ETOH-W)
REFERENCES:
JA, 33, 249, 1980

24120-8127

NAME:	CP-50833, 21-HYDROXY-25-DEMETHYL-25-METHYLTHIOPROTOSTREPTOVARICIN
PO:	S.NIGELLUS-AFRICANUS
CT:	ANSAMYCIN, STREPTOVARICIN T., ACIDIC
FORMULA:	C36H47NO10S
EA:	(N, 2)(S, 5)
MW:	685
PC:	ORANGE, POW.
UV:	MEOH-NAOH: (237, 482,)(299, 301,)(412, 123,)
SOL-GOOD:	MEOH, CHL
SOL-POOR:	HEX
TO:	(S.AUREUS, .2)(S.LUTEA,)(E.COLI, 1.56)
IS-EXT:	(MIBK, , FILT.)
IS-CHR:	(SEPHADEX LH-20, MEOH)(SILG, ETOAC)
IS-CRY:	(PREC., ETOAC, HEX)

REFERENCES:

USP 4225674; 4247462

24120-8128

NAME:	ACTAMYCIN
PO:	S.SP.
CT:	ANSAMYCIN, STREPTOVARICIN T., ACIDIC
FORMULA:	C39H45NO10
EA:	(N, 2)
MW:	687
PC:	ORANGE, POW.
UV:	ETOH-HCL: (303, ,)(284, ,)(346, ,)(443, ,)
TO:	(G.POS.,)

REFERENCES:

TL, 1145, 1981; *JA*, 34, 605, 1981; *CC*, 768, 1980; *Proc. 10th Int. IUPAC Symp. Nat. Prod.*, 1976, C-22

24130-6757

NAME: HERBIMYCIN-A, HERBIMYCIN, AM-3672
PO: S.HYGROSCOPICUS
CT: NEUTRAL, ANSAMYCIN
FORMULA: C30H42N2O9
EA: (N, 5)
MW: 574
PC: YELLOW, CRYST.
OR: (+137, CHL)
UV: MEOH: (270, , 20090)(392.5, , 1650)
SOL-GOOD: ACET, CHL, ETOAC, DMSO, DMFA
SOL-FAIR: BENZ, MEOH, ETOH
SOL-POOR: W, ET2O, HEX
TO: (S.LUTEA, 200)(C.ALB., 200)(S.CEREV., 200)
IS-EXT: (ETOAC, 3, EVAP.FILT.)
IS-CHR: (SILG, ETOAC)
IS-CRY: (CRYST., MEOH)
REFERENCES:
JA, 32, 255, 1979; 33, 781, 1980; *TL*, 4323, 1979; JP 80/13063; 81/87512; *CA*, 93, 130603

24130-7101

NAME: 17-O-DEMETHYLGELDANAMYCIN
PO: S.HYGROSCOPICUS
CT: ANSAMYCIN, ACIDIC
FORMULA: C28H38N2O9
EA: (N, 5)
MW: 546
PC: PURPLE, CRYST.
TO: (G.POS.,)
TV: ANTITUMOR
REFERENCES:
JP 79/135300; *CA*, 92, 126908

24130-8067

NAME: HERBIMYCIN-B
PO: S.HYGROSCOPICUS
CT: ANSAMYCIN
FORMULA: C28H38N2O8
EA: (N, 5)
MW: 530
PC: YELLOW, CRYST.
OR: (+109, CHL)
UV: MEOH: (272, , 20000)(395, , 1900)
SOL-GOOD: MEOH, ETOH, ACET, DMSO, DMFA
SOL-FAIR: ETOAC, CHL, BENZ
SOL-POOR: W, ET2O, HEX
TO: (PHYT.FUNGI, 100)(HERBICID,)
LD50: (19, IP)
TV: TMV
IS-EXT: (ETOAC, , EVAP.FILT.)
IS-ABS: (CARBON, ETOAC)
IS-CRY: (CRYST., ETOAC)
REFERENCES:

JA, 33, 1114, 1980; *Nippon Kogaku Kaishi*, 5, 892, 1981; JP 81/87595; *CA*, 95, 24781, 185560

24130-8789

NAME: ANSATRIENIN-A
PO: S.COLLINUS
CT: ANSAMYCIN
FORMULA: C36H48N2O8
EA: (N, 4)
MW: 636
PC: YELLOW, POW.
OR: (+109.7, CHL)
UV: MEOH: (261, , 35800)(270, , 45500)(281, , 35500)(388, , 1500)
UV: MEOH-NAOH: (263, , 38000)(268, , 50000)(278, , 39000)(477, , 1700)
SOL-GOOD: MEOH, CHL, BASE
SOL-POOR: W, HEX
QUAL: (NINH., -)(FECL3, -)
TO: (S.CEREV., 50)(FUNGI, 5)(PHYT.FUNGI, 6)
IS-FIL: ORIG.
IS-EXT: (ACET, , MIC.)(ETOAC, , W)
IS-CHR: (SILG, BENZ-ETOAC)
IS-CRY: (PREC.ETOAC, HEX)
REFERENCES:

Zbl. Bakt. Parasit., C2, 122, 1981

24130-8790

NAME: ANSATRIENIN-B
PO: S.COLLINUS
CT: ANSAMYCIN
FORMULA: C36H50N2O8
EA: (N, 4)
MW: 638
PC: WH., POW.
OR: (+334.2, CHL)
UV: MEOH: (261, , 32500)(271, , 41500)(280, , 32200)(304, , 3500)
UV: MEOH-NAOH: (477, ,)
SOL-GOOD: MEOH, ETOAC
SOL-FAIR: CHL
SOL-POOR: W, HEX
QUAL: (NINH., -)(FECL3, -)
TO: (S.CEREV., 50)(FUNGI, 5)(PHYT.FUNGI,)
IS-CHR: (SILG, ETOAC)
REFERENCES:
Zbl. Bakt. Parasit., C2, 122, 1981

24210-8129

NAME: C-15003 PND-1, N-DEMETHYLANSAMITOCIN-P-1
PO: NOC.SP.
CT: ANSA-MACROLACTAM, MAYTANSIN T., NEUTRAL
FORMULA: C29H37N2O9CL
EA: (N, 5)(CL, 6)
MW: 593
PC: WH., POW.
OR: (-55.8, ETOH)
UV: MEOH: (232, , 31500)(239, , 32000)(279, 3780)(288, , 3700)
SOL-GOOD: MEOH, CHL, PYR, THF, DMSO
SOL-POOR: W, HEX
TO: (PROTOZOA,)(FUNGI,)
LD50: (1.5|.5, IP)
TV: P-388
IS-EXT: (ETOAC, , FILT.)(ACET-W, , MIC.)
IS-ION: (DIAION HP-10, MEOH-W)
IS-CHR: (SILG, CHL-MEOH)
IS-CRY: (PREC., ETOAC, HEX)(CRYST., ETOAC-ET2O)
REFERENCES:
JA, 34, 489, 496, 1981; EP 19934, 25898

24210-8130

NAME: C-15003 PND-2, N-DEMETHYLANSAMITOCIN-P-2
PO: NOC.SP.
CT: ANSA-MACROLACTAM, MAYTANSIN T., NEUTRAL
FORMULA: C30H39N2O9CL
EA: (N, 5)(CL, 6)
MW: 607
PC: WH., POW.
OR: (-56.3, ETOH)
UV: MEOH: (232, , 31000)(239, , 32000)(279, , 3800)(288, , 3760)
SOL-GOOD: MEOH, CHL, PYR, THF, DMSO
SOL-POOR: W, HEX
TO: (FUNGI,)(PROTOZOA,)
LD50: (1.5|.5, IP)
TV: P-388
REFERENCES:
JA, 34, 489, 496, 1981; EP 19934, 25898

24210-8131

NAME: C-15003 PND-3, N-DEMETHYLANSAMITOCIN-P-3
PO: NOC.SP.
CT: ANSA-MACROLACTAM, MAYTANSIN T., NEUTRAL
FORMULA: C31H41N2O9CL
EA: (N, 5)(CL, 5)
MW: 621
PC: WH., POW.
OR: (-57.1, ETOH)
UV: MEOH: (239, , 33000)(279, , 3880)(288, , 3790)
SOL-GOOD: MEOH, CHL, PYR, THF, DMSO
SOL-POOR: W, HEX
TO: (FUNGI,)(PROTOZOA,)
LD50: (1.5|.5, IP)
TV: P-388
IS-FIL: 4
IS-ION: (DX-50X2-NH4, NH4OH)(CG-50-H,)
IS-CHR: (SP-SEPHADEX C-25, PH4.7 BUFF.)(BIOGEL P-2, W)
IS-CRY: (LIOF.,)
REFERENCES:
JA, 34, 489, 496, 1981; EP 19934, 25898

24210-8132

NAME: C-15003 PND-4, N-DEMETHYLANSAMITOCIN-P-4
PO: NOC.SP.
CT: ANSA-MACROLACTAM, MAYTANSIN T., NEUTRAL
FORMULA: C32H41N2O9CL
EA: (N, 4)(CL, 5)
MW: 635
PC: WH., POW.
OR: (-56.6, ETOH)
UV: MEOH: (232, , 31500)(239, , 31100)(279, , 3690)(288, , 3760)
SOL-GOOD: MEOH, CHL, THF, PYR, DMSO
SOL-POOR: W, HEX
TO: (FUNGI,)(PROTOZOA,)
LD50: (1.5|.5, IP)
TV: P-388
IS-ION: (DX-50X2-NH4, NH4OH)(CG-50-H,)
IS-CHR: (SP-SEPHADEX C-25, PH4.7 BUFF.)(BIOGEL P-2, W)
IS-CRY: (LIOF.,)
REFERENCES:
JA, 34, 489, 496, 1981; EP 19934, 25898

24210-8506

NAME: C-15003-PHM-1, ANSAMITOCIN-PHM-1, 26-HYDROXYANSAMITOCIN-P1
PO: NOC.SP.
CT: ANSA-MACROLACTAM, MAYTANSIN T., NEUTRAL
FORMULA: C30H39N2O10CL
EA: (N, 4)(CL, 6)
MW: 622
PC: WH., POW.
UV: MEOH: (232, , 25800)(249, , 23900)(280, , 4210)(288, , 4250)
SOL-GOOD: PYR, DMFA, MEOH, THF
SOL-POOR: HEX, W
TO: (PROTOZOA, 64)
IS-EXT: (ETOAC, , FILT.)
IS-CHR: (SILG, CHL-MEOH)(SILG, HEX-ETOAC)
REFERENCES:
JA, 34, 489, 496, 1981; EP 26338

24210-8507

NAME: C-15003-PHM-2, ANSAMITOCIN-PHM-2
PO: NOC.SP.
CT: ANSA-MACROLACTAM, MAYTANSIN T., NEUTRAL
FORMULA: C31H41N2O10CL
EA: (N, 4)(CL, 6)
MW: 636
PC: WH., POW.
UV: MEOH: (232, , 24500)(249, , 23600)(280, 4200)(288, , 4240)
SOL-GOOD: PYR, DMFA, DMSO, CHL, THF
SOL-POOR: HEX, W
TO: (PROTOZOA, 64)
IS-EXT: (ETOAC, , FILT.)
IS-CHR: (SILG, CHL-MEOH)(SILG, HEX-ETOAC)
REFERENCES:
JA, 34, 489, 496, 1981; EP 26338

24210-8508

NAME: C-15003-PHM-3, ANSAMITOCIN-PHM-3
PO: NOC.SP.
CT: ANSA-MACROLACTAM, MAYTANSIN T., NEUTRAL
FORMULA: C32H43N2O10CL
EA: (N, 4)(CL, 6)
MW: 650
PC: WH., CRYST.
OR: (-148, ETOH)
UV: MEOH: (232, , 25600)(249, , 23700)(280, , 4190)(288, , 4250)
SOL-GOOD: PYR, DMFA, DMSO, MEOH, CHL, THF
SOL-POOR: HEX, W
TO: (PROTOZOA, 32)
IS-EXT: (ETOAC, , FILT.)
IS-CHR: (SILG, CHL-MEOH)(SILG, HEX-ETOAC)
REFERENCES:
JA, 34, 489, 496, 1981; EP 26338

24210-8509

NAME: C-15003-PHM-4, ANSAMITOCIN-PHM-4
PO: NOC.SP.
CT: ANSA-MACROLACTAM, MAYTANSIN T., NEUTRAL
FORMULA: C33H45N2O10CL
EA: (N, 4)(CL, 5)
MW: 664
PC: WH., CRYST.
OR: (-149, ETOH)
UV: MEOH: (232, , 25400)(249, , 23600)(280, , 4160)(288, , 4220)
SOL-GOOD: PYR, DMFA, DMSO, CHL, MEOH
SOL-POOR: HEX, W
TO: (PROTOZOA, 32)
IS-EXT: (ETOAC, , FILT.)
IS-CHR: (SILG, CHL-MEOH)(SILG, HEX-ETOAC)
REFERENCES:

JA, 34, 489, 496, 1981; EP 26338

24210-8510

NAME: C-15003-PHO-3, ANSAMITOCIN-PHO-3, 15-HYDROXYASAMITOCIN-P3, DESACETYLMAYTANBUTACIN
PO: NOC.SP., S.SP.+ANSAMITOCIN-P-3, CHAINIA SP.+ANSAMITOCIN P-3
CT: ANSA-MACROLACTAM, MAYTANSIN T., NEUTRAL
FORMULA: C32H43N2O10CL
EA: (N, 4)(CL, 6)
MW: 650
PC: WH., CRYST.
OR: (-96, ETOH)
UV: ETOH: (233, , 26600)(252, , 23100)(281, , 4520)(289, , 4520)
SOL-GOOD: PYR, DMFA, DMSO, THF, CHL, MEOH, ACET, ETOAC
SOL-POOR: HEX, W
TO: (PROTOZOA, 16)
IS-EXT: (ETOAC, , FILT.)
IS-CHR: (SILG, CHL-MEOH)(SILG, HEX-ETOAC)
REFERENCES:

JA, 34, 489, 496, 1981; *JOC*, 42, 2349, 1977; EP 28683

24210-8511

NAME: EPI-C-15003-PHO-3, EPI-ANSAMITOCIN-PHO-3
PO: NOC.SP., S.SP.+ANSAMITOCIN-P-3
CT: ANSA-MACROLACTAM, MAYTANSIN T., NEUTRAL
FORMULA: C32H43N2O10CL
EA: (N, 4)(CL, 6)
MW: 650
PC: WH., CRYST.
UV: ETOH: (233, , 28100)(253, , 25400)(281, , 5220)(289, , 5220)
SOL-GOOD: DMFA, PYR, DMSO, CHL, MEOH
SOL-POOR: HEX, W
TO: (PROTOZOA, 32)
IS-EXT: (ETOAC, , FILT.)
IS-CHR: (SILG, CHL-MEOH)(SILG, HEX-ETOAC)
REFERENCES:
JA, 34, 489, 496, 1981; *JOC*, 42, 2349, 1977; EP 28683

24210-8512

NAME: C-15003-PND-4-B"HY, ANSAMITOCIN-PND-4B"HY
PO: NOC.SP.
CT: ANSA-MACROLACTAM, MAYTANSIN T., NEUTRAL
FORMULA: C32H43N2O10CL
EA: (N, 4)(CL, 6)
MW: 650
PC: WH., POW.
UV: MEOH: (232, , 31500)(239, , 32000)(279, , 3800)(288, , 3700)
TO: (PROTOZOA, 32)
IS-EXT: (ETOAC, , FILT.)
IS-CHR: (SILG, CHL-MEOH)(SILG, HEX-ETOAC)
REFERENCES:
JA, 34, 489, 496, 1981

24210-8513

NAME: C-15003-P-4-B"HY, ANSAMITOCIN-P-4-B"HY
PO: NOC.SP.
CT: ANSA-MACROLACTAM, MAYTANSIN T., NEUTRAL
FORMULA: C33H45N2O10CL
EA: (N, 4)(CL, 5)
MW: 664
PC: WH., CRYST.
UV: MEOH: (231, , 30100)(251, , 27500)(280, , 5650)(288, , 5630)
TO: (PROTOZOA, 16)
IS-EXT: (ETOAC, , FILT.)
IS-CHR: (SILG, CHL-MEOH)(SILG, HEX-ETOAC)
REFERENCES:
JA, 34, 489, 496, 1981

24210-8514

NAME: C-15003-P-G"HY, ANSAMITOCIN-P-4-G"HY
PO: NOC.SP.
CT: ANSA-MACROLACTAM, MAYTANSIN T., NEUTRAL
FORMULA: C33H45N2O10CL
EA: (N, 4)(CL, 5)
MW: 664
PC: WH., POW.
UV: MEOH: (232, , 30000)(252, , 27300)(280, , 5630)(288, , 5610)
TO: (PROTOZOA, 16)
IS-EXT: (ETOAC, , FILT.)
IS-CHR: (SILG, CHL-MEOH)(SILG, HEX-ETOAC)
REFERENCES:
JA, 34, 489, 496, 1981

24210-8515

NAME: C-15003-QND-O
PO: NOC.SP.
CT: ANSA-MACROLACTAM, MAYTANSIN T., NEUTRAL
FORMULA: C27H35N2O7CL
EA: (N, 5)(CL, 7)
MW: 534
PC: WH., POW.
UV: MEOH: (231, , 30300)(240, , 30000)(251, , 25800)(279, , 3600)(288, , 3560)
TO: (PROTOZOA, 32)
IS-EXT: (ETOAC, , FILT.)
IS-CHR: (SILG, CHL-MEOH)(SILG, HEX-ETOAC)
REFERENCES:
JA, 34, 489, 496, 1981

24210-8516

NAME: C-15003-DECL-QND-O
PO: NOC.SP.
CT: ANSA-MACROLACTAM, MAYTANSIN T., NEUTRAL
FORMULA: C27H36N2O7
EA: (N, 6)
MW: 500
PC: WH., POW.
UV: MEOH: (218, , 39800)(243, , 35400)(280, , 2800)(288, , 2500)
TO: (PROTOZOA, 64)
IS-EXT: (ETOAC, , FILT.)
IS-CHR: (SILG, CHL-MEOH)(SILG, HEX-ETOAC)
REFERENCES:
JA, 34, 489, 496, 1981

24220-6720

NAME: RUBRADIRIN-C
PO: S.ACHROMOGENES-RUBRADIRIS
CT: ACIDIC
FORMULA: C40H33N3014
EA: (N, 5)
MW: 779
PC: RED, CRYST.
UV: MEOH-H2SO4: (203, , 86205)(220, , 41644)(271, , 26239)(320, , 30400)(530, ,)
UV: MEOH-NAOH: (302, , 35818)(410, , 14100)
SOL-GOOD: BASE, DMFA, DMSO
SOL-FAIR: MEOH, ETOH, THF, CHL
SOL-POOR: W, BENZ, TOL, HEX
TO: (S.AUREUS,)(B.SUBT.,)
IS-FIL: 4
IS-EXT: (ETOAC, , FILT.)
IS-CHR: (SILG, CHL-MEOH)
IS-CRY: (PREC., ETOAC, HEX)(CRYST., DMSO)
REFERENCES:
USP 4107296; *CA*, 90, 119747

24220-7102

NAME: CP-48926
PO: ACTINOPLANES DECCANENSIS
CT: ANSA-MACROLACTAM, RUBRADIRIN T., ACIDIC
EA: (C, 56)(H, 5)(N, 5)(O, 33)
PC: RED, POW.
OR: (+413, CHL)
UV: MEOH: (211, 615,)(304, 339,)(500, 17,)
SOL-GOOD: CHL, ETOAC
SOL-POOR: W, HEX
TO: (S.AUREUS,)(B.SUBT.,)
IS-EXT: (ME.I.BU.KETONE, 5.5, WB.)(W, 10, ME.I.BU.KETONE)
IS-CHR: (SILG, CHL-MEOH)
REFERENCES:
USP 4169887

24220-7103

NAME: CP-48927
PO: ACTINOPLANES DECCANENSIS
CT: ANSA-MACROLACTAM, RUBRADIRIN T., BASIC
EA: (C, 55)(H, 5)(N, 5)(O, 35)
PC: RED, POW.
OR: (+513, CHL)
UV: MEOH: (303, 367,)(500, 20,)
SOL-GOOD: CHL, ETOAC
SOL-POOR: W, HEX
TO: (S.AUREUS,)(B.SUBT.,)
REFERENCES:
USP 4169887

24300-8172

NAME: STUBOMYCIN
PO: S.SP.
CT: LACTONE-LACTAM
FORMULA: C29H35NO5
EA: (N, 3)
MW: 477
PC: WH., CRYST.
OR: (+246, DMSO)
UV: MEOH: (300, 930,)
UV: MEOH-HCL: (300, 930,)
UV: MEOH-NAOH: (300, 930,)
SOL-GOOD: DMSO, PYR, DMFA
SOL-FAIR: MEOH
SOL-POOR: W, CHL, ET2O
QUAL: (FECL3, +)(DNPH, +)(NINH., -)(EHRL., -)
TO: (B.SUBT., .4)(S.LUTEA, .4)(S.AUREUS, 3.1)
(FUNGI, 6.3)(PHYT.FUNGI, 3.1)
LD50: (500, IP)
TV: HELA, EHRLICH, P-388
IS-EXT: (ETOAC, , FILT.)(MEOH, 2, MIC.)
IS-ION: (DIAION HP-20, MEOH)
IS-CHR: (SEPHADEX LH-20, MEOH)(SILG, CH2CL2)
IS-CRY: (PREC., ETOAC, HEX)(CRYST., MEOH)
REFERENCES:
JA, 34, 259, 1981

Quinone and Similar Antibiotics

31
TETRACYCLIC COMPOUNDS AND ANTHRAQUINONES

311 Tetracyclines and Similar Antibiotics
312 Anthracyclines
313 Anthraquinone Derivatives

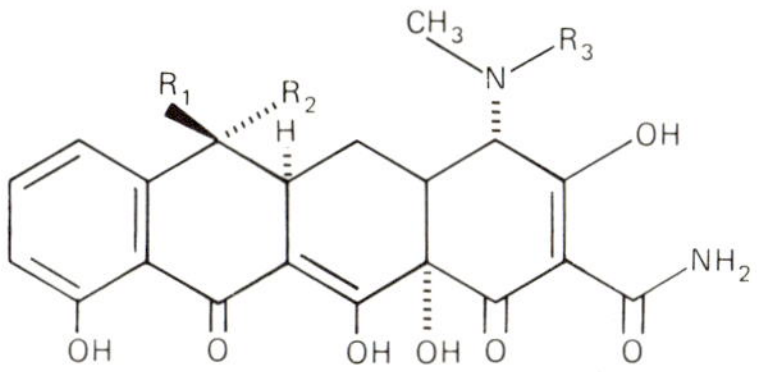

	R_1	R_2	R_3
6-Deoxytetracycline	H	CH_3	CH_3
4-N-Demethyl-4-N-ethyl-6-demethyltetracycline	OH	H	C_2H_5

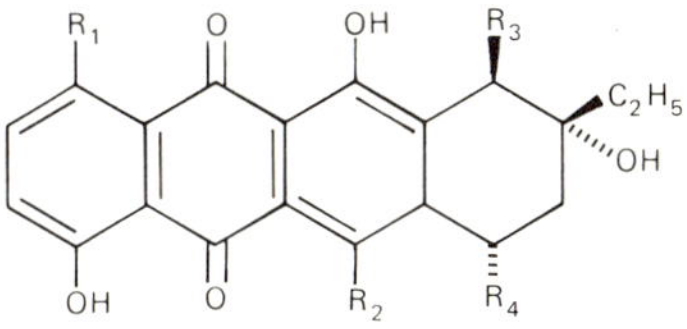

	R_1	R_2	R_3	R_4
Iremycin	H	OH	O-Roa	H
Roseorubicin A	H	OH	O-(Roa)$_2$-dF-(Rod)$_2$	H
Roseorubicin B	H	OH	O-Roa-Roa	H
Violamycin A_1	OH	OH	OH	O-Roa
Violamycin A_2	OH	H	OH	O-Roa
Violamycin A_3	OH	OH	$COOCH_3$	O-Roa
Violamycin A_4	H	OH	$COOCH_3$	O-Roa
Violamycin A_5	OH	OH	$COOCH_3$	O-Roa
Violamycin A_6	H	OH	OH	O-Roa
Violamycin B_1	OH	OH	OH	O-Roa-dF-Rod
Violamycin B_2	OH	H	OH	O-Roa-dF-Rod
Violamycin B_3	OH	OH	$COOCH_3$	O-Roa-dF-Rod
Violamycin B_4	H	OH	$COOCH_3$	O-Roa-dF-Rod
Violamycin B_5	OH	OH	$COOCH_3$	O-Roa-dF-Rod
Violamycin B_6	H	OH	OH	O-Roa-dF-Rod
G-5 (11-OH-aclacinomycin)	H	OH	$COOCH_3$	O-Roa-dF-LcinA
G-6 (11-OH-cinerubin)	OH	OH	$COOCH_3$	O-Roa-dF-LcinA
G-7	H	OH	OH	O-Roa-dF-LcinA
G-8	H	H	O-Rod-dF-LcinA	H

Note — Roa: L-rhodosamine; dF: 2-deoxy-L-fucose; Rod: L-rhodinose; LcinA: L-cinerulose A

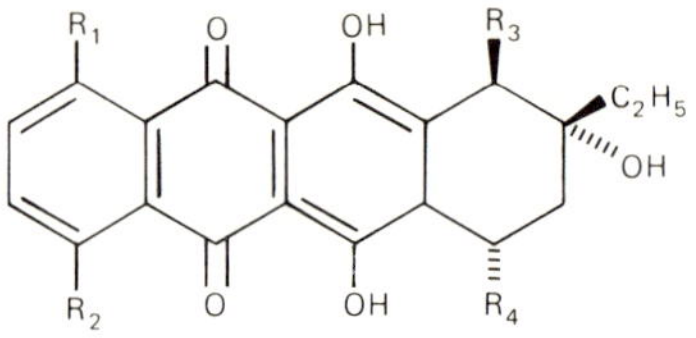

	R_1	R_2	R_3	R_4
G-2	H	OH	H	O-Roa-dF-LcinA
G-3	H	OCH_3	$COOCH_3$	O-Roa-dF-LcinA
G-9	OH	OH	OH	O-Roa-dF-LcinA
MA-144-KH	H	OH	$COOCH_3$	O-Daun
Alcindoromycin	OH	OH	$COOCH_3$	O-demRoa-dF-dF

Note — Daun: L-daunosamine; demRoa: N-monodemethyl-L-rhodosamine

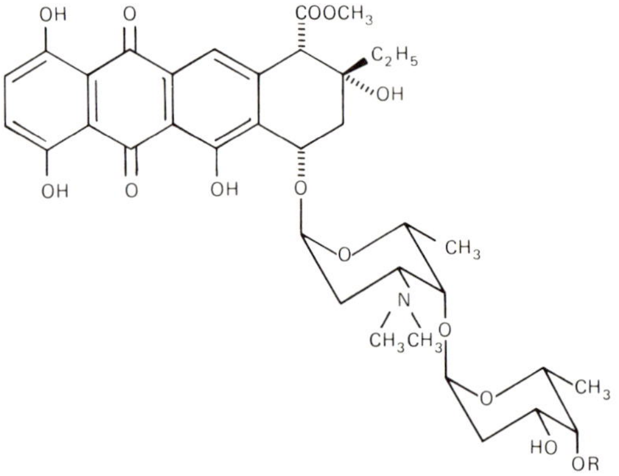

	R
Collinemycin (10-epi-musettamycin)	H
Mimimycin(10-epi-marcellomycin)	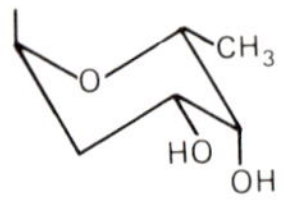

X:

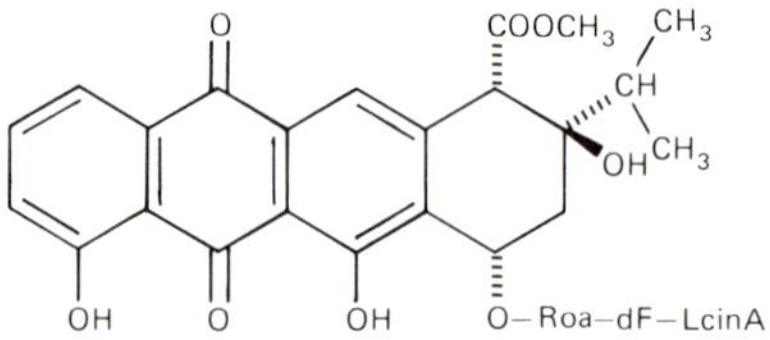

13-Methylaclacinomycin A

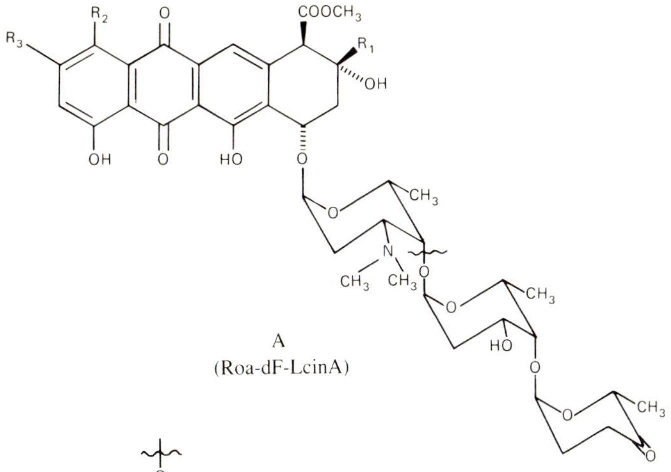

A
(Roa-dF-LcinA)

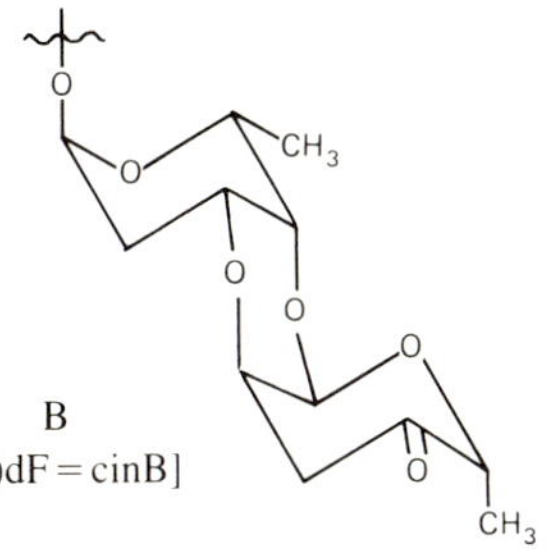

B
[(Roa)dF=cinB]

	R_1	R_2	R_3	Str.*
Auramycin A	CH_3	H	H	A
Auramycin B	CH_3	H	H	B
Sulfurmycin A	CH_2COCH_3	H	H	A
Sulfurmycin B	CH_2COCH_3	H	H	B
1-Hydroxy-auramycin A	CH_3	OH	H	A
1-Hydroxy-auramycin B	CH_3	OH	H	B
1-Hydroxy-sulfurmycin A	CH_2COCH_3	OH	H	A
1-Hydroxy-sulfurmycin B	CH_2COCH_3	OH	H	B
2-Hydroxy-aclacinomycin A	C_2H_5	H	OH	A
13-Methyl-aclacinomycin A[x]	||||$CH(CH_3)_2$	H	H	A

* Str.: structure of the sugar part in the A and B series

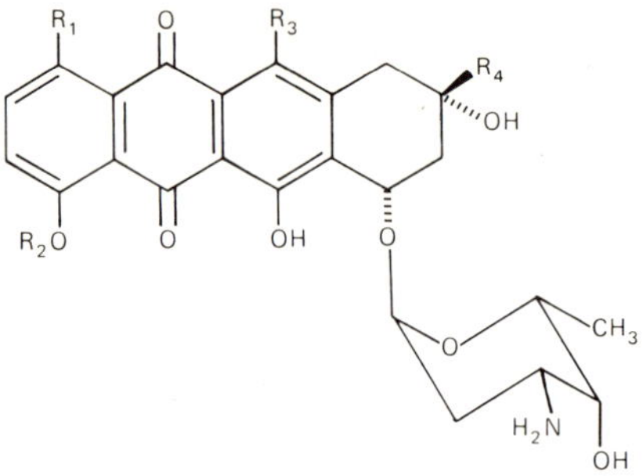

	R_1	R_2	R_3	R_4
11-Deoxydaunomycin	H	CH_3	H	$COCH_3$
11-Deoxyadriamycin	H	CH_3	H	$COCH_2OH$
11-Deoxy-13-dihydrodaunomycin	H	CH_3	H	$CH(OH)CH_3$
11-Deoxy-13-deoxodaunomycin	H	CH_3	H	C_2H_5
1-Hydroxy-13-dihydrodaunomycin	OH	CH_3	OH	$CH(OH)CH_3$
11-Deoxycarminomycin	H	H	H	$COCH_3$
11-Deoxy-13-deoxocarminomycin	H	H	H	C_2H_5
11-Deoxy-14-hydroxycarminomycin	H	H	H	$COCH_2OH$
GP-III	OH	CH_3	OH	$CH(OH)CH_3$
13-Deoxycarminomycin	H	H	OH	C_2H_5
Feudomycin A	H	CH_3	OH	C_2H_5
Feudomycin B	H	CH_3	OH	CH_2COCH_3

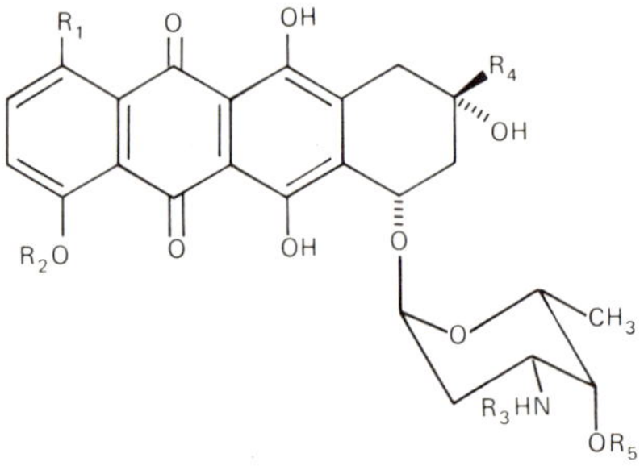

	R_1	R_2	R_3	R_4	R_5
N-Formyl-1-hydroxy-13-dihydrodaunomycin	OH	CH_3	CHO	$CH(OH)CH_3$	H
Rubeomycin A (D-326 III,Carminomycin 2)	H	H	H	$COCH_3$	X_1
Rubeomycin A_1(D-326 IV, Carminomycin 3)	H	H	H	$COCH_3$	X_2
Rubeomycin B (D-326 I, tentative)	H	H	H	$CH(OH)CH_3$	X_1
Rubeomycin B_1(D-326 II, tentative)	H	H	H	$CH(OH)CH_3$	X_2

Structures of rubeomycin A, A_1, B and B_1:

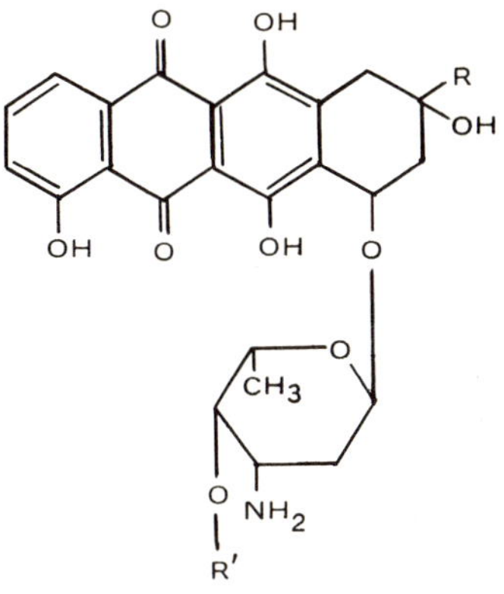

Structure of D326 complex.

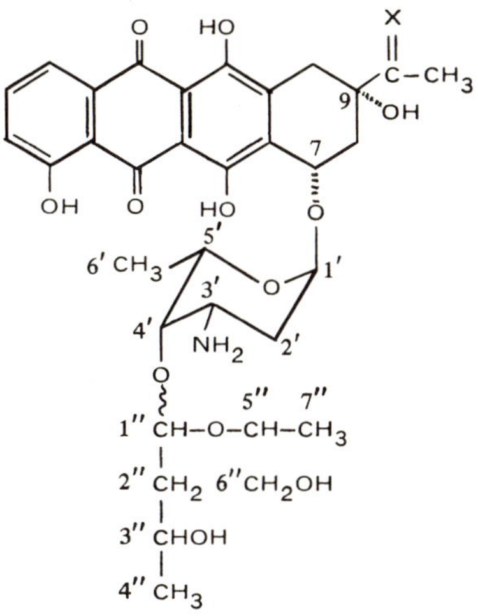

	Aglycone	Sugar
D326 I	X = OH,H	4′-epimer A
D326 II	X = OH,H	4′-epimer B
D326 III	X = O	4′-epimer A
D326 IV	X = O	4′-epimer B

X_1, X_2 =

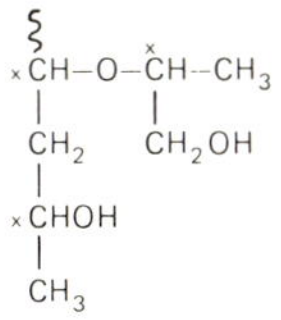

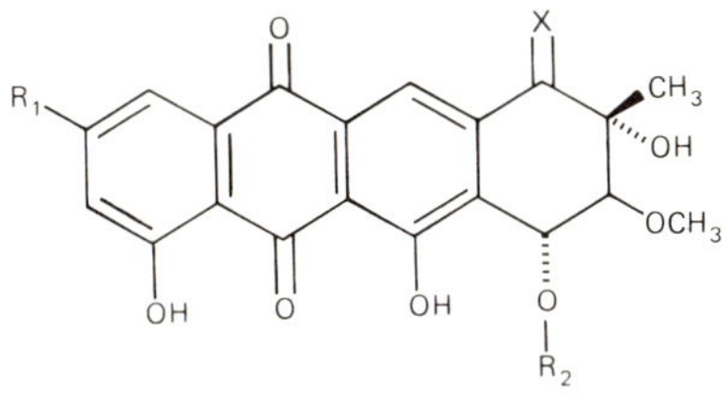

	R_1	R_2	R_3
10-Dihydrosteffimycin A	OCH_3	Y	H; X:H,OH
10-Dihydrosteffimycin B	OCH_3	Y	CH_3; X:H,OH
Steffimycinol	OCH_3	H	O
SM-173 A (Aranciamycin)	H	Y	H; X: = O

Y:

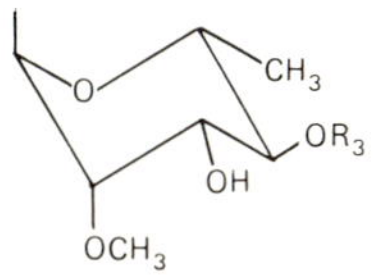

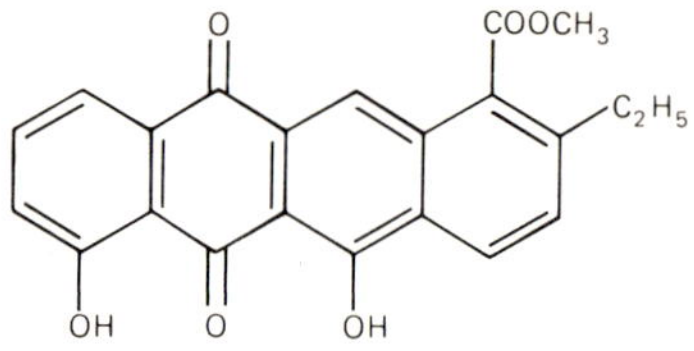

Galirubinone B_1 (MA-144-F)

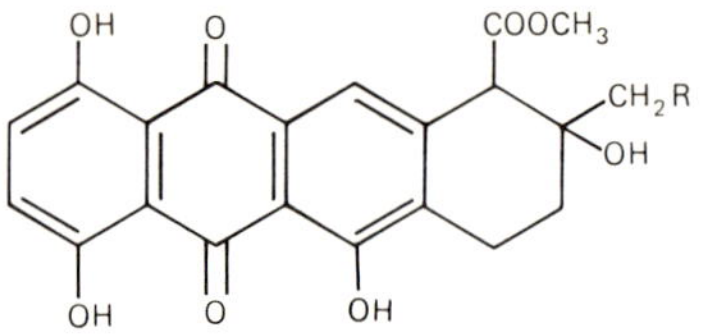

DC-44 A; R = $COCH_3$
DC-44 B; R = H

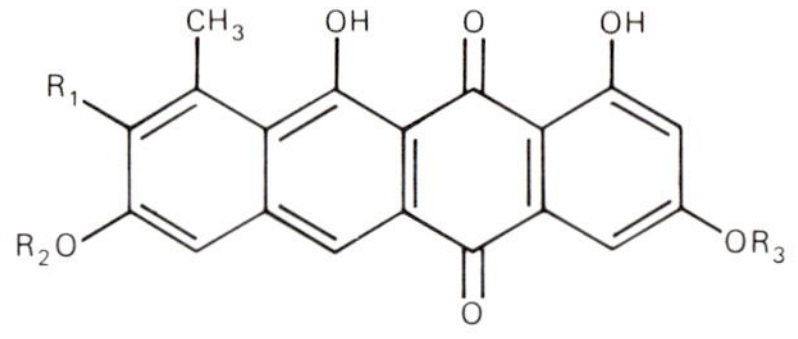

	R_1	R_2	R_3
Tetracenomycin A_2	$COOCH_3$	CH_3	CH_3
Tetracenomycin B_1	$COOCH_3$	H,CH_3 (tentative)	
Tetracenomycin B_2	H	CH_3,H (tentative)	
Tetracenomycin D	H	H	H

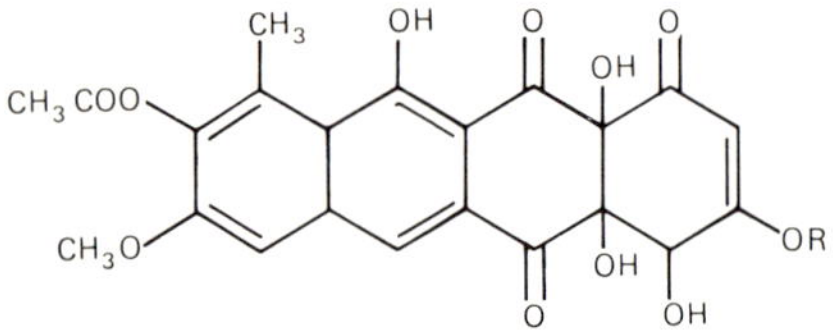

	R
Tetracenomycin C	CH_3
3-Demethyoxy-3-ethoxy-tetracenomycin C	C_2H_5

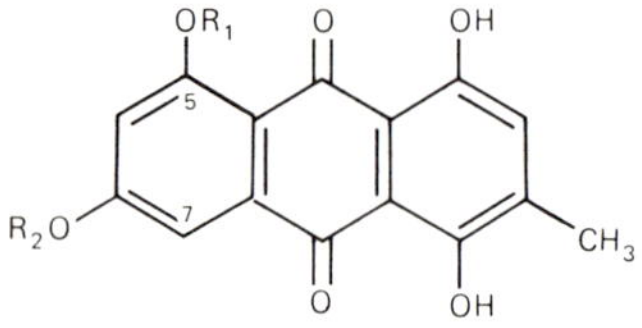

	R_1	R_2
Catenarin-5-methylether	CH_3	
Erythroglaucin	H	CH_3

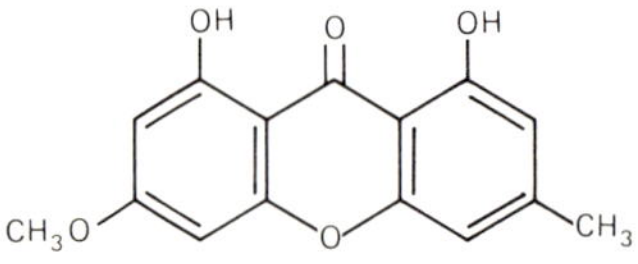

Physcion-9-anthrone

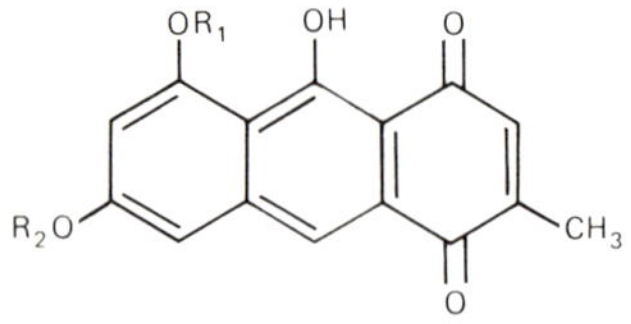

	R_1	R_2
Viocristin	CH_3	H
Isoviocristin	H	CH_3

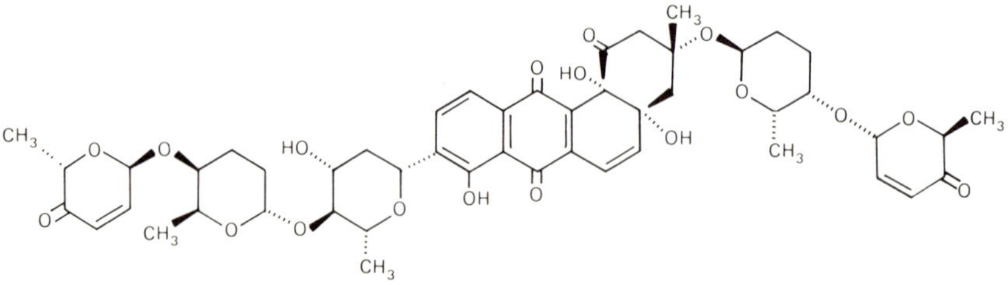

P-1894-B
(Vinenomycin A_1)

31111-7104

NAME: <u>6-DEOXYTETRACYCLINE</u>
PO: S.HYDROGENANS
CT: TETRACYCLINE T., AMPHOTER
FORMULA: C22H24N2O7
EA: (N, 6)
MW: 428
PC: YELLOW
TO: (G.POS.,)(G.NEG.,)
REFERENCES:
JP 76/57887; *CA*, 85, 190733

31111-7926

NAME: <u>4-N-DEMETHYL-4-N-ETHYL-6-DEMETHYLTETRACYCLINE</u>
PO: S.AUREOFACIENS+ETHIONINE
CT: TETRACYCLINE T., AMPHOTER
FORMULA: C22H24N2O8
EA: (N, 6)
MW: 444
PC: YELLOW, POW.
TO: (G.POS.,)(G.NEG.,)
IS-EXT: (ETOAC, , WB.)
REFERENCES:
USP 3364123

31130-7657

NAME: <u>VEROTETRONE</u>
PO: S.AUREOFACIENS
CT: TETRACYCLINE L., AMPHOTER
FORMULA: C19H9NO9
EA: (N, 3.5)
MW: 395
PC: RED, POW.
OR: (0,)
UV: MEOH: (296, ,)(420, ,)
SOL-GOOD: MEOH, CHL
SOL-POOR: HEX
TV: EHRLICH, ANTITUMOR
IS-EXT: (ETOAC, , MIC.)
IS-CHR: (SILG, CHL-ETOAC)
IS-CRY: (PREC., ETOAC, HEX)
REFERENCES:
Folia Micr., 25, 289, 1980

31211-6721

NAME: IREMYCIN, A-43615
IDENTICAL: G"-RHODOMYCIN-I
PO: S.VIOLACEUS+S.HYGROSCOPICUS, S.VIOLACEUS-IREMYCETICUS
CT: ANTHRACYCLINE, RHODOMYCIN T., BASIC, AMPHOTER
FORMULA: C38H33NO9
EA: (N, 2)
PC: RED, BROWN, CRYST.
UV: (535, 273,)
UV: C.HEX: (238, 595,)(254, 522,)(295, 142,)(469, 222,)(489, 266,)(498, 327,)
SOL-GOOD: MEOH, CHL, DIOXAN
SOL-FAIR: MEOH
SOL-POOR: W, HEX
TO: (G.POS.,)
TV: ANTITUMOR, ANTIVIRAL
IS-EXT: (BUOH, 7.5, FILT.)(MEOH, 7, MIC.)(CHL, , FILT.)(CHL-MEOH, , MIC.)
IS-CHR: (SILG, CHL-MEOH-ACOH-W)(SILG, CHL-ACET-MEOH)(SEPHADEX LH-20, MEOH-CHL)
IS-CRY: (PREC., CHL, HEX)(CRYST., MEOH)
REFERENCES:

Z. Allg. Mikr., 20, 529, 533, 1980; *JA*, 33, 1457, 1980; *Proc. Antib. Symp., Weimar*, 1979, A-26; DDR P 140672

31211-6759

NAME: ROSEORUBICIN-A
PO: ACT.ROSEOVIOLACENS, ACT.VIOLASCENS, ACT.VIOLARUS, ACT.VIOLACEOCHROMOGENES, S.PURPURASCENS
CT: BASIC, AMPHOTER, ANTHRACYCLINE, RHODOMYCIN T.
FORMULA: C54H78N2O18
EA: (N, 2.5)
MW: 1042
PC: RED, POW.
UV: MEOH: (237, 354,)(256, 300,)(295, 80,)(495, 152,)(530, 117,)(580, 26,)
SOL-GOOD: MEOH, CHL
SOL-POOR: W
TO: (S.AUREUS, .4)(S.LUTEA, .4)(B.SUBT., .2)
TV: L-1210
IS-FIL: 7.4
IS-EXT: (ACET, , MIC.)(CHL, 7.4, FILT.)(CHL, 7, W)
IS-CHR: (SEPHADEX LH-20, CHL-MEOH)
IS-CRY: (PREC., CHL, HEX)
REFERENCES:

JA, 32, 420, 1979; JP 80/47699; *CA*, 93, 112305

31211-6760

NAME: ROSEORUBICIN-B
IDENTICAL: G"-RHODOMYCIN-II
PO: ACT.ROSEOVIOLACENS, ACT.VIOLASCENS, ACT.VIOLARUS, ACT.VIOLACEOCHROMOGENES, S.PURPURASCENS
CT: BASIC, AMPHOTER, ANTHRACYCLINE, RHODOMYCIN T.
FORMULA: C36H48N2O11
EA: (N, 4)
MW: 684
PC: RED, POW.
UV: MEOH: (234, 481,)(256, 436,)(295, 107,) (495, 211,)(530, 172,)(575, 47,)
SOL-GOOD: MEOH, CHL
SOL-POOR: W
TO: (S.AUREUS, 3.1)(B.SUBT., 3.1)(S.LUTEA, 12.5)
TV: L-1210
IS-EXT: (BUOH, 7, W)
IS-CHR: (SEPHADEX LH-20, MEOH)
IS-CRY: (PREC., MEOH, HEX)
REFERENCES:

JA, 32, 420, 1979; JP 80/47699; *CA*, 93, 112305

31211-7927

NAME: RHODOMYCIN-COMPLEX
PO: S.GRISEORUBER
CT: ANTHRACYCLINE, RHODOMYCIN T., BASIC, AMPHOTER
EA: (N,)
PC: RED, POW.
TO: (B.SUBT.,)
IS-EXT: (MEOH, , MIC.)(BUOH, , FILT.)
IS-CHR: (SEPHADEX LH-20, MEOH)
REFERENCES:

Folia Micr., 25, 464, 1980

31211-7928

NAME: VIOLAMYCIN-A2
PO: S.VIOLACEUS
CT: ANTHRACYCLINE, RHODOMYCIN T., BASIC, AMPHOTER
FORMULA: C28H33NO10
EA: (N, 2.5)
MW: 543
PC: RED, POW.
UV: MEOH: (495, ,)(516, ,)(529, ,)
SOL-GOOD: MEOH, CHL
SOL-FAIR: BENZ
SOL-POOR: W
TO: (G.POS.,)(MYCOPLASMA SP.,)(G.NEG.,)
TV: L-1210, ANTIVIRAL
IS-FIL: 5.5
IS-EXT: (BUOH, 5.5, FILT.)(MEOH, , MIC.)
IS-CHR: (SEPHADEX LH-20, MEOH)
REFERENCES:

Z. Allg. Mikr., 14, 503, 551, 1974; *Antib.*, 966, 1975; *Proc. Antib. Symp., Weimar*, 1979, A-18, A-19, A-20, A-21, A-22, A-23, A-24, A-25; DDR P 100494; DT 2243554

31211-7929

NAME: VIOLAMYCIN-A3
PO: S.VIOLACEUS
CT: ANTHRACYCLINE, RHODOMYCIN T., BASIC, AMPHOTER
FORMULA: C30H35NO12
EA: (N, 2)
MW: 601
PC: RED, POW.
UV: MEOH: (495, ,)(524, ,)(551, ,)(364, ,)
SOL-GOOD: MEOH, CHL
SOL-FAIR: BENZ
SOL-POOR: W
TO: (G.POS.,)(MYCOPLASMA SP.,)(G.NEG.,)
TV: L-1210, ANTIVIRAL
IS-FIL: 5.5
IS-EXT: (BUOH, 5.5, FILT.)(MEOH, , MIC.)
IS-CHR: (SEPHADEX LH-20, MEOH)
REFERENCES:

Z. Allg. Mikr., 14, 503, 551, 1974; *Antib.*, 966, 1975; *Proc. Antib. Symp., Weimar*, 1979, A-18, A-19, A-20, A-21, A-22, A-23, A-24, A-25; DDR P 100494; DT 2243554

31211-7930

NAME:	VIOLAMYCIN-A4
PO:	S.VIOLACEUS
CT:	ANTHRACYCLINE, RHODOMYCIN T., BASIC, AMPHOTER
FORMULA:	C30H35NO11
EA:	(N, 2.5)
MW:	585
PC:	RED, POW.
UV:	MEOH: (495, ,)(516, ,)(529, ,)
SOL-GOOD:	MEOH, CHL
SOL-FAIR:	BENZ
SOL-POOR:	W
TO:	(G.POS.,)(MYCOPLASMA SP.,)(G.NEG.,)
TV:	L-1210, ANTIVIRAL
IS-FIL:	5.5
IS-EXT:	(BUOH, 5.5, FILT.)(MEOH, , MIC.)
IS-CHR:	(SEPHADEX LH-20, MEOH)

REFERENCES:

Z. Allg. Mikr., 14, 503, 551, 1974; *Antib.*, 966, 1975; *Proc. Antib. Symp., Weimar*, 1979, A-18, A-19, A-20, A-21, A-22, A-23, A-24, A-25; DDR P 100494; DT 2243554

31211-7931

NAME:	VIOLAMYCIN-A5
PO:	S.VIOLACEUS
CT:	ANTHRACYCLINE, RHODOMYCIN T., BASIC, AMPHOTER
FORMULA:	C30H35NO10
EA:	(N, 2.5)
MW:	569
PC:	RED, POW.
UV:	MEOH: (495, ,)(524, ,)(551, ,)(564, ,)
SOL-GOOD:	MEOH, CHL
SOL-FAIR:	BENZ
SOL-POOR:	W
TO:	(G.POS.,)(MYCOPLASMA SP.,)(G.NEG.,)
TV:	L-1210, ANTIVIRAL
IS-FIL:	5.5
IS-EXT:	(BUOH, 5.5, FILT.)(MEOH, , MIC.)
IS-CHR:	(SEPHADEX LH-20, MEOH)

REFERENCES:

Z. Allg. Mikr., 14, 503, 551, 1974; *Antib.*, 966, 1975; *Proc. Antib. Symp., Weimar*, 1979, A-18, A-19, A-20, A-21, A-22, A-23, A-24, A-25; DDR P 100494; DT 2243554

31211-7932

NAME: VIOLAMYCIN-A6
PO: S.VIOLACEUS
CT: ANTHRACYCLINE, RHODOMYCIN T., BASIC, AMPHOTER
FORMULA: C28H33NO10
EA: (N, 2.5)
MW: 543
PC: RED, POW.
UV: MEOH: (495, ,)(516, ,)(529, ,)
SOL-GOOD: MEOH, CHL
SOL-FAIR: BENZ
SOL-POOR: W
TO: (G.POS.,)(MYCOPLASMA SP.,)(G.NEG.,)
TV: L-1210, ANTIVIRAL
IS-FIL: 5.5
IS-EXT: (BUOH, 5.5, FILT.)(MEOH, , MIC.)
IS-CHR: (SEPHADEX LH-20, MEOH)
REFERENCES:

Z. Allg. Mikr., 14, 503, 551, 1974; *Antib.*, 966, 1975; *Proc. Antib. Symp., Weimar*, 1979, A-18, A-19, A-20, A-21, A-22, A-23, A-24, A-25; DDR P 100494; DT 2243554

31211-7933

NAME: VIOLAMYCIN-B2, VIOLAMYCIN-B-I2
PO: S.VIOLACEUS
CT: ANTHRACYCLINE, RHODOMYCIN T., BASIC, AMPHOTER
FORMULA: C40H53NO15
EA: (N, 2)
MW: 787
PC: RED, POW.
UV: MEOH: (495, ,)(516, ,)(529, ,)
SOL-GOOD: MEOH, CHL
SOL-FAIR: BENZ
SOL-POOR: W
TO: (G.POS.,)(MYCOPLASMA SP.,)(G.NEG.,)
TV: L-1210, ANTIVIRAL
IS-FIL: 5.5
IS-EXT: (BUOH, 5.5, FILT.)(MEOH, , MIC.)
IS-CHR: (SEPHADEX LH-20, MEOH)
REFERENCES:

Z. Allg. Mikr., 14, 503, 551, 1974; *Antib.*, 966, 1975; *Proc. Antib. Symp., Weimar*, 1979, A-18, A-19, A-20, A-21, A-22, A-23, A-24, A-25; DDR P 100494; DT 2243554

31211-7934

NAME: VIOLAMYCIN-B3
PO: S.VIOLACEUS
CT: ANTHRACYCLINE, RHODOMYCIN T., BASIC, AMPHOTER
FORMULA: C42H55NO17
EA: (N, 1.5)
MW: 845
PC: RED, POW.
UV: MEOH: (495, ,)(524, ,)(551, ,)(564, ,)
SOL-GOOD: MEOH, CHL
SOL-FAIR: BENZ
SOL-POOR: W
TO: (G.POS.,)(MYCOPLASMA SP.,)(G.NEG.,)
TV: L-1210, ANTIVIRAL
IS-FIL: 5.5
IS-EXT: (BUOH, 5.5, FILT.)(MEOH, , MIC.)
IS-CHR: (SEPHADEX LH-25, MEOH)
REFERENCES:

Z. Allg. Mikr., 14, 503, 551, 1974; *Antib.*, 966, 1975; *Proc. Antib. Symp., Weimar*, 1979, A-18, A19, A-20, A-21, A-22, A-23, A-24, A-25; DDR P 100494; DT 2243554

31211-7935

NAME: VIOLAMYCIN-B4
PO: S.VIOLACEUS
CT: ANTHRACYCLINE, RHODOMYCIN T., BASIC, AMPHOTER
FORMULA: C42H55NO16
EA: (N, 1.5)
MW: 829
PC: RED, POW.
UV: MEOH: (495, ,)(516, ,)(529, ,)
SOL-GOOD: MEOH, CHL
SOL-FAIR: BENZ
SOL-POOR: W
TO: (G.POS.,)(MYCOPLASMA SP.,)(G.NEG.,)
TV: L-1210, ANTIVIRAL
IS-FIL: 5.5
IS-EXT: (BUOH, 5.5, FILT.)(MEOH, , MIC.)
IS-CHR: (SEPHADEX LH-20, MEOH)
REFERENCES:

Z. Allg. Mikr., 14, 503, 551, 1974; *Antib.*, 966, 1975; *Proc. Antib. Symp., Weimar*, 1979, A-18, A-19, A-20, A-21, A-22, A-23, A-24, A-25; DDR P 100494; DT 2243554

31211-7936

NAME: VIOLAMYCIN-B5
PO: S.VIOLACEUS
CT: ANTHRACYCLINE, RHODOMYCIN T., BASIC, AMPHOTER
FORMULA: C42H55NO15
EA: (N, 1.5)
MW: 813
PC: RED, POW.
UV: MEOH: (495, ,)(524, ,)(551, ,)(564, ,)
SOL-GOOD: MEOH, CHL
SOL-FAIR: BENZ
SOL-POOR: W
TO: (G.POS.,)(MYCOPLASMA SP.,)(G.NEG.,)
TV: L-1210, ANTIVIRAL
IS-FIL: 5.5
IS-EXT: (BUOH, 5.5, FILT.)(MEOH, , MIC.)
IS-CHR: (SEPHADEX LH-20, MEOH)
REFERENCES:

Z. Allg. Mikr., 14, 503, 551, 1974; *Antib.*, 966, 1975; *Proc. Antib. Symp., Weimar,* 1979, A-18, A-19, A-20, A-21, A-22, A-23, A-24, A-25; DDR P 100494; DT 2243554

31211-7937

NAME: VIOLAMYCIN-B6
PO: S.VIOLACEUS
CT: ANTHRACYCLINE, RHODOMYCIN T., BASIC, AMPHOTER
FORMULA: C40H53NO15
EA: (N, 2)
MW: 787
PC: RED, POW.
UV: MEOH: (495, ,)(516, ,)(529, ,)
SOL-GOOD: MEOH, CHL
SOL-FAIR: BENZ
SOL-POOR: W
TO: (G.POS.,)(MYCOPLASMA SP.,)(G.NEG.,)
TV: L-1210, ANTIVIRAL
IS-FIL: 5.5
IS-EXT: (BUOH, 5.5, FILT.)(MEOH, , MIC.)
IS-CHR: (SEPHADEX LH-20, MEOH)
REFERENCES:

Z. Allg. Mikr., 14, 503, 551, 1974; *Antib.*, 966, 1975; *Proc. Antib. Symp., Weimar,* 1979, A-18, A-19, A-20, A-21, A-22, A-23, A-24, A-25; DDR P 100494; DT 2243554

31211-8133

NAME: G-6, 11-HYDROXYCINERUBIN-A
PO: S.GALILAEUS+E"-PYRROMYCINONE
CT: ANTHRACYCLINE, RHODOMYCIN T., BASIC, AMPHOTER
FORMULA: C42H53NO17
EA: (N, 1.5)
MW: 843
PC: RED, POW.
UV: MEOH: (243, 585,)(585, 229,)(632, 261,)
UV: MEOH-HCL: (240, 508,)(295, 90,)(521, 184,)(547, 175,)(560, 187,)(605, 48,)
SOL-GOOD: MEOH, CHL, TOL
SOL-POOR: W
TO: (S.AUREUS, .4)(B.SUBT., .2)(S.LUTEA, .2)(C.ALB., 50)
TV: L-1210
IS-EXT: (ACET, , MIC.)(CHL, , W)(TOL, 7, W)
IS-ION: (DX-50X2-NH4, NH4OH)(CG-50-H,)
IS-CHR: (SEPHADEX LH-20, MEOH)(SP-SEPHADEX C-25, PH4.7 BUFF.)(BIOGEL P-2, W)
IS-CRY: (PREC., CHL-MEOH, HEX)(LIOF.,)
REFERENCES:
JA, 33, 1331, 1341, 1980; JP 80/120786

31211-8134

NAME: G-7, 10-DECARBOMETHOXY-10.11-DIHYDROXYACLACINOMYCIN-A
PO: S.GALILAEUS+B"-RHODOMYCINONE
CT: ANTHRACYCLINE, RHODOMYCIN T., BASIC, AMPHOTER
FORMULA: C40H51NO15
EA: (N, 2)
MW: 785
PC: RED, POW.
UV: MEOH: (235, , 302)(252, 315,)(292, 98,)(495, 183,)(528, 187,)(580, 30,)
UV: MEOH-HCL: (235, , 302)(252, 315,)(292, 98,)(495, 183,)(528, 187,)(580, 30,)
UV: MEOH-NAOH: (241, 595,)(285, 108,)(565, 220,)
SOL-GOOD: MEOH, CHL, TOL
SOL-POOR: W
TO: (S.AUREUS, 6.2)(B.SUBT., 1.6)(S.LUTEA, .8)(C.ALB., 50)
TV: L-1210
REFERENCES:
JA, 33, 1331, 1341, 1980; JP 80/120786

31211-8135

NAME: G-7, 4"-DEHYDRORHODOMYCIN-Y
PO: S.GALILAEUS+G"-RHODOMYCINONE
CT: ANTHRACYCLINE, RHODOMYCIN T., BASIC, AMPHOTER
FORMULA: C40H51NO14
EA: (N, 2)
MW: 769
PC: RED, POW.
UV: MEOH: (236, 431,)(254, 384,)(295, 98,)(495, 196,)(528, 130,)
UV: MEOH-HCL: (236, 431,)(254, 384,)(295, 98,)(495, 196,)(52, , 130,)
UV: MEOH-NAOH: (242, 591,)(290, 108,)(558, 222,)(592, 200,)
SOL-GOOD: CHL, MEOH, TOL
SOL-POOR: W
TO: (S.AUREUS, 12.5)(B.SUBT., 3.1)(S.LUTEA, 6.2)
TV: L-1210
REFERENCES:
JA, 33, 1331, 1980; JP 80/120786

31211-8136

NAME: G-9, 10-DECARBOMETHOXY-10-HYDROXYCINERUBIN-A
PO: S.GALILAEUS+B"-RHODOMYCINONE
CT: ANTHRACYCLINE, RHODOMYCIN T., BASIC, AMPHOTER
FORMULA: C40H51NO15
EA: (N, 1.5)
MW: 785
PC: RED, POW.
UV: MEOH: (234, 496,)(256, 256,)(290, 132,)(490, 132,)
UV: MEOH-HCL: (234, 496,)(256, 256,)(290, 132,)(490, 132,)
UV: MEOH-NAOH: (235, 433,)(296, 88,)(560, 141,)(597, 127,)
SOL-GOOD: MEOH, CHL, TOL
SOL-POOR: W
TO: (G.POS.,)
TV: L-1210
REFERENCES:
JA, 33, 1331, 1980; JP 80/120786

31212-7106

NAME: ALCINDOROMYCIN, N-MONODEMETHYLMARCELLOMYCIN
PO: ACTINOSPORANGIUM SP.
CT: ANTHRACYCLINE, CINERUBIN T., BASIC, AMPHOTER
FORMULA: C41H53NO17
EA: (N, 1.5)
PC: RED
OR: (+13, CHL)
UV: CHL: (258, , 21200)(288, , 9400)(484, , 14100)(495, , 15100)(515, , 8480)
TO: (G.POS.,)(S.AUREUS, 8)
TV: ANTITUMOR, L-1210
REFERENCES:
JACS, 101, 7041, 1979; *Lloydia*, 43, 242, 1980; USP 4039736

31212-7107

NAME: COLLINEMYCIN, 10-EPI-MUSETTAMYCIN
PO: ACTINOSPORANGIUM SP.
CT: ANTHRACYCLINE, CINERUBIN T., BASIC, AMPHOTER
FORMULA: C36H45NO14
EA: (N, 2)
MW: 715
PC: RED
UV: CHL: (257, , 22600)(288, , 9300)(296, , 9090)(484, , 13800)(494, , 14700)(514, , 11200)(528, , 9600)
TO: (G.POS.,)(S.AUREUS, 8)
TV: L-1210
REFERENCES:
JACS, 101, 7041, 1979; *Lloydia*, 43, 242, 1980; USP 4039736

31212-7108

NAME: MIMIMYCIN, 10-EPI-MARCELLOMYCIN
PO: ACTINOSPORANGIUM SP.
CT: ANTHRACYCLINE, CINERUBIN T., BASIC, AMPHOTER
FORMULA: C42H55NO17
EA: (N, 1.5)
MW: 845
PC: RED
UV: CHL: (258, , 23300)(296, , 9300)(495, , 15400)(515, , 1700)(529, , 9380)
TO: (G.POS.,)(S.AUREUS, 16)
TV: L-1210
REFERENCES:
JACS, 101, 7041, 1979; *Lloydia*, 43, 242, 1980; USP 4039736

31212-7938

NAME: TRYPANOMYCIN COMPLEX
PO: S.DIASTATOCHROMOGENES
CT: ANTHRACYCLINE, CINERUBIN T., BASIC, AMPHOTER
PC: RED, BROWN, POW.
UV: MEOH: (495, ,)
SOL-GOOD: W, MEOH, CHL
SOL-POOR: HEX
TO: (G.POS.,)(G.NEG.,)(MYCOPLASMA SP.,)(C.ALB.,)(PROTOZOA,)
LD50: (60, IV)(31, IP)
TV: ANTITUMOR
IS-EXT: (BUOH, , FILT.)(MEOH, , MIC.)
REFERENCES:
AAC, 1, 385, 1972; DT 2009116

31212-8137

NAME: G-5, 11-HYDROXYACLACINOMYCIN-A
PO: S.GALILAEUS+E"-RHODOMYCINONE
CT: ANTHRACYCLINE, CINERUBIN T., BASIC, AMPHOTER
FORMULA: C42H53NO16
EA: (N, 1.5)
MW: 827
PC: RED, POW.
UV: MEOH: (235, 513,)(255, 307,)(292, 102,) (495, 118,)(527, 118,)
UV: MEOH-NAOH: (242, 555,)(287, 99,)(566, 223,) (605, 194,)
SOL-GOOD: MEOH, CHL
SOL-POOR: W
TO: (S.AUREUS, .8)(B.SUBT., .4)(S.LUTEA, .2) (C.ALB., 50)
TV: L-1210
REFERENCES:
JA, 33, 1331, 1341, 1980; JP 80/120786

31212-8518

NAME: 1-HYDROXYAURAMYCIN-A
PO: S.MELANOGENES, S.GALILAEUS
CT: ANTHRACYCLINE, CINERUBIN T., BASIC, AMPHOTER
FORMULA: C41H51NO16
EA: (N, 2)
MW: 813
PC: RED, POW.
OR: (+93.1, CHL)
UV: MEOH: (234, 500,)(292, 103,)(493, 150,)
(511, 130,)(526, 140,)(570, 80,)
UV: MEOH-NAOH: (241, 535,)(565, 210,)(602, 170,)
SOL-GOOD: MEOH, CHL
SOL-POOR: W, HEX
TO: (B.SUBT., .78)(S.AUREUS, .78)(S.LUTEA, .39)
TV: L-1210
IS-EXT: (CHL, , FILT.)(ACET, , MIC.)
REFERENCES:
JA, 34, 912, 1981

31212-8519

NAME: 1-HYDROXYAURAMYCIN-B
PO: S.MELANOGENES, S.GALILAEUS
CT: ANTHRACYCLINE, CINERUBIN T., BASIC, AMPHOTER
FORMULA: C41H49NO16
EA: (N, 2)
MW: 811
PC: RED, POW.
OR: (+89.4, CHL)
UV: MEOH: (234, 555,)(256, 325,)(292, 105,)
(493, 185,)(526, 150,)(570, 80,)
UV: MEOH-NAOH: (242, 595,)(565, 240,)(602, 200,)
SOL-GOOD: MEOH, CHL
SOL-POOR: W, HEX
TO: (S.AUREUS, .78)(B.SUBT., .78)(S.LUTEA, .39)
TV: L-1210
IS-EXT: (CHL, , FILT.)(ACET, , MIC.)
REFERENCES:
JA, 34, 912, 1981

31212-8520

NAME: 1-HYDROXYSULFURMYCIN-A
PO: S.MELANOGENES, S.GALILAEUS
CT: ANTHRACYCLINE, CINERUBIN T., BASIC, AMPHOTER
FORMULA: C43H53NO17
EA: (N, 2)
MW: 855
PC: RED, POW.
OR: (+57.88, CHL)
UV: MEOH: (234, 570,)(256, 345,)(292, 115,) (493, 175,)(511, 165,)(526, , 160)(570, 75,)
UV: MEOH-NAOH: (241, 530,)(292, 170,)(565, 230,)(602, 180,)
SOL-GOOD: MOEH, CHL
SOL-POOR: W, HEX
TO: (B.SUBT., .78)(S.LUTEA, .78)(S.AUREUS, .78)
TV: L-1210
IS-EXT: (CHL, , FILT.)(ACET, , MIC.)
REFERENCES:

JA, 34, 912, 1981

31212-8521

NAME: 1-HYDROXYSULFURMYCIN-B
PO: S.MELANOGENES, S.GALILAEUS
CT: ANTHRACYCLINE, CINERUBIN T., BASIC, AMPHOTER
FORMULA: C43H51NO17
EA: (N, 2)
MW: 853
PC: RED, POW.
OR: (+60.33, CHL)
UV: MEOH: (234, 545,)(256, 340,)(290, 110,) (493, 175,)(511, 160,)(526, 160,)(570, 85,)
UV: MEOH-NAOH: (241, 540,)(293, 165,)(565, 190,)(602, 190,)
SOL-GOOD: MEOH, CHL
SOL-POOR: W, HEX
TO: (S.AUREUS, .78)(B.SUBT., .78)(S.LUTEA, .78)
TV: L-1210
REFERENCES:

JA, 34, 912, 1981

31213-8068

NAME: <u>MA-144-KH</u>, 7-O-DAUNOSAMINYLAKLAVINONE
PO: S.GALILAEUS
CT: ANTHRACYCLINE, AKLAVIN T., BASIC, AMPHOTER
FORMULA: C28H31NO10
EA: (N, 2.5)
MW: 541
PC: RED, POW.
TO: (G.POS.,)
TV: L-1210
REFERENCES:

JA, 33, 1158, 1980

31213-8138

NAME: <u>G-2</u>, 10-DECARBOMETHOXYACLACINOMYCIN-A
PO: S.GALILAEUS+10-DECABOMETHOXYAKLAVINONE
CT: ANTHRACYCLINE, AKLAVIN T., BASIC, AMPHOTER
FORMULA: C40H51NO13
EA: (N, 2)
MW: 753
PC: YELLOW, POW.
OR: (-44.7, CHL)
UV: CHL: (262, , 24300)(282, , 11200)(293, , 11600)(436, , 12700)
UV: MEOH: (230, 687,)(260, 483,)(295, 212,) (433, 239,)
UV: MEOH-NAOH: (217, 1230,)(235, 570,)(287, 18)(520, 245,)
SOL-GOOD: MEOH, CHL
SOL-POOR: W
TO: (G.POS.,)
TV: L-1210
IS-EXT: (ACET, , MIC.)(CHL, , W)
REFERENCES:

JA, 33, 1331, 1341, 1980; JP 80/120786

31213-8139

NAME:	G-3, 4-O-METHYLACLACINOMYCIN-A
PO:	S.GALILAEUS+4-O-METHYLAKLAVINONE
CT:	ANTHRACYCLINE, AKLAVIN T., BASIC, AMPHOTER
FORMULA:	C43H56NO15
EA:	(N, 1.5)
MW:	826
PC:	YELLOW, POW.
OR:	(-38, CHL)
UV:	CHL: (258, , 23300)(288, , 11400)(419, , 10900)
UV:	MEOH: (229, 527,)(258, 324,)(418, 13, ,)
UV:	MEOH-NAOH: (217, 871,)(25
TV:	L-1210

REFERENCES:

JA, 33, 1331, 1341, 1980; JP 80/120786; JP 80/108889; *CA*, 94, 47703

31213-8140

NAME:	AURAMYCIN-A
PO:	S.GALILAEUS
CT:	ANTHRACYCLINE, AKLAVIN T., BASIC, AMPHOTER
FORMULA:	C41H51NO15
EA:	(N, 2)
MW:	797
PC:	YELLOW, POW.
OR:	(-8, CHL)
SOL-GOOD:	MEOH, CHL
SOL-POOR:	W, HEX
TO:	(B.SUBT., 3.12)(S.LUTEA, .1)(S.AUREUS, .78)
LD50:	(100, IP)
TV:	P-388
IS-EXT:	(MEOH, , MIC.)(CHL-W, , MIC.)(CHL-MEOH, , FILT.)
IS-CHR:	(SEPHADEX LH-20, CHL-MEOH)(SILG, CHL-MEOH)
IS-CRY:	(DRY, MEOH)

REFERENCES:

JA, 34, 608, 1981; EP 19302

31213-8141

NAME: AURAMYCIN-B
PO: S.GALILAEUS
CT: ANTHRACYCLINE, AKLAVIN T., BASIC, AMPHOTER
FORMULA: C41H49NO15
EA: (N, 2)
MW: 795
PC: YELLOW, POW.
OR: (-8, CHL)
SOL-GOOD: MEOH, CHL
SOL-POOR: W, HEX
TO: (B.SUBT., .78)(S.LUTEA, .05)(S.AUREUS, .39)
LD50: (100, IP)
TV: P-388
REFERENCES:

JA, 34, 608, 1981; EP 19302

31213-8142

NAME: SULFURMYCIN-A
PO: S.GALILAEUS
CT: ANTHRACYCLINE, AKLAVIN T., BASIC, AMPHOTER
FORMULA: C43H53NO16
EA: (N, 1.5)
MW: 839
PC: YELLOW, POW.
OR: (-23.3, CHL)
SOL-GOOD: MEOH, CHL
SOL-POOR: W, HEX
TO: (B.SUBT., .78)(S.LUTEA, .1)(S.AUREUS, .2)
LD50: (100, IP)
TV: P-388
IS-EXT: (MEOH, , MIC.)(CHL-MEOH, , FILT.)
IS-CHR: (SEPHADEX LH-20, CHL-MEOH)
IS-CRY: (PREC., CHL, HEX)
REFERENCES:

JA, 34, 608, 1981; EP 19302

31213-8143

NAME:	SULFURMYCIN-B
PO:	S.GALILAEUS
CT:	ANTHRACYCLINE, AKLAVIN T., BASIC, AMPHOTER
FORMULA:	C43H51NO16
EA:	(N, 1.5)
MW:	837
PC:	YELLOW, POW.
OR:	(-21.5, CHL)
SOL-GOOD:	MEOH, CHL
SOL-POOR:	W, HEX
TO:	(B.SUBT., .78)(S.LUTEA, .1)(S.AUREUS, .39)
LD50:	(100, IP)
TV:	P-388

REFERENCES:

JA, 34, 608, 1981; EP 19302

31213-8522

NAME:	13-METHYLACLACINOMYCIN-A
PO:	S.GALILAEUS
CT:	ANTHRACYCLINE, AKLAVIN T., BASIC, AMPHOTER
FORMULA:	C43H55NO15
EA:	(N, 2)
MW:	825
PC:	YELLOW, POW.
UV:	MEOH: (229, 531,)(258, 310,)(289, 121,) (430, 158,)
TO:	(G.POS.,)
TV:	L-1210
IS-EXT:	(TOL, 7, FILT.)(W, 3.5, TOL)
IS-CRY:	(PREC., TOL, HEX)

REFERENCES:

JA, 34, 770, 1981

31213-8523

NAME: 2-HYDROXYACLACINOMYCIN-A
PO: S.GALILAEUS, S.CINEREORUBER+2-HYDROXYAKLAVINON
CT: ANTHRACYCLINE, AKLAVIN T., BASIC, AMPHOTER
FORMULA: C42H53NO16
EA: (N, 2)
MW: 828
PC: ORANGE, YELLOW, POW.
OR: (+42.3, MEOH)
UV: MEOH: (222, 375,)(256, 235,)(295, 207,)
(450, 110,)
UV: MEOH-HCL: (226, 448,)(254, 238,)(268, 245,)
(291, 251,)(440, 158,)
UV: MEOH-NAOH: (297, 250,)(540, 143,)
SOL-GOOD: MEOH, CHL
SOL-POOR: HEX, W
TO: (G.POS.,)
LD50: (50, IP)
TV: L-1210
IS-EXT: (CHL-MEOH, , WB.)(ACET, , MIC.)(CHL, 7, W)
IS-CHR: (SEPHADEX LH-20, MEOH)(SILG, CHL)
REFERENCES:
JA, 34, 916, 959, 1495, 1981

31214-6722

NAME: 11-DEOXYDAUNORUBICIN, 11-DEOXYDAUNOMYCIN
PO: MIC.PEUCETICA
CT: ANTHRACYCLINE, DAUNOMYCIN T., BASIC, AMPHOTER
FORMULA: C27H29NO9
EA: (N, 3)
MW: 511
PC: RED, POW.
OR: (+139, MEOH)
UV: 7: (235, 617,)(262, 424,)(426, 166,)
UV: MEOH: (228, 743,)(260, 450,)(418, 199,)
UV: NAOH: (510, 128,)
SOL-GOOD: MEOH, CHL, DIOXAN, ACID, PYR
SOL-POOR: ET2O, HEX, W
TO: (S.AUREUS, 62)(S.LUTEA, 12.5)(B.SUBT., 50)
(E.COLI, 25)
TV: HELA, P-388, L-1210
IS-FIL: 4
IS-EXT: (CHL-MEOH, 8.5, FILT.)(ACET-HCL, , MIC.)
IS-CHR: (SILG, CHL-MEOH-W)
REFERENCES:
JA, 33, 1462, 1468, 1980; *JACS*, 102, 1462, 1980; *Proc. Antib. Symp., Weimar*, 1979, A-3; DT 2904186; Belg. P 874032

31214-6723

NAME: 11-DEOXYDOXORUBICIN, 11-DEOXYADRIAMYCIN
PO: MIC.PEUCETICA
CT: ANTHRACYCLINE, DAUNOMYCIN T., BASIC, AMPHOTER
FORMULA: C27H29NO10
EA: (N, 3)
MW: 527
PC: RED, POW.
OR: (+111, MEOH)
UV: 7: (235, 600,)(262, 406,)(426, 161,)
UV: MEOH: (228, 645,)(260, 420,)(418, 193,)
UV: NAOH: (510, 120,)
SOL-GOOD: MEOH, CHL, PYR, DIOXAN, ACID
SOL-POOR: ET2O, HEX, W
TO: (E.COLI, 50)(S.AUREUS, 125)(S.LUTEA, 100)(B.SUBT., 100)
TV: HELA, P-388, L-1210
IS-FIL: 4
IS-EXT: (CHL-MEOH, 8.5, FILT.)(ACET-HCL, , MIC.)
IS-CHR: (SILG, CHL-MEOH-W)
REFERENCES:

JA, 33, 1462, 1468, 1980; *JACS*, 102, 1462, 1980; *Proc. Antib. Symp., Weimar*, 1979, A-3; DT 2904186; Belg. P 874032

31214-6724

NAME: 11-DEOXY-13-DIHYDRODAUNOMYCIN, 11-DEOXY-RP-20798
PO: MIC.PEUCETICA
CT: ANTHRACYCLINE, DAUNOMYCIN T., BASIC, AMPHOTER
FORMULA: C27H31NO9
EA: (N, 3)
MW: 513
PC: RED, POW.
OR: (+107, MEOH)
UV: 7: (235, 605,)(262, 400,)(426, 160,)
UV: MEOH: (228, 640,)(260, 410,)(418, 179,)
UV: NAOH: (510, 118,)
SOL-GOOD: MEOH, CHL, PYR, ACOH, ACID
SOL-POOR: ET2O, HEX, W
TO: (S.LUTEA, 100)(S.AUREUS, 1000)(B.SUBT., 1000)(E.COLI, 100)
TV: HELA
IS-FIL: 4
IS-EXT: (CHL-MEOH, 8.5, FILT.)(ACET-HCL, , MIC.)
IS-CHR: (SILG, CHL-MEOH-W)
REFERENCES:

JA, 33, 1462, 1468, 1980; *JACS*, 102, 1462, 1980; *Proc. Antib. Symp., Weimar*, 1979, A-13; DT 2904186; Belg. P 874032

31214-6725

NAME: 11-DEOXY-13-DEOXODAUNOMYCIN, 11-DEOXY-13-DEOXO-DAUNORUBICIN
PO: MIC.PEUCETICA
CT: ANTHRACYCLINE, DAUNOMYCIN T., BASIC, AMPHOTER
FORMULA: C27H31NO8
EA: (N, 3)
MW: 497
PC: RED, POW.
OR: (+122, MEOH)
UV: 7: (235, 550,)(262, 400,)(426, 140,)
UV: MEOH: (228, 610,)(260, 395,)(418, 171,)
SOL-GOOD: MEOH, CHL, PYR, ACOH, ACID
SOL-POOR: ET2O, HEX, W
TO: (S.AUREUS, 250)(B.SUBT., 100)(S.LUTEA, 23)(E.COLI, 50)
TV: HELA
IS-FIL: 4
IS-EXT: (CHL-MEOH, 8.5, FILT.)(ACET-HCL, , MIC.)
IS-CHR: (SILG, CHL-MEOH-W)
REFERENCES:
JA, 33, 1462, 1468, 1980; *JACS*, 102, 1462, 1980; *Proc. Antib. Symp., Weimar*, 1979, A-13; DT 2904186; Belg. P 874032

31214-7658

NAME: OODM, GP-I, 1-HYDROXY-13-DIHYDRODAUNOMYCIN
PO: S.COELUREORUBIDUS+E"-PYRROMYCINON, S.PEUCETICUS-CARNEUS+E"-PYRROMYCINON, S.PEUCETICUS-CAESIUS+E"-ISORHODO-MYCINON
CT: ANTHRACYCLINE, DAUNOMYCIN T., BASIC, AMPHOTER
FORMULA: C27H31NO11
EA: (N, 2.5)
MW: 545
PC: RED, POW.
UV: MEOH: (240, 720,)(285, 123,)(520, 239,)(546, 216,)(600, 42,)
SOL-GOOD: MEOH, CHL
SOL-POOR: W, HEX
TO: (B.SUBT.,)
TV: L-1210
IS-EXT: (ACET, , MIC.)(BUOH, 7, W)
IS-CHR: (SEPHADEX LH-20, MEOH)
REFERENCES:
JA, 33, 1150, 1980; EP 12159; JP 80/43012; *CA*, 93, 112302

31214-7659

NAME: N-FORMYL-1-HYDROXY-13-DIHYDRODAUNOMYCIN, FOODM, GP-II
PO: S.COELUREORUBIDUS+E"-PYRROMYCINON
CT: ANTHRACYCLINE, DAUNOMYCIN T., BASIC, AMPHOTER
FORMULA: C28H31NO12
EA: (N, 2.5)
MW: 573
PC: RED, POW.
UV: MEOH: (240, 760,)(285, 123,)(520, 258,) (546, 241,)(600, 61,)
SOL-GOOD: MEOH, CHL
SOL-POOR: W, HEX
TO: (B.SUBT.,)
TV: L-1210
IS-EXT: (CHL, , FILT.)(CHL, 6.5, W)
IS-CHR: (SEPHADEX LH-20, TOL-MEOH)(SILG, CHL-MEOH)
REFERENCES:
JA, 33, 1150, 1980; EP 12159; JP 80/43012; *CA*, 93, 112302

31214-7660

NAME: 11-DEOXYCARMINOMYCIN, COMP Y
PO: S.PEUCETICUS-CAESIUS
CT: ANTHRACYCLINE, DAUNOMYCIN T., BASIC, AMPHOTER
FORMULA: C26H27NO9
EA: (N, 2.5)
MW: 548
PC: ORANGE, CRYST.
OR: (+150, MEOH)
UV: MEOH: (228, 670,)(258, 440,)(430, 208,)
UV: MEOH-NAOH: (520, 150)
SOL-GOOD: MEOH, CHL, ACID, PYR, DIOXAN
SOL-POOR: ET2O, HEX, C.HEX, W
TO: (S.AUREUS, 25)(B.SUBT., 12.5)(S.LUTEA, 3.12) (E.COLI, 6.25)
TV: L-1210, P-388
IS-FIL: 4
IS-EXT: (MEOH, 4, MIC.)(CHL, 9, W)
IS-CHR: (SILG, CHL-MEOH-W)
IS-CRY: (PREC., CHL, HEX)(CRYST., MEOH-HCL)(CRYST., MEOH-BUOH)
REFERENCES:
BP 2036021

31214-7661

NAME: 11-DEOXY-13-DEOXOCARMINOMYCIN, COMP Z
PO: S.PEUCETICUS-CAESIUS
CT: ANTHRACYCLINE, DAUNOMYCIN T., BASIC, MACROLIDE
FORMULA: C26H29NO8
EA: (N, 2.5)
MW: 534
PC: ORANGE, CRYST.
OR: (+134, MEOH)
UV: MEOH: (228, 660,)(258, 430,)(430, 216,)
UV: MEOH-NAOH: (520, 150,)
SOL-GOOD: MEOH, CHL, ACID, PYR, DIOXAN
SOL-POOR: ET2O, HEX, W
TO: (S.AUREUS, 100)(B.SUBT., 25)(S.LUTEA, 6.25)
(E.COLI, 25)
TV: L-1210, P-388
REFERENCES:
BP 2036021

31214-7662

NAME: 11-DEOXY-14-HYDROXYCARMINOMYCIN, COMP W
PO: S.PEUCETICUS-CAESIUS
CT: ANTHRACYCLINE, DAUNOMYCIN T., BASIC, AMPHOTER
FORMULA: C26H27NO10
EA: (N, 2.5)
MW: 564
PC: ORANGE, CRYST.
OR: (+130, MEOH)
UV: MEOH: (228, 650,)(258, 435,)(430, 217,)
UV: MEOH-NAOH: (520, 152,)
SOL-GOOD: MEOH, CHL, ACID, PYR, DIOXAN
SOL-POOR: ET2O, HEX, W
TO: (S.AUREUS, 25)(B.SUBT., 50)(S.LUTEA, 12.6)
(E.COLI, 12.5)
TV: L-1210, P-388
IS-FIL: 4
IS-EXT: (MEOH, 4, MIC.)(CHL, 9, W)
IS-CHR: (SILG, CHL-MEOH-W)
IS-CRY: (CRYST., MEOH-BUOH)
REFERENCES:
BP 2036021

31214-7663

NAME: COMP X
PO: S.PEUCETICUS-CAESIUS
CT: ANTHRACYCLINE, DAUNOMYCIN T., BASIC, AMPHOTER
PC: ORANGE, RED
UV: MEOH: (228, ,)(258, ,)(430, ,)
SOL-GOOD: MEOH, CHL
SOL-POOR: ET2O, HEX, W
TO: (B.SUBT.,)(S.AUREUS,)(S.LUTEA,)
TV: L-1210
IS-EXT: (MEOH, , MIC.)(CHL, 9, W)
IS-CHR: (SILG, CHL-MEOH-W)
REFERENCES:
BP 2036021

31214-8069

NAME: GP-III
PO: S.COELUREORUBIDUS+E"-PYRROMYCINON
CT: ANTHRACYCLINE, DAUNOMYCIN T., BASIC, AMPHOTER
EA: (N,)
PC: RED, POW.
TO: (B.SUBT.,)
IS-EXT: (ACET, , MIC.)(CHL, , FILT.)(W, 3.1, TOL)(CHL, 7, W)
IS-CHR: (SEPHADEX LH-20, CHL-MEOH)(SEPHADEX LH-20, TOL-MEOH)
REFERENCES:
JA, 33, 1150, 1980

31214-8144

NAME: 13-DEOXYCARMINOMYCIN, DOC
PO: S.PENCETICUS-CARMINATUS
CT: ANTHRACYCLINE, DAUNOMYCIN T., BASIC, AMPHOTER
FORMULA: C26H29NO9
EA: (N, 3)
MW: 499
PC: RED, POW.
OR: (+275, MEOH)
UV: MEOH: (236, 620,)(256, 520,)(293, 164,)
(495, 260,)(529, 186,)
SOL-GOOD: MEOH, CHL, DIOXAN, CH2CL2
SOL-FAIR: ETOAC
SOL-POOR: BENZ, HEX, W
TO: (S.AUREUS, 12.5)(S.LUTEA, 1.56)(B.SUBT., 6.25)
(E.COLI, 3.12)
TV: P-388, HELA
IS-FIL: 4
IS-EXT: (ACET-HCL, 4, MIC.)(CHL, 9, W)(W, 2, BUOH)
IS-CHR: (SILG, CHL-MEOH-W)(CEL, BUOH-ETOAC)
IS-CRY: (PREC., ACET, HEX)(CRYST., MEOH-HCL)
REFERENCES:
DT 3012665

31214-8173

NAME: RUBEOMYCIN-A, FA-1180-A
IDENTICAL: CARMINOMYCIN-2, DF-4466-A, D-326-III
PO: ACTINOMADURA ROSEOVIOLACEA-BIWAKOENSIS
CT: ANTHRACYCLINE, DAUNOMYCIN T., BASIC, AMPHOTER
FORMULA: C33H41NO13
EA: (N, 2)
MW: 659
PC: RED, CRYST.
OR: (+147, CHL)(+120.6, CHL)
UV: MEOH: (237, 980,)(255, 760,)(292, 230,)
(493, 390,)(530, 270,)
SOL-GOOD: MEOH, CHL
SOL-POOR: HEX
TO: (B.SUBT., 1.56)(S.AUREUS, .8)
TV: P-388, YOSHIDA
IS-EXT: (ACET, , MIC.)(W, 2, CHL)(CHL, 8, W)
IS-CHR: (SILG, CHL-TOL-MEOH)
IS-CRY: (CRYST., CHL)
REFERENCES:
JA, 34, 938, 1981; EP 25713

31214-8174

NAME: RUBEOMYCIN-A1, FA-1180-A1
IDENTICAL: CARMINOMYCIN-3, DF-4466-B, D-326-IV
PO: ACTINOMADURA ROSEOVIOLACEA-BIWAKOENSIS
CT: ANTHRACYCLINE, DAUNOMYCIN T., BASIC, AMPHOTER
FORMULA: C33H41NO13
EA: (N, 2)
MW: 659
PC: RED, POW.
OR: (+185, CHL)(+170.4, CHL)
UV: MEOH: (234, 457,)(254, 351,)(292, 101,)
(492, 172,)(527, 132,)(580, 30,)
UV: MEOH-HCL: (237, 605,)(255, 464,)(292, 161,)
(493, 260,)(530, 196,)
SOL-GOOD: MEOH, CHL
SOL-POOR: HEX
TO: (B.SUBT., 1.56)(S.AUREUS, .4)
TV: YOSHIDA
IS-EXT: (ACET, , MIC.)(W, 2, CHL)(CHL, 8, W)
IS-CHR: (SILG, CHL-TOL-MEOH)
IS-CRY: (CRYST., CHL)
REFERENCES:

JA, 34, 938, 1981; EP 25713

31214-8175

NAME: RUBEOMYCIN-B, FA-1180-B
PO: ACTINOMADURA ROSEOVIOLACEA-BIWAKOENSIS
CT: ANTHRACYCLINE, DAUNOMYCIN T., BASIC, AMPHOTER
FORMULA: C33H43NO13
EA: (N, 2)
MW: 661
PC: RED, POW.
OR: (+88, MEOH-CHL)(-70, MEOH-CHL)(+118, CHL)
UV: MEOH: (237, 850,)(255, 660,)(295, 200,)
(492, 350,)(528, 250,)
SOL-GOOD: MEOH, CHL
SOL-POOR: HEX
TO: (B.SUBT., .4)(S.AUREUS, 3.12)
TV: P-388, YOSHIDA
IS-EXT: (ACET, , MIC.)(W, 2, CHL)(CHL, 8, W)
IS-CHR: (SILG, CHL-TOL-MEOH)
IS-CRY: (CRYST., CHL)
REFERENCES:

JA, 34, 938, 1981; EP 25713

31214-8176

NAME: RUBEOMYCIN-B1, FA-1180-B1
PO: ACTINOMADURA ROSEOVIOLACEA-BIWAKOENSIS
CT: ANTHRACYCLINE, DAUNOMYCIN T., BASIC, AMPHOTER
EA: (N, 2)
MW: 661
PC: RED, CRYST.
OR: (+165, CHL-MEOH)(+151, CHL)
UV: MEOH: (234, 603,)(253, 487,)(293, 133,)
(491, 249,)(529, 186,)
UV: MEOH-HCL: (237, 582,)(255, 466,)(292, 137,)
(493, 251,)(530, 182,)
SOL-GOOD: MEOH, CHL
SOL-POOR: HEX
TO: (B.SUBT., .4)(S.AUREUS, 50)
TV: P-388, YOSHIDA
IS-EXT: (ACET, , MIC.)(W, 2, CHL)(CHL, 8, W)
IS-CHR: (SILG, CHL-TOL-MEOH)
IS-CRY: (CRYST., CHL)
REFERENCES:
JA, 34, 938, 1981; EP 25713

31214-8524

NAME: FEUDOMYCIN-A, 13-DEOXYDAUNOMYCIN
PO: S.COELUREORUBIDUS
CT: ANTHRACYCLINE, DAUNOMYCIN T., BASIC, AMPHOTER
FORMULA: C27H31NO9
EA: (N, 3)
MW: 513
PC: RED, POW.
OR: (+243, MEOH)
UV: MEOH: (235, 688,)(253, 510,)(290, 170,)
(475, 227,)(49M, 235,)(530, 135,)
SOL-GOOD: MEOH, BENZ
SOL-POOR: W, HEX
TO: (G.POS.,)
TV: L-1210
IS-EXT: (ACET, , MIC.)(CHL, , FILT.)(CHL, 7, W)
IS-CHR: (SEPHADEX LH-20, MEOH)
IS-CRY: (PREC., CHL, HEX)
REFERENCES:
JA, 34, 783, 1981

31214-8525

NAME: FEUDOMYCIN-B
PO: S.COELUREORUBIDUS
CT: ANTHRACYCLINE, DAUNOMYCIN T., BASIC, AMPHOTER
FORMULA: C28H31NO10
EA: (N, 3)
MW: 541
PC: RED, POW.
OR: (+146, MEOH)
UV: MEOH: (235, 631,)(253, 469,)(290, 170,)
(475, 214,)(498, 224,)(532, 130,)
SOL-GOOD: MEOH, BENZ
SOL-POOR: W, HEX
TO: (G.POS.,)
TV: L-1210
REFERENCES:
JA, 34, 783, 1981

31214-8526

NAME: D-326-I, 4-HYDROXYBAUMYCINOL-A2
PO: ACTINOMADURA SP.
CT: ANTHRACYCLINE, DAUNOMYCIN T., BASIC, AMPHOTER
FORMULA: C33H43NO13
EA: (N, 2)
MW: 661
PC: RED, POW.
OR: (+72, CHL-MEOH)
UV: MEOH: (235, 482,)(254, 384,)(294, 111,)
(492, 189,)(526, , 138)(575, 21,)
SOL-GOOD: MEOH, BENZ
SOL-POOR: W, HEX
TV: L-1210
IS-EXT: (ACET, , MIC.)(CHL, , W)
IS-CHR: (SEPHADEX LH-20, CHL-MEOH)(SILG, CHL)
REFERENCES:
JA, 34, 774, 1981

31214-8527

NAME: D-326-II, 4-HYDROXYBAUMYCINOL-A1
PO: ACTINOMADURA SP.
CT: ANTHRACYCLINE, DAUNOMYCIN T., BASIC, AMPHOTER
FORMULA: C33H43NO13
EA: (N, 2)
MW: 661
PC: RED, POW.
OR: (-34, CHL-MEOH)
UV: MEOH: (235, 551,)(255, 431,)(295, 120,)
(492, 221,)(527, 159,)(575, 16,)
SOL-GOOD: MEOH, BENZ
SOL-POOR: W, HEX
TV: L-1210
REFERENCES:
JA, 34, 744, 1981

31214-8528

NAME: D-326-III, 4-HYDROXYBAUMYCIN-A2
IDENTICAL: CARMINOMYCIN-2, RUBEOMYCIN-A, DF-4466-A
PO: ACTINOMADURA SP.
CT: ANTHRACYCLINE, DAUNOMYCIN T., BASIC, AMPHOTER
FORMULA: C33H41NO13
EA: (N, 2)
MW: 659
PC: RED, POW.
OR: (-1, CHL-MEOH)
UV: MEOH: (235, 581,)(255, 430,)(294, 125,)
(492, 221,)(527, 155,)(575, 15,)
SOL-GOOD: MEOH, BENZ
SOL-POOR: W, HEX
TV: L-1210
REFERENCES:
JA, 34, 744, 1981

31214-8529

NAME: D-326-IV, 4-HYDROXYBAUMYCIN-A1
IDENTICAL: CARMINOMYCIN-3, RUBEOMYCIN-A1, DF-4466-B
PO: ACTINOMADURA SP.
CT: ANTHRACYCLINE, DAUNOMYCIN T., BASIC, AMPHOTER
FORMULA: C33H41NO13
EA: (N, 2)
MW: 659
PC: RED, POW.
OR: (+48, CHL-MEOH)
UV: MEOH: (235, 674,)(255, 423,)(292, 128,)
(492, 223,)(526, 156,)(575, 18,)
SOL-GOOD: MEOH, BENZ
SOL-POOR: W, HEX
TV: L-1210
REFERENCES:
JA, 34, 744, 1981

31214-8791

NAME: DF-4466-A, 4466-A
IDENTICAL: RUBOMYCIN-A, CARMINOMYCIN-2
PO: ACTINOMADURA SP.
CT: ANTHRACYCLINE, DAUNOMYCIN T., BASIC, AMPHOTER
FORMULA: C33H41NO13
EA: (N, 2)
MW: 659|36
PC: RED, POW.
TO: (G.POS.,)
TV: ANTITUMOR
REFERENCES:
JP 81/90098; *CA*, 96, 33344

31214-8792

NAME: DF-4466-B, 4466-B
IDENTICAL: RUBOMYCIN-A1, CARMINOMYCIN-3
PO: ACTINOMADURA SP.
CT: ANTHRACYCLINE, DAUNOMYCIN T., BASIC, MACROLIDE
FORMULA: C33H41NO13
EA: (N, 2)
MW: 659
PC: RED, POW.
TO: (G.POS.,)
TV: ANTITUMOR
REFERENCES:
JP 81/90098; *CA*, 96, 33344

31221-8530

NAME: VIRIPLANIN
PO: AMPULLARIELLA REGULARIS
CT: ANTHRACYCLINE, NOQALAMYCIN T., BASIC, AMPHOTER
FORMULA: C21|1H30|1NO10|1.N
EA: (N, 3)
MW: 2000
EW: 500|50
PC: ORANGE, RED, CRYST.
UV: MEOH: (237, 338,)(260, 156,)(290, 59,)(480, 88,)
UV: MEOH-HCL: (237, 389,)(260, 151,)(290, 61,)(480, 99,)
UV: MEOH-NAOH: (241, 322,)(285, 64,)(320, 30,)(553, 86,)
SOL-GOOD: MEOH, CHL, DMFA, DMSO, BASE
SOL-POOR: W, HEX, ET2O
TO: (B.SUBT.,)
TV: HERPES
IS-FIL: 7
IS-EXT: (BUOH, 7, FILT.)(MEOH, 7, MIC.)(CHL, 7, W)
IS-CHR: (SILG, CHL-MEOH-NH4OH)
IS-CRY: (DRY, MEOH)
REFERENCES:

DT 2946523

31222-7516

NAME: SM-173-A
IDENTICAL: ARANCIAMYCIN
PO: S.CHROMOFUSCUS
CT: ANTHRACYCLINE, ACIDIC, STEFFIMYCIN T.
FORMULA: C27H28O12
MW: 544
PC: ORANGE, RED, POW.
TO: (S.AUREUS,)(MYCOB.SP.,)(FUNGI,)
REFERENCES:

CA, 92, 144895; 93, 130479

31222-7939

NAME: 10-DIHYDROSTEFFIMYCIN-A, U-58875
PO: ACTINOPLANES UTAHENSIS+STEFFIMYCIN, CHAETOMIUM SP.+STEFFIMYCIN
CT: ANTHRACYCLINE, STEFFIMYCIN T.
FORMULA: C28H32O13
MW: 576
PC: ORANGE, CRYST.
UV: ETOH: (227, , 26120)(268, , 17510)(434, , 12670)
TO: (B.SUBT.,)(S.AUREUS,)(S.LUTEA,)
IS-EXT: (CH2CL2, , FILT.)
IS-CHR: (SILG, CHL-MEOH)
IS-CRY: (CRYST., ACET)
REFERENCES:
JA, 33, 819, 1980; USP 3824305; 4209611, 4264726

31222-7940

NAME: 10-DIHYDROSTEFFIMYCIN-B, U-58874
PO: ACTINOPLANES UTAHENSIS+STEFFIMYCIN, CHAETOMIUM SP.+STEFFIMYCIN
CT: ANTHRACYCLINE, STEFFIMYCIN T.
FORMULA: C29H34O13
MW: 590
PC: ORANGE, CRYST.
UV: ETOH: (227, , 29750)(267, , 18700)(285, , 15930)(430, , 12500)
TO: (B.SUBT.,)(S.AUREUS,)(S.LUTEA,)
IS-EXT: (CH2CL2, , FILT.)
IS-CHR: (SILG, CHL-MEOH)
IS-CRY: (CRYST., ACET)
REFERENCES:
JA, 33, 819, 1980; USP 3824305, 4209611, 4264726

31223-8842

NAME: A-12918
PO: ACTINOPLANES SP.
CT: BASIC, QUINONE T.
EA: (C, 63)(H, 7)(N, 4)
PC: ORANGE, CRYST.
OR: (+34, MEOH)
SOL-GOOD: MEOH, BENZ, DMSO
SOL-POOR: HEX
QUAL: (FECL3, -)(NINH., -)
REFERENCES:
JP, 81/103195; *CA*, 96, 50695

31230-6761

NAME: STEFFIMYCINOL
PO: S.PEUCETICUS-CAESIUS+STEFFIMYCINON, S.NOGALATER+STEFFIMYCINON
CT: ACIDIC, ANTHRACYCLINONE
FORMULA: C21H20O9
MW: 416
PC: ORANGE, RED, POW.
TO: (B.SUBT.,)
REFERENCES:
JA, 28, 838, 1975; USP 4077844

31230-7941

NAME: GALIRUBINONE-B1, BISANHYDROAKLAVINONE
IDENTICAL: MA-144-F
PO: S.GALILAEUS
CT: ANTHRACYCLINONE
FORMULA: C22H16O6
MW: 376
PC: ORANGE, CRYST.
UV: C.HEX: (244, ,)(257, ,)(263, ,)(290, ,)
(447, ,)(475, ,)
IS-EXT: (MEOH, 7, MIC.)
REFERENCES:
Naturwiss., 52, 539, 1965; *Ber.*, 100, 2561, 1967; *JA*, 30, 683, 1977

31230-8793

NAME: DC-44-A
PO: S.ARGENTEOGRISEUS
CT: ANTHRACYCLINONE L., ACIDIC
FORMULA: C21H18O7
MW: 382
PC: RED, CRYST.
TO: (S.AUREUS, 25)(B.SUBT., 50)
TV: ANTITUMOR
IS-EXT: (ACET, , MIC.)
IS-ION: (DIAION HP-20,)
IS-CRY: (CRYST., ET2O)
REFERENCES:
JP 81/51435; *CA*, 95, 148696

31230-8794

NAME: DC-44-B
PO: S.ARGENTEOGRISEUS
CT: ANTHRACYCLINONE L., ACIDIC
FORMULA: C23H20O8
MW: 424
PC: RED, CRYST.
TO: (S.AUREUS, 100)(B.SUBT., 50)
TV: ANTITUMOR
IS-CRY: (CRYST., MEOH-ET2O)
REFERENCES:
JP 81/51435; *CA*, 95, 148696

31240-6727

NAME:	<u>TETRACENOMYCIN-C</u>
PO:	S.GLAUCESCENS
CT:	ACIDIC, ANTHRACYCLINONE L.
FORMULA:	C23H20O11
MW:	472
PC:	YELLOW, RED, CRYST.
OR:	(+22, DIOXAN)
UV:	CHL: (391, ,)(410, ,)
UV:	MEOH-HCL: (236, , 25000)(287, , 47600)(386, , 12500)(403, , 13200)
UV:	MEOH-NAOH: (255, , 32000)(436, , 13000)
SOL-GOOD:	ACET, ETOAC, DIOXAN, NAOH
SOL-FAIR:	CHL
SOL-POOR:	W, HEX
QUAL:	(FECL3, +)
TO:	(CORYNEBACTERIUM SP., 1.56)(STREPTOMYCES SP., .78)
TV:	P-388, CYTOTOXIC
IS-EXT:	(CHL, , W)
IS-ION:	(XAD-2, MEOH)
IS-CHR:	(SILG-OXALIC ACID, CHL-ACET)(SEPHADEX LH-20, CHL)
IS-CRY:	(PREC., ACET, HEX)(CRYST., MEOH)

REFERENCES:

Arch. Mikr., 121, 111, 1979; *Actinomycetes, Zbl. Bakt. Suppl.*, 11, 465, 1981

31240-8795

NAME:	<u>3-DEMETHYOXY-3-ETHOXYTETRACENOMYCIN-C</u>
PO:	S.GLAUCESCENS
CT:	ANTHRACYCLINONE L., ACIDIC
FORMULA:	C24H22O11
MW:	486
PC:	YELLOW, POW.
UV:	CHL: (391, ,)(410, ,)
UV:	MEOH-HCL: (236, ,)(287, ,)(386, ,)(403, ,)
UV:	MEOH-NAOH: (255, ,)(436, ,)
TO:	(ARTHROBACTER SP.,)(STREPTOMYCES SP.,)
TV:	P-388
IS-EXT:	(ETOAC, 5.7, FILT.)
IS-CHR:	(SILG, CHL-ACET)(SEPHADEX LH-20, CHL)
IS-CRY:	(PREC., ETOAC, HEX)

REFERENCES:

JA, 34, 1067, 1981

31240-8796

NAME: TETRACENOMYCIN-A2
PO: S.GLAUCESCENS
CT: ANTHRACYCLINONE L., ACIDIC
FORMULA: C23H1808
MW: 422
PC: RED, POW.
SOL-FAIR: MEOH, BENZ
SOL-POOR: W
TO: (ARTHROBACTER SP.,)
REFERENCES:
Actinomycetes, Zbl. Bakt. Suppl., 11, 465, 1981

31240-8797

NAME: TETRACENOMYCIN-B1
PO: S.LAUCESCENS
CT: ANTHRACYCLINONE L., ACIDIC
FORMULA: C20H1406
MW: 350
PC: RED, POW.
SOL-POOR: W
TO: (ARTHROBACTER SP.,)
REFERENCES:
Actinomycetes, Zbl. Bakt. Suppl., 11, 465, 1981

31240-8798

NAME: TETRACENOMYCIN-B2
PO: S.GLAUCESCENS
CT: ANTHRACYCLINONE L., ACIDIC
FORMULA: C22H1608
MW: 408
PC: RED, POW.
SOL-POOR: W
TO: (ARTHROBACTER SP.,)
REFERENCES:
Actinomycetes, Zbl. Bakt. Suppl., 11, 465, 1981

31240-8799

NAME: <u>TETRACENOMYCIN-D</u>
PO: S.GLAUCESCENS
CT: ANTHRACYCLINONE L., ACIDIC
FORMULA: C19H12O6
MW: 336
PC: RED, POW.
SOL-POOR: W
TO: (ARTHROBACTER SP.,)
REFERENCES:
Actinomycetes, Zbl. Bakt. Suppl., 11, 465, 1981

31200-7105

NAME: <u>MYCETIN-C</u>
PO: S.VIOLACEUS
CT: ANTHRACYCLINE L.
PC: RED
TO: (G.POS.,)
REFERENCES:
Antib., 771, 1965; *Mikrob.*, 35, 1053, 1966

31310-7110

NAME: <u>CATENARIN-5-METHYLETHER</u>
PO: ASP.GLAUCUS, ASP.CRISTATUS
CT: ANTHRAQUINONE, ACIDIC
FORMULA: C16H12O6
MW: 300
TO: ANTIMICROBIAL
REFERENCES:
Abst. Ann. Meet. ASM, 1979, O-25

31310-7111

NAME: <u>PHYSCION-9-ANTHRONE</u>
PO: ASP.GLAUCUS, ASP.CRISTATUS
CT: ANTHRAQUINONE, ACIDIC
FORMULA: C16H14O4
MW: 270
PC: CRYST., YELLOW
UV: CHL: (239, , 9800)(319, , 13800)
TO: ANTIMICROBIAL
REFERENCES:
Arch. Mikr., 126, 223, 231, 1980; *Holzforsch.*, 21, 89, 1967; *Abst. Ann. Meet. ASM*, 1979, O-25

31310-7112

NAME: ERYTHROGLAUCIN, CATENARIN-7-METHYLETHER
PO: ASP.GLAUCUS, ASP.CRISTATUS, ASP.RUBER, XANTHORIA ELEGANS
CT: ANTHRAQUINONE, ACIDIC
FORMULA: C16H12O6
MW: 300
PC: RED, CRYST.
UV: ETOH: (255, , 13800)(276, , 12900)(304, , 7850)(490, , 11000)(524, , 7400)
UV: MEOH: (231, , 27500)(255, , 14800)(275, , 13550)(302, , 8350)(489, , 11500)(523, , 7950)
TO: (B.SUBT.,)
REFERENCES:

Bioch. J., 28, 1640, 1934; *Mycol.*, 56, 185, 1964; *Abst. Ann. Meet. ASM*, 1979, 0-25; *Arch. Mikr.*, 126, 223, 231, 1980

31330-7562

NAME: P-1894-B
IDENTICAL: VINENOMYCIN-A1, OS-4742-A1
PO: S.ALBOGRISEOLUS
CT: ACIDIC, BENZANTHRAQUINONE
FORMULA: C49H58O18
EA: (C, 63)(H, 6)(O, 30)
MW: 934, 922
PC: ORANGE, RED, CRYST.
OR: (+155, MEOH)
UV: MEOH: (218, 432,)(318, 59,)(440, 71,)
UV: MEOH-HCL: (218, 427,)(318, 62,)(435, 70,)
UV: MEOH-NAOH: (228, 297,)(282, 144,)(530, 77,)
SOL-GOOD: MEOH, BENZ, DIOXAN, DMSO, PYR, BASE
SOL-POOR: W, HEX, ACID
QUAL: (NINH., -)
IS-EXT: (ETOAC, 3, FILT.)
IS-CHR: (SILG, CHL)(SILG, CHL-ETOAC)
IS-CRY: (PREC., CHL, HEX)(CRYST., CHL-TOL)
REFERENCES:

CC, 154, 1981; *Chem. Ph. Bull.*, 29, 1788, 1981; DT 2931792; BP 2028132; *JA*, 34, 1355, 1981

31350-7424

NAME: DC-14
PO: S.OLIVACEUS
CT: BASIC, PLURAMYCIN T.
FORMULA: C43H50N2O13
EA: (N, 3.5)
MW: 802
PC: ORANGE, RED
UV: MEOH: (243, ,)(420, ,)
TO: (S.AUREUS, 1)(B.SUBT., 1)(E.COLI, 32)(K.PNEUM., 8)
LD50: (8, IP)
TV: S-180, P-388
REFERENCES:
JP 79/141703; *CA*, 92, 144957

31350-7942

NAME: KIDAMYCIN-E, 289-E
PO: S.PHAEOVERTICILLATUS-TAKATSUKIENSIS+ANTHRAQUINOUSULPHONIC ACID
CT: PLURAMYCIN T., BASIC
EA: (C, 67)(H, 7)(N, 4)
PC: YELLOW, CRYST.
OR: (+477, CHL)
UV: MEOH: (244, 790,)(430, 160,)
SOL-GOOD: MEOH, ETOAC
SOL-POOR: W, HEX
TO: (G.POS.,)(G.NEG.,)
LD50: (.3|.05, IV)(.75, IP)
TV: EHRLICH
IS-EXT: (CHL, 8.3, FILT.)
REFERENCES:
JP 73/28079

31370-7943

NAME: VARIACYCLOMYCIN-A
PO: S.OLIVOVARIABILIS
CT: ANTHRAQUINONE DERIV.
FORMULA: C27H23NO11
EA: (N, 2.5)
MW: 537
PC: RED, POW.
TO: (G.POS.,)
IS-EXT: (ETOAC, , FILT.)
REFERENCES:
Antib., 486, 1974

31380-7664

NAME: VIOCRISTIN
PO: ASP.CRISTATUS
CT: ACIDIC
FORMULA: C16H12O5
MW: 284
PC: VIOLET, POW.
UV: ETOH: (253, , 39000)(340, , 5500)(497, , 8150)(517, , 8145)(527, , 7950)
UV: ETOH-NAOH: (214, , 52500)(244, , 26900)(279, , 27600)(350, , 7080)(580, , 8900)
SOL-FAIR: ETOH, ETOAC, DIOXAN, PYR, DMSO
SOL-POOR: CHL
TO: (B.SUBT., 8)(P.VULG.,)
TV: EHRLICH
IS-EXT: (ACET, , MIC.)
IS-CHR: (SILG, CHL-MEOH)
REFERENCES:
Arch. Mikr., 126, 223, 231, 1980

31380-7665

NAME: ISOVIOCRISTIN
PO: ASP.CRISTATUS
CT: ACIDIC
FORMULA: C16H12O5
MW: 284
PC: VIOLET, POW.
SOL-FAIR: ETOH, ETOAC, PYR
TO: (B.SUBT.,)
REFERENCES:
Arch. Mikr., 126, 223, 231, 1980

32
NAPHTOQUINONE DERIVATIVES

321 Simple Naphtoquinones
322 Condensed Naphtoquinones
323 Naphtoquinone-Like Compounds

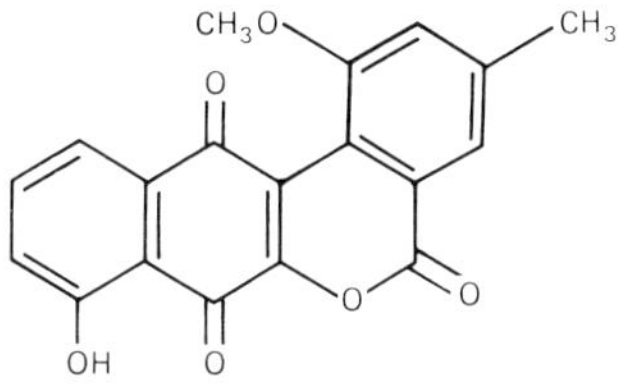

WS-5995 A

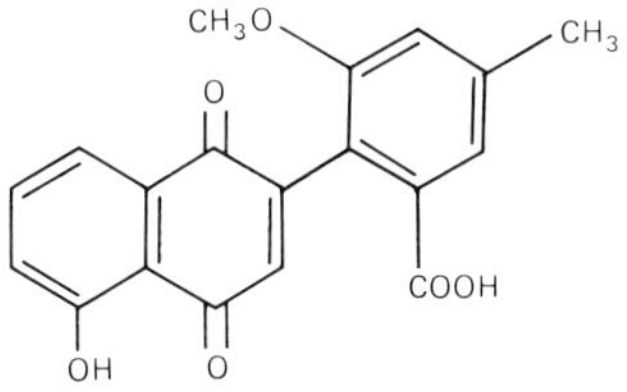

WS-5995 B

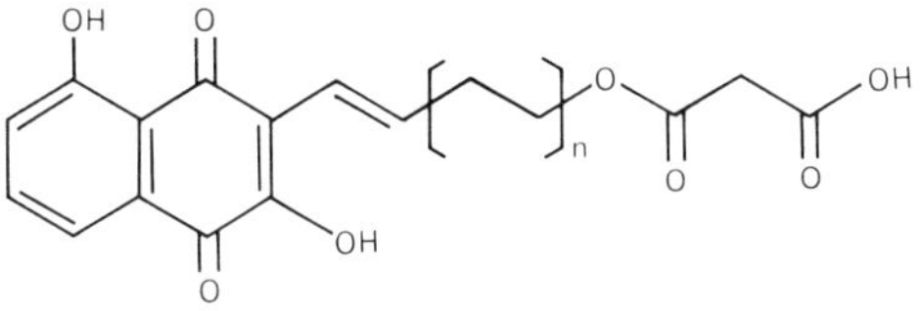

	n
Trichione	1
Homotrichione	2

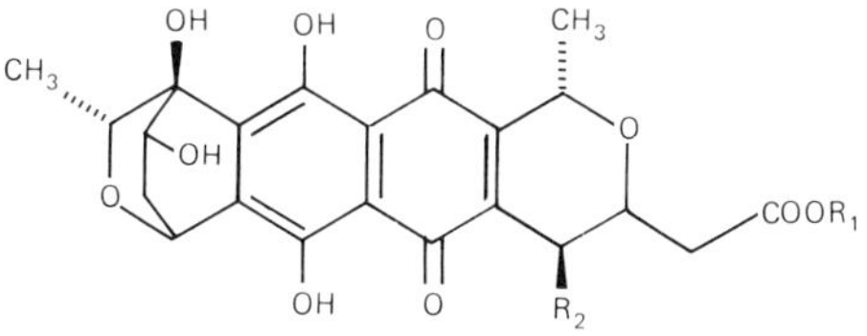

	R_1	**R_2**
Granaticinic Acid	H	OH
Granatomycin A	CH_3	H

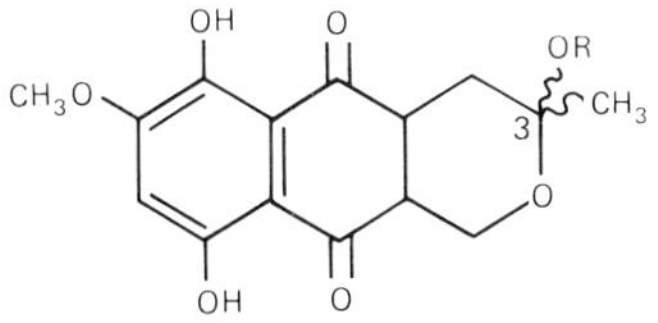

	R
Dihydrofusarubin	H
3-O-Ethyldihydrofusarubin	C_2H_5

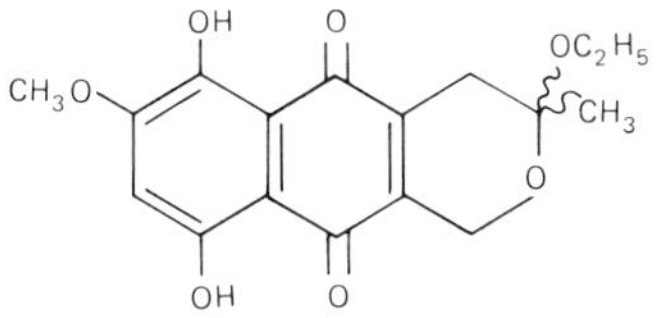

O-Ethylfusarubin

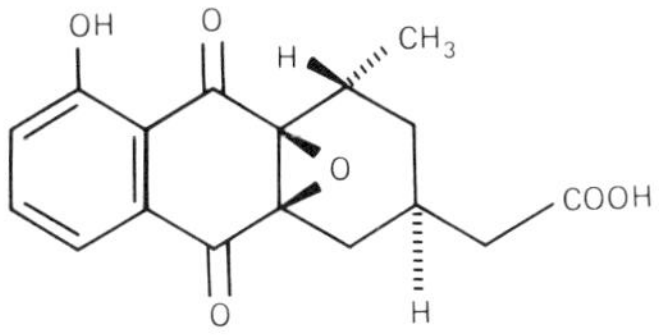

Nanaomycin E

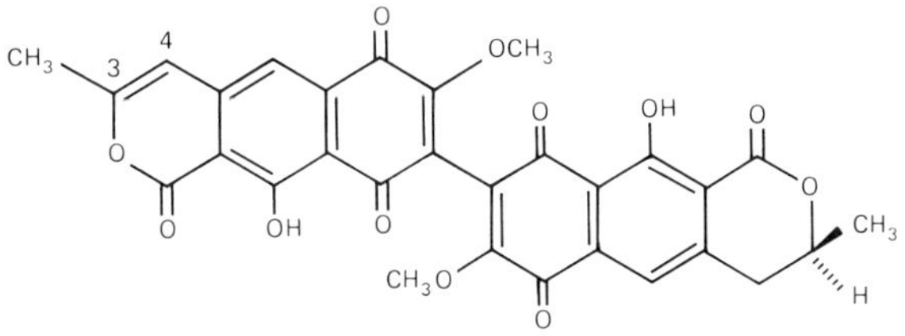

3,4-Dehydroxanthomegnin

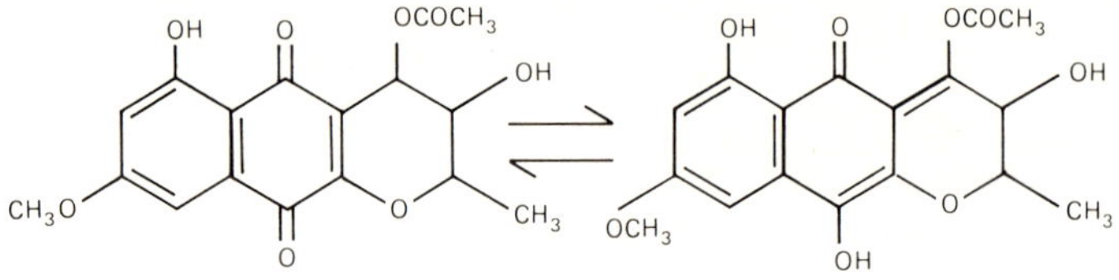

Gunacin

32120-7944

NAME:	WS-5995-A
PO:	S.AURANTICOLOR
CT:	P-NAPHTOQUINONE, ACIDIC
FORMULA:	C19H12O6
MW:	336
PC:	ORANGE, YELLOW, CRYST.
UV:	MEOH: (242, ,)(303, ,)(426, ,)
UV:	THF: (242, 946,)(303, 343,)(434, 260,)
SOL-FAIR:	MEOH, ETOH, THF
SOL-POOR:	ACET, HEX, W
QUAL:	(FECL3, +)
TO:	(PROTOZOA,)
LD50:	NONTOXIC
IS-FIL:	ORIG.
IS-EXT:	(ETOAC, , FILT.)
IS-CHR:	(SILG, BENZ)
IS-CRY:	(CRYST., THF)

REFERENCES:

TL, 4359, 1980; *JA*, 33, 1103, 1107, 1980; JP 81/2574

32120-8070

NAME:	WS-5995-B
PO:	S.AURANTICOLOR
CT:	P-NAPHTOQUINONE, ACIDIC
FORMULA:	C19H14O6
MW:	338
PC:	YELLOW, CRYST.
UV:	ETOH: (275, 280,)(408, 150,)
SOL-GOOD:	MEOH, CHL
SOL-POOR:	BENZ, HEX, W
QUAL:	(FECL3, +)
TO:	(P.VULG., 100)(PROTOZOA,)
IS-EXT:	(ETOAC, , EVAP.FILT.)
IS-CHR:	(SILG, BENZ-ETOAC)
IS-CRY:	(CRYST., THF)

REFERENCES:

JA, 33, 1103, 1107, 1980

32120-8531

NAME: TRICHIONE
PO: TRICHIA FLORIFORMIS, METATRICHIA VESPARIUM
CT: P-NAPHTOQUINONE, ACIDIC
FORMULA: C18H14O9
MW: 374
PC: ORANGE, POW.
TO: ANTIBIOTIC
REFERENCES:
Pure Appl. Chem., 53, 1233, 1981

32120-8532

NAME: HOMOTRICHIONE
PO: TRICHIA FLORIFORMIS, METATRICHIA VESPARIUM
CT: P-NAPHTOQUINONE, ACIDIC
FORMULA: C20H18O9
MW: 402
PC: ORANGE, POW.
TO: ANTIBIOTIC
REFERENCES:
Pure Appl. Chem., 53, 1233, 1981

32221-7945

NAME: PROACTINORHODIN
PO: S.COELICOLOR, S.VIOLACEORUBER
CT: NAPHTOQUINONE DERIV., ACTINORHODIN T., ACIDIC
FORMULA: C16H16O7.N
PC: PINK, CRYST.
REFERENCES:
Ber., 88, 778, 1955

32222-7666

PO: S.PRUNICEUS
CT: NAPHTOQUINONE DERIV., RUBROMYCIN T., ACIDIC
EA: (C, 49)(H, 6)(O, 45)
PC: RED, POW.
OR: (+24.2,)
UV: MEOH: (255, ,)(315,)(365, ,)(510, ,)
SOL-GOOD: MEOH, CHL
SOL-POOR: W
TO: (G.POS.,)
IS-CRY: (PREC., CHL, HEX)
REFERENCES:
Egypt J. Microb., 13, 37, 1979; *CA*, 92, 194063

32223-6728

NAME: GRANATICINIC ACID
PO: S.SP.
CT: ACIDIC
FORMULA: C22H22O11
MW: 462
PC: RED, POW.
TO: (S.AUREUS, 50)(B.SUBT., 50)(S.LUTEA, 500)
IS-FIL: 2.5
IS-EXT: (CHL, 2.5, FILT.)
IS-CHR: (SEPHADEX LH-20, ACET)
IS-CRY: (PREC., ACET, HEX)
REFERENCES:

Monatsh., 110, 531, 1979

32223-7946

NAME: GRANATOMYCIN-A, METHYL-DIHYDROGRANATICIN
PO: S.LATERITUS
CT: NAPHTOQUINONE DERIV., GRANATICIN T., ACIDIC
FORMULA: C23H24O10
MW: 460
PC: RED, CRYST.
UV: MEOH: (272, ,)(495, ,)(520, ,)(556, ,)
UV: MEOH-NAOH: (360, ,)(585, ,)(625, ,)
SOL-GOOD: MEOH, BENZ
SOL-POOR: W
TO: (P.VULG.,)(S.AUREUS, 975)(E.COLI, 10)(PS.AER., 10)(MYCOB.SP., .6)(FUNGI, 10)(B.SUBT., .15)
LD50: (28, IP)
TV: ANTIVIRAL
IS-FIL: ORIG.
IS-EXT: (BUOH, , FILT.)(BUOH, , MIC.)
IS-CHR: (SEPHADEX LH-20, MEOH)(SILG, CHL)
REFERENCES:

Z. Allg. Mikr., 20, 543, 1980; DDR P 147115

32224-6762

NAME: <u>O-ETHYLFUSARUBIN</u>
PO: FUS.SOLANI
CT: NAPHTOQUINONE DERIV., NAPHTAZARIN, ACIDIC
FORMULA: C17H18O7
MW: 334
PC: ORANGE, POW.
UV: MEOH: (301, ,)(472, ,)(498, ,)(535, ,)
TO: (S.LUTEA, 20)(B.SUBT., 75)(S.AUREUS, 75)
(E.COLI, 75)(C.ALB., 75)(S.CEREV., 75)
IS-FIL: 2
IS-EXT: (CHL, , MIC.)(CHL, 2, EVAP.FILT.)
IS-CHR: (SILG, CHL)
IS-CRY: (DRY, CHL)
REFERENCES:

JA, 32, 679, 685, 1979; 33, 1376, 1980; *Can. J. Chem.*, 56, 1593, 1978

32224-6763

NAME: <u>DIHYDROFUSARUBIN</u>
PO: FUS.SOLANI
CT: NAPHTOQUINONE DERIV., NAPHTAZARIN, ACIDIC
FORMULA: C15H16O7
MW: 308
PC: YELLOW, POW.
UV: MEOH: (250, ,)(272, ,)(300, ,)(388, ,)
TO: (S.LUTEA, 20)(B.SUBT., 50)(S.AUREUS, 50)
(E.COLI, 75)(C.ALB., 30)(S.CEREV., 50)(FUNGI, 40)
IS-FIL: 2
IS-EXT: (CHL, 2, EVAP.FILT.)
IS-CHR: (SILG, CHL)
IS-CRY: (DRY, CHL)
REFERENCES:

JA, 32, 679, 685, 1979; 33, 1376; 1980; *Can. J. Chem.*, 56, 1593, 1978

32224-6764

NAME: 3-O-ETHYLDIHYDROFUSARUBIN
PO: FUS.SOLANI
CT: ACIDIC, NAPHTOQUINONE DERIV., NAPHTAZARIN
FORMULA: C17H20O7
MW: 336
PC: YELLOW, POW.
UV: MEOH: (250, ,)(272, ,)(300, ,)(388, ,)
TO: (S.LUTEA, 20)(S.AUREUS, 40)(B.SUBT., 40)
(E.COLI, 75)(C.ALB., 100)(S.CEREV., 75)(FUNGI, 40)
IS-EXT: (CHL, 2, EVAP.FILT.)
REFERENCES:
JA, 32, 679, 685, 1979; 33, 1376, 1980; *Can. J. Chem.*, 56, 1593, 1978

32225-6765

NAME: NANAOMYCIN-E
PO: S.ROSA-NOTOENSIS
CT: ACIDIC, NAPHTOQUINONE DERIV., KALAFUNGIN T.
FORMULA: C16H14O7
MW: 318
PC: ORANGE, CRYST.
OR: (+89, MEOH)
UV: MEOH: (236, ,)(364, ,)(232, , 15000)(276, , 2930)(361, , 3870)
SOL-GOOD: MEOH, ETOAC
SOL-POOR: HEX
QUAL: (FECL3, +)(EHRL., -)(NINH., -)
TO: (S.AUREUS, 25)(B.SUBT., 11.5)(E.COLI, 50)
(P.VULG., 25)
LD50: (60, IP)
IS-EXT: (ETOAC, 4.5, FILT.)
IS-CHR: (SILG, BENZ-ETOAC)
IS-CRY: (CRYST., CH2CL2-HEX)
REFERENCES:
JA, 32, 442, 1979; BP 2015525; 2015526

32226-8177

NAME: 3.4-DEHYDROXANTHOMEGNIN
PO: NANNIZZIA CAJETANI, MICROSPORUM COOKEI
CT: NAPHTOQUINONE DERIV., PYRANONAPHTOQUINONE, ACIDIC
FORMULA: C30H20O12
MW: 572
PC: RED, POW.
UV: MEOH: (237, ,)(385, ,)
UV: MEOH-NAOH: (394, ,)(550, ,)
TO: (B.SUBT., 100)
REFERENCES:
Coll., 46, 1210, 1981; *Ber.*, 112, 957, 1979; *Exp.*, 26, 803, 1970

32227-7113

NAME: GUNACIN
PO: USTILAGO SP.
CT: NAPHTOQUINONE DERIV., ACIDIC
FORMULA: C17H16O8
MW: 348
PC: ORANGE, POW.
UV: ETOH: (228, ,)(267, ,)(314|3, ,)(438|8, ,)
UV: ETOH-NAOH: (238, ,)(290, ,)(315|5, ,) (533|12, ,)
SOL-GOOD: MEOH, CHL
SOL-POOR: W, HEX
TO: (S.AUREUS, .025)(E.COLI, 1.25)(P.VULG., 1.25) (K.PNEUM., 50)(PS.AER., 50)(C.ALB., 10)
LD50: (12, IV)(16, IP)
TV: HELA, ANTIVIRAL
IS-EXT: (CHL, 3, FILT.)
IS-CHR: (SILG, CHL-MEOH)
IS-CRY: (DRY, MEOH)
REFERENCES:
JA, 32, 1104, 1979

32228-8633

NAME: FREDERICAMYCIN-A, FCRC-A-48-A
PO: S.GRISEUS
PC: GREEN, POW.
TO: (G.POS.,)(FUNGI,)
TV: P-388, L-1210
REFERENCES:

32320-6766

NAME: A-5945
PO: S.SP.
CT: BASIC, XANTHOMYCIN T.
EA: (C, 57)(H, 7)(N, 8)
MW: 340
PC: ORANGE, POW., RED, BROWN
OR: (+282, MEOH)(+282, ETOH)
UV: HCL: (289, 121,)(405, 27,)
UV: MEOH: (271, 114,)(315, 50,)(450, 27,)
UV: W: (283, 149,)(408, 59,)
SOL-GOOD: MEOH, ETOH, ACET, CHL, W
SOL-FAIR: ET2O
SOL-POOR: ET2O, HEX
TO: (S.AUREUS, .006)(B.SUBT., .013)(S.LUTEA, .006)
(E.COLI, .05)(K.PNEUM., .05)
LD50: (2.78, IP)
IS-EXT: (CHL, 8.5, FILT.)(W, 2, CHL)
IS-CHR: (CM-SEPHADEX,)(SEPHADEX LH-20, MEOH)
IS-CRY: (PREC., ETOH, ACET)
REFERENCES:
JP 79/41393; *CA*, 91, 122146

32320-8533

NAME: XANTHOMYCIN-LIKE
PO: S.SP.
CT: XANTHOMYCIN T., BASIC
EA: (C, 51)(H, 6)(N, 8)
PC: RED, ORANGE, CRYST.
UV: HCL: (265, 311,)(358, 21,)
UV: NAOH: (270, 201,)(504, 29,)
SOL-GOOD: MEOH, BENZ
SOL-POOR: W
STAB: (ACID, +)(BASE, -)
TO: (S.AUREUS, .02)(B.SUBT., .1)(S.LUTEA, .02)
(E.COLI, .8)(P.VULG., 12.5)(K.PNEUM., 12.5)
(SHYG., 25)(PS.AER., 100)
LD50: (1.5, IV)
TV: P-388, S-180, EHRLICH
IS-EXT: (ETOAC, 7.7, W)(W, 1.5, ETOAC)
IS-ION: (IRC-50-H, HCL)
IS-CRY: (LIOF., W)
REFERENCES:
JA, 34, 856, 1981

33
BENZOQUINONES

331 Simple Benzoquinones
332 Condensed Benzoquinones

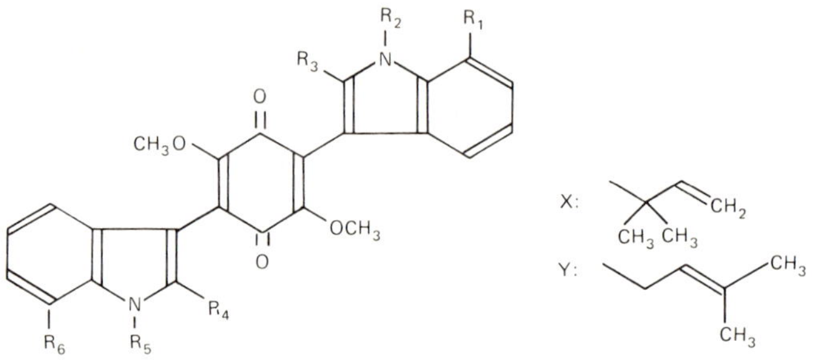

Asterriquinone	R_1	R_2	R_3	R_4	R_5	R_6
A_1	H	X	H	H	X	H
A_2	Y	H	H	X	H	Y
A_3	H	H	X	H	X	H
A_4	H	H	X	X	H	Y
B_1	Y	H	H	X	H	H
B_2	H	H	H	X	H	Y
B_3	H	H	H	H	X	H
B_4	H	H	X	X	H	H
C_1	H	H	H	X	H	H
C_2	H	H	H	H	H	X
D	H	H	H	H	H	H

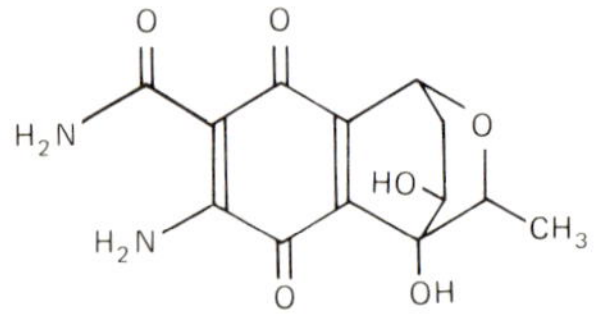

U-58431

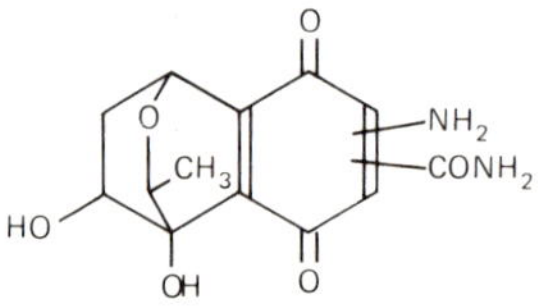

Sarubicin A

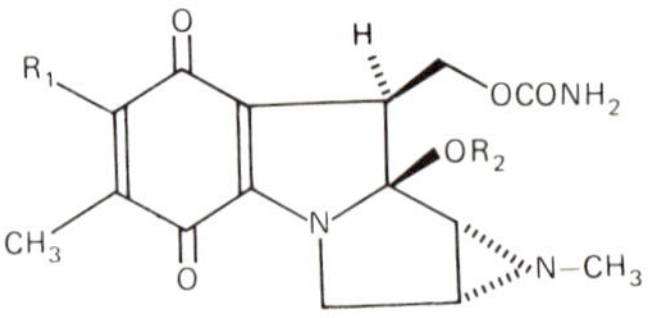

	R_1	R_2
7-Demethoxy-7-aminomitomycin B	NH_2	H
7-Demethoxy-7-amino-9a-O-methylmitomycin B	NH_2	CH_3
9a-O-Methylmitomycin B	CH_3O	CH_3

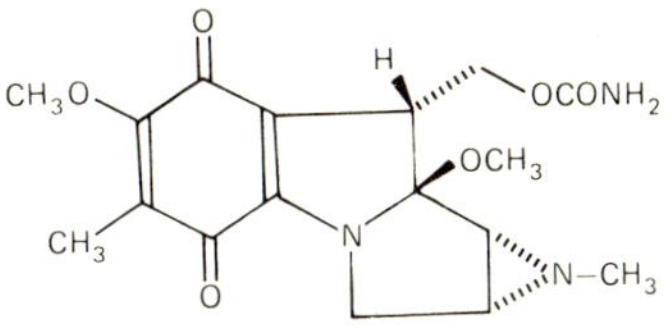

1a-N-Methylmitomycin A

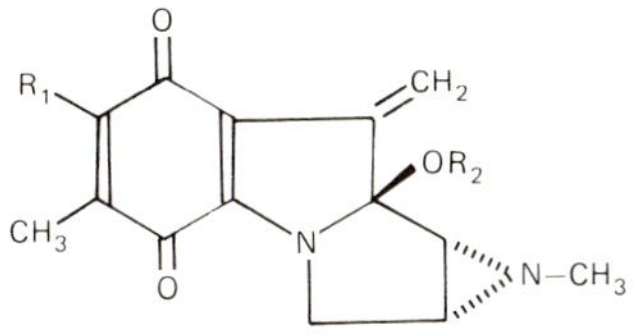

	R_1	R_2
Mitomycin G (10-Decarbamoyloxy-9-dehydroporfiromycin)	NH_2	CH_3
Mitomycin H (10-Decarbamoyloxy-9-dehydro-N-methylimitomycin A)	CH_3O	CH_3
10-Decarbamoyloxy-9-dehydromitomycin B	CH_3O	H

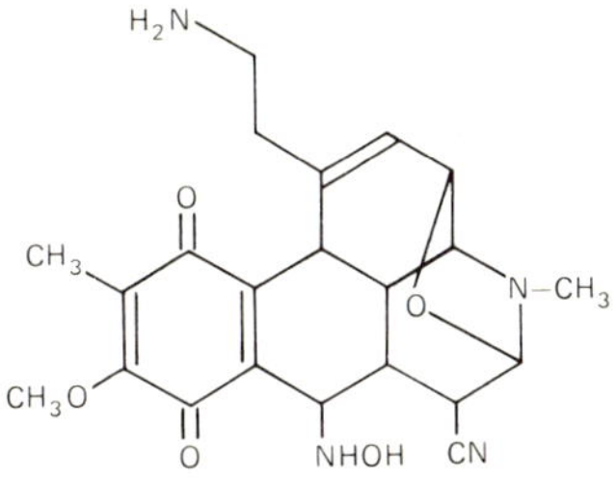

49-A

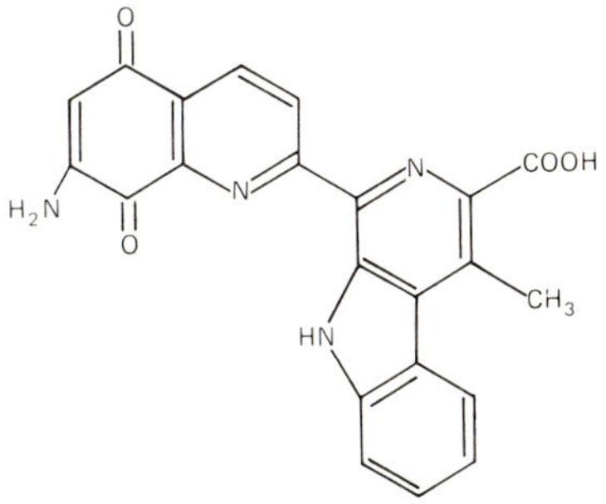

Lavendomycin
(K-82-A)

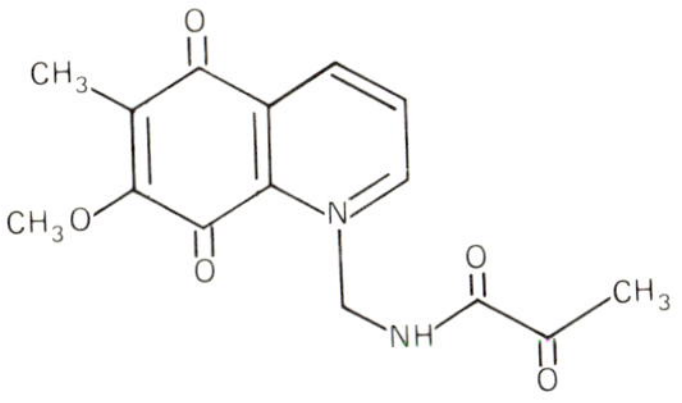

Mimocin

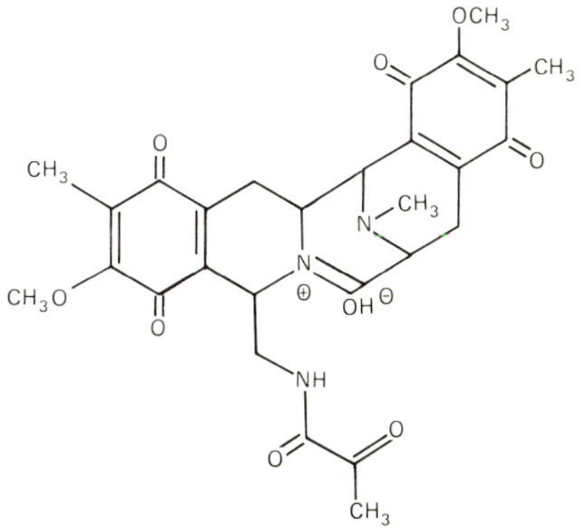

Saframycin S

33140-8145

NAME: P1
IDENTICAL: PLEUROTIN
PO: HOHENBUEHELIA GEOGENIUS
CT: BENZOQUINONE DERIV., NEUTRAL
FORMULA: C21H22O5
MW: 354
PC: YELLOW, CRYST.
UV: MEOH: (248, , 101)
SOL-POOR: W
QUAL: (DNPH, +)
LD50: (25, IP)
TV: L-1210, EHRLICH
IS-EXT: (CHL, , FILT.)
IS-CHR: (SILG, BENZ-ET2O)
IS-CRY: (CRYST., CHL-CCL4)
REFERENCES:
Arzn. Forsch., 31, 293, 1981

33140-8534

NAME: ASTERRIQUINONE A-1, DIMETHYLASTERRIQUINONE
PO: ASP.TERREUS-AFRICANUS
CT: BENZOQUINONE DERIV., NEUTRAL
FORMULA: C34H34N2O4
EA: (N, 5)
MW: 534
PC: PURPLE, RED, CRYST.
OR: (0,)
UV: ETOH: (295, ,)(299, ,)(490, ,)
SOL-GOOD: MEOH, ET2O
SOL-POOR: W
TV: ANTITUMOR
IS-EXT: (HEX, , MIC.)
IS-CHR: (SILG, BENZ)(AL, BENZ-ETOAC)
REFERENCES:
Chem. Ph. Bull., 29, 961, 1008, 1981; JP 81/32456; *CA*, 95, 130971

33140-8535

NAME:	ASTERRIQUINONE A-2
PO:	ASP.TERREUS-AFRICANUS
CT:	BENZOQUINONE DERIV., NEUTRAL
FORMULA:	C39H42N2O4
EA:	(N, 5)
MW:	602
PC:	PURPLE, CRYST.
OR:	(0,)
UV:	ETOH: (283, ,)(292, ,)(374, ,)(505, ,)
SOL-GOOD:	MEOH, ET2O
SOL-POOR:	W
TV:	ANTITUMOR
IS-EXT:	(HEX, , MIC.)
IS-CHR:	(SILG, BENZ)(AL, BENZ-ETOAC)
REFERENCES:	

Chem. Ph. Bull., 29, 961, 991, 1008, 1981; JP 81/32456; *CA*, 95, 130971

33140-8536

NAME:	ASTERRIQUINONE A-3
PO:	ASP.TERREUS-AFRICANUS
CT:	BENZOQUINONE DERIV., NEUTRAL
FORMULA:	C34H34N2O4
EA:	(N, 5)
MW:	534
PC:	PURPLE, POW.
OR:	(0,)
UV:	ETOH: (284, ,)(292, ,)(376, ,)(516, ,)
SOL-GOOD:	MEOH, ET2O
SOL-POOR:	W
TV:	ANTITUMOR
IS-EXT:	(HEX, , MIC.)
IS-CHR:	(SILG, BENZ)(AL, BENZ-ETOAC)
REFERENCES:	

Chem. Ph. Bull., 29, 961, 991, 1008, 1981; JP 81/32456; *CA*, 95, 130971

33140-8537

NAME:	<u>ASTERRIQUINONE A-4</u>
PO:	ASP.TERREUS-AFRICANUS
CT:	BENZOQUINONE DERIV., NEUTRAL
FORMULA:	C39H42N2O4
EA:	(N, 5)
MW:	602
PC:	PURPLE, RED, POW.
OR:	(0,)
UV:	ETOH: (284, ,)(291, ,)(390, ,)(497, ,)
SOL-GOOD:	MEOH, ET2O
SOL-POOR:	W
TV:	ANTITUMOR
IS-EXT:	(HEX, , MIC.)
IS-CHR:	(SILG, BENZ)(AL, BENZ-ETOAC)

REFERENCES:

Chem. Ph. Bull., 29, 961, 991, 1008, 1981; JP 81/32456; *CA*, 95, 130971

33140-8538

NAME:	<u>ASTERRIQUINONE B-1</u>
PO:	ASP.TERREUS-AFRICANUS
CT:	BENZOQUINONE DERIV., NEUTRAL
FORMULA:	C34H34N2O4
EA:	(N, 5)
MW:	534
PC:	PURPLE, CRYST.
OR:	(0,)
UV:	ETOH: (285, ,)(292, ,)(382, ,)(522, ,)
SOL-GOOD:	MEOH, BENZ
SOL-POOR:	W
TV:	ANTITUMOR
IS-EXT:	(ET2O, , MIC.)
IS-CHR:	(SILG, BENZ-ETOAC)(AL, BENZ-ETOAC)
IS-CRY:	(PREC., ET2O, HEX)

REFERENCES:

Chem. Ph. Bull., 29, 961, 991, 1008, 1981; JP 81/32456; *CA*, 95, 130971

33140-8539

NAME: ASTERRIQUINONE B-2
PO: ASP.TERREUS, AFRICANUS
CT: BENZOQUINONE DERIV., NEUTRAL
FORMULA: C34H34N2O4
EA: (N, 5)
MW: 534
PC: PURPLE, CRYST.
OR: (0,)
UV: ETOH: (283, ,)(292, ,)(372, ,)(518, ,)
SOL-GOOD: MEOH, BENZ
SOL-POOR: W
TV: ANTITUMOR
IS-EXT: (ET2O, , MIC.)
IS-CHR: (SILG, BENZ-ETOAC)(AL, BENZ-ETOAC)
IS-CRY: (PREC.ET2O, HEX)
REFERENCES:

Chem. Ph. Bull., 29, 961, 991, 1008, 1981; JP 81/32456; *CA*, 95, 130971

33140-8540

NAME: ASTERRIQUINONE B-3
PO: ASP.TERREUS-AFRICANUS
CT: BENZOQUINONE DERIV., NEUTRAL
FORMULA: C29H26N2O4
EA: (N, 6)
MW: 466
PC: RED, PURPLE, POW.
OR: (0,)
UV: ETOH: (284, ,)(240, ,)(484, ,)
SOL-GOOD: MEOH, BENZ
SOL-POOR: W
TV: ANTITUMOR
IS-EXT: (ET2O, , MIC.)
IS-CHR: (SILG, BENZ-ETOAC)(AL, BENZ-ETOAC)
IS-CRY: (PREC.ET2O, HEX)
REFERENCES:

Chem. Ph. Bull., 29, 961, 991, 1008, 1981; JP 81/32456; *CA*, 95, 130971

33140-8541

NAME: ASTERRIQUINONE B-4
PO: ASP.TERREUS-AFRICANUS
CT: BENZOQUINONE DERIV., NEUTRAL
FORMULA: C34H34N2O4
EA: (N, 5)
MW: 534
PC: RED, PURPLE, CRYST.
OR: (0,)
UV: ETOH: (284, ,)(292, ,)(385, ,)(502, ,)
SOL-GOOD: MEOH, BENZ
SOL-POOR: W
TV: ANTITUMOR
IS-EXT: (ET2O, , MIC.)
IS-CHR: (SILG, BENZ-ETOAC)(AL, BENZ-ETOAC)
IS-CRY: (PREC., ET2O, HEX)
REFERENCES:
Chem. Ph. Bull., 29, 961, 991, 1008, 1981; JP 81/32456; *CA*, 95, 130971

33140-8542

NAME: ASTERRIQUINONE C-1
PO: ASP.TERREUS-AFRICANUS
CT: BENZOQUINONE DERIV., NEUTRAL
FORMULA: C29H26N2O4
EA: (N, 6)
MW: 466
PC: PURPLE, CRYST.
OR: (0,)
UV: ETOH: (283, ,)(291, ,)(267, ,)(510, ,)
TV: ANTITUMOR
IS-EXT: (ET2O, , MIC.)
IS-CHR: (SILG, BENZ-ETOAC)(AL, BENZ-ETOAC)
IS-CRY: (PREC., ET2O, HEX)
REFERENCES:
Chem. Ph. Bull., 29, 961, 991, 1008, 1981; JP 81/32456; *CA*, 95, 130971

33140-8543

NAME:	ASTERRIQUINONE-C-2
PO:	ASP.TERREUS-AFRICANUS
CT:	BENZOQUINONE DERIV., NEUTRAL
FORMULA:	C29H26N2O4
EA:	(N, 6)
MW:	466
PC:	PURPLE, CRYST.
OR:	(0,)
UV:	ETOH: (285, ,)(290, ,)(485, ,)
TV:	ANTITUMOR
IS-EXT:	(ET2O, , MIC.)
IS-CHR:	(SILG, BENZ-ETOAC)(AL, BENZ-ETOAC)
IS-CRY:	(PREC., ET2O, HEX)

REFERENCES:

Chem. Ph. Bull., 29, 961, 991, 1008, 1981; JP 81/32456; *CA*, 95, 130971

33140-8544

NAME:	ASTERRIQUINONE-D
PO:	ASP.TERREUS-AFRICANUS
CT:	BENZOQUINONE ERIV., NEUTRAL
FORMULA:	C24H18N2O4
EA:	(N, 7)
MW:	398
PC:	PURPLE, POW.
OR:	(0,)
UV:	ETOH: (284, ,)(290, ,)(488, ,)
TV:	ANTITUMOR
IS-EXT:	(ET2O, , MIC.)
IS-CHR:	(SILG, BENZ-ETOAC)(AL, BENZ-ETOAC)
IS-CRY:	(PREC., ET2O, HEX)

REFERENCES:

Chem. Ph. Bull., 29, 961, 991, 1008, 1981; JP 81/32456; *CA*, 95, 130971

33150-6729

NAME: SARUBICIN-A, SARCINAMYCIN-A
IDENTICAL: U-58431
PO: S.VIOLACEORUBER
CT: BASIC, GRANATICIN T.
FORMULA: C13H14N2O6
EA: (N, 9)
MW: 294
PC: RED, CRYST.
UV: CHL: (262, , 12600)(498, , 1580)
SOL-GOOD: MEOH, BENZ
SOL-POOR: W
TO: (S.LUTEA, 10)(E.COLI, 50)
IS-FIL: 4
IS-EXT: (BUOH, 4, FILT.)
IS-CHR: (SILG, ETOAC)(AL, CHL-BUOH)
IS-CRY: (CRYST., ETOAC)
REFERENCES:
JA, 33, 787, 1980; *Z. Chem.*, 20, 1980; *Tetr.*, 37, 1961, 1981; *Proc. Antib. Symp., Weimar*, 1979, B-31

33150-7947

NAME: U-58431
IDENTICAL: SARUBICIN-A
PO: S.HELICUS
CT: GRANATICIN T., AMPHOTER
FORMULA: C13H14N2O6
EA: (N, 9)
MW: 294
PC: ORANGE, RED, CRYST.
OR: (+9, MEOH)
UV: MEOH: (216, ,13171)(472, ,1517)
UV: MEOH-HCL: (261, ,)(472, ,)
UV: MEOH-NAOH: (261, ,)(472, ,)
SOL-GOOD: MEOH, BUOH
SOL-POOR: W, HEX
TO: (S.AUREUS, 125)(DIPLOCOCCUS PNEUMONIAE, 1)
(E.COLI, 250)(P.VULG., 250)(K.PNRUM., 125)
IS-FIL: 3
IS-EXT: (BUOH, ,FILT.+AMMONIUM SULPHATE)
IS-ION: (XAD-4, MEOH-W)
IS-CHR: (SILG, CHL-MEOH)
REFERENCES:
JA, 33, 919, 1980

33211-7114

NAME: 7-DEMETHOXY-7-AMINOMITOMYCIN-B
PO: S.CAESPITOSUS
CT: MITOMYCIN T., BASIC
FORMULA: C15H18N4O5
EA: (N, 17)
MW: 334
PC: BLUE, VIOLET, CRYST.
UV: MEOH: (217, ,)(361, ,)(558, ,)
TO: (G.POS.,)
TV: ANTITUMOR
IS-EXT: (CHL, , W+NACL)
IS-ION: (DIAION HP-20, MEOH)
IS-CHR: (SILG, CHL-MEOH)
IS-CRY: (CRYST., BENZ-MEOH)
REFERENCES:

Progr. AAC, 1969, 112; JP 79/122797; *CA*, 92, 179073

33211-7667

NAME: 7-DEMETHOXY-7-AMINO-9A-O-METHYLMITOMYCIN-B
PO: S.CAESPITOSUS
CT: MITOMYCIN T., BASIC
FORMULA: C16H20N4O5
EA: (N, 16)
MW: 348
PC: PURPLE, CRYST.
UV: MEOH: (217, , 17800)(361, , 19100)(558, , 215)
TO: (G.POS.,)
TV: ANTITUMOR
IS-EXT: (CHL, , NACL-W)
IS-ION: (DIAION HP-20, MEOH-W)
IS-CHR: (SILG, CHL-MEOH)(SILG, ETOAC-ACET)(AL, CHL-MEOH)
IS-CRY: (CRYST., CHL-ACET)
REFERENCES:

Progr. AAC, 1969, 112; JP 79/122797; *CA*, 92, 179073

33211-7668

NAME: <u>9A-O-METHYLMITOMYCIN-B</u>
PO: S.CAESPITOSUS
CT: MITOMYCIN T., BASIC
FORMULA: C17H21N3O6
EA: (N, 12)
MW: 363
PC: RED, VIOLET, CRYST.
UV: MEOH: (220, , 12050)(324, , 7580)(520, , 850)
SOL-GOOD: MEOH, CHL
TO: (G.POS.,)
TV: ANTITUMOR
IS-FIL: ORIG.
IS-EXT: (CHL, , W-NACL)
IS-ION: (DIAION HP-20, MEOH)
IS-CHR: (SILG, CHL-MEOH)
IS-CRY: (CRYST., BENZ-MEOH)
REFERENCES:

JP 80/45322; *CA*, 93, 166099

33211-7669

NAME: <u>1A-N-METHYLMITOMYCIN-A</u>
PO: S.CAESPITOSUS
CT: MITOMYCIN T., BASIC
FORMULA: C17H21N3O6
EA: (N, 12)
MW: 363
PC: RED, CRYST.
UV: MEOH: (219, , 12550)(324, , 10200)(539, , 1230)
TV: ANTITUMOR
REFERENCES:

JP 80/45322; *CA*, 93, 166099

33211-7670

NAME: MITOMYCIN-G, 7-AMINO-9A-O-METHYL-10-DECARBAMOYLOXY-9, -DEHYDRO-7-DEMETHYLMITOMYCIN-B, 10-DECARBAMOYLOXY-9-DEHYDROPORFIROMYCIN
PO: S.CAESPITOSUS
CT: MITOMYCIN T., BASIC
FORMULA: C15H17N3O3
EA: (N, 14)
MW: 287
PC: CRYST., GREEN
UV: MEOH: (222, , 10500)(289, , 10700)(373, , 17800)(602, , 234)
TO: (S.AUREUS, 50)(B.SUBT., .19)(K.PNEUM., 12.5)(P.VULG., 50)
TV: ANTITUMOR, KB
IS-EXT: (CHL, , W+NACL)(CHL, FILT.)
IS-ION: (DIAION HP-20, MEOH)(DIAION HP-20, MEOH-W)
IS-CHR: (SILG, CHL-MEOH)(SILG, ACET-CHL)(AL, CHL)
REFERENCES:
JA, 34, 243, 1152, 1981; *Proc. 23rd Symp. Chem. Nat. Prod., Nagoya*, 1980; JP 80/15408; 80/118396; *CA*, 93, 43923; 94, 82164

33211-7671

NAME: MITOMYCIN-K, 9A-O-METHYL-10-DECARBAMOYLOXY-9-DEHYDROMITOMYCIN-B, 10-DECARBAMOYLOXY-9-DEHYDRO-N-METHYLMITOMYCIN-A
PO: S.CAESPITOSUS
CT: MITOMYCIN T., BASIC
FORMULA: C16H18N2O4
EA: (N, 9)
MW: 302
PC: CRYST., PURPLE
TO: (S.AUREUS, 6.25)(B.SUBT., .39)(K.PNEUM., 12.5)(P.VULG., 50)
TV: ANTITUMOR, KB
IS-EXT: (CHL, , W+NACL)(CHL, FILT.)
IS-ION: (DIAION HP-20, MEOH)(DIAION HP-20, MEOH-W)
IS-CHR: (SILG, CHL-MEOH)(SILG, ACET-CHL)(AL, CHL)
REFERENCES:
JA, 34, 243, 1152, 1981; *Proc. 23rd Symp. Chem. Nat. Prod., Nagoya*, 1980; JP 80/15408; 80/118396; *CA*, 93, 43923; 94, 82164

33211-7672

NAME: MITOMYCIN-H, 10-DECARBAMOYLOXY-9-DEHYDROMITOMYCIN-B
PO: S.CAESPITOSUS
CT: MITOMYCIN T., BASIC
FORMULA: C15H16N2O4
EA: (N, 10)
MW: 288
PC: CRYST., BLUE, PURPLE
UV: MEOH: (226, , 12000)(291, , 11700)(578, , 1100)
TO: (S.AUREUS, .78)(B.SUBT., .1)(E.COLI, 50) (K.PNEUM., 12.5)(P.VULG., 6.25)
LD50: (210, IP)
TV: ANTITUMOR, KB, S-180
IS-EXT: (CHL, , W+NACL)(CHL, FILT.)
IS-ION: (DIAION HP-20, MEOH)(DIAION HP-20, MEOH-W)
IS-CHR: (SILG, CHL-MEOH)(SILG, CHL-ACET)(SILG, ACET-CHL)(AL, CHL)
IS-CRY: (CRYST., ACET)
REFERENCES:

JA, 34, 243, 1152, 1981; *Proc. 23rd Symp. Chem. Nat. Prod., Nagoya*, 1980; JP 80/15408; 80/118396; *CA*, 93, 43923; 94, 82164

33212-7115

NAME: 49-A
PO: S.SP., S.FLAVOGRISEUS
CT: MITOMYCIN L., BASIC
FORMULA: C22H26N4O5
EA: (N, 13)
MW: 426
PC: ORANGE, CRYST.
UV: MEOH: (267, 273,)(360, 20,)
SOL-GOOD: MEOH, CHL, ETOAC
SOL-POOR: W, HEX, ACET
TO: (B.SUBT., .05)(S.AUREUS, .005)(S.LUTEA, .0025) (E.COLI, .15)(K.PNEUM., .31)(SHYG., .08) (PS.AER., .15)
IS-FIL: 4.5
IS-EXT: (ETOAC, 8.5, W)(W, 1, ETOAC)
IS-ION: (XAD-2, MEOH)
IS-CHR: (SILG, MEOH)(SILG, BENZ-MEOH)(SILG, CHL-MEOH)
IS-CRY: (CRYST., CHL-ACET)
REFERENCES:

JP, 79/126793; *CA*, 93, 68655

33220-7517

NAME: K-82-A
IDENTICAL: LAVENDAMYCIN
PO: S.LAVENDULAE
CT: ACIDIC, STREPTONIGRIN T.
EA: (C, 63)(H, 4)(N, 12)
PC: RED, CRYST.
OR: (+201, MEOH)
UV: MEOH: (230, 790,)(250, 787,)(280, 525,) (390, 306,)
UV: MEOH-NAOH: (231, 710,)(300, ,)(380, 260,) (500, ,)
UV: NAOH: (230, 692,)(388, 241,)
SOL-GOOD: BASE, DMFA
SOL-FAIR: MEOH, CHL
SOL-POOR: BENZ, HEX, W, ACID
QUAL: (FEHL., -)(FECL3, -)(NINH., -)(BIURET, -) (PAULY, -)(SAKA., -)(EHRL., -)
TO: (FUNGI,)(C.ALB.,)(S.AUREUS, .5)(S.LUTEA, .2) (B.SUBT., .2)(E.COLI, 2)(PS.AER., 5)
TV: ANTITUMOR
IS-EXT: (ETOAC, 2, FILT.)
IS-CHR: (SILG,)(SEPHADEX LH-20,)
IS-CRY: (PREC., MEOH, ET2O)
REFERENCES:
JA, 33, 1231, 1980; JP 80/2624; *CA*, 92, 179074

33220-8545

NAME: AB-111
PO: ACTINOPLANES SP.
CT: STREPTONIGRIN T.
EA: (C, 54)(H, 4)(N, 9)(ASH, 7)
PC: RED, BROWN, POW.
UV: MEOH: (258, ,)(388, ,)
SOL-GOOD: MEOH, DMSO, PYR, W
SOL-FAIR: BUOH, ETOAC
SOL-POOR: CHL, HEX
QUAL: (FECL3, +)(NINH., -)
TO: (S.AUREUS, .39)(E.COLI, 3.13)(MYCOB.TUB., .3)
IS-EXT: (ETOAC, , FILT.)
REFERENCES:
JP 81/18594

33220-8800

NAME:	LAVENDAMYCIN
IDENTICAL:	K-82-A
PO:	S.LAVENDULAE
CT:	STREPTONIGRIN T., AMPHOTER
FORMULA:	C22H14N4O4
EA:	(N, 12)
MW:	398
PC:	RED, POW.
UV:	MEOH: (234, 492,)(246, 498,)(391, 211,)
UV:	MEOH-HCL: (252, 474,)(277, 360,)(385, 190,)
UV:	MEOH-NAOH: (245, 941,)(309, 423,)(390, ,)
SOL-GOOD:	MEOH, CHL
SOL-POOR:	W
TV:	ANTITUMOR

REFERENCES:

TL, 4595, 1981

33240-7673

NAME:	MIMOCIN
PO:	S.LAVENDULAE
CT:	BENZOQUINONE DERIV.
FORMULA:	C15H14N2O5
EA:	(N, 9)
MW:	302
PC:	YELLOW, CRYST.
UV:	MEOH: (243, ,)(322, ,)
TO:	(B.SUBT.,)(C.ALB.,)
IS-CRY:	(CRYST., ET2O)

REFERENCES:

TL, 3207, 1980

33250-7674

NAME: SAFRAMYCIN-S, DECYANOSAFRAMYCIN-A
PO: S.LAVENDULAE
CT: SAFRAMYCIN T., BASIC
FORMULA: C28H31N3O9
EA: (N, 8)
MW: 537
PC: YELLOW, POW.
OR: (+32.5, MEOH)
UV: MEOH: (268, , 15200)
SOL-GOOD: MEOH, CHL
SOL-FAIR: ET2O
SOL-POOR: HEX, W
QUAL: (NINH., -)(FECL3, -)(EHRL., -)
TO: (S.AUREUS,)(B.SUBT.,)(S.LUTEA,)(G.NEG.,) (FUNGI,)
TV: L-1210, EHRLICH
IS-EXT: (CHL, 7, FILT.)(W, 1, ETOAC)(CH2CL2, 8, W)
IS-CHR: (SILG, BENZ-ETOAC)
REFERENCES:
JA, 33, 951, 1980; *Abst. AAC*, 19, 853, 1979; *Gann*, 71, 790, 1980; *J. Pharm. Dyn.*, 4, 282, 1981

33260-6730

NAME: DNACTIN-A1, C-14482-A1
PO: NOC.SP.
CT: BASIC, QUINONE T.
EA: (C, 59)(H, 6)(N, 17)
MW: 460
PC: RED, BROWN, CRYST.
OR: (+150, ETOH)
UV: MEOH: (214, 503,)(281, 209,)(498, 46,)
SOL-GOOD: CHL, MEOH, DMSO
SOL-FAIR: ETOH, BUOH, ETOAC, W
SOL-POOR: HEX
QUAL: (SAKA., -)(EHRL., -)
STAB: (HEAT, -)(ACID, +)(BASE, -)
TO: (B.SUBT., 1)(S.AUREUS, 1)(S.LUTEA, 1)(E.COLI, 2)(PS.AER., 2)(P.VULG., 10)
LD50: (5, IV)
TV: P-388
REFERENCES:
JA, 33, 1437, 1443, 1980; DT 2833689; BP 2001954; EP 26338; *Agr. Biol. Ch.*, 45, 2013, 1981

33260-7919

NAME: DNACTIN-B1, C-14482-B1
PO: NOC.SP.
CT: QUINONE T., BASIC
EA: (C, 56)(H, 6)(N, 14)
MW: 400|50
PC: RED, BROWN, CRYST.
OR: (0, ETOH)
UV: MEOH: (213, 592,)(283, 227,)(496, 50.1,)
SOL-GOOD: MEOH, DMSO
SOL-FAIR: W, ETOAC, ET2O
SOL-POOR: HEX
QUAL: (NINH., -)(SAKA., -)(EHRL., -)
STAB: (ACID, -)(BASE, -)
TO: (B.SUBT., .05)(S.AUREUS, .01)(S.LUTEA, .2)
(E.COLI, .2)(PS.AER., .2)
LD50: (.94|.3, IV)
IS-FIL: 5
IS-EXT: (BUOH, 8, W)(W, 2.5, BUOH)(CHL, 8, W)(W, 2, CHL)
IS-ION: (DIAION HP-10, MEOH-W)
IS-CHR: (XAD-2, MEOH-W)
IS-CRY: (CRYST., ACET-HEX)
REFERENCES:

JA, 33, 1437, 1443, 1980; Holl. P. 80/627; DT 3003359

33260-7920

NAME: DNACTIN-B2, C-14482-B2
PO: NOC.SP.
CT: QUINONE T., BASIC
EA: (C, 57)(H, 7)(N, 13)
PC: RED, BROWN, CRYST.
UV: MEOH: (214, 555,)(283, 207,)(499, 55.8,)
SOL-GOOD: MEOH, ETOH, DMSO, CHL
SOL-FAIR: W, ETOAC, ET2O
SOL-POOR: HEX
QUAL: (NINH., -)(SAKA., -)
TO: (S.AUREUS,)(B.SUBT.,)(E.COLI,)(P.VULG.,)
(PS.AER.,)
IS-FIL: 5
IS-EXT: (BUOH, 8, W)(W, 2.5, BUOH)(CHL, 8, W)
IS-ION: (DIAION HP-10, MEOH-W)
IS-CHR: (XAD-2, MEOH-W)
REFERENCES:

JA-, 33, 1437, 1443, 1980; Holl. P 80/627; DT 3003359

33260-7921

NAME:	DNACTIN-B3, C-14482-B3
PO:	NOC.SP.
CT:	QUINONE T., BASIC
EA:	(C, 59)(H, 7)(N, 14)
PC:	BROWN, RED, POW.
UV:	MEOH: (214, 620,)(283, 251,)(492, 55.6,)
SOL-GOOD:	MEOH, ETOH, DMSO, CHL
SOL-FAIR:	ETOAC, ET2O, W
SOL-POOR:	HEX
QUAL:	(NINH., -)(SAKA., -)
TO:	(S.AUREUS,)(B.SUBT.,)

REFERENCES:

JA, 33, 1437, 1443, 1980; Holl. P 80/627; DT 3003359

34
QUINONE-LIKE COMPOUNDS

341 Semiquinone Antibiotics
342 Other Quinone-Like Compounds

Structures of the deflectins.

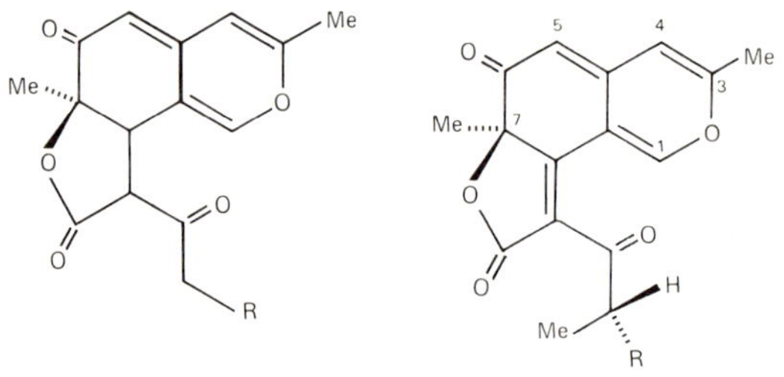

1a R = n-C_6H_{13} 2a R = n-C_8H_{17}
1b R = n-C_8H_{17} 2b R = n-$C_{10}H_{21}$
1c R = n-$C_{10}H_{21}$

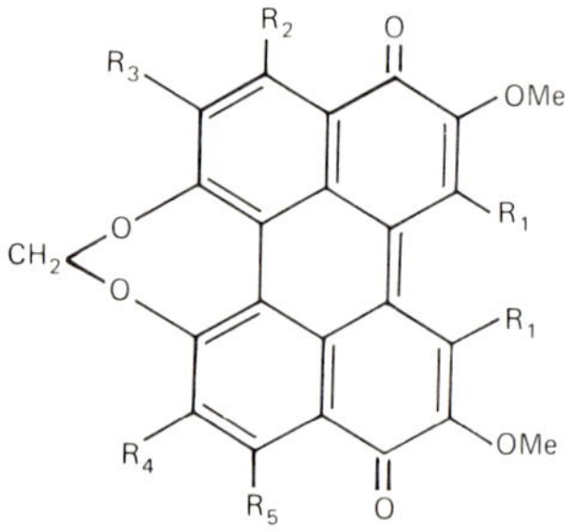

(I) (R_1 = — CH_2 — CH(OH) — CH_3,R_2 = R_4 = OH, R_3 = R_5 = H)
(III) (R_1 = — CH_2 — CH(OH) — CH_3,R_3 = R_4 = OH,R_2 = R_5 = H)

I. Cercosporin
III: Neocercosporin

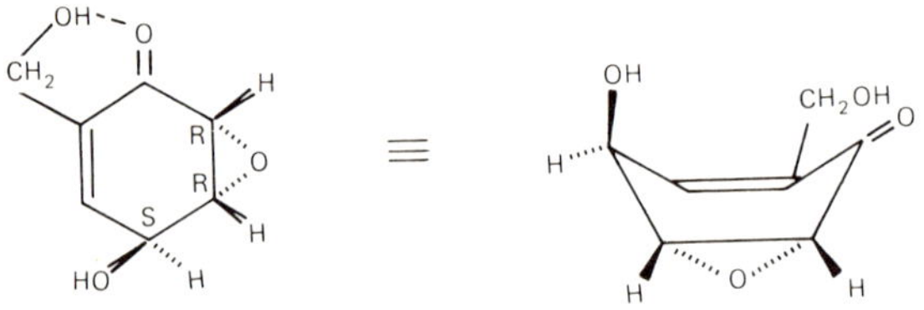

(+)-Isoepoxidon

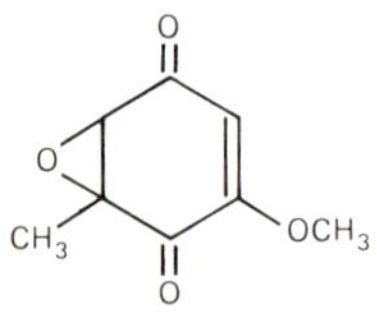

Coriloxin

34110-8801

NAME: DEFLECTIN-1A
PO: ASP.DEFLECTUS
CT: SEMIQUINONE, QUINONE METHIDE, NEUTRAL
FORMULA: C21H2405
MW: 356
PC: WH., CRYST.
OR: (+431, ETOAC)
UV: MEOH: (209, ,)(219, ,)(261, ,)(340, ,)
SOL-GOOD: MEOH, CHL
SOL-POOR: W, HEX
TO: (G.POS.,)
TV: EHRLICH
IS-EXT: (ACET, , MIC.)(CHL, , W)
IS-CHR: (SILG, CHL-MEOH)
IS-CRY: (CRYST., MEOH)
REFERENCES:

JA, 34, 923, 1981

34110-8802

NAME: DEFLECTIN-1B, DEFLECTIN-A
PO: ASP.DEFLECTUS
CT: SEMIQUINONE, QUINONE METHIDE, NEUTRAL
FORMULA: C23H2805
MW: 384
PC: WH., CRYST.
OR: (+405, ETOAC)
UV: MEOH: (209, , 7900)(219, , 7200)(261, , 11500)(340, , 10700)
UV: MEOH-NAOH: (210, ,)(265, ,)(300, ,)(367, ,)(530, ,)
SOL-GOOD: MEOH, CHL
SOL-POOR: W, HEX
TO: (B.SUBT., 10)(S.AUREUS,)(P.VULG., 50)
TV: EHRLICH
REFERENCES:

JA, 34, 923, 1981

34110-8803

NAME: DEFLECTIN-1C
PO: ASP.DEFLECTUS
CT: SEMIQUINONE, QUINONE METHIDE, NEUTRAL
FORMULA: C25H32O5
MW: 412
PC: WH., CRYST.
UV: MEOH: (209, ,)(219, ,)(261, ,)(340, ,)
SOL-GOOD: MEOH, CHL
SOL-POOR: W, HEX
TO: (G.POS.,)
REFERENCES:
JA, 34, 923, 1981

34110-8804

NAME: DEFLECTIN-2A, DEFLECTIN-B
PO: ASP.DEFLECTUS
CT: SEMIQUINONE, QUINONE METHIDE, NEUTRAL
FORMULA: C24H30O5
MW: 398
PC: WH., CRYST.
OR: (+397, ETOAC)
UV: MEOH: (208, , 8600)(218, , 7800)(261, , 12700)(338, , 11600)
UV: MEOH-NAOH: (210, , 10800)(265, , 11900)(300, , 5700)(367, , 10600)(530, , 9100)
SOL-GOOD: MEOH, CHL
SOL-POOR: HEX, W
TO: (B.SUBT., 5)(S.AUREUS,)(P.VULG., 20)
TV: EHRLICH
IS-EXT: (ACET, , MIC.)
IS-CHR: (SILG, CHL)
IS-CRY: (CRYST., MEOH)
REFERENCES:
JA, 34, 923, 1981

34110-8805

NAME: <u>DEFLECTIN-2B</u>
PO: ASP.DEFLECTUS
CT: SEMIQUINONE, QUINONE METHIDE, NEUTRAL
FORMULA: C26H34O5
MW: 426
PC: WH., CRYST.
OR: (+346, ETOAC)
UV: MEOH: (208, ,)(218, ,)(261, ,)(338, ,)
UV: MEOH-NAOH: (210, ,)(265, ,)(300, ,)(367, ,)(530, ,)
SOL-GOOD: MEOH, CHL
SOL-POOR: HEX, W
TO: (G.POS.,)
TV: EHRLICH
REFERENCES:
JA, 34, 923, 1981

34133-6767

NAME: <u>NEOCERCOSPORIN</u>, NEOSPORIN
PO: CERCOSPORA KIKUCHII
CT: ACIDIC, SEMIQUINONE
FORMULA: C29H26O10
MW: 534
PC: VIOLET, RED, CRYST.
UV: MEOH: (222, , 49000)(270, , 31500)(475, , 23500)(562, , 8500)
STAB: (LIGHT, +)(HEAT, +)
TO: (B.SUBT., .1)(S.LUTEA, 1)(E.COLI, .1)(PS.AER., .5)(P.VULG., 2.5)
IS-EXT: (ET2O, , MIC.)
IS-CHR: (CAHPO4, CHL)
IS-CRY: (CRYST., ETOH)
REFERENCES:
J. Ph. Soc. Jap., 98, 1553, 1978; 99, 20, 1979; JP 79/73746; *CA*, 91, 14461, 138856

34133-6769

NAME: CERCOSPORIN
PO: CERCOSPORA BETICOLA
CT: ACIDIC, SEMIQUINONE
FORMULA: C29H26010
MW: 534
PC: RED, CRYST.
OR: (+470, CHL)
UV: ETOH: (205, , 51900)(223, , 49100)(267|7, , 31400)(325, , 3500)(473, , 23600)(525, , 11400)(565, , 8600)
UV: MEOH: (223, 940,)(260, 638,)(271, 652,)(275, 650,)(470, 498,)
UV: NAOH: (250, , 48300)(288, , 49500)(297, , 48300)(480, , 23300)(595, , 9000)(690, , 13400)
SOL-GOOD: BASE, PYR, DIOXAN, MEOH, CHL
SOL-FAIR: BENZ, ET20
SOL-POOR: ACID, W, HEX
QUAL: (FECL3, +)
STAB: (HEAT, +)(LIGHT, -)
TO: (B.SUBT.,)(S.AUREUS,)(S.LUTEA,)(E.COLI,)(C.ALB.,)(S.CEREV.,)
LD50: PHYTOTOXIC
IS-EXT: (ET20, , MIC.)(ETOAC, , WB.)
IS-CHR: (CAHP04, CHL)
IS-CRY: (CRYST., ETOH)
REFERENCES:

Phytopath. Z., 44, 295, 1962; *JACS*, 79, 5727, 1957; *Agr. Biol. Ch.*, 36, 1707, 1972; 39, 287, 1975; *Trans. Brit. Mycol. Soc.*, 69, 496, 1978; 70, 77, 1978; *CC*, 1463, 1971; *Curr. Trends Life Sci.*, 6, 49, 1979; *J. Ph. Soc. Jap.*, 100, 900, 1980; *CA*, 93, 217663

34210-6768

NAME: +-ISOEPOXYDON, U-III
PO: P.URTICAE
CT: QUINONE L., NEUTRAL, EPOXYDON T
FORMULA: C7H804
MW: 156
PC: WH., CRYST.
OR: (+206, MEOH)
UV: MEOH: (212, , 5756)(240, , 3806)(329, , 367)
SOL-GOOD: ETOAC
TO: (B.SUBT.,)
IS-EXT: (ETOAC, 2, FILT.)
IS-CRY: (CRYST., 1.2-DICHLOROETAN)
REFERENCES:

Bioch. J., 182, 445, 1979

34210-8146

NAME: CORILOXIN
PO: CORIOLUS VERNICIPES
CT: NEUTRAL, EPOXYDON T., QUINONE L.
FORMULA: C8H10O4
MW: 170
PC: WH., CRYST.
UV: MEOH: (263, ,)
IS-EXT: (ACET, , MIC.)
IS-CHR: (SILG, BENZ-ETOAC)
REFERENCES:

JP 80/89274

30
OTHER LESS KNOWN QUINONE ANTIBIOTICS

30000-6700

NAME:	SF-1971
PO:	S.PHAEOPURPUREUS
CT:	NEUTRAL, ACIDIC, QUINONE T.
FORMULA:	C27H24010
EA:	(C, 65)(H, 5)(O, 30)
MW:	500, 508
PC:	YELLOW, CRYST., BROWN, RED, ORANGE
OR:	(-65, MEOH)
UV:	MEOH: (215, ,)(310, ,)(420, ,)
UV:	MEOH-HCL: (215, ,)(310, ,)(420, ,)
UV:	MEOH-NAOH: (280, ,)(386, ,)(530, ,)
SOL-GOOD:	MEOH, ETOAC
SOL-FAIR:	CHL, BENZ
SOL-POOR:	W
QUAL:	(NINH., -)
STAB:	(BASE, -)(ACID, +)
TO:	(G.POS.,)(G.NEG.,)(SHYG., 3.12)(MYCOPLASMA,)
LD50:	(7|3, IP)

REFERENCES:

JP 78/124691; *CA*, 90, 184891

30000-6758

NAME:	XK-99
PO:	MIC.MELANOSPORA-COMAENSIS
CT:	QUINONE T.
EA:	(C, 58)(H, 5)(N, 10)(O, 27)
MW:	390
PC:	RED, BROWN, POW.
OR:	(-360, MEOH)
UV:	MEOH: (385, ,)
UV:	MEOH-HCL: (252, ,)(385, ,)
UV:	MEOH-NAOH: (375|5, ,)
SOL-GOOD:	MEOH, ETOH, BUOH
SOL-POOR:	ETOAC, HEX
TO:	(S.AUREUS, .7)(B.SUBT., 10.5)(S.LUTEA, 5.3) (E.COLI, .7)(PS.AER., 1.4)(K.PNEUM., 2.7) (SHYG., 2.7)(MYCOB.SP., .03)
TV:	S-180
IS-EXT:	(MEOH, , MIC.)(ET2O, , W)
IS-CHR:	(HP-10, MEOH-NH4OH)(SEPHADEX LH-20, MEOH)
IS-CRY:	(PREC., MEOH, HEX)

REFERENCES:

USP 4162305; JP 78/121701; *CA*, 90, 119749

30000-7653

NAME:	SF-2012
PO:	S.VINACEUS
CT:	QUINONE T., ACIDIC
FORMULA:	C22H21O10CL
EA:	(CL, 8)
MW:	492, 445
PC:	RED, POW.
UV:	MEOH: (232, 670,)(310, 134,)(505, 84,)(542, 80,)
UV:	MEOH-HCL: (232, 270,)(310, 166,)(500, 116,)
UV:	MEOH-NAOH: (232, 830,)(390, 172,)(570, 224,)
SOL-GOOD:	PYR, DMFA
SOL-FAIR:	MEOH, CHL
SOL-POOR:	W, HEX
QUAL:	(FECL3, +)(NINH., -)
TO:	(B.SUBT., .39)(S.AUREUS, .19)(S.LUTEA, .78)
IS-EXT:	(ETOAC, , FILT.)
IS-ION:	(XAD-2, MEOH-W)
IS-CHR:	(SEPHADEX LH-20, MEOH)

REFERENCES:

JP 80/48394; *CA*, 93, 184274

30000-7654

NAME:	C-15462-A
PO:	S.LONGISPOROFLAVUS
CT:	QUINONE T., NEUTRAL
FORMULA:	C19H16O5
MW:	324
PC:	YELLOW, CRYST.
OR:	(-205, PYR)
UV:	MEOH: (242, 705,)(290, 275,)(360, 140,)
UV:	MEOH-NAOH: (230, 1090,)(318, 510,)(385, 110,)(472, 165,)
SOL-GOOD:	PYR, DMSO
SOL-FAIR:	MEOH, CHL
SOL-POOR:	BENZ, HEX, W
QUAL:	(FECL3, +)
TO:	(S.LUTEA, 50)(B.SUBT., 20)(S.AUREUS, 2) (MYCOB.TUB., 12.5)(TREPANOMA SP., .03)
IS-FIL:	ORIG.
IS-EXT:	(ETOAC, 7.5, FILT.)
IS-CHR:	(SILG-OXALIC ACID, HEX-ETOAC)

REFERENCES:

JP 79/132501; *CA*, 93, 6122

30000-7655

NAME:	<u>C-15462-V</u>
PO:	S.LONGISPOROFLAVUS
CT:	QUINONE T., NEUTRAL
FORMULA:	C20H18O5
MW:	338
PC:	YELLOW, POW.
OR:	(-175, PYR)
UV:	MEOH: (240, 1030,)(285, 380,)(313, 230,) (336, 245,)
UV:	MEOH-NAOH: (230, 1560,)(340, 760,)(560, 46,)
SOL-GOOD:	PYR, DMSO
SOL-FAIR:	MEOH, CHL
SOL-POOR:	BENZ, HEX, W
QUAL:	(FECL3, +)
TO:	(B.SUBT., 100)(S.AUREUS, 50)(MYCOB.TUB., 3.1)

REFERENCES:

JP 79/132501; *CA*, 93, 6122

30000-7656

NAME:	<u>VETICILLOMYCIN</u>
PO:	STV.SP.
CT:	QUINONE T., NEUTRAL
FORMULA:	C23H25NO6
EA:	(C, 62)(H, 6)(N, 4)(O, 23)
MW:	411
PC:	RED, CRYST.
OR:	(+560, CHL)
UV:	MEOH: (210, ,)(265, ,)(510, ,)
UV:	NAOH: (212, ,)(275, ,)(555, ,)
SOL-GOOD:	MEOH, BENZ
SOL-FAIR:	ET2O, HEX
SOL-POOR:	W
QUAL:	(FECL3, +)
TO:	(S.AUREUS, 25)(S.LUTEA, 50)(B.SUBT., 6.25) (FUNGI,)
LD50:	(500|250, IP)(500|250, IV)
TV:	L-1210, CYTOTOXIC
IS-EXT:	(CHL, ,)
IS-ION:	(HP-20, MEOH)(HP 20, MEOH-W)
IS-CHR:	(SILG, CHL-MEOH)
IS-CRY:	(PREC., CHL, HEX)

REFERENCES:

JP 80/48395; *CA*, 93, 184273

30000-7922

NAME: CRATERIFERMYCIN-A
PO: S.CRATERIFER-ANTIBIOTICUS
CT: QUINONE T., ACIDIC
PC: PINK, POW.
UV: MEOH: (270, ,)
SOL-GOOD: BASE, MEOH, CHL
SOL-POOR: W, ACID, HEX
TO: (B.SUBT.,)(S.AUREUS,)(S.LUTEA,)(G.NEG.,)
IS-EXT: (CHL, 7, FILT.)
REFERENCES:
Rev. Inst. Antib., 12, 81, 1972; 13, 3, 1973

30000-7923

NAME: CRATERIFERMYCIN-B
PO: S.CRATERIFER-ANTIBIOTICUS
CT: QUINONE T., ACIDIC
PC: PINK, POW.
UV: MEOH: (275, ,)(320, ,)
SOL-GOOD: BASE, MEOH, CHL
SOL-POOR: W, HEX, ACID
TO: (B.SUBT.,)(S.AUREUS,)(S.LUTEA,)(G.NEG.,)
IS-EXT: (CHL, 7, FILT.)
REFERENCES:
Rev. Inst. Antib., 12, 81, 1972; 13, 3, 1973

30000-7924

NAME: CRATERIFERMYCIN-C
PO: S.CRATERIFER-ANTIBIOTICUS
CT: QUINONE T., ACIDIC
PC: PINK, POW.
UV: MEOH: (320, ,)
SOL-GOOD: BASE, MEOH, CHL
SOL-POOR: W, ACID, HEX
TO: (B.SUBT.,)(S.AUREUS,)(G.NEG.,)
REFERENCES:
Rev. Inst. Antib., 12, 81, 1972; 13, 3, 1973

30000-7925

NAME: CRATERIFERMYCIN-D
PO: S.CRATERIFER-ANTIBIOTICUS
CT: QUINONE T., ACIDIC
PC: PINK, POW.
UV: MEOH: (244, ,)(290, ,)(332, ,)
SOL-GOOD: BASE, MEOH, CHL
SOL-POOR: W, ACID, HEX
TO: (S.AUREUS,)(B.SUBT.,)
REFERENCES:

Rev. Inst. Antib., 12, 81, 1972; 13, 3, 1973

Amino Acid and Peptide Antibiotics

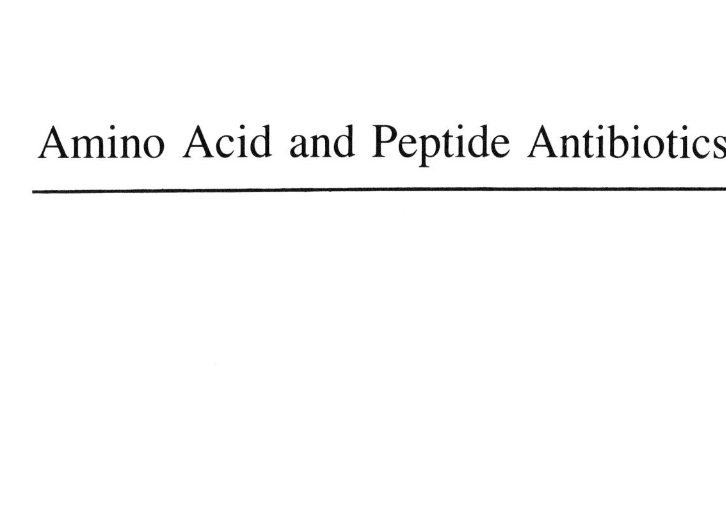

41
AMINO ACID DERIVATIVES

411 Simple Amino Acids
412 Cyclic Amino Acid Derivatives (Including Beta Lactams)
413 Diketipiperazine Derivatives

$CH_2{=}C(CH_3){-}CH_2{-}CH_2{-}CH(NH_2){-}COOH$

L-2-Amino-5-methyl-5-hexenoic acid

$HC{\equiv}C{-}CH(NH_2){-}COOH$

FR-900130
(L-2-Amino-3-butynoic acid)

$O_2N{-}NH(CH_2CH_2NHCO)_nCH_2CH_2{-}CH(NH_2){-}COOH$

	n
Compound No. 7682	0
Compound No. 7683	1

$CH_3{-}CH(Cl){-}CH_2{-}CH(NH_2){-}COOH$

γ-Chloronorvaline
(AL-719)

γ-Hydroxynorvalinelactone
(AL-719 Y)

$H_2N{-}O{-}CH_2CH(NH_2){-}COOH$

β-Aminoxy-D-alanine

$H_2N{-}CH_2{-}CH_2{-}O{-}CH_2{-}CH(NH_2){-}COOH$

I-677
(L-4-Oxalysine)

L-2(1-Methylcyclopropyl)glycine

	R_1	R_2	R_3
Oganomycin A	H	SO_3H	H
Oganomycin B	H	H	H
S-3907C/3	OCH_3	SO_3H	SO_3H
S-3907C/4B	OCH_3	H	SO_3H
Cephamycin A (S-3907C/2)	OCH_3	SO_3H	H

PA-32413-I

	R_1	R_2
PS-3 (Epithienamycin C)	OH	CH_2–CH_2–NHOCOCH$_3$
PS-4 (Epithienamycin D)	OH	CH=CH–NHCOCH$_3$
PS-5	H	CH=CH–NHCOCH$_3$
PS-6 (Dihydro-PS-8)	CH_3	CH_2–CH_2–NHCOCH$_3$
PS-7	H	CH_2–CH_2–NHCOCH$_3$
PS-8	CH_3	CH=CH–NHCOCH$_3$
NS-5 (Deshydroxythienamycin)	H	CH=CH–NH_2

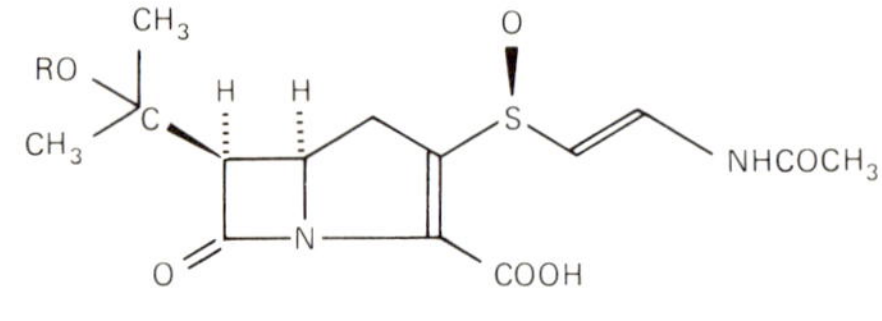

	R
Carpetimycin A (C-19393-H_2)	H
Carpetimycin B (C-19393-S_2)	SO_3H

17927-D

	R
Asperenomycin A (PA-31088 IV)	CH=CH–NHCOCH$_3$
Asperenomycin B (PA-39504 X_1)	CH_2–CH_2–NHCOCH$_3$

	R
Thienamycin	CH_3
Northienamycin	H

2-(3-Alanyl)clavam

(−)-2-(2-Hydroxyethyl)clavam

Isosulfazecin

Sulfazecin

Monobactam I

Monobactam III

	R_1	R_2	R_3	M
Monobactam VII	H	H	CH_3O	Na
Monobactam VIII	OH	H	CH_3O	K
Monobactam IX	OH	H	H	K
Monobactam X	OSO_3Na	OH	CH_3O	Na
Monobactam XI	OSO_3Na	OSO_3Na	CH_3O	Na

Cairomycin A
(tentative)

Absolute configuration of epicorazine B

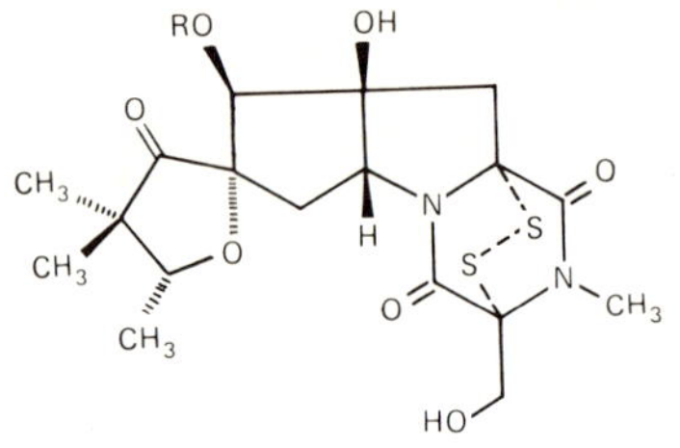

	R
Sirodesmin PL	– CO – CH_3
Deacetylsirodesmin PL	H

Bisdethio-bis-methylthiogliotoxin:

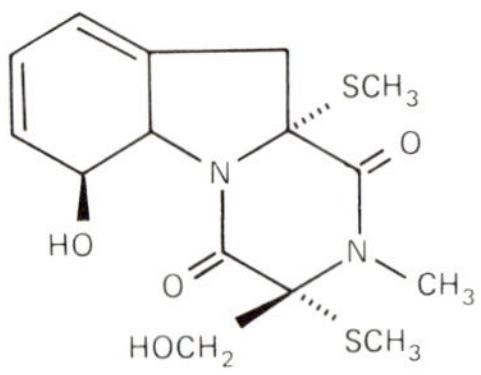

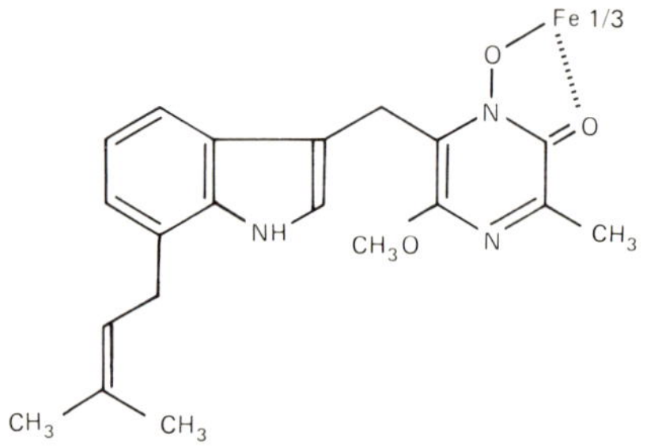

Asterchrome

41121-7117

NAME:	L-2-AMINO-5-METHYL-5-HEXENOIC ACID, AMHA
PO:	S.SP.
CT:	AMINO ACID, AMPHOTER
FORMULA:	C7H13NO2
EA:	(N, 10)
MW:	143
EW:	147
PC:	WH., CRYST.
OR:	(+34, HCL)
UV:	W: (200, ,)
SOL-GOOD:	ACID, BASE
SOL-FAIR:	W
SOL-POOR:	MEOH, HEX
QUAL:	(NINH., +)
TO:	(E.COLI,)
IS-FIL:	2
IS-ION:	(DX-50X4-H, NH4OH)(DX-50X8, PH3.1 BUFF)
IS-CRY:	(CRYST., W)

REFERENCES:

JA, 32, 1118, 1979

41121-7518

NAME:	FR-900130, L-2-AMINO-3-BUTYNOIC ACID
PO:	S.CATENULAE
CT:	AMINO ACID, AMPHOTER
FORMULA:	C4H5NO2
EA:	(N, 14)
MW:	99
PC:	YELLOW, WH., POW., HYGROSCOPIC
SOL-GOOD:	W
SOL-FAIR:	MEOH
SOL-POOR:	ETOH, HEX
QUAL:	(NINH., +)
STAB:	(ACID, +)(BASE, -)(HEAT, -)
TO:	(S.AUREUS, 39)(B.SUBT., 39)(E.COLI, 156)
IS-FIL:	2
IS-ION:	(DUOLIT C-20-H, H2SO4)(DUOLIT C-20-H, NH4OH)
IS-CHR:	(SEPHADEX G-15, W)
IS-CRY:	(LIOF.,)

REFERENCES:

JA, 33, 125, 132, 1980

41121-7682

PO: AGARICUS SUBRUTILESCENS
CT: AMINO ACID, AMPHOTER
FORMULA: C4H9N3O4
EA: (N, 26)
MW: 177
PC: WH.
OR: (+18, W)(+38, HCL)
UV: 1: (231, , 5800)
UV: 13: (230, , 7800)
UV: 6: (230, , 7500)
SOL-GOOD: W
TO: (B.SUBT., 10)(S.LUTEA, 10)(E.COLI, 30)(P.VULG., 100)
IS-EXT: (ETOH-W, , MIC.)
IS-CHR: (IR-120-4,)(DX-1X4-AC,)
REFERENCES:

JP 80/45624; *CA*, 93, 68666

41121-7683

PO: AGARICUS SUBRUTILESCENS
CT: AMINO ACID, AMPHOTER
FORMULA: C7H14N4O5
EA: (N, 24)
MW: 234
PC: WH.
UV: 1: (233, , 6400)
UV: 13: (230, , 9600)
UV: 6: (231, , 8200)
TO: (B.SUBT., 100)(S.LUTEA, 30)(E.COLI, 200) (P.VULG., 200)
REFERENCES:

JP 80/45624; *CA*, 93, 68666

41121-8148

NAME: G"-CHLORONORVALINE, AL-719
PO: S.GRISEOSPOREUS
CT: AMINO ACID, AMPHOTER
FORMULA: C5H10NO2CL
EA: (N, 9)(CL, 23)
MW: 151
PC: WH., CRYST.
OR: (+70, HCL)
UV: W: (200, ,)
SOL-GOOD: W
SOL-POOR: MEOH, CHL, HEX
QUAL: (NINH., +)
STAB: (ACID, +)(BASE, -)(HEAT, -)
TO: (PS.AER., 100)(B.SUBT.,)(K.PNEUM.,)
TV: ANTIVIRAL
IS-FIL: 3
IS-ION: (DX-50X2-H, NH4OH)
IS-CHR: (DEAE-SEPHADEX A-25, W)(SEPHADEX G-10, W)
IS-CRY: (LIOF.,)(CRYST., W-MEOH)
REFERENCES:
JA, 33, 1249, 1980

41122-8546

NAME: 1-667
IDENTICAL: L-4-OXALYSIN
PO: S.ROSEOVIRIDOFUSCUS
CT: AMINO ACID, AMPHOTER
FORMULA: C5H12N2O3
EA: (N, 19)
MW: 148
PC: WH., CRYST.
UV: W: (200, ,)
SOL-GOOD: W
QUAL: (NINH., +)
TV: S-180, ANTITUMOR
REFERENCES:
Acta Micr. Sinica, 21, 218, 1981; *CA*, 95, 95394

41123-6772

NAME: B"-AMINOXY-D-ALANINE
PO: S.SP.
CT: AMINO ACID, AMPHOTER
FORMULA: C3H8N2O3
EA: (N, 22)
MW: 120
PC: WH., CRYST.
OR: (-19.5, HCL)
UV: W: (200, ,)
SOL-GOOD: W
SOL-POOR: ETOH, HEX
QUAL: (NINH., +)
STAB: (HEAT, +)
TO: (S.AUREUS, 500)(E.COLI, 16)(PS.AER., 500)
(B.SUBT.,)
IS-ION: (IR-120-H, NH4OH)
REFERENCES:
Bull. Coll. Agric. Veter. Med., Nihon Univ., 36, 1, 1979; *JOC*, 27, 2957, 1962; *CA*, 90, 199159

41123-8085

NAME: L-2-1-METHYLCYCLOPROPYL-GLYCINE, PA-4046-I
PO: MIC.MIYAKANENSIS
CT: AMINO ACID, AMPHOTER
FORMULA: C6H11NO2
EA: (N, 11)
MW: 129
PC: WH., CRYST.
OR: (+120.6, HCL)
UV: W: (200, ,)
SOL-GOOD: W, HEX
SOL-POOR: ACET, HEX
QUAL: (NINH., +)
STAB: (ACID, +)(BASE, +)(HEAT, +)
TO: (E.COLI, 2)
LD50: (75|25, SC)
IS-ION: (DX-2X2-AC, ACOH)(DX-50X2-NH4, NH4OH)
IS-CRY: (CRYST., ACET-W)
REFERENCES:
JA, 34, 370, 1981; JP 81/68648; *CA*, 95, 185551

41123-8149

NAME: <u>G"-HYDROXYNORVALINELACTONE</u>, AL-719-Y
PO: S.GRISEOSPOREUS
CT: AMINO ACID, BASIC
FORMULA: C5H11NO2
EA: (N, 10)
MW: 117
PC: WH., CRYST.
OR: (+4, W)
UV: W: (200, ,)
SOL-GOOD: W
SOL-FAIR: MEOH
SOL-POOR: BUOH, HEX
QUAL: (NINH., +)
TV: ANTIVIRAL
IS-FIL: 3
IS-ION: (DX-50X2-H, NH4OH)
IS-CRY: (LIOF.,)
REFERENCES:
JA, 33, 1249, 1980

41211-7681

NAME: <u>"PENICILLIN LIKE FACTOR"</u>
PO: TRICHOPHYTON MENTAGROPHYTES
TO: (G.POS.,)(B.SUBT.,)
REFERENCES:
J. Appl. Bact., 48, 359, 1980

41212-7440

NAME: <u>XK-201-IV</u>
PO: S.SP.
CT: ACIDIC, G-LACTAM
EA: (C,)(H,)(N,)(S,)
MW: 470
PC: WH., POW.
OR: (-10, W)
UV: W: (265, ,)
SOL-GOOD: W
QUAL: (NINH., +)
TO: (P.VULG., 1)(PHYT.BACT., 1)
IS-ABS: (CARBON, ACET-W)
REFERENCES:
JP 79/138592; *CA*, 92, 109112

41212-7949

NAME: "COMPOUND A", OGANOMYCIN-A
PO: S.OGANOENSIS
CT: B"-LACTAM, CEPHALOSPORIN T., AMPHOTER
FORMULA: C24H27N3013S2
EA: (N, 6)(S, 10)
MW: 629
PC: WH., POW.
UV: 6.5: (218, ,)(282, ,)
SOL-GOOD: W
SOL-FAIR: MEOH
SOL-POOR: BUOH, HEX
QUAL: (NINH., +)
TO: (B.SUBT., 12.3)(S.AUREUS, 100)(E.COLI, 6.25)
(SHYG., 6.25)(P.VULG., .39)(K.PNEUM., 25)
IS-ION: (DIAION HP-20, ACET-W)(IRA-68-CL, NANO3-NAOAC)
IS-CHR: (AVICEL, BUOH-ACOH-W)
REFERENCES:
BP 2037764; *JA*, 34, 1507, 1981

41212-7950

NAME: OGANOMYCIN-B, "COMPOUND-C"
PO: S.OGANOENSIS
CT: B"-LACTAM, CEPHALOSPORIN T., AMPHOTER
FORMULA: C24H27N3010S2
EA: (N, 7)(S, 6)
MW: 581
PC: WH., POW.
UV: 6.5: (307.5, ,)
SOL-GOOD: W
SOL-FAIR: MEOH
SOL-POOR: BUOH, HEX
QUAL: (NINH., +)
TO: (B.SUBT., 3.13)(S.AUREUS, 50)(E.COLI, 3.13)
(SHYG., 6.25)(P.VULG., .78)(K.PNEUM., 3.13)
IS-ION: (DIAION HP-20, ACET-W)(IRA-68-CL, NANO3-NAOAC)
IS-CHR: (AVICEL, BUOH-ACOH-W)
REFERENCES:
BP 2037764; *JA*, 34, 1507, 1981

41212-8086

NAME: PA-32413-I
PO: S.CLAVULIGERUS
CT: B"-LACTAM, CEPHALOSPORIN T., ACIDIC
FORMULA: C18H25N5O10S
EA: (N, 14)(S, 6)
MW: 503
PC: WH., POW.
UV: W: (240, ,)(266, ,)
TO: (E.COLI,)
REFERENCES:
JP 80/143996; *CA*, 94, 137796

41212-8547

NAME: S-3907C 3
PO: S.SP., S.ROCHEI
CT: B"-LACTAM, CEPHALOSPORIN T., ACIDIC
FORMULA: C25H29N3O17S2
EA: (N, 6)(S, 9)
MW: 707
PC: WH., POW.
UV: MEOH: (288, 230,)
UV: MEOH-HCL: (289, 230,)
UV: MEOH-NAOH: (284, 230,)
SOL-GOOD: W
TO: (P.VULG.,)
REFERENCES:
EP 28511

41212-8548

NAME: S-3907C 4B
PO: S.SP., S.ROCHEI
CT: B"-LACTAM, CEPHALOSPORIN T., ACIDIC
FORMULA: C25H29N3O14S
EA: (N, 6)(S, 5)
MW: 627
PC: WH., POW.
UV: 7: (290, 233,)
UV: MEOH: (292, 230,)
UV: MEOH-NAOH: (353, ,)
TO: (P.VULG.,)
REFERENCES:
EP 28511

41214-6773

NAME: PS-6, DIHYDRO-PS-8
PO: S.CREMEUS-AURATILIS, S.FULVOVIRIDIS, S.FLAVOGRISEUS, S.OLIVACEUS
CT: B"-LACTAM, THIENAMYCIN T., ACIDIC
FORMULA: C14H20N204S
EA: (N, 9)(S, 10)
MW: 312
PC: WH., POW.
OR: (+55, W)
UV: W: (300, , 9003)
SOL-GOOD: W
SOL-FAIR: ACID
SOL-POOR: ACET, BENZ, ETOAC
QUAL: (EHRL., +)(NINH., -)
TO: (S.AUREUS, .33)(B.SUBT., 1.39)(K.PNEUM., 12.5) (S.LUTEA, .39)(P.VULG., 12.5)(E.COLI, 6.25) (PS.AER., 50)
LD50: NONTOXIC
IS-ION: (DIAION PA-306, NACL)
IS-ABS: (CARBON, ACET-W)
IS-CHR: (DIAION HP-20, ACET-W)(QAE-SEPHADEX, NACL)
REFERENCES:

JA, 33, 1128, 1138, 1980; EP 1567; JP 79/154598; 79/59295; 79/92983; 80/33431

41214-6774

NAME: PS-7
PO: S.CREMEUS-AURATILIS, S.FULVOVIRIDIS, S.FLAVOGRISEUS, S.OLIVACEUS
CT: B"-LACTAM, THIENAMYCIN T., ACIDIC
FORMULA: C13H16N204S
EA: (N, 9)(S, 10)
MW: 296
PC: WH., POW.
OR: (+62, W)
UV: W: (226, , 14300)(308, , 13900)
SOL-GOOD: W
SOL-FAIR: ACID
SOL-POOR: ACET, BENZ, ETOAC
QUAL: (EHRL., +)(NINH., -)
TO: (S.AUREUS, .19)(B.SUBT., .39)(S.LUTEA, .2) (E.COLI, .78)(K.PNEUM., 3.13)(PS.AER., 6.25) (P.VULG., 12.5)
LD50: NONTOXIC
REFERENCES:

JA, 33, 1128, 1138, 1980; EP 1567; JP 79/154598; 79/59295; 79/92983; 80/33431

41214-7118

NAME: SF-2050
PO: S.SP.
CT: B"-LACTAM, THIENAMYCIN T., ACIDIC
EA: (N, 6)(S, 13)
PC: WH., POW.
OR: (-46, W)
UV: 7: (295, ,)
UV: W: (295, ,)
SOL-GOOD: W, MEOH
SOL-POOR: ETOAC, BENZ
QUAL: (NINH., -)
STAB: (ACID, -)(BASE, -)
TO: (B.SUBT., 10)(S.AUREUS, 10)(E.COLI, 20)
(K.PNEUM., 20)
IS-ABS: (CARBON, ACET-W)
IS-CHR: (DEAE-SEPHADEX, PH7 BUFF)
REFERENCES:
DT 2905066; BP 2014129

41214-7119

NAME: SF-2050-B
PO: S.SP.
CT: B"-LACTAM, THIENAMYCIN T., ACIDIC
EA: (N,)(S,)
PC: WH., POW.
UV: 6.5: (228, ,)(305, ,)
UV: W: (228, ,)(305, ,)
SOL-GOOD: MEOH, W
SOL-POOR: ETOAC, BENZ
QUAL: (NINH., -)
STAB: (ACID, -)(BASE, -)
TO: (B.SUBT., 2)(S.AUREUS, 10)(E.COLI, 10)
(K.PNEUM., 10)
REFERENCES:
DT 2905066; BP 2014129

41214-7684

NAME: <u>17927-D</u>
PO: S.FULVOVIRIDIS, S.ARGENTEOLUS, S.OLIVACEUS, S.FLAVOGRISEUS
CT: B"-LACTAM, THIENAMYCIN T., ACIDIC
FORMULA: C13H20N2O5S
EA: (N, 9)(S, 10)
MW: 316
PC: WH.
UV: W: (200, ,)
SOL-GOOD: W
SOL-FAIR: MEOH
SOL-POOR: ETOAC, CHL, BENZ
QUAL: (NINH., -)
REFERENCES:
JP 80/24129; 80/29909; *CA*, 93, 130554, 130561

41214-7685

NAME: <u>PS-3</u>
IDENTICAL: 890-A3, EPITHIENAMYCIN-C, MM-22381, 890-A1
PO: S.CREMEUS-AURATILIS
CT: B"-LACTAM, THIENAMYCIN T., ACIDIC
FORMULA: C13H18N2O5S
EA: (N, 9)(S, 10)
MW: 314
PC: WH.
UV: 7.5: (298, ,)
SOL-GOOD: W
TO: (S.AUREUS, .13)(E.COLI, 1)(P.VULG., 2) (K.PNEUM., 2)
REFERENCES:
JP 80/54899

41214-7686

NAME: <u>PS-4</u>
IDENTICAL: MM-22383, EPITHIENAMYCIN-D, 890-A5
PO: S.CREMEUS-AURATILIS
CT: B"-LACTAM, THIENAMYCIN T., ACIDIC
FORMULA: C13H16N2O5S
EA: (N, 9)(S, 10)
MW: 312
PC: WH.
UV: W: (229, ,)(309, ,)
SOL-GOOD: W
TO: (S.AUREUS,)(B.SUBT.,)(P.VULG.,)(K.PNEUM.,)
REFERENCES:
JP 79/92983

41214-7687

NAME: NS-5, DESACETYL-PS-5, DESHYDROXYTHIENAMYCIN
PO: S.OLIVACEUS+PS-5, S.CATTLEYA
CT: B"-LACTAM, THIENAMYCIN T., ACIDIC, AMPHOTER
FORMULA: C11H16N2O3S
EA: (N, 11)(S, 12)
MW: 256
PC: WH.
UV: 7: (293, ,)
UV: W: (297, ,)
SOL-GOOD: W
QUAL: (NINH., +)
TO: (S.AUREUS,)(P.VULG.,)
LD50: NONTOXIC
IS-ION: (XAD-2, MEOH-W)(CM-SEPHADEX C-25,)(CG-400, W)
IS-CHR: (SEPHADEX LH-20, MEOH)
REFERENCES:
JA, 34, 341, 1981; *EP*, 2058; *JA*, 33, 543, 550, 1980; JP 80/42536; 80/72191; *CA*, 94, 3920

41214-7951

NAME: C-19393-H2, S-19393-H2
IDENTICAL: CARPETIMYCIN-A
PO: S.SAPROPHYTICUS, S.GRISEUS-CRYOPHILUS+COC12
CT: B"-LACTAM, THIENAMYCIN T., ACIDIC
FORMULA: C14H18N2O6S
EA: (N, 7)(S, 6)
MW: 342
EW: 374|52
PC: WH., POW.
OR: (-134, W)(-141, W)
UV: W: (242, 395,)(289, 314,)(244.5, , 14300)
(290, , 11000)
SOL-GOOD: W, MEOH
SOL-FAIR: BUOH, ACET
SOL-POOR: ETOAC, HEX
QUAL: (EHRL., +)(NINH., -)(FECL3, -)(SAKA., -)
STAB: (ACID, -)(BASE, -)
TO: (S.AUREUS, .63)(B.SUBT., .31)(S.LUTEA, .31)
(E.COLI, .08)(K.PNEUM., .31)(P.VULG., 5)
(PS.AER., 10)
IS-ION: (DX-1X2-CL, NACL)(HP-20, MEOH-W)(XAD-2, W)
IS-ABS: (CARBON, I.BUOH-W)
IS-CHR: (QAE-SEPHADEX A-25-CL, NACL)
IS-CRY: (LIOF.,)
REFERENCES:
JA, 33, 1414, 1425, 1980; 34, 206, 211, 1981; Holl. P 80/628

41214-7952

NAME:	C-19393-S2
IDENTICAL:	CARPETIMYCIN-B
PO:	S.SAPROPHYTICUS, S.GRISEUS-CRYOPHILUS+COC12
CT:	B"-LACTAM, THIENAMYCIN T., ACIDIC
FORMULA:	C14H18N2O9S2
EA:	(N, 7)(S, 14)
MW:	422
EW:	478|50
PC:	WH., POW.
OR:	(-152, W)
UV:	W: (240, 296,)(285, 245,)(243.5, , 15900) (288, , 13200)
SOL-GOOD:	W, MEOH, DMSO, ACOH
SOL-FAIR:	ETOH, BUOH, PYR
SOL-POOR:	ACET, HEX
QUAL:	(EHRL., +)(NINH., -)(FECL3, -)(SAKA., -)
STAB:	(ACID, -)(BASE, -)(HEAT, +)
TO:	(S.AUREUS, 12.5)(S.LUTEA, 12.5)(B.SUBT., 12.5) (E.COLI, 12.5)(K.PNEUM., 12.5)(P.VULG., 100)
IS-ION:	(DX-1X2-CL, NACL)(HP-20, MEOH-W)(XAD-2, W)
IS-ABS:	(CARBON, I.BUOH-W)
IS-CHR:	(QAE-SEPHADEX A-25-CL, NACL)
IS-CRY:	(LIOF.,)

REFERENCES:

JA, 33, 1414, 1425, 1980; 34, 206, 211, 1981; Holl. P 80/628

41214-8150

NAME: CARPETIMYCIN-A, KA-6643-A
IDENTICAL: C-19393-H2
PO: S.SP.
CT: B"-LACTAM, THIENAMYCIN T., ACIDIC
FORMULA: C14H18N2O6S
EA: (N, 8)(S, 9)
MW: 342
PC: WH., POW.
OR: (-27, W)
UV: W: (240, 369,)(288, 300,)
SOL-GOOD: W
SOL-POOR: ACET, HEX
STAB: (ACID, -)(BASE, -)
TO: (S.AUREUS, .39)(B.SUBT., .2)(K.PNEUM., .2) (E.COLI, .05)(P.VULG., .39)(PS.AER., 6.25)
IS-ION: (DIAION PA-306-CL, PH7 BUFF-NACL)(IRA-68-CL, NACL)
IS-ABS: (CARBON, ACET-W)
IS-CHR: (QAE-SEPHADEX A-25, W)
IS-CRY: (LIOF.,)
REFERENCES:

JA, 33, 1388, 1980; 34, 818, 1981; *Abst. AAC*, 20, 165, 1980; BP 2047700; Holl. P 80/2210

41214-8151

NAME: CARPETIMYCIN-B, KA-6643-B
IDENTICAL: C-19393-S2
PO: S.SP.
CT: B"-LACTAM, THIENAMYCIN T., ACIDIC
FORMULA: C14H18N2O9S2
EA: (N, 6)(S, 15)
MW: 422
PC: WH., POW.
OR: (-145, W)
UV: W: (240, 357,)(285, 305,)
SOL-GOOD: W
SOL-POOR: ACET, HEX
STAB: (ACID, -)(BASE, -)
TO: (S.AUREUS, 6.25)(B.SUBT., 6.25)(E.COLI, 1.56) (K.PNEUM., 6.25)(P.VULG., 12.5)(PS.AER., 25)
IS-ION: (DIAION PA-306, NACL)(IRA-458-CL, NACL)
IS-CHR: (SEPHADEX)
REFERENCES:

JA, 33, 1388, 1980; 34, 818, 1981; *Abst. AAC*, 20, 165, 1980; BP 2047700; Holl. P 80/2210

41214-8152

NAME:	ASPERENOMYCIN-A, PA-31088-IV
PO:	S.TOKUNONENSIS, S.ARGENTEOLUS
CT:	B"-LACTAM, THIENAMYCIN T., ACIDIC
FORMULA:	C14H16N206S
EA:	(N, 8)(S, 8)
MW:	341
PC:	WH., YELLOW, POW.
OR:	(-210.8, PH7 BUFF)
UV:	7: (241, , 21472)(280, ,)(320, ,)
UV:	W: (241, 562,)
SOL-GOOD:	W, MEOH, DMSO
SOL-POOR:	ACET, ET2O
QUAL:	(EHRL., +)(NINH., -)
TO:	(S.AUREUS, 3.13)(E.COLI, 3.13)(K.PNEUM., 1.56) (P.VULG., 6.25)(PS.AER., 50)
IS-ION:	(IRA-68-CL, NACL)(DIAION HP-20, EDTA-W)
IS-ABS:	(CARBON, ACET-W)
IS-CRY:	(LIOF.,)

REFERENCES:

JA, 34, 909, 1981; EP 17246; JP 81/13628

41214-8178

NAME:	NORTHIENAMYCIN
PO:	S.CATTLEYA
CT:	B"-LACTAM, THIENAMYCIN T., ACIDIC
FORMULA:	C10H14N2O4S
EA:	(N, 11)(S, 12)
MW:	258
PC:	WH., POW.
TO:	(S.AUREUS,)
IS-ION:	(DX-1X2-CO3, W+CO2)(DX-1X2-CL, W)
IS-CHR:	(XAD-2, W)

REFERENCES:

USP 4247640

41214-8549

NAME: PS-8
PO: S.CREMEUS-AURATILIS
CT: B"-LACTAM, THIENAMYCIN T., ACIDIC
FORMULA: C15H20N2O4S
EA: (N, 8)(S, 10)
MW: 324
PC: WH., POW.
UV: W: (223, , 10800)(308, , 9720)
SOL-GOOD: W
TO: (S.AUREUS, 1.56)(P.VULG., 12)
LD50: NONTOXIC
REFERENCES:
JP 81/25183; *CA*, 95, 95460

41214-8550

NAME: ASPERENOMYCIN-B, PA-39504-X1
PO: S.TOKUNONENSIS, S.ARGENTEOLUS
CT: B"-LACTAM, THIENAMYCIN T., ACIDIC
FORMULA: C14H18N2O6S
EA: (N, 8)(S, 9)
MW: 342
PC: WH., POW.
UV: W: (240, ,)(318, ,)
SOL-GOOD: W
TO: (S.AUREUS, 6.25)(E.COLI, 3.13)(K.PNEUM., 6.25)
IS-ION: (DX-1X2-CL, NACL)
IS-CHR: (DIAION-HP-20, MEOH)
IS-CRY: LIOF.
REFERENCES:
EP 30719

41215-7953

NAME: 2-3-ALANYL-CLAVAM
PO: S.CLAVULIGERUS
CT: B"-LACTAM, CALVULANIC ACID T., AMPHOTER
FORMULA: C8H12N2O4
EA: (N, 15)
MW: 190
PC: WH., CRYST.
OR: (-137, W)
SOL-GOOD: W
SOL-POOR: ETOH, HEX
TO: (B.SUBT., .25)(E.COLI, 03)(K.PNEUM., 62.5)
(S.CEREV., 125)(C.ALB., 500)
IS-FIL: 6.5
IS-ION: (XAD-2, W)
IS-CHR: (SEPHADEX LH-20, MEOH-W)(AG-50X4-NA, W)
IS-CRY: (CRYST., ETOH-W)
REFERENCES:
USP 4202819; *CA*, 93, 130567

41215-8179

NAME: 2-2-HYDROXYETHYL-CLAVAM
PO: S.ANTIBIOTICUS-ANTIBIOTICUS
CT: B"-LACTAM, CLAVULANIC ACID T.
FORMULA: C7H11NO3
EA: (N, 9)
MW: 157
PC: WH., OIL
OR: (-141, CHL)(-202, DMSO)
QUAL: (EHRL., +)
STAB: (ACID, -)(HEAT, -)
TO: (G.POS.,)(G.NEG.,)(FUNGI,)
REFERENCES:
TL, 2539, 1981

41216-6895

NAME:	SULFAZECIN, G-6302
IDENTICAL:	MONOBACTAM-I
PO:	PS.ACIDOPHILA
CT:	ACIDIC, B"-LACTAM, MONOBACTAM
FORMULA:	C12H20N409S
EA:	(N, 13)(S, 8)
MW:	396
EW:	400|20
PC:	WH., CRYST.
OR:	(+94, W)(+82, W)
UV:	W: (200, ,)
SOL-GOOD:	W, MEOH, DMSO
SOL-FAIR:	ETOH, ACET, PYR
SOL-POOR:	ETOAC, HEX
QUAL:	(NINH., +)(EHRL., +)(SAKA., -)
STAB:	(BASE, -)(ACID, +)(HEAT, +)
TO:	(E.COLI, 25)(SHYG., 25)(K.PNEUM., 12.5) (B.SUBT., 100)(S.AUREUS, 200)(E.COLI, 6.25) (P.VULG., 6.25)(PS.AER., 800)
LD50:	(10000, IV)
IS-FIL:	4
IS-ION:	(DX-1X2-CL, NACL)
IS-ABS:	(CARBON, ACET-W)
IS-CHR:	(DEAE-SEPHADEX A-25, PH6.6 PUFF)
IS-CRY:	(PREC., W, ACET-ET20)

REFERENCES:

Nature, 289, 590, 1981; *JA*, 34, 621, 1981; *J. Ferm. Techn.*, 59, 263, 1981; DT 2855949; BP 2011380; Belg. P 873217

41216-7771

NAME:	ISOSULFAZECIN, SB-72310
PO:	PS.MESOACIDOPHILA
CT:	ACIDIC, B"-LACTAM, MONOBACTAM
FORMULA:	C12H20N4O9S
EA:	(N, 14)(S, 8)
MW:	396
EW:	400|20
OR:	(+.5, W)(+8.5, W-NAOH)(+4.5, W)
UV:	W: (200, ,)
SOL-GOOD:	W, MEOH, DMSO
SOL-FAIR:	ETOH, PYR, ACET
SOL-POOR:	ETOAC, HEX
QUAL:	(NINH., +)(EHRL., +)(SAKA., -)
STAB:	(ACID, +)(BASE, -)(HEAT, +)
TO:	(S.LUTEA,)(S.AUREUS, 200)(B.SUBT., 100) (E.COLI, 50)(P.VULG., 100)(PS.AER., 1600)
LD50:	(NONTOXIC,)(10000, IV)
IS-ION:	(DX-1X2-CL, NACL)
IS-ABS:	(CARBON, ACET-W)(CARBON, MEOH-W)
IS-CHR:	(DEAE-SEPHADEX A-25, NACL)
IS-CRY:	(PREC., W, ACET)

REFERENCES:

Nature, 289, 590, 1981; *JA*, 34, 1081, 1981; Holl. P. 79/7191

41216-8551

NAME:	MONOBACTAM-I, SQ-26445, EM-5210
IDENTICAL:	SULFAZECIN
PO:	GLUCONOBACTER SP., ACETOBACTER PASTEURIANUS, GLUCONOBACTER OXYDANS, ACETOBACTER ACETI
CT:	B"-LACTAM, MONOBACTAM, ACIDIC
FORMULA:	C12H20N4O9S
EA:	(N, 14)(S, 8)
MW:	396
PC:	WH., POW.
SOL-GOOD:	W
QUAL:	(NINH., +)
TO:	(E.COLI, 25)(K.PNEUM., 50)(PS.AER., 50)
LD50:	NONTOXIC
IS-ION:	(DX-1X2-CL, NACL)
IS-ABS:	(CARBON, MEOH-W)
IS-CHR:	(SEPHADEX LH-20, MEOH)(SEPHADEX G-10, W)

REFERENCES:

Nature, 291, 489, 1981; *Abst. 12th Int. Congr. Chemoth., Florence*, 1981, 55; Belg. P 887428

41216-8552

NAME: MONOBACTAM-III, SQ-26180, E-5117
PO: CHROMOBACTERIUM VIOLACEUM
CT: B"-LACTAM, MONOBACTAM, ACIDIC
FORMULA: C6H10N2O6S
EA: (N, 12)(S, 13)
MW: 238
PC: WH., CRYST.
OR: (+94.3, W)
SOL-GOOD: W
TO: (S.AUREUS, 50)(PS.AER.,)(K.PNEUM.,)
LD50: NONTOXIC
IS-ION: (DX-1X2-CL, NACL)
IS-ABS: (CARBON, MEOH-W)
IS-CHR: (SEPHADEX LH-20, MEOH)(SEPHADEX G-10, W)
REFERENCES:

Nature, 291, 489, 1981; *Abst. 12th Int. Congr. Chemoth., Florence*, 1981, 55; Belg. P 887428

41216-8553

NAME: MONOBACTAM-VII
PO: AGROBACTERIUM RADIOBACTER
CT: B"-LACTAM, MONOBACTAM, ACIDIC
FORMULA: C15H19N3O7S
EA: (N, 11)(S, 8)
MW: 377
PC: WH.
TO: (S.AUREUS,)(E.COLI,)(K.PNEUM.,)
LD50: NONTOXIC
IS-ION: (DX-1X2-CL, NACL)
IS-ABS: (CARBON, MEOH-W)
IS-CHR: (SEPHADEX LH-20, MEOH)(SEPHADEX G-10, W)
REFERENCES:

Nature, 291, 489, 1981; *Abst. 12th Int. Congr. Chemoth., Florence*, 1981, 55; Belg. P 887428

41216-8554

NAME: MONOBACTAM-VIII
PO: AGROBACTERIUM RADIOBACTER
CT: B''-LACTAM, MONOBACTAM, ACIDIC
FORMULA: C15H19N3O8S
EA: (N, 10)(S, 8)
MW: 393
PC: WH.
TO: (S.AUREUS,)(G.NEG.,)
LD50: NONTOXIC
IS-ION: (DX-1X2-CL, NACL)
IS-ABS: (CARBON, MEOH-W)
IS-CHR: (SEPHADEX LH-20, MEOH)(SEPHADEX G-10, W)
REFERENCES:

Nature, 291, 489, 1981; *Abst. 12th Int. Congr. Chemoth., Florence*, 1981, 55; Belg. P 887428

41216-8555

NAME: MONOBACTAM-IX
PO: AGROBACTERIUM RADIOBACTER
CT: B''-LACTAM, MONOBACTAM, ACIDIC
FORMULA: C14H17N3O7S
EA: (N, 11)(S, 9)
MW: 363
PC: WH.
TO: (S.AUREUS,)(G.NEG.,)
LD50: NONTOXIC
IS-ION: (DX-1X2-CL, NACL)
IS-ABS: (CARBON, MEOH-W)
IS-CHR: (SEPHADEX LH-20, MEOH)(SEPHADEX G-10, W)
REFERENCES:

Nature, 291, 489, 1981; *Abst. 12th Int. Congr. Chemoth., Florence*, 1981, 55; Belg. P 887428

41216-8556

NAME: MONOBACTAM-X
PO: AGROBACTERIUM RADIOBACTER
CT: B"-LACTAM, MONOBACTAM, ACIDIC
FORMULA: C15H19N3O12S2
EA: (N, 8)(S, 13)
MW: 497
PC: WH.
TO: (S.AUREUS,)(G.NEG.,)
LD50: NONTOXIC
IS-ION: (DX-1X2-CL, NACL)
IS-ABS: (CARBON, MEOH-W)
IS-CHR: (SEPHADEX LH-20, MEOH)(SEPHADEX G-10, W)
REFERENCES:

Nature, 291, 489, 1981; *Abst. 12th Int. Congr. Chemoth., Florence*, 1981, 55; Belg. P 887428

41216-8557

NAME: MONOBACTAM-XI
PO: AGROBACTERIUM RADIOBACTER
CT: B"-LACTAM, MONOBACTAM, ACIDIC
FORMULA: C15H19N3O15S3
EA: (N, 7)(S, 16)
MW: 577
PC: WH.
TO: (S.AUREUS,)(G.NEG.,)
LD50: NONTOXIC
IS-ION: (DX-1X2-CL, NACL)
IS-ABS: (CARBON, MEOH-W)
IS-CHR: (SEPHADEX LH-20, MEOH)(SEPHADEX G-10, W)
REFERENCES:

Nature, 291, 489, 1981; *Abst. 12th Int. Congr. Chemoth., Florence*, 1981, 55; Belg. P 887428

41310-8558

NAME: CAIROMYCIN-A
PO: S.SP.
CT: DIKETOPIPERAZINE DERIV., ACIDIC
FORMULA: C9H14N2O4
EA: (N, 13)
MW: 214
PC: YELLOW, BROWN, POW.
UV: MEOH: (200, ,)
SOL-GOOD: ACET, ETOAC, CHL, BENZ
SOL-FAIR: BUOH
SOL-POOR: HEX, W
QUAL: (EHRL., +)(FEHL., -)(BIURET, -)(NINH., -)
(SAKA., -)(FECL3, -)
TO: (S.AUREUS, .3)(B.SUBT., .3)(S.LUTEA, .3)
(E.COLI, 25)(K.PNEUM., 25)(PS.AER., 50)(C.ALB.,
50)(FUNGI, 50)
LD50: (10, IP)
IS-EXT: (CHL-ETOAC, 7, FILT.)
IS-CHR: (SILG, CHL-MEOH)
IS-CRY: (PREC., ETOAC, HEX)
REFERENCES:

AAC, 19, 941, 1981

41321-6775

NAME: EPICORAZINE-B
PO: EPICOCCUM NIGRUM
CT: EPIDIKETOOLIGOTHIAPIPERAZINE, GLIOTOXIN T.,
NEUTRAL
FORMULA: C18H16N2O6S2
EA: (N, 6)(S, 16)
MW: 420
PC: WH., CRYST.
OR: (-320, MEOH)
UV: MEOH: (220, , 15300)
SOL-GOOD: MEOH, CHL, ACCN, DMFA
SOL-POOR: W
TO: (S.AUREUS, 30)
IS-FIL: ORIG.
IS-EXT: (CHL, , FILT.)
IS-CHR: (SILG, CHL-MEOH)
IS-CRY: (CRYST., CHL-HEX)
REFERENCES:

JA, 31, 1099, 1106, 1978; *Acta Cryst., Ser. B*, 33, 1474, 1977; *Bull. Soc. Pharm. Bordeaux*, 120, 23, 1981; *CA*, 95, 164335

41321-7122

NAME: BISDETHIO-BIS-METHYLTHIOGLIOTOXIN
PO: GLIOCLADIUM DELIQUESCENS
CT: EPIDIKETOOLIGOTHIAPIPERAZINE, GLIOTOXIN T., NEUTRAL
FORMULA: C15H20N2O4S2
EA: (N, 8)(S, 19)
MW: 330
PC: WH., POW.
TV: ANTIVIRAL
IS-EXT: (ETOAC, , FILT.)
REFERENCES:
JCS Perkin I, 119, 1980

41325-7120

NAME: SIRODESMIN-PL
PO: PHOMA LINGAM
CT: EPIDIKETOOLIGOTHIAPIPERAZINE, SIRODESMIN T., NEUTRAL
FORMULA: C20H26N2O8S2
EA: (N, 6)(S, 13)
MW: 486
PC: WH., CRYST.
OR: (-224, MEOH)
UV: MEOH: (208, , 9900)(228, , 4800)
SOL-GOOD: MEOH, ETOAC
SOL-POOR: W
TO: (FUNGI,)
IS-EXT: (ETOAC, , FILT.)
IS-CHR: (SILG, ETOAC-MEOH)(SILG, CHL-ETOAC-MEOH)
REFERENCES:
CR Ser. D, 284, 927, 1977; *Nouveau J. Chim.*, 1, 327, 1977; *JCS Perkin* I, 113, 1739, 1980

41325-7121

NAME: DEACETYLSIRODESMIN-PL
PO: PHOMA LINGAM
CT: EPIDIKETOOLIGOTHIAPIPERAZINE, SIRODESMIN T. NEUTRAL
FORMULA: C18H24N2O7S2
EA: (N, 6)(S, 14)
MW: 444
PC: WH.
OR: (-385, MEOH)
UV: MEOH: (204, , 6000)(228, , 3200)
SOL-GOOD: MEOH, CHL
SOL-POOR: W
TO: (FUNGI,)
IS-EXT: (ETOAC, , FILT.)
REFERENCES:

CR Ser. D, 284, 927, 1977; *Nouveau J. Chim.*, 1, 327, 1977; *JCS Perkin I*, 113, 1739, 1980

41330-8559

NAME: ASTERCHROME
PO: ASP.TERREUS+FECL3
CT: ASPERGILLIC ACID T., NEUTRAL
FORMULA: C20H22N3O3.FE-3
EA: (N, 11)(FE, 5)
MW: 351, 1112
PC: RED, CRYST.
OR: (0,)
UV: MEOH: (223, , 151360)(282, , 24550)(292, , 21900)(347, , 26300)(450, , 4790)
SOL-GOOD: MEOH, BENZ
SOL-POOR: W
TO: (G.POS.,)
IS-EXT: (FLEX, , MIC.)(CHL, , MIC.)
IS-CHR: (SILG, BENZ-ETOAC)
IS-CRY: (CRYST., CHL-HEX)
REFERENCES:

Chem. Ph. Bull., 29, 1510, 1981

42
HOMOPEPTIDE ANTIBIOTICS

421 Oligopeptides
422 Linear Homopeptides
423 Cyclic Homopeptides

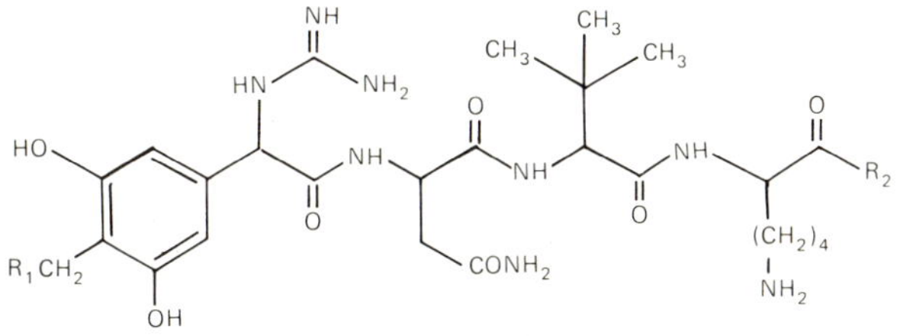

	R_1	R_2
Pheganomycin	OH	OH
Deoxypheganomycin D	OH	NHCH(COOH)CH_2COOH
Pheganomycin D	H	NHCH(COOH)CH_2COOH
Pheganomycin-DR	OH	L-Asp-L-Arg
Pheganomycin-DGPT	OH	L-Asp-gly-L-pro-L-thr

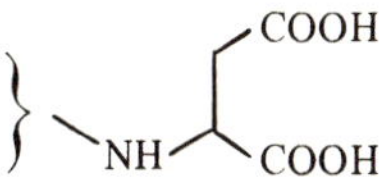

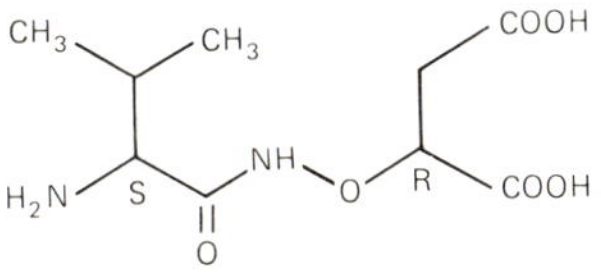

Malioxamycin

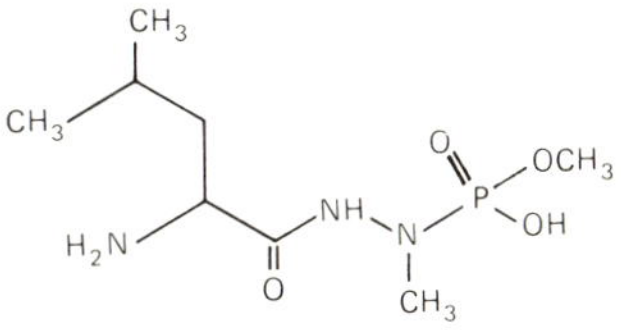

FR-900137

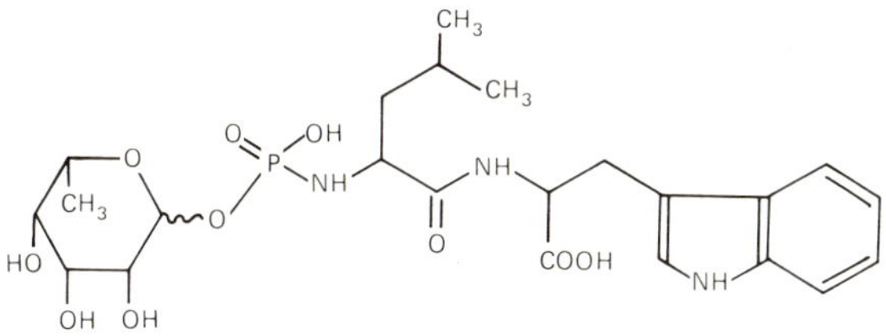

AJ-9406

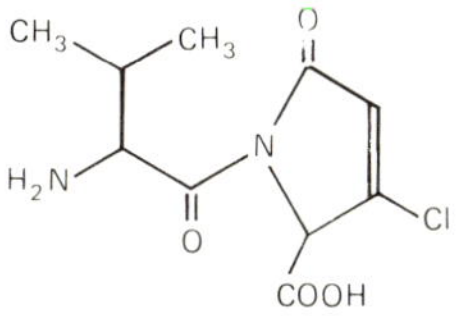

FR-900148

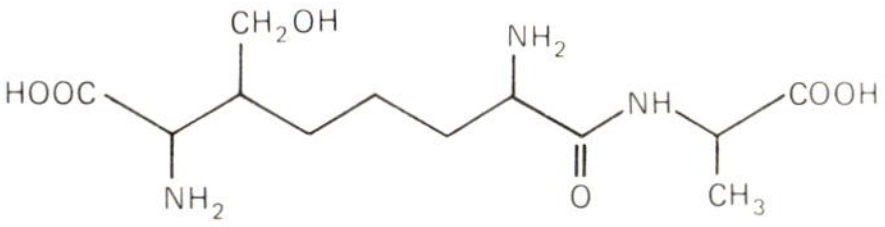

N-(2,6-Diamino-6-hydroxymethylpimelyl)-L-Alanine

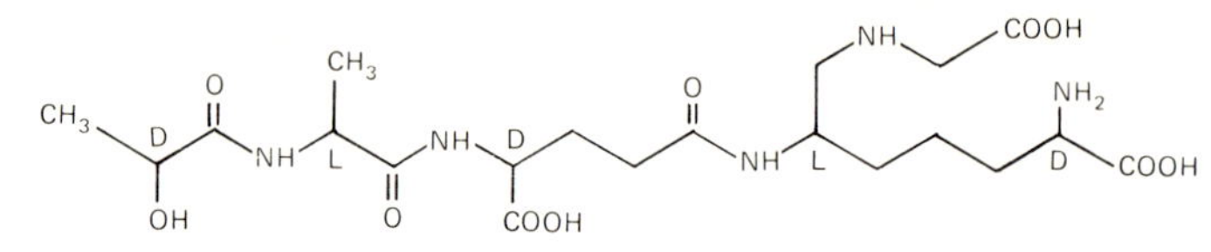

FK-156 (FR-900156)

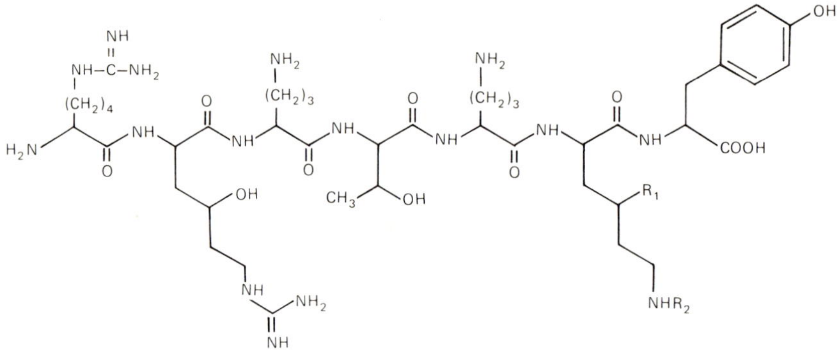

	R_1	R_2
K-582 A	H	H
K-582 B	OH	$-C(=NH)NH_2$

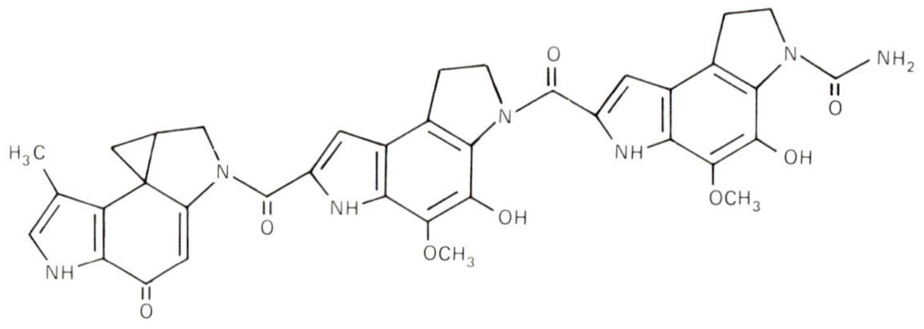

CC-1065 (Rachelmycin)

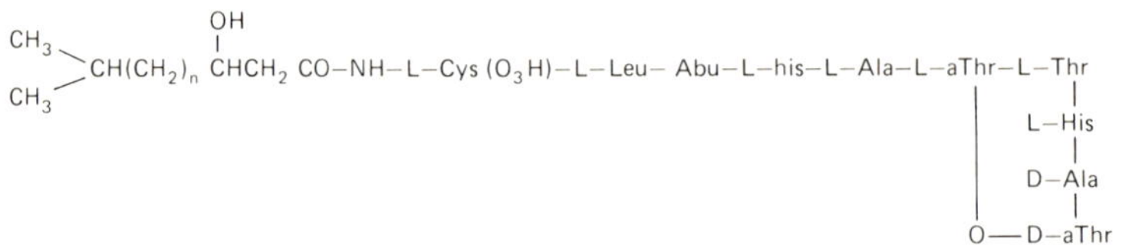

$(CH_3)_2CH(CH_2)_n$ CH(OH)CH_2 CO–NH–L–Cys (O_3H)–L–Leu– Abu–L–his–L–Ala–L–aThr–L–Thr–L–His–D–Ala–D–aThr–O (lactone to aThr)

	n
Imacidin A	5
Imacidin B	6
Imacidin C	7
Imacidin D	8
Imacidin E	9

Imacidinic acids are the open lactones of Imacidins

Ac–aib–pro–aib–ala–aib–aib–gln–leu–aib–gly–aib–aib–aib–pro–val–aib–aib–gln–gln–Leuol

Hypelcin A
Leuol : L-leucinol = $(CH_3)_2CHCH_2CH(NH_2)CH_2OH$

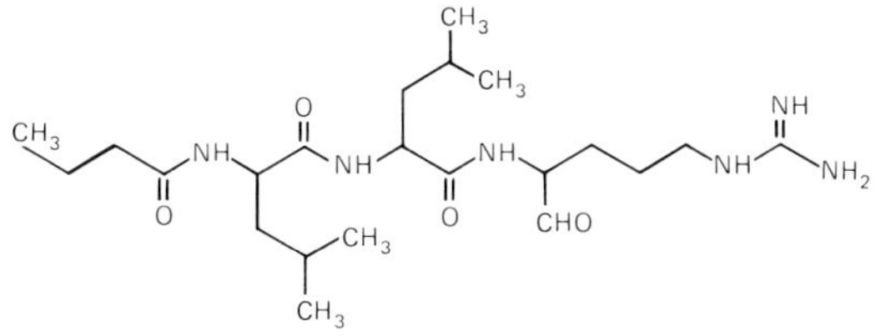

Leupeptin PrLL

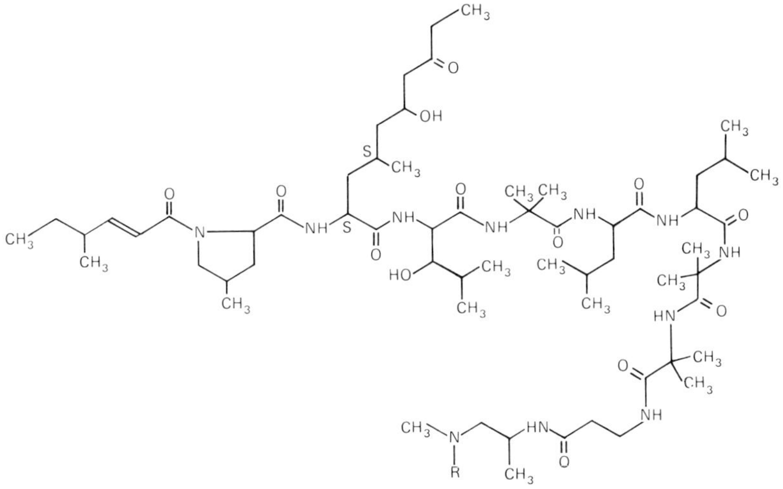

	R
1907 II	H
1907 VIII (P-168, Leucinostatin)	CH_3

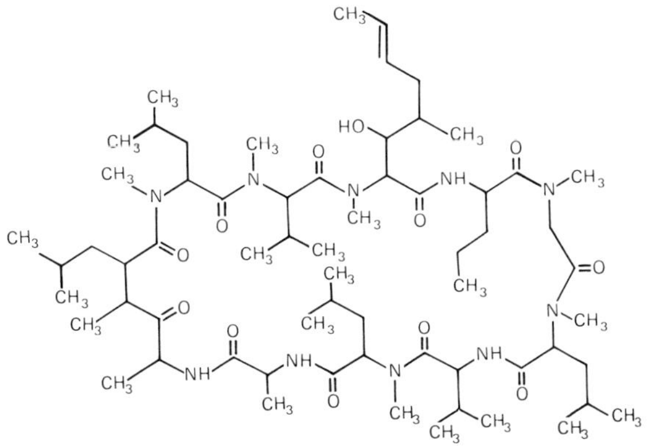

Cyclosporin G

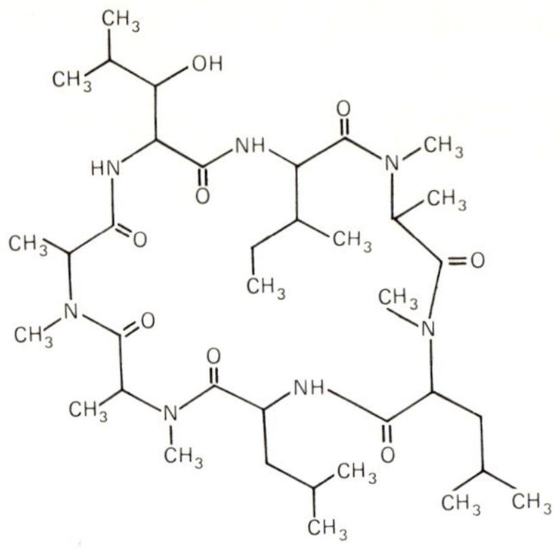

Ternatin

42120-6776

NAME:	PHEGANOMYCIN
PO:	S.CIRRATUS
CT:	OLIGOPEPTIDE, AMPHOTER
FORMULA:	C26H42N8O9
EA:	(N, 17)
MW:	610
PC:	WH., POW.
OR:	(+3, W)
UV:	W: (288, 27.2,)
SOL-GOOD:	W
SOL-POOR:	MEOH, HEX
QUAL:	(NINH., +)(PAULY, +)(EHRL., -)(SAKA., +)
TO:	(MYCOB.SP., 100)
LD50:	NONTOXIC
IS-ION:	(XAD-2, MEOH-W)
IS-CHR:	(CARBON, ACET-HCL)(CM-SEPHADEX C-25, NACL) (SEPHADEX G-10, W)

REFERENCES:

Proc. 15th Symp. Peptide Chem., Osaka, 1978, 121; JP 78/108943; 78/111034; *CA*, 90, 101941; 91, 89513

42120-6777

NAME:	PHEGANOMYCIN-D
PO:	S.CIRRATUS
CT:	OLIGOPEPTIDE, AMPHOTER
FORMULA:	C30H47N9O12
EA:	(N, 17)
MW:	725
PC:	WH., POW.
OR:	(-36, W)
UV:	W: (288, 224,)
SOL-FAIR:	W
SOL-POOR:	MEOH, HEX
QUAL:	(NINH., +)(PAULY, +)(SAKA., +)(EHRL., -)
TO:	(S.AUREUS, 100)(MYCOB.SP., 12.5)(K.PNEUM., 50)
LD50:	NONTOXIC

REFERENCES:

Proc. 15th Symp. Peptide Chem., Osaka, 1978, 121; JP 78/108943; 78/111034; *CA*, 90, 101941; 91, 89513

42120-6778

NAME:	PHEGANOMYCIN-DR
PO:	S.CIRRATUS
CT:	OLIGOPEPTIDE, AMPHOTER
FORMULA:	C36H59N13O13
EA:	(N, 20)
MW:	882
PC:	WH., POW.
OR:	(-10, W)
UV:	W: (288, 17.6,)
SOL-GOOD:	W
SOL-POOR:	MEOH, HEX
QUAL:	(NINH., +)(PAULY, +)(SAKA., -)(EHRL., -)
TO:	(MYCOB.SP., 3.12)(G.NEG., 24)
LD50:	NONTOXIC

REFERENCES:

Proc. 15th Symp. Peptide Chem., Osaka, 1978, 121; JP 78/108943; 78/111034; *CA*, 90, 101941; 91, 89513

42120-6779

NAME:	PHEGANOMYCIN-DGPT
PO:	S.CIRRATUS
CT:	OLIGOPEPTIDE, AMPHOTER
FORMULA:	C41H64N12O16
EA:	(N, 17)
MW:	981
PC:	WH., POW.
OR:	(-53, W)
UV:	W: (288, 16.2,)
SOL-GOOD:	W
SOL-POOR:	MEOH, HEX
QUAL:	(NINH., +)(PAULY, +)(EHRL., -)(SAKA., -)
TO:	(E.COLI, 100)(MYCOB.SP., 6.25)(SHYG., 12) (P.VULG., 100)(K.PNEUM., 50)
LD50:	NONTOXIC

REFERENCES:

Proc. 15th Symp. Peptide Chem., Osaka, 1978, 121; JP 78/108943; 78/111034; *CA*, 90, 101941; 91, 89513

42120-6780

NAME: DEOXYPHEGANOMYCIN-D
PO: S.CIRRATUS
CT: OLIGOPEPTIDE, AMPHOTER
FORMULA: C30H47N9O11
EA: (N, 18)
MW: 709
PC: WH., POW.
OR: (-7, W)
UV: W: (277, 14.5,)(283, 14.3,)
SOL-FAIR: W
SOL-POOR: MEOH, HEX
TO: (MYCOB.SP., 6.25)(G.NEG., 25)(K.PNEUM., 25)
LD50: NONTOXIC
REFERENCES:

Proc. 15th Symp. Peptide Chem., Osaka, 1978, 121; JP 78/108943; 78/111034; *CA*, 90, 101941; 91, 89513

42120-7439

NAME: MALIOXAMYCIN
PO: S.LYDICUS
CT: ACIDIC, OLIGOPEPTIDE, AMPHOTER
FORMULA: C9H16N2O6
EA: (N, 10)
MW: 248
PC: WH., CRYST.
OR: (+37.4, W)
UV: MEOH: (200, ,)
SOL-GOOD: W, MEOH, ACET, DMFA, DMSO
SOL-POOR: ETOAC, HEX, PHENOL
QUAL: (NINH., +)(FECL3, +)
TO: (E.COLI, 50)(K.PNEUM., 100)(S.AUREUS,)
(B.SUBT.,)(PS.AER.,)
IS-FIL: 7
IS-ION: (DIAION-SK-H, NH4OH)
IS-CHR: (SEPHADEX G-15, W)(SP-SEPHADEX C-25-H, W)
(SEPHADEX G-15, BUOH-ACOH-W)
IS-CRY: (CRYST., W)(LIOF.,)
REFERENCES:

JA, 33, 1214, 1220, 1980; JP 79/141702; *CA*, 92, 126905

42120-7688

NAME: AJ-9406, N-6-DEOXY-L-TALOSYLOXY-HYDROXYPHOSPHINYL-L-LEUCYL-L-TRYPTOPHANE
PO: S.MOZUNENSIS
CT: OLIGOPEPTIDE, ACIDIC
FORMULA: C23H34N3O10P
EA: (N, 9)(P, 5)
MW: 543
PC: WH., POW.
OR: (-25.6,)
TO: (PS.AER.,)
REFERENCES:
JP 80/69597; *CA*, 93, 219326; WO P 80/1069; JP 81/77296; *CA*, 96, 4952

42120-7689

NAME: FR-900137
PO: S.UZENENSIS
CT: OLIGOPEPTIDE, ACIDIC
FORMULA: C8H20N3O4P
EA: (N, 17)(P, 12)
MW: 253
EW: 270
PC: WH., POW.
OR: (+31.6, W)
UV: HCL: (205, , 1000)(270, , 300)
UV: NAOH: (205, , 1000)(280, , 180)
UV: W: (205, , 1000)
SOL-GOOD: W, MEOH
SOL-FAIR: ETOH
SOL-POOR: ACET, CHL, HEX
TO: (S.AUREUS, 100)(B.SUBT., 3)(E.COLI, 1.5)
(P.VULG., 3)(K.PNEUM., 50)(PS.AER., 6)
LD50: NONTOXIC
IS-ION: (DEAE-SEPHADEX-OH, NH4OH)(DIAION HP-20, W)
IS-ABS: (CARBON, MEOH-W)
IS-CHR: (CM-SEPHADEX C-25, PH6.5 BUFF)(G-15, W)
IS-CRY: (PREC., MEOH, EVAP.FILT.)(LIOF.,)
REFERENCES:
JA, 33, 272, 280, 1980

42120-7690

NAME: FR-900148
PO: S.XANTHOCIDICUS
CT: OLIGOPEPTIDE, ACIDIC, AMPHOTER
FORMULA: C10H13N2O4CL
EA: (N, 11)(CL, 14)
MW: 260
EW: 330
PC: WH., POW.
OR: (+77.1, W)
UV: W: (200, ,)
SOL-GOOD: W, MEOH
SOL-FAIR: ETOH
SOL-POOR: ACET, HCL
QUAL: (NINH., +)(EHRL., -)
STAB: (ACID, -)(BASE, -)
TO: (S.AUREUS, 3.1)(B.SUBT., 200)(S.LUTEA, 50)
(E.COLI, 1.6)(SHYG., 50)(K.PNEUM., 800)
(PS.AER., 25)
LD50: (5000|1500, IV)
IS-ION: (DEAE-SEPHADEX, NACL)
IS-ABS: (CARBON, ACET-W)
IS-CHR: (CARBON, W)(CM-SEPHADEX H, W)(G-15, W)
IS-CRY: (LIOF.,)
REFERENCES:

JA, 33, 259, 267, 1980; JP 81/86196; *CA*, 95, 148705

42120-8087

NAME: N-2.6-DIAMINO-6-HYDROXYMETHYLPIMELYL-ALANINE
PO: MIC.CHALCEA
CT: OLIGOPEPTIDE, AMPHOTER
FORMULA: C11H21N3O6
EA: (N, 14)
MW: 291
PC: WH., POW.
OR: (+8.7, W)
UV: W: (200, ,)
SOL-GOOD: W
SOL-POOR: ACET, HEX
QUAL: (NINH., +)
STAB: (ACID, +)(BASE, +)
TO: (E.COLI,)
IS-ION: (DX-1X2-AC, ACOH)(DX-50X2-NH4, NH4OH)
IS-CRY: (PREC., W, ACET)
REFERENCES:

JA, 34, 374, 1981

42120-8560

NAME: CAIROMYCIN-C
PO: S.SP.
CT: OLIGOPEPTIDE, AMPHOTER
FORMULA: C23H38N6O6
EA: (N, 17)
MW: 494
PC: RED, BROWN, POW.
UV: MEOH: (235, ,)(275, ,)
SOL-GOOD: CHL, ETOAC, ACET
SOL-FAIR: BUOH
SOL-POOR: W, HEX
QUAL: (BIURET, +)(NINH., +)(EHRL., +)(FEHL., -)
(SAKA., -)(FECL3, -)
TO: (S.AUREUS, 1.5)(S.LUTEA, 1.5)(B.SUBT., 1.5)
(E.COLI, 25)(K.PNEUM., 25)(PS.AER., 50)
(MYCOB.TUB., .5)(C.ALB., 50)(FUNGI, 50)
LD50: (10, IP)
REFERENCES:

AAC, 19, 941, 1981

42120-8561

NAME: FK-156, FR-900156
PO: S.SP.
CT: OLIGOPEPTIDE
FORMULA: C20H33N5O11
EA: (N, 13)
MW: 519
TV: INTERFERON INDUCER
REFERENCES:

EP 25842; *JACS*, 103, 7026, 1981; *TL*, 693, 1982; *Abst. AAC*, 21, 414, 1981

42140-6703

NAME: CC-1065, NSC-298223
IDENTICAL: RACHELMYCIN
PO: S.ZELENSIS
CT: INDOLE DERIV., OLIGOPEPTIDE L.
FORMULA: C37H33N708
EA: (C, 61)(H, 5)(N, 13)
MW: 703
PC: CRYST., AMBER
UV: DIOXAN: (230, ,)(258, ,)(364, , 49100)
SOL-GOOD: DMFA, DMSO, ACET, ETOAC, CHL, DIOXAN
SOL-FAIR: CHL, THF
SOL-POOR: HEX, MEOH
TO: (S.AUREUS, .0015)(B.SUBT., .012)(S.LUTEA, .012)
(E.COLI, .32)(P.VULG., .08)(PS.SER., .08)
(K.PNEUM., .08)(C.ALB., .3)(S.CEREV., .04)
LD50: (.1, IP)
TV: L-1210, P-388
IS-EXT: (ACET, , MIC.)(CH2CL2, , W)(CH2CL2, 10, FILT.)
IS-CHR: (SILG, CHL-MEOH-NH4OH)
IS-CRY: (PREC., CH2C12, HEX)(CRYST., MEOH)
UTILITY: ON CLINICAL TRIAL
REFERENCES:
Proc. Am. Ass. Cancer Res., 19, 99, 1978; *JA*, 31, 1211, 1978; 33, 902, 1980; 34, 1119, 1981; *JACS*, 103, 5621, 7629, 1981; USP 4169888; DT 2841361

42140-6770

NAME: RACHELMYCIN, NSC-219877
IDENTICAL: CC-1065
PO: S.SP.
CT: PEPTIDE, OLIGOPEPTIDE L.
FORMULA: C37H33N708
EA: (N, 14)
MW: 703
TV: P-388, MELANOMA
REFERENCES:
Recent Results Cancer Res., 63, 49, 1978; USP 4301248

42220-6781

NAME: K-582-A, K-582M-A, MYRORIDIN-K
PO: METARRHIZIUM ANISOPLIAE
CT: PEPTIDE, EDEIN T., BASIC, AMPHOTER
EA: (C, 46)(H, 8)(N, 20)
MW: 1250|50
PC: WH., POW.
OR: (+.2, W)
UV: MEOH: (278, 1175,)
UV: W: (276, ,)
SOL-GOOD: MEOH, W
SOL-FAIR: ETOH
SOL-POOR: BUOH, HEX
QUAL: (NINH., +)(PAULY, +)(SAKA., +)(BIURET, +)
STAB: (ACID, +)(BASE, -)
TO: (C.ALB., .2)(S.CEREV., .2)
LD50: (132|12, IV)(6000, PEROS)
TV: EHRLICH, S-180, SN-36, ANTIVIRAL, INFL, POLIO, NDV
IS-EXT: (W, 2, BUOAC)
IS-ION: (IRC-50-NHA, HCL)
IS-CHR: (BIOGEL D-2, NACL)(CM-SEPHADEX C-25, NACL-PH6 PUFF)
IS-CRY: (PREC., PENTACHLOROPHENOL, W)(PREC., W, ETOH) (PREC., MEOH, ACET)
UTILITY: ON CLINICAL TRIAL
REFERENCES:

JA, 33, 533, 1980; *Sendai Shi Eisi*, 8, 173, 176, 1979; BP 2011425; DT 2856410; JP 81/75098; *CA*, 93, 68587, 125433, 130483; *CA*, 95, 202069

42220-6782

NAME:	K-582-B, K-582M-B
PO:	METARRHIZIUM ANISOPLIAE
CT:	PEPTIDE, EDEIN T., BASIC, AMPHOTER
EA:	(C, 41)(H, 7)(N, 21)
MW:	1210|10
PC:	WH., POW.
OR:	(+.6, W)
UV:	MEOH: (278, 1175,)
UV:	W: (276, ,)
SOL-GOOD:	W, MEOH
SOL-FAIR:	ETOH
SOL-POOR:	BUOH, HEX
QUAL:	(NINH., +)(BIURET, +)(SAKA., +)(PAULY, +)
STAB:	(ACID, +)(BASE, -)
TO:	(C.ALB., .4)(S.CEREV., .4)(P.VULG., 40)
LD50:	(30|5, IV)(50|7, IP)(110|15, SC)(2700|300, PEROS)
TV:	EHRLICH, S-180, SN-36, ANTIVIRAL, POLIO, INFL, NDV

REFERENCES:

JA, 33, 533, 1980; *Sendai Shi Eisi*, 8, 173, 176, 1979; BP 2011425; DT 2856410; JP 81/75098; *CA*, 93, 68587, 125433,130483; *CA*, 95, 202069

42220-7691

NAME:	TATUMINE
PO:	B.BREVIS
CT:	PEPTIDE, EDEIN T.
EA:	(N,)
MW:	335
PC:	WH., POW.
UV:	W: (206, ,)
SOL-GOOD:	W, PROH
QUAL:	(NINH., +)
STAB:	(HEAT, +)
TV:	SA, ANTITUMOR
IS-ION:	(DX-50X8-H, NH4OH)
IS-CHR:	(CM-CEL, I.PROH-NH4OH)(G-15, W)

REFERENCES:

JA, 33, 359, 1980; *Cell. Biol.*, 70, 131A, 1976

42230-8153

NAME: "PENTADECAPEPTIDE"
PO: STV.GRISEOVERTICILLATUM
CT: PEPTIDE, AMPHOTER
EA: (N,)
PC: WH., CRYST.
OR: (-72, HCL)
UV: W: (251, 2.68,)(257, 2.90,)(263, 2.29,)
SOL-GOOD: W, MEOH
SOL-FAIR: ETOAC
SOL-POOR: HEX
QUAL: (SAKA., +)(PAULY, -)
TO: (B.SUBT., .78)
LD50: NONTOXIC
TV: LEWIS, MELANOMA
IS-EXT: (BUOH, 7, W)
IS-ION: (IRC-50-H, ACET-HCL)(SP-SEPHADEX C-25-NA, NACL)
IS-ABS: (CARBON, ACET-W)
IS-CRY: (CRYST., W)
REFERENCES:

EP 19072

42240-6789

NAME: IMACIDIN-B
PO: S.OLIVACEUS
CT: PEPTIDE, NEUTRAL
FORMULA: C55H90N14O18S
EA: (N, 15)(S, 2.5)
MW: 1267
PC: WH., POW.
SOL-GOOD: ACID, BASE, BUOH-W
SOL-POOR: BUOH, W
QUAL: (NINH., +)
STAB: (BASE, -)
TO: (STREPTOMYCES SP.,)
IS-FIL: 5.1
IS-EXT: (MEOH, , MIC.)(BUOH, 5, FILT.)
IS-ION: (XAD-2, MEOH-W)
IS-CHR: (DX-50X2, PYR-ACOH)(SEPHADEX LH-20, W-BUOH)
REFERENCES:

Arch. Mikr., 122, 219, 1979; *Liebigs Ann.*, 28, 1982

42240-6790

NAME: IMACIDIN-C
PO: S.OLIVACEUS
CT: PEPTIDE, NEUTRAL
FORMULA: C56H92N14O18S
EA: (N, 15)(S, 2.5)
MW: 1281
PC: WH., POW.
SOL-GOOD: ACID, BASE, BUOH-W
SOL-POOR: BUOH, W
QUAL: (NINH., +)
STAB: (BASE, -)
TO: (STREPTOMYCES SP.,)
REFERENCES:
Arch. Mikr., 122, 219, 1979; *Liebigs Ann.*, 28, 1982

42240-6791

NAME: IMACIDINIC ACID-C
PO: S.OLIVACEUS
CT: PEPTIDE, NEUTRAL, ACIDIC
FORMULA: C56H94N14O18S
EA: (N, 15)(S, 2.5)
MW: 1283
PC: WH., POW.
SOL-GOOD: ACID, BASE
SOL-POOR: BUOH, W
QUAL: (NINH., +)
STAB: (BASE, -)
TO: (STREPTOMYCES SP.,)
REFERENCES:
Arch. Mikr., 122, 219, 1979; *Liebigs Ann.*, 28, 1982

42250-6930

NAME: HYPELCIN-A
PO: HYPOCREA PELTATA
CT: PEPTIDE, NEUTRAL, PEPTAIBOPHOL
FORMULA: C89H153N23O24
EA: (N, 16)
MW: 1927
PC: WH., POW.
OR: (-9.75, MEOH)
QUAL: (NINH., -)
TO: (FUNGI,)
IS-EXT: (CH2CL2, , MIC.)
IS-CHR: (SEPHADEX LH-20, MEOH)
REFERENCES:
CC, 413, 1979; *Exp.*, 36, 590, 1980

42250-6931

NAME: HYPELCIN-B
PO: HYPOCREA PELTATA
CT: PEPTIDE, PEPTAIBOPHOL
FORMULA: C89H152N22O25
EA: (N, 16)
MW: 1928
PC: WH., POW.
OR: (-7.25, MEOH)
QUAL: (NINH., -)
TO: (FUNGI,)
REFERENCES:
CC, 413, 1979; *Exp.*, 36, 590, 1980

42260-8154

NAME: CEREXIN-B2
PO: B.CEREUS
CT: PEPTIDE, CEREXIN T., AMPHOTER
FORMULA: C65H99N15O18
EA: (N, 15)
MW: 1377
PC: WH., POW.
OR: (+,)
UV: MEOH: (275, ,)(282.5, ,)(290.5, ,)
SOL-GOOD: DMSO, DMFA, BASE
SOL-FAIR: MEOH
SOL-POOR: ETOH, HEX
QUAL: (NINH., +)(EHRL., +)
TO: (B.SUBT., 6)(S.AUREUS, 6)
IS-EXT: (BUOH, 7, W)
IS-CHR: (HPLC, ACCN-W)
REFERENCES:
JA, 32, 313, 1979

42260-8155

NAME:	CEREXIN-B3, CEREXIN-B4
PO:	B.CEREUS
CT:	PEPTIDE, CEREXIN T., AMPHOTER
FORMULA:	C66H101N15O18
EA:	(N, 15)
MW:	1391
PC:	WH., POW.
OR:	(+,)
UV:	MEOH: (275, ,)(282.5, ,)(290.5, ,)
SOL-GOOD:	DMSO, DMFA, BASE
SOL-FAIR:	MEOH
SOL-POOR:	ETOH, HEX
QUAL:	(NINH., +)(EHRL., +)
TO:	(B.SUBT., 6)(S.AUREUS, 6)
REFERENCES:	

JA, 32, 313, 1979

42260-8156

NAME:	CEREXIN-D2
PO:	B.CEREUS
CT:	PEPTIDE, CEREXIN T., AMPHOTER
FORMULA:	C65H99N15O17
EA:	(N, 15)
MW:	1361
PC:	WH., POW.
OR:	(+,)
UV:	MEOH: (275, ,)(282.5, ,)(290.5, ,)
SOL-GOOD:	DMSO, DMFA, BASE
SOL-FAIR:	MEOH
SOL-POOR:	ETOH, HEX
QUAL:	(NINH., +)(EHRL., +)
TO:	(B.SUBT., 6)(S.AUREUS, 6)
REFERENCES:	

JA, 32, 313, 1979

42260-8157

NAME:	CEREXIN-D3, CEREXIN-D4
PO:	B.CEREUS
CT:	PEPTIDE, CEREXIN T., AMPHOTER
FORMULA:	C66H101N15017
EA:	(N, 15)
MW:	1375
PC:	WH., POW.
OR:	(+,)
UV:	MEOH: (275, ,)(282.5, ,)(290.5, ,)
SOL-GOOD:	DMSO, DMFA, BASE
SOL-FAIR:	MEOH
SOL-POOR:	ETOH, HEX
QUAL:	(NINH., +)(EHRL., +)
TO:	(B.SUBT., 6)(S.AUREUS, 6)

REFERENCES:

JA, 32, 313, 1979

42260-8203

NAME:	LEUPEPTIN-PRLL
PO:	S.ROSEUS, S.CHARTREUSIS, S.ALLERNERULI, S.SP.
CT:	PEPTIDE, BASIC
FORMULA:	C21H40N604
EA:	(N, 18)
MW:	440
PC:	WH., POW.
OR:	(-56, MEOH)(-73, MEOH)
UV:	MEOH: (200, ,)
SOL-GOOD:	W, MEOH, BUOH, DMSO
SOL-POOR:	ACET, HEX
QUAL:	(SAKA., +)(DNPH, +)(NINH., -)(FECL3, -)
LD50:	(118, IV)(1405, SC)(1550, PEROS)
TV:	YOSHIDA
IS-EXT:	(BUOH, ,)
IS-ION:	(LEWATIT-CNP, MEOH-HCL)
IS-CHR:	(DX-1X2-CL, W)
IS-CRY:	(LIOF.,)

REFERENCES:

JA, 22, 283, 558, 1969; 31, 95, 1978; 32, 523, 1979; 33, 1172, 1980; *Chem. Ph. Bull.*, 17, 1896, 1902, 1969; *Cancer Res.*, 32, 1725, 1972; 40, 2539, 1980; JP 70/17159

42270-8158

NAME: P-168
IDENTICAL: 1907-VIII
PO: PAECILOMYCES LILACINUS
CT: PEPTIDE, BASIC
FORMULA: C62H111N11O13
EA: (N, 12)
MW: 1217
PC: WH., CRYST.
OR: (-27, MEOH)
UV: MEOH: (200, ,)(220, , 23000)
SOL-GOOD: MEOH, BENZ
SOL-POOR: W, HEX
TO: (S.CEREV.,)(FUNGI,)(C.ALB.,)(PHYT.FUNGI,)
(S.AUREUS,)(B.SUBT.,)(S.LUTEA,)(MYCOB.SP.,)
IS-EXT: (ETOAC, , FILT.)
IS-CHR: (SILG, ETOAC-MEOH)(AL, ETOAC-BENZ)(AL, BENZ-ETOAC)
REFERENCES:
Agr. Biol. Ch., 44, 3029, 3033, 1980; 45, 1023, 1981

42270-8159

NAME: 1907-II
PO: PAECILOMYCES LILACINUS
CT: PEPTIDE, BASIC
FORMULA: C61H109N11O13
EA: (N, 13)
MW: 1203
PC: WH., POW.
OR: (-38.1, MEOH)
SOL-GOOD: MEOH, BENZ
SOL-POOR: HEX, W
QUAL: (NINH., +)(BIURET, +)(SAKA., -)(EHRL., -)
TO: (B.SUBT., 6.25)(S.AUREUS, 3.12)(S.LUTEA, .78)
(PS.AER., 100)(C.ALB., 6.25)(S.CEREV., 50)
(FUNGI, 1.56)(PHYT.FUNGI, .78)
REFERENCES:
Agr. Biol. Ch., 44, 3037, 1980

42270-8160

NAME: 1907-VIII
IDENTICAL: P-168
PO: PAECILOMYCES LILACINUS
CT: PEPTIDE, BASIC
FORMULA: C62H111N11013
EA: (N, 13)
MW: 1217
PC: WH., POW.
OR: (-38, MEOH)
SOL-GOOD: MEOH, BENZ
SOL-POOR: W, HEX
QUAL: (BIURET, +)(NINH., -)(SAKA., -)(EHRL., -)
TO: (B.SUBT., 12.5)(S.AUREUS, 6.25)(S.LUTEA, .78)
(PS.AER., 100)(C.ALB., 6.25)(S.CEREV., 50)
REFERENCES:
Agr. Biol. Ch., 44, 3033, 3037, 1980

42350-7692

NAME: CYCLOSPORIN-G
PO: TOLYPOCLADIUM INFLATUM, TRICHODERMA POLYSPORUM
CT: CYCLOPEPTIDE, CYCLOSPORIN T., NEUTRAL
FORMULA: C63H113N11O12
EA: (N, 12)
MW: 1215
PC: WH., CRYST.
OR: (-245, CHL)(-191, MEOH)
UV: MEOH: (200, ,)
SOL-GOOD: MEOH, CHL
TV: CYTOTOXIC
IS-EXT: (BUOAC, , FILT.)
IS-CHR: (SEPHADEX LH-20, MEOH)(SILG, HEX-ACET)
REFERENCES:
Belg. P 879402

42300-6783

NAME:	TERNATIN
PO:	DIDYMOCLADIUM TERNATUM, CALDOTHRICUM TERNATUM
CT:	NEUTRAL, CYCLOPEPTIDE
FORMULA:	C37H67N7O8
EA:	(N, 13)
MW:	737
PC:	WH., CRYST.
OR:	(-39.9, ETOH)(-34.3, ETOH)
UV:	ETOH: (211, , 9500)
SOL-GOOD:	MEOH, ET2O
SOL-FAIR:	HEX
SOL-POOR:	W
TO:	(PHYT.FUNGI,)(S.CEREV.,)
IS-EXT:	(CHL, 7, FILT.)(CH2CL2-HEX, 7, FILT.)
IS-CHR:	(SILG, CHL-ETOAC)
IS-CRY:	(CRYST., BENZ-HEX)(CRYST., CCL4-HEX)

REFERENCES:

SU P 302914, 405235, 508151, 517198

43
HETEROMER PEPTIDES

431 Lipopeptide Antibiotics
432 Thiapeptides
433 Chelate-Forming Peptides

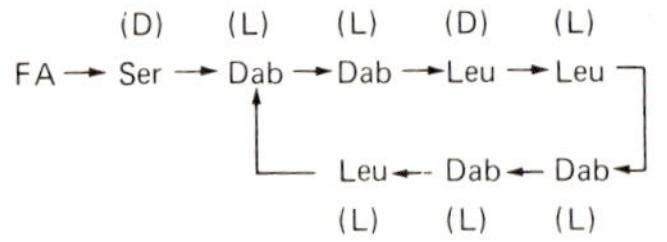
(D) (L) (L) (D) (L)
FA → Ser → Dab → Dab → Leu → Leu
Leu ← Dab ← Dab
(L) (L) (L)

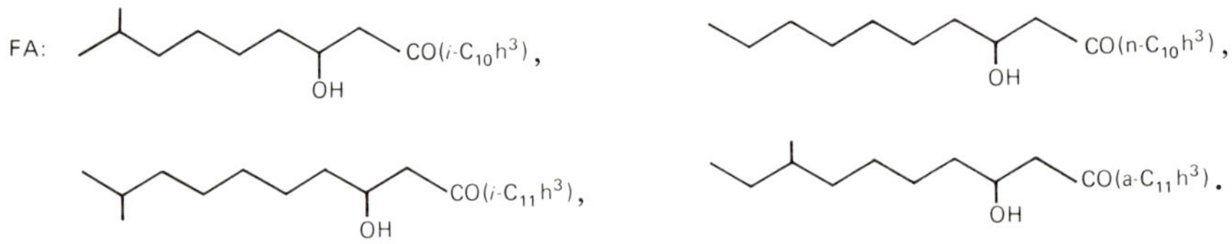
FA:
CO(i-C10h3),
CO(n-C10h3),
CO(i-C11h3),
CO(a-C11h3).
OH

The fatty acid component in Octapeptins A_4 and B_4 is $iC_{11}h^3$

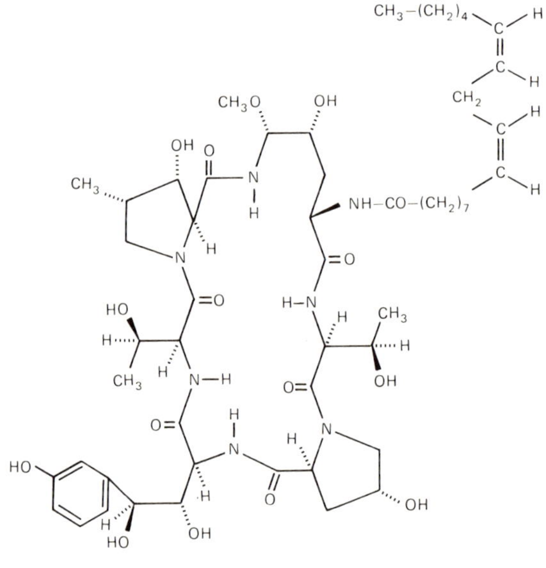
CH3−(CH2)4
NH−CO−(CH2)7
CH3O
OH
HO

A-30912 H

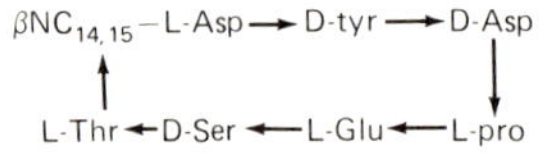
βNC14,15 − L-Asp → D-tyr → D-Asp
L-Thr ← D-Ser ← L-Glu ← L-pro

Bacillomycin-D

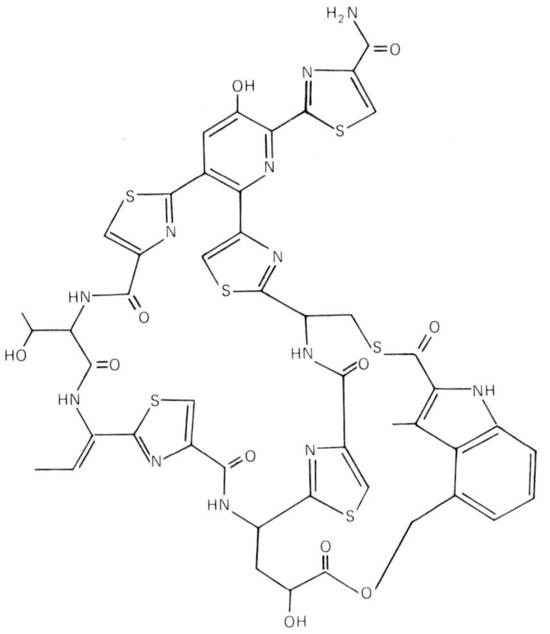

RP-35665

The structures of Thiostrepton B and Siomycin D_1 are in the first part of this Supplement.

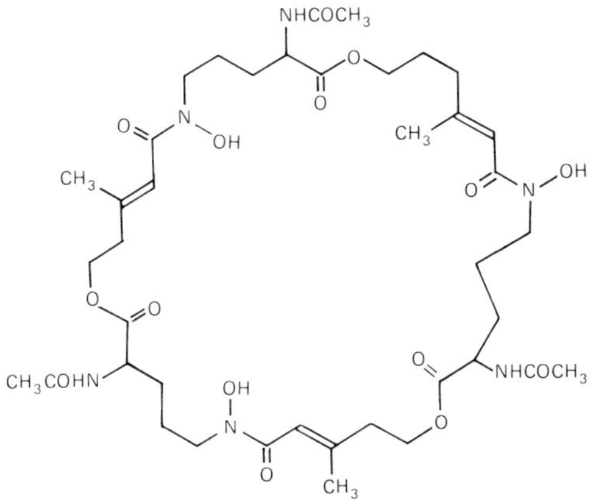

V (Desferritriacetylfusigen-like)

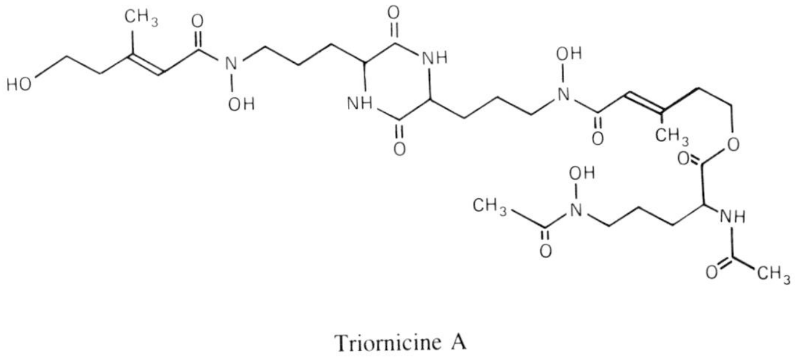

Triornicine A

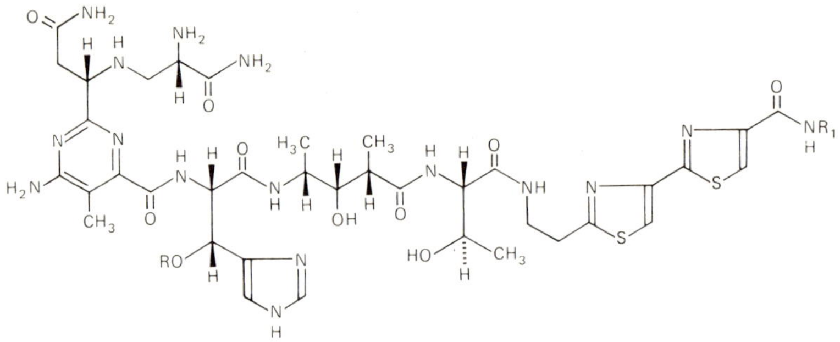

R = 2-O-(3-O--carbamoyl-α-D-mannopyranosyl)-α-L--gulopyranosyl

3 (S)-1'-Phenyletylamino propylaminobleomycin R_1: —$NH(CH_2)_3NHCH(CH_3)$—phenyl

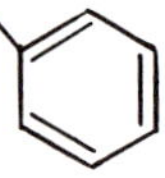

Bleomycin B_1' R_1: H

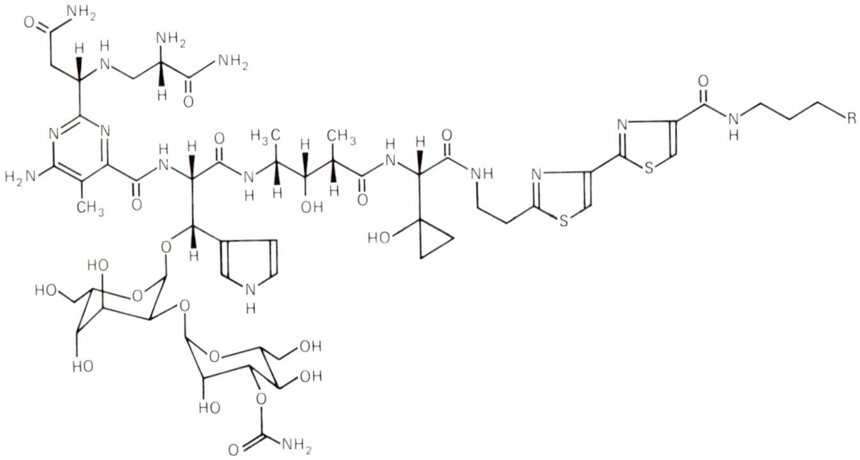

Cleomycin 1	$-S^+(CH_3)_2$
Cleomycin 2 (Cleomycin B_2)	$CH_2NH-C(=NH)-NH_2$
Cleomycin 6	$NH(CH_2)_4NH_2$
Cleomycin 11	$NH(CH_2)_3NH-C_4H_9$
Cleomycin 20	$N(CH_3)-(CH_2)_3NH_2$
Cleomycin 33	**R** $-NH-CH(CH_3)$—phenyl

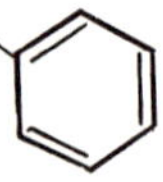

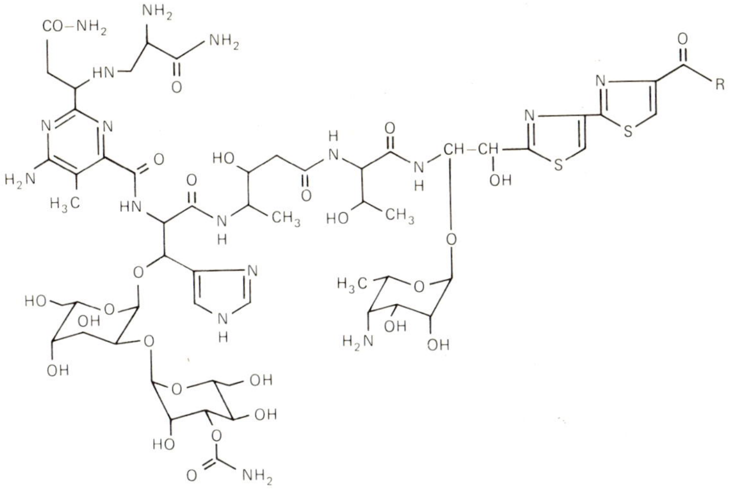

Tallysomycin A

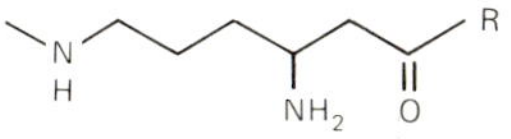

Tallysomycin B R

Name of new tallysomycins produced	**Terminal amine structure**
S_{1a}, S_{1b}	$-NH-(CH_2)_3-NH_2$
S_{2a}, S_{2b}	$-NH-(CH_2)_3-S^+(CH_3)_2$
S_{3b}	$-NH-(CH_2)_2-NH_2$
S_{4b}	$-NH-CH_2-CH(OH)-CH_2-NH_2$
S_{5b}	$-NH-(CH_2)_2-NH-CH_2-CH(OH)-CH_3$
S_{6a}, S_{6b}	$-NH-(CH_2)_3-NH-CH_2-CH_2OH$
S_{7b}	$-NH-(CH_2)_3-N(CH_3)_2$
S_{8a}, S_{8b}	$-NH-(CH_2)_3-N(CH_2-CH_2OH)_2$
S_{9b}	$-NH-(CH_2)_2-NH-CH_2-CH_2OH$
S_{10a}, S_{10b}	$-NH-(CH_2)_4-NH_2$
S_{11a}, S_{11b}	$-NH-(CH_2)_3-NH-CH_3$
S_{12b}	$-NH-(CH_2)_3-N$ (morpholine ring, O) 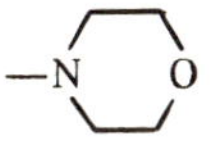
S_{13b}	$-NH-(CH_2)_3-N$ (piperidine ring, CH_3) 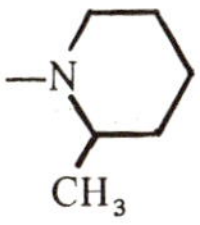
S_{14a}, S_{14b}	$-NH-(CH_2)_3-NH-CH(CH_3)-$ 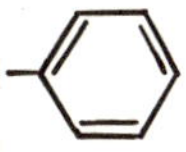 (phenyl)

43110-7954

NAME: PARVULIN-B
PO: S.PARVULUS-PARVULI
CT: LIPOPEPTIDE, AMPHOMYCIN T., ACIDIC, AMPHOTER
EA: (C, 49)(H, 7)(N, 10)
MW: 1200
PC: WH., POW.
TO: (G.POS.,)
IS-FIL: 3.5
IS-EXT: (ACET, 4, MIC.)
IS-CHR: (SILG, CHL-MEOH)
REFERENCES:
USP 3798129; Hung. P 157984

43110-7955

NAME: PARVULIN-C
PO: S.PARVULUS-PARVULI
CT: LIPOPEPTIDE, AMPHOMYCIN T., ACIDIC, AMPHOTER
EA: (C, 51)(H, 2)(N, 11)
MW: 1200
PC: WH., POW.
TO: (G.POS.,)
IS-FIL: 3.5
IS-EXT: (ACET, 4, MIC.)
IS-CHR: (SILG, CHL-MEOH)
REFERENCES:
USP 3798129; Hung. P 157984

43110-8180

NAME: 33-A
PO: S.ALBOGRISEOLUS
CT: LIPOPEPTIDE, AMPHOMYCIN T., ACIDIC, AMPHOTER
EA: (N, 12)
PC: WH., POW.
UV: MEOH: (277, 56.5,)
SOL-GOOD: MEOH
SOL-POOR: HEX
TO: (S.AUREUS,)(B.SUBT.,)
IS-ION: (XAD-2, ACET-BUOH-W)
IS-CHR: (SILG, CHL-MEOH)
IS-CRY: (PREC., BUOH, HEX)(CRYST., ETOAC)
REFERENCES:
Antib., 883, 1980

43122-6792

NAME: AM-157
PO: B.SP.
CT: LIPOPEPTIDE, OCTAPEPTIN T., BASIC
EA: (C, 48)(H, 8)(N, 18)
MW: 678|34
PC: WH., POW.
OR: (-65.5, W)
UV: W: (200, ,)
SOL-GOOD: W, MEOH, ETOH, PROH, BUOH
SOL-POOR: ACET, HEX
QUAL: (NINH., +)(SAKA., -)(PAULY, -)(EHRL., -)
TO: (S.AUREUS, 200)(B.SUBT., 200)(PS.AER., 6.25)
(E.COLI, 100)(K.PNEUM., 400)
LD50: (25, IV)
IS-EXT: (BUOH, 10, FILT.)(W, 4, BUOH)
IS-CHR: (CG-50-NH4, NH4OH)
IS-CRY: (LIOF.,)
REFERENCES:

JP 78/141203; *CA*, 90, 150293

43122-6793

NAME: OCTAPEPTIN-A4
PO: B.CIRCULANS
CT: LIPOPEPTIDE, OCTAPEPTIN T., BASIC
FORMULA: C49H93N13O10
EA: (N, 18)
MW: 1023
PC: WH.
TO: (G.POS.,)(G.NEG.,)(FUNGI,)
REFERENCES:

J. Chrom., 173, 313, 1979

43122-6794

NAME: OCTAPEPTIN-B4
PO: B.CIRCULANS
CT: LIPOPEPTIDE, OCTAPEPTIN T., BASIC
FORMULA: C52H91N13O10
EA: (N, 17)
MW: 1057
PC: WH.
TO: (G.POS.,)(G.NEG.,)(FUNGI,)
REFERENCES:

J. Chrom., 173, 313, 1979

43122-7123

NAME: OCTAPEPTIN-D
PO: B.SP.
CT: LIPOPEPTIDE, OCTAPEPTIN T., BASIC
FORMULA: C48H90N12O10, C47H88N12O11, C48H90N12O11
EA: (N, 15)
MW: 1076, 1090
PC: WH., POW.
OR: (-41.7, HCL)
UV: W: (200, ,)
SOL-GOOD: W, MEOH
SOL-POOR: ACET, CHL
QUAL: (NINH., +)(SAKA., -)(PAULY, -)(EHRL., -)
TO: (K.PNEUM., 6.25)(PS.AER., 12.5)(SHYG., 3.13)
LD50: (30|20, IP)
IS-FIL: 5
IS-EXT: (BUOH-MEOH, 5, WB.)(BUOH, 8, W)(W, 2, BUOH)
IS-ION: (IRC-50-NA, HCL)
IS-CHR: (SILG, CHL-ETOH-NH4OH)(SEPHADEX LH-20, MEOH-W)
IS-CRY: (PREC., MEOH-HCL, ACET)
REFERENCES:

JA, 33, 182, 186, 1980; JP 81/87596; *CA*, 95, 185561

43123-6795

NAME: PERMETIN-A, NLF-II, PERMYCIN-A
PO: B.CIRCULANS
CT: LIPOPEPTIDE, POLYPEPTIN T., BASIC
FORMULA: C54H92N12O12
EA: (N, 15)
MW: 1100
PC: WH., POW.
UV: MEOH: (258|12, ,)
SOL-GOOD: W, MEOH
SOL-POOR: ACET, ET2O
QUAL: (NINH., +)
TO: (S.AUREUS, 6.25)(B.SUBT., 3.13)(E.COLI, 12.5)
(K.PNEUM., 12.5)(PS.AER., 12.5)
LD50: (36, IP)(2100, PEROS)
IS-EXT: (BUOH, 8, FILT.)
IS-CHR: (SEPHADEX LH-20, MEOH)(CM-CEL, MEOH-PH7.2 PUFF-NACL)
REFERENCES:

JA, 32, 115, 121, 1979; JP 79/132548; 80/98115; *CA*, 92, 179072; USP 4294754

43123-7124

NAME: PERMETIN-B, PELMYCIN, PERMYCIN-B
PO: B.CIRCULANS
CT: LIPOPEPTIDE, POLYPEPTIN T., BASIC
FORMULA: C55H94N12O12
EA: (N, 15)
MW: 1114
PC: WH., POW.
TO: (G.POS.,)(G.NEG.,)(FUNGI,)
IS-EXT: (BUOH, , FILT.)
IS-CHR: (SEPHADEX G-25, W)
REFERENCES:
JA, 32, 115, 121, 1979; JP 79/132548; 80/98115; *CA*, 92, 179072; USP 4294754

43124-8181

NAME: TRIDECAPTIN-CA"2
PO: B.POLYMYXA
CT: LIPOPEPTIDE, TRIDECAPTIN T., BASIC, AMPHOTER
FORMULA: C73H115N17O21
EA: (N, 15)
MW: 1565
PC: WH., POW.
SOL-GOOD: ACID, BASE
SOL-POOR: W, ACET, ET2O
QUAL: (NINH., +)(EHRL., +)(SAKA., -)(PAULY, -)
TO: (B.SUBT., 6)(S.AUREUS, 6)(E.COLI, 3)(K.PNEUM., 6)
REFERENCES:
JA, 32, 313, 1979

43124-8182

NAME: TRIDECAPTIN-CB"1
PO: B.POLYMYXA
CT: LIPOPEPTIDE, TRIDECAPTIN T., AMPHOTER, BASIC
FORMULA: C75H119N17O21
EA: (N, 15)
MW: 1593
PC: WH., POW.
SOL-GOOD: ACID, BASE
SOL-POOR: W, ACET, ET2O
QUAL: (NINH., +)(EHRL., +)(SAKA., -)(PAULY, -)
TO: (B.SUBT., 6)(S.AUREUS, 6)(E.COLI, 3)(K.PNEUM., 25)
REFERENCES:
JA, 32, 313, 1979

43124-8183

NAME:	TRIDECAPTIN-BB", TRIDECAPTIN-BG"
PO:	B.POLYMYXA
CT:	LIPOPEPTIDE, TRIDECAPTIN T., AMPHOTER, BASIC
FORMULA:	C66H109N17O20
EA:	(N, 16)
MW:	1459
PC:	WH., POW.
SOL-GOOD:	ACID, BASE
SOL-FAIR:	MEOH-W
SOL-POOR:	W, ACET, ET2O
QUAL:	(NINH., +)(EHRL., +)(SAKA., -)(PAULY, -)
TO:	(B.SUBT., 12)(S.AUREUS, 12)(E.COLI, 6) (K.PNEUM., 25)

REFERENCES:

JA, 32, 313, 1979

43124-8184

NAME:	TRIDECAPTIN-BD"
PO:	B.POLYMYXA
CT:	LIPOPEPTIDE, TRIDECAPTIN T., AMPHOTER, BASIC
FORMULA:	C65H107N17O20
EA:	(N, 16)
MW:	1445
PC:	WH., POW.
SOL-GOOD:	ACID, BASE
SOL-POOR:	W, ACET, ET2O
QUAL:	(NINH., +)(EHRL., +)(SAKA., -)(PAULY, -)
TO:	(B.SUBT., 12)(S.AUREUS, 25)(E.COLI, 6) (K.PNEUM., 25)

REFERENCES:

JA, 32, 313, 1979

43130-8185

NAME: A-30912-H, A-42355
PO: ASP.NINDULANS-ROSEUS
CT: PEPTIDE, ECHINOCANDIN T.
FORMULA: C53H83N7016
EA: (N, 9)
MW: 1073
PC: WH., POW.
OR: (-40, MEOH)
UV: MEOH: (223, , 13100)(275, , 2100)
UV: MEOH-HCL: (223, , 13000)(275, , 2000)
UV: MEOH-NAOH: (245, , 14700)(290, , 3500)
SOL-GOOD: MEOH, ETOAC, BASE, DMFA, DMSO
SOL-POOR: ET20, HEX
TO: (C.ALB., 1.25)(FUNGI, .078)
IS-CHR: (SILG, MEOH-ACCN-W)
IS-CRY: (DRY,)
REFERENCES:
BP 2065130; Belg. P 883592, 883593; EP 21684

43140-6787

NAME: MANILOSPORIN-C1
PO: B.SUBTILIS-MANILOSPORA
CT: LIPOPEPTIDE, BACILLOMYCIN T.
EA: (C, 56)(H, 7)(N, 13)
PC: WH., POW.
OR: (+31.15, PYR)
UV: MEOH: (227, ,)(272, ,)
TO: (FUNGI, 10)(PHYT.FUNGI, 10)
LD50: (130, IP)(600, PEROS)
IS-EXT: (CHL-ETOH, ,)(MEOH, ,)
IS-CHR: (SEPHADEX LH-20, CHL-ETOH)
REFERENCES:
DT 2732467, 2806813

43140-6788

NAME:	MANILOSPORIN-C2
PO:	B.SUBTILIS-MANILOSPORA
CT:	LIPOPEPTIDE, BACILLOMYCIN T.
FORMULA:	C30H48N6012, C35H56N7014
EA:	(C, 52)(H, 9)(N, 13)
MW:	750
PC:	WH., POW.
OR:	(+22.1, PYR)
UV:	MEOH: (227, ,)(272, ,)
SOL-GOOD:	MEOH
TO:	(FUNGI, 10)(PHYT.FUNGI, 5)
LD50:	(130, IP)(600, PEROS)

REFERENCES:

DT 2732467, 2806813

43140-8074

NAME:	BACILLOMYCIN-D
IDENTICAL:	"RAUBITSCHEK SUBSTANCE"
PO:	B.SUBTILIS
CT:	PEPTIDE, BACILLOMYCIN T., ACIDIC
FORMULA:	C48H74N9016
EA:	(N, 12)
MW:	1060|32, 1039
PC:	WH., POW.
OR:	(-,)
UV:	(200, ,)
SOL-GOOD:	BASE
SOL-FAIR:	PYR, ETOH-W
SOL-POOR:	MEOH, HEX, W
QUAL:	(PAULY, +)(NINH., -)(EHRL., -)
TO:	(FUNGI,)
IS-CHR:	(SEPHADEX LH-20, HEX-CHL-MEOH)
IS-CRY:	(CRYST., CHL-MEOH)

REFERENCES:

JA, 33, 1146, 1980; *Eur. J. Bioch.*, 118, 323, 1981

43100-6784

NAME:	MANILOSPORIN-A
PO:	B.SUBTILIS-MANILOSPORA
CT:	LIPOPEPTIDE
EA:	(C, 60)(H, 10)(N, 10)
PC:	WH., POW.
OR:	(-34.62, MEOH)
UV:	MEOH: (227, ,)(272, ,)
SOL-GOOD:	MEOH
TO:	(S.AUREUS, 5)(S.LUTEA, 10)
LD50:	(125|25, IP)
IS-EXT:	(ETOH, , PREC.)(CHL, ,)
IS-CHR:	(SILG, CHL-ETOH)
IS-CRY:	(PREC., ACID, FILT.)(PREC., ETOH, W)

REFERENCES:

DT 2732467, 2806813

43100-6785

NAME:	MANILOSPORIN-B1
PO:	B.SUBTILIS-MANILOSPORA
CT:	LIPOPEPTIDE
FORMULA:	C56H96N8016, C63H108N9018
EA:	(C, 57)(H, 9)(N, 10)
MW:	1250
PC:	WH., POW.
OR:	(+23.81, MEOH)
UV:	MEOH: (227, ,)(272, ,)
SOL-GOOD:	MEOH
TO:	(E.COLI, 2.5)(K.PNEUM., 50)(P.VULG., 1) (PS.AER., 25)(S.LUTEA, 10)
LD50:	(400, IP)(1000, PEROS)
IS-EXT:	(ETOH, ,)(CHL-ETOH, ,)
IS-CHR:	(SEPHADEX LH-20, CHL-MEOH)
IS-CRY:	(PREC., ACID, FILT.)

REFERENCES:

DT 2732467, 2806813

43100-6786

NAME: MANILOSPORIN-B2
PO: B.SUBTILIS-MANILOSPORA
CT: LIPOPEPTIDE
FORMULA: C56H98N7014, C64H112N8016
EA: (C, 61)(N, 8)(H, 9)
MW: 1209
PC: WH., POW.
OR: (-23.15, MEOH)
UV: MEOH: (227, ,)(272, ,)
SOL-GOOD: MEOH
TO: (S.LUTEA, 10)(S.AUREUS, 25)(E.COLI, 10)
(K.PNEUM., 50)(PS.AER., 50)(P.VULG., 2.5)
LD50: (150, IP)
REFERENCES:
DT 2732467, 2806813

43100-7678

NAME: HERBICOLIN-A
PO: ERWINIA HERBICOLA
CT: LIPOPEPTIDE, NEUTRAL
EA: (O,)(C, 46)(H, 7)(N, 12)(S, 2)
MW: 2250|750
PC: WH., CRYST.
UV: MEOH: (200, ,)
SOL-GOOD: MEOH, BUOH, W
SOL-FAIR: ETOAC, CHL, ACET
SOL-POOR: ET2O, HEX
QUAL: (SAKA., +)(NINH., -)
TO: (FUNGI,)(PHYT.FUNGI,)(C.ALB.,)(PROTOZOA,)
(ALGAE,)
IS-ION: (XAD-2, MEOH)
IS-CHR: (SEPHADEX LH-20, MEOH)
IS-CRY: (CRYST., MEOH)
REFERENCES:
JA, 33, 353, 1980; *Mykosen*, 23, 290, 1980; EP 26485; *CA*, 93, 215949

43100-7679

NAME: HERBICOLIN-B
PO: ERWINIA HERBICOLA
CT: LIPOPEPTIDE, NEUTRAL
EA: (O,)(C, 46)(H, 7)(N, 12)(S, 2)
MW: 2250|750
PC: WH.
UV: MEOH: (200, ,)
SOL-GOOD: MEOH, BUOH, W
SOL-FAIR: ACET
SOL-POOR: ET2O, HEX
QUAL: (NINH., -)
TO: (PHYT.FUNGI,)(FUNGI,)(C.ALB.,)
REFERENCES:

JA, 33, 353, 1980; *Mykosen*, 23, 290, 1980; EP 26485; *CA*, 93, 215949

43211-6796

NAME: RP-35665
PO: S.ACTUOSUS+ACID
CT: THIAZOLYL-PEPTIDE, THIOSTREPTON T., NEUTRAL
FORMULA: C48H40N12O11S6
EA: (N, 14)(S, 15)(ASH, 6)
PC: YELLOW, POW.
OR: (+61, PYR)
UV: MEOH: (320, 313,)(405, 107,)
SOL-GOOD: DMFA, ACOH, PYR
SOL-POOR: W, ETOH, HEX
TO: (S.AUREUS, .001)(S.LUTEA, .00046)(B.SUBT., .0023)
LD50: NONTOXIC
REFERENCES:

Belg. P 869492

43211-6801

NAME: A-6984-B, PEPTIDE-B
PO: ACTINOPLANES UTAHENSIS
CT: THIAZOLYL-PEPTIDE, THIOSTREPTON T.
EA: (C, 48)(H, 3)(N, 15)(S, 16)
MW: 1250|150
PC: WH., YELLOW, POW.
OR: (+33, PYR)
UV: MEOH: (218, 710,)(325, 292,)(416, 124,)
(440, 105,)
UV: MEOH-HCL: (218, 794,)(231, 246,)(351, 237,)
UV: MEOH-NAOH: (215, 288,)(414, 136,)
SOL-GOOD: PYR, DMSO
SOL-POOR: W, BENZ, HEX
QUAL: (FECL3, -)(NINH., -)(FEHL., -)
TO: (S.AUREUS, .0063)(S.LUTEA, .0031)(B.SUBT., .2)
(E.COLI, .2)(G.POS.,)
LD50: NONTOXIC
IS-EXT: (ACET, , MIC.)(ETOAC, 7, FILT.)
IS-CHR: (SILG, CHL-MEOH)
REFERENCES:

JP 79/32401; *CA*, 91, 18344

43211-7125

NAME: A-6984-C, PEPTIDE-C
PO: ACTINOPLANES UTAHENSIS
CT: THIAZOLYL-PEPTIDE, THIOSTREPTON T., NEUTRAL
EA: (C, 49)(H, 4)(N, 14)(S, 14)
MW: 1250|150
PC: YELLOW, CRYST.
OR: (+75, PYR)
UV: MEOH: (218, 689,)(324, 222,)(412, 93,)
UV: MEOH-HCL: (218, 715,)(300, 190,)(333, 215,)
(351, 211,)
UV: MEOH-NAOH: (232, 553,)(295, 248,)(408, 151,
)
SOL-GOOD: PYR, DMSO
SOL-POOR: W, BENZ, HEX
QUAL: (FECL3, -)(NINH., -)(FEHL., -)
TO: (S.AUREUS, .025)(B.SUBT., .025)(S.LUTEA, .0063)
LD50: NONTOXIC
IS-EXT: (ACET, , MIC.)(ETOAC, 7, FILT.)
IS-CHR: (SILG, CHL-MEOH)
REFERENCES:

JP 79/32402; *CA*, 91, 18345

43211-7126

NAME: A-6984-D, PEPTIDE-D
PO: ACTINOPLANES UTAHENSIS
CT: THIAZOLYL-PEPTIDE, THIOSTREPTON T., NEUTRAL
EA: (C, 47)(H, 3)(N, 13)(S, 14)
MW: 1250|150
PC: YELLOW, CRYST.
OR: (+53, PYR)
UV: MEOH: (219, 641,)(327, 267,)(414, 141,)
UV: MEOH-HCL: (219, 715,)(300, 215,)(333, 247,) (352, 242,)
UV: MEOH-NAOH: (232, 577,)(298, 268,)(410, 177,)
SOL-GOOD: PYR, DMSO
SOL-POOR: W, BENZ, HEX
QUAL: (FECL3, -)(NINH., -)(FEHL., -)
TO: (S.AUREUS, .05)(B.SUBT., .05)(S.LUTEA, .0125)
IS-EXT: (ACET, , MIC.)(ETOAC, 7, FILT.)
IS-CHR: (SILG, CHL-MEOH)
REFERENCES:
JP 79/32402; *CA*, 91, 18345

43211-7127

NAME: A-10947
PO: ACTINOPLANES SP.
CT: THIAZOLYL-PEPTIDE, THIOSTREPTON T., AMPHOTER
EA: (C, 49)(H, 4)(N, 10)(S, 10)
EW: 1272
PC: WH., YELLOW, CRYST.
OR: (+64.1, CHL)
UV: MEOH: (217, 476,)(261, 260,)(302, 191,) (372, 78,)(409, 51,)
SOL-GOOD: ACOH, PYR, DMSO, CHL
SOL-FAIR: MEOH, ETOH, BUOH, ETOAC
SOL-POOR: BENZ, HEX, W
QUAL: (SAKA., -)(FEHL., -)(FECL3, -)(BIURET, -) (NINH., -)
TO: (S.AUREUS, .1)(B.SUBT., .05)(S.LUTEA, .05)
LD50: (1000|500, IP)
IS-EXT: (ACET, , MIC.)(ETOAC, 3.5, W)
IS-CHR: (SILG, CHL-MEOH)
IS-CRY: (PREC., ETOAC, HEX)(CRYST., ETOAC)
REFERENCES:
JA, 32, 967, 1979; JP 79/157502; *CA*, 92, 126912

43211-7129

NAME: SYRIAMYCIN
PO: S.VIOLACEONIGER
CT: PEPTIDE L., NEUTRAL
EA: (N,)(S,)
PC: YELLOW
TO: (S.AUREUS, .01)(B.SUBT., .01)
TV: EHRLICH, S-180
REFERENCES:

Ain Shams Med. J., 29, 305, 1978; *Ind. J. Exp. Biol.*, 17, 721, 1979; *CA*, 91, 191294

43211-7427

NAME: S-54832-A-I
PO: MIC.GLOBOSA
CT: THIAZOLYL-PEPTIDE, THIOSTREPTON T., AMPHOTER
EA: (C, 50)(H, 4)(N, 13)(O, 22)(S, 11)
EW: 1521
PC: YELLOW, POW.
OR: (+118.7, PYR)
UV: MEOH: (219, 73,)(287, 26,)(364, 14,)
SOL-GOOD: CHL, DMFA, ACCN, DIOXAN, PYR, DMSO
SOL-FAIR: MEOH, ETOH
SOL-POOR: W, HEX
TO: (S.AUREUS, .03)(B.SUBT., .01)(S.LUTEA, .01)
IS-EXT: (ETOAC, , FILT.)(MEOH, , MIC.)
IS-CHR: (SEPHADEX LH-20, CH2CL2-MEOH)(SILG, CH2CL2-MEOH-W)
IS-CRY: (DRY,)
REFERENCES:

BP 2022573; 200120, DT 2825618; *CA*, 90, 150294

43211-7428

NAME: S-54832-A-II
PO: MIC.GLOBOSA
CT: THIAZOLYL-PEPTIDE, THIOSTREPTON T.
EA: (C, 50)(H, 4)(N, 12)(S, 12)(O, 22)
PC: YELLOW, POW.
UV: MEOH: (218, 80,)(293, 25,)(362, 14.5,)
SOL-GOOD: CHL, DMFA, ACCN, DIOXAN, PYR, DMSO
SOL-FAIR: MEOH, ETOH
SOL-POOR: W, HEX
TO: (S.AUREUS, .01)
IS-CHR: (SILG, CH2CL2-MEOH-W)
REFERENCES:

BP 2022573, 200120; DT 2825618; *CA*, 90, 150294

43211-7429

NAME: S-54832-A-III
PO: MIC.GLOBOSA
CT: THIAZOLYL-PEPTIDE, THIOSTREPTON T.
EA: (C, 42)(H, 4)(N, 12)(S,)
PC: YELLOW, POW.
UV: MEOH: (218, 73,)(292, 23,)(363, 14,)
SOL-GOOD: CHL, DMFA, DMSO, PYR, DIOXAN, ACCN
SOL-FAIR: MEOH
SOL-POOR: W, HEX
TO: (S.AUREUS, .019)
IS-CHR: (SILG, CH2CL2-MEOH-W)
REFERENCES:

BP 2022573, 200120; DT 2825618; *CA*, 90, 150294

43211-7430

NAME: S-54832-A-IV
PO: MIC.GLOBOSA
CT: THIAZOLYL-PEPTIDE, THIOSTREPTON T., AMPHOTER
EA: (C, 49)(H, 4)(N, 13)(O, 22)(S, 12)
EW: 1554
PC: YELLOW, POW.
UV: MEOH: (219, 76,)(292, 26,)(360, 14,)
SOL-GOOD: CHL, DMFA, DMSO, ACCN, PYR, DIOXAN
SOL-FAIR: MEOH, ETOH
SOL-POOR: W, HEX
TO: (S.AUREUS, .019)(S.LUTEA, .01)(B.SUBT., 1)
IS-CHR: (SILG, CHL-MEOH-W)
REFERENCES:

BP 2022573, 200120; DT 2825618; *CA*, 90, 150294

43211-7693

NAME: CINROPEPTIN, TSINROPEPTIN
PO: ACT.CINERACEUS, S.CINERACEUS
CT: THIAZOLYL-PEPTIDE, THIOSTREPTON T., NEUTRAL
FORMULA: C50H63N11O12S6
EA: (C, 50)(H, 5)(N, 13)(S, 16)
MW: 1200, 1072
PC: WH., YELLOW, POW.
OR: (-169.4, CHL)
UV: MEOH: (218, 620,)(232, 632,)(304, 100,)
(364, 98,)
SOL-GOOD: CHL, BENZ, DMFA, ACOH, BASE, PYR, DMSO
SOL-FAIR: MEOH, ETOH
SOL-POOR: ET2O, HEX, W, ACID
QUAL: (NINH., -)
TO: (S.AUREUS, .015)(B.SUBT., .007)(S.LUTEA, .03)
(E.COLI, 50)(P.VULG., 100)(PS.AER., 50)
(K.PNEUM., 25)
IS-EXT: (ACET, , MIC.)
IS-CHR: (SILG, ETOAC)
IS-CRY: (PREC., CHL, HEX)
REFERENCES:
Antib., 323, 403, 1981; SU P 555663, 563027; *CA*, 92, 109102

43211-8186

NAME: THIOSTREPTON-B
PO: S.AZUREUS, S.LAURENTII
CT: THIAZOLYL-PEPTIDE, THIOSTREPTON T., AMPHOTER
FORMULA: C66H78N16O16S5
EA: (N, 16)(S, 10)
MW: 1510
PC: WH., YELLOW, POW.
UV: MEOH: (200, ,)
SOL-GOOD: CHL, PYR
SOL-FAIR: MEOH
SOL-POOR: W
TO: (S.AUREUS, .05)(B.SUBT., .05)
REFERENCES:
Proc. 17th Symp. Peptide Chem., 1979, 19—24; *CA*, 94, 4233; *JA*, 34, 124, 1981

43211-8187

NAME: SIOMYCIN-D1
PO: S.SIOYAENSIS
CT: THIAZOLYL-PEPTIDE, THIOSTREPTON T., AMPHOTER
FORMULA: C70H79N19O18S5
EA: (N, 16)(S, 10)
MW: 1633, 1700|100
PC: WH., CRYST.
OR: (-69.9, DIOXAN)
UV: ETOH: (200, ,)(250, ,)(280, ,)
SOL-GOOD: CHL, DMFA
TO: (S.AUREUS, .05)(B.SUBT., .05)
IS-CHR: (SILG, CHL-MEOH)
IS-CRY: (CRYST., CHL-MEOH)
REFERENCES:
JA, 33, 1563, 1980

43211-8188

NAME: PLANOTHIOCIN-A
PO: ACTINOPLANES SP.
CT: THIAZOLYL-PEPTIDE, THIOSTREPTON T., ACIDIC
EA: (C, 49)(H, 4)(N, 13)(S, 14)
MW: 1438
PC: WH., CRYST.
OR: (+19.6, PYR)
UV: MEOH: (217, 625,)(330, 31, ,)
UV: MEOH-HCL: (330, 323,)
UV: MEOH-NAOH: (234, 589,)(328, , 231)
SOL-GOOD: CHL-MEOH, PYR, DMSO, ACOH
SOL-POOR: ACET, HEX, W
QUAL: (FECL3, -)(NINH., -)
STAB: (ACID, +)(BASE, -)
TO: (S.AUREUS, .08)(S.LUTEA, .004)(B.SUBT., .031)
LD50: NONTOXIC
IS-EXT: (ACET, , MIC.)(ETOAC, 3, W)
IS-CHR: (SILG, CHL-MEOH-ACOH)
IS-CRY: (PREC., ACET, HEX)(CRYST., CHL-MEOH)
REFERENCES:
DT 3028594

43211-8189

NAME:	PLANOTHIOCIN-B
PO:	ACTINOPLANES SP.
CT:	THIAZOLYL-PEPTIDE, THIOSTREPTON T., ACIDIC
EA:	(C, 49)(H, 3)(N, 13)(S, 14)
MW:	1350
PC:	WH., CRYST.
OR:	(+95.4, PYR)
UV:	MEOH: (218, 540,)(330, 277,)
UV:	MEOH-HCL: (219, 566,)(331, 302,)
UV:	MEOH-NAOH: (297, 264,)(329, 262,)
SOL-GOOD:	CHL-MEOH, PYR, DMSO, ACOH
SOL-POOR:	ACET, HEX, W
QUAL:	(FECL3, -)(NINH., -)
STAB:	(ACID, +)(BASE, -)
TO:	(S.AUREUS, .016)(B.SUBT., .063)(S.LUTEA, .008)
LD50:	NONTOXIC
IS-CHR:	(SILG, ETOAC-MEOH-ACOH)
IS-CRY:	(CRYST., CHL-MEOH)

REFERENCES:

DT 3028594

43211-8190

NAME:	PLANOTHIOCIN-C
PO:	ACTINOPLANES SP.
CT:	THIAZOLYL-PEPTIDE, THIOSTREPTON T., ACIDIC
EA:	(C, 49)(H, 4)(N, 13)(S, 15)
MW:	1350|150
PC:	WH., POW.
OR:	(+28.4, PYR)
UV:	MEOH: (218, 721,)(330, 327,)
UV:	MEOH-NAOH: (299, 271,)(329, 241,)
SOL-GOOD:	CHL-MEOH, PYR, DMSO, ACOH
SOL-POOR:	ACET, HEX, W
QUAL:	(FECL3, -)(NINH., -)
STAB:	(ACID, +)(BASE, -)
TO:	(S.AUREUS, .08)(S.LUTEA, .002)(B.SUBT., .031)
LD50:	NONTOXIC
IS-CHR:	(SILG, ETOAC-MEOH-ACOH)

REFERENCES:

DT 3028594

43211-8191

NAME: PLANOTHIOCIN-D
PO: ACTINOPLANES SP.
CT: THIAZOLYL-PEPTIDE, THIOSTREPTON T., ACIDIC
EA: (C, 47)(H, 4)(N, 13)(S, 16)
MW: 1350|150
PC: WH., POW.
OR: (+31.4, PYR)
UV: MEOH: (217, 764,)(330, 343,)
UV: MEOH-NAOH: (300, 282,)(329, 248,)
SOL-GOOD: CHL-MEOH, PYR, DMSO, ACOH
SOL-POOR: ACET, HEX, W
QUAL: (FECL3, -)(NINH., -)
STAB: (ACID, +)(BASE, -)
TO: (S.AUREUS, .004)(B.SUBT., .016)(S.LUTEA, .002)
LD50: NONTOXIC
IS-CHR: (SILG, ETOAC-MEOH-ACOH)
REFERENCES:
DT 3028594

43211-8192

NAME: PLANOTHIOCIN-E
PO: ACTINOPLANES SP.
CT: THIAZOLYL-PEPTIDE, THIOSTREPTON T., ACIDIC
EA: (C, 48)(H, 13)(N, 13)(S, 14)
MW: 1350|150
PC: WH., POW.
OR: (+45.5, PYR)
UV: MEOH: (217, 556,)(329, 207,)
UV: MEOH-NAOH: (299, 2472,)(329, 207,)
SOL-GOOD: CHL-MEOH, PYR, DMSO, ACOH
SOL-POOR: ACET, HEX, W
QUAL: (FECL3, -)(NINH., -)
STAB: (ACID, +)(BASE, -)
TO: (S.AUREUS, .016)(B.SUBT., .031)(S.LUTEA, .008)
LD50: NONTOXIC
IS-CHR: (SILG, ETOAC-MEOH-ACOH)
REFERENCES:
DT 3028594

43211-8193

NAME:	PLANOTHIOCIN-F
PO:	ACTINOPLANES SP.
CT:	THIAZOLYL-PEPTIDE, THIOSTREPTON T., ACIDIC
EA:	(C, 50)(H, 4)(N, 15)(S, 15)
MW:	1350|150
PC:	WH., POW.
OR:	(+107.8, PYR)
UV:	MEOH: (217, 744,)(330, 333,)
UV:	MEOH-NAOH: (240, 515,)(315, 291,)
SOL-GOOD:	CHL-MEOH, PYR, DMSO, ACOH
SOL-POOR:	ACET, HEX, W
QUAL:	(FECL3, -)(NINH., -)
STAB:	(ACID, +)(BASE, -)
TO:	(S.AUREUS, .031)(B.SUBT., 1)(S.LUTEA, .008)
LD50:	NONTOXIC
IS-CHR:	(SILG, CHL-MEOH)

REFERENCES:

DT 3028594

43211-8194

NAME:	PLANOTHIOCIN-G
PO:	ACTINOPLANES SP.
CT:	THIAZOLYL-PEPTIDE, THIOSTREPTON T., ACIDIC
EA:	(C, 49)(H, 3)(N, 14)(S, 16)
MW:	1350|150
PC:	WH., POW.
OR:	(+122.4, PYR)
UV:	MEOH: (217, 770,)(330, 35,)
UV:	MEOH-NAOH: (240, 51.,)(299, 247,)(329, 284)
SOL-GOOD:	CHL-MEOH, PYR, DMSO, ACOH
SOL-POOR:	ACET, HEX, W
QUAL:	(FECL3, -)(NINH., -)
STAB:	(ACID, +)(BASE, -)
TO:	(S.AUREUS, .016)(B.SUBT., .063)(S.LUTEA, .002)
LD50:	NONTOXIC
IS-CHR:	(SILG, CHL-MEOH)

REFERENCES:

DT 3028594

43213-8195

NAME: JINGSIMYCIN
PO: S.HYGROSCOPICUS-JINGGANSIS
CT: THIAZOLYL-PEPTIDE
EA: (C, 41)(H, 5)(N, 14)(S, 13)
MW: 2100
PC: YELLOW, POW.
OR: (+42, W)
UV: MEOH: (222, ,)(270, ,)
SOL-GOOD: MEOH, DMFA
QUAL: (DNPH, -)
TO: (FUNGI, 2.5)
LD50: (2000|1000, IP)
REFERENCES:
Acta Micr. Sinica, 20, 191, 1980

43200-8562

NAME: A-7413-D
PO: ACTINOPLANES SP.
CT: PEPTIDE
EA: (N,)(S,)
TO: (B.SUBT.,)
REFERENCES:
DT 2703938

43200-8601

NAME: L-31-A
PO: ACTINOMADURA SP.
CT: PEPTIDE, ACIDIC
EA: (C, 52)(H, 6)(N, 14)(O, 21)(S, 6)(CL, 1)
PC: WH., YELLOW, POW.
OR: (-34, W)
UV: MEOH: (280, 45.2,)
SOL-GOOD: MEOH, BUOH, DMFA, ACOH, BASE
SOL-POOR: ETOAC, HEX, W
QUAL: (NINH., +)(EHRL., +)(FECL3, -)
STAB: (ACID, +)(BASE, -)
TO: (B.STEAROTHERMOPHILUS, 125)(S.AUREUS,)(E.COLI,)(PS.AER.,)
LD50: (500|200, IP)
IS-FIL: 2
IS-EXT: (BUOH, 2, FILT.)(ACET, , MIC.)(BUOH, 2, W)
IS-CHR: (CEL, BUOH-ACOH-W)
IS-CRY: (PREC., BUOH, ETOAC)(DRY, ETOAC)
REFERENCES:
BP 2069500

43200-8602

NAME: L-31-B
PO: ACTINOMADURA SP.
CT: PEPTIDE, ACIDIC
EA: (C, 51)(H, 16)(N, 15)(S, 6)(O, 21)(CL, 1)
PC: WH., YELLOW, POW.
OR: (-26, W)
UV: MEOH: (-280, 44.8,)
SOL-GOOD: MEOH, BUOH, DMFA, ACOH, BASE
SOL-POOR: ETOAC, HEX, W
QUAL: (NINH., +)(EHRL., +)(FECL3, -)
STAB: (ACID, +)(BASE, -)
TO: (B.STEAROTHERMOPHILUS, 125)(S.AUREUS,)(E.COLI,)(PS.AER.,)
LD50: (200|50, IP)
REFERENCES:
BP 2069500

43311-6898

NAME: 216
PO: S.SP.
CT: SIDEROMYCIN, ALBOMYCIN T.
EA: (C, 39)(H, 5)(N, 14)(S, 3)(FE, 5)
PC: ORANGE, POW.
UV: W: (278, ,)(420|10, ,)
SOL-GOOD: W, DMSO
SOL-FAIR: MEOH
SOL-POOR: ETOH, HEX
QUAL: (NINH., -)
STAB: (ACID, -)(BASE, +)(HEAT, -)
TO: (E.COLI, .005)(PS.AER., 3.12)(K.PNEUM., .01)
(B.SUBT., .02)(S.AUREUS, .001)(S.LUTEA, 1)
(PS.AER., .39)(SHYG., .0012)
LD50: (3000|1000, IP)
IS-ION: (IRA-400,)
REFERENCES:

43311-7959

NAME: SF-2012-L
PO: S.VINACEUS
CT: SIDEROMYCIN, ALBOMYCIN T., BASIC
EA: (C, 48)(H, 7)(N, 16)(FE, 4)(O, 21)
MW: 1450|250
PC: RED, BROWN, POW.
UV: MEOH: (291|1, 94,)(421|3, 23,)
UV: W: (291|1, ,)(421|3, ,)
SOL-GOOD: MEOH, W
SOL-POOR: ACET, BENZ
QUAL: (NINH., -)
STAB: (ACID, +)(BASE, -)
TO: (B.SUBT., .09)(S.AUREUS, .19)
IS-ION: (DIAION HP-20, MEOH)(IRC-50-H, HCL)
REFERENCES:

JP 80/79396; *CA*, 93, 219315

43315-6797

NAME: V, DESFERRITRIACETYLFUSIGEN-LIKE
PO: ASP.NIDYLANS
CT: SIDERAMINE
EA: (C, 55)(H, 7)(N, 10)
PC: WH., POW.
SOL-GOOD: W, MEOH, CHL
QUAL: (FECL3, +)
TO: (S.AUREUS, 5)(B.SUBT., 5)(P.VULG., 5)
IS-EXT: (CHL, 4, FILT.)
IS-CHR: (SEPHADEX LH-20, MEOH)(SILG, ETOAC-MEOH)
IS-CRY: (DRY, ETOAC)
REFERENCES:

JA, 31, 1110, 1978

43315-7694

NAME: TRIORNICINE-A
PO: EPICOCUM PURPURASCENS
CT: SIDERAMINE
FORMULA: C31H50N6O12
EA: (N, 12)
MW: 698, 660|50
EW: 710|75
PC: POW., RED, BROWN
OR: (-22.1, MEOH)
UV: W: (195, , 40740)(250, , 12880)(430, , 1862)
SOL-GOOD: W
TV: EHRLICH
IS-EXT: (BENZYLALCOHOL, , FILT.+FECL3+AMMONIUMSULPHATE)
(W, , ET2O)
REFERENCES:

Biochem., 20, 2432, 2436, 1981; *Abst. 179th ACS Meet.*, 1980, ORGN-6

43320-7128

NAME: 3-S-1'-PHENYLETYLAMINO-PROPILAMINO-BLEOMYCIN
PO: S.VERTICILLUS+N-S-1'-PHENYLETYL-1.3-DIAMINOPROPANE
CT: GLYCOPEPTIDE, BLEOMYCIN T.
EA: (C,)(H,)(N,)(S,)
TO: (B.SUBT.,)
TV: ANTITUMOR
IS-ION: (IRC-50-H, HCL)
REFERENCES:

JP 79/41387; *CA*, 91, 138847

43320-7573

NAME: PINGYANMYCIN
IDENTICAL: BLEOMYCIN-A5
PO: S.PINGYANGENSIS, S.VERTICILLATUS-PINGYANGENSIS
CT: GLYCOPEPTIDE, BLEOMYCIN T., BASIC
EA: (N,)(S,)
TO: (G.POS.,)(G.NEG.,)
TV: S-180, S-37, WALKER-256, LEWIS, MELANOMA
REFERENCES:

Yao Hsueh Hsueh Pao, 15, 609, 1980; *Chung-Hua Ch. Liu Tsa Chin*, 1, 161, 1979; *CA*, 92, 104424; 95, 18396; *Acta Micr. Sinica*, 20, 94, 1980

43320-7956

NAME: BLEOMYCIN-B1′
PO: S.VERTICILLUS
CT: GLYCOPEPTIDE, BLEOMYCIN T., BASIC
FORMULA: C50H70N16O21S2
EA: (N, 17)(S, 5)
MW: 1294
PC: BLUE, POW.
UV: MEOH: (240, ,)(295, ,)
SOL-GOOD: W
TO: (G.POS.,)(G.NEG.,)
TV: L-1210
IS-ION: (IRC-50-H, HCL)
IS-CHR: (CM-SEPHADEX C-25,)
REFERENCES:
JA, 26, 396, 1973

43320-7957

NAME: CLEOMYCIN-B2
PO: S.VERTICILLATUS
CT: GLYCOPEPTIDE, BLEOMYCIN T., BASIC
FORMULA: C56H81N19O21S2
EA: (N, 18)(S, 4)
MW: 1419
PC: WH., POW.
UV: MEOH: (244, ,)(294, ,)
SOL-GOOD: W
TO: (G.POS.,)(G.NEG.,)
TV: ANTITUMOR
REFERENCES:
JA, 33, 1079, 1980; *TL*, 4925, 1980; DT 3027326

43320-7958

NAME: MC-637-SY-1
PO: S.FLAVOVIRIDIS
CT: GLYCOPEPTIDE, BLEOMYCIN T., BASIC
EA: (N,)(S,)(CU,)
PC: BLUE, POW.
UV: HCL: (240, ,)(295, ,)
UV: W: (240, ,)(295, ,)
SOL-GOOD: W
TO: (G.POS.,)(G.NEG.,)
TV: ANTITUMOR
IS-ION: (IRC-50-NA, HCL)
REFERENCES:
JP 73/22687

43320-8196

NAME:	CLEOMYCIN-A2, CLEOMYCIN-1
PO:	S.VERTICILLUS+3-AMINOPROPYL-DIMETHYLSULPHONIUM BROMIDE
CT:	GLYCOPEPTIDE, BLEOMYCIN T., BASIC
FORMULA:	C56H84N1MO21S3
EA:	(N, 16)(S, 7)
MW:	1426
PC:	WH., POW.
UV:	HCL: (290, , 16700)
UV:	W: (290, 112,)
SOL-GOOD:	W
QUAL:	(EHRL., +)(PAULY, +)
TO:	(G.POS.,)(G.NEG.,)(MYCOB.SP.,)
TV:	ANTITUMOR
IS-FIL:	7
IS-ION:	(IRC-50-H, HCL)(XAD-2, MEOH-HCL)
IS-CHR:	(AL, MEOH-W)(SEPHADEX G-25, W)(CM-SEPHADEX C-25-NA, NACL)

REFERENCES:

DT 3027326

43320-8197

NAME:	CLEOMYCIN-2, 4-GUANIDINOBUTYLAMINOCLEOMYCIN
PO:	S.VERTICILLUS+AGMATIN-SULPHATE
CT:	GLYCOPEPTIDE, BLEOMYCIN T., BASIC
FORMULA:	C56H84N21O21S2
EA:	(N, 20)(S, 4)
MW:	1450
PC:	WH., POW.
UV:	HCL: (290, , 16300)
UV:	W: (290, 109,)
SOL-GOOD:	W
QUAL:	(EHRL., +)(PAULY, +)
TO:	(G.POS.,)(G.NEG.,)(MYCOB.SP.,)
TV:	ANTITUMOR

REFERENCES:

DT 3027326

43320-8198

NAME: CLEOMYCIN-6, 3-4-AMINOBUTYLAMINO-PROPYLAMINOCLEOMYCIN
PO: S.VETICILLUS+SPERMIDINE
CT: GLYCOPEPTIDE, BLEOMYCIN T., BASIC
FORMULA: C58H89N20O21S2
EA: (N, 17)(S, 4)
MW: 1465
PC: WH., POW.
UV: HCL: (290, 98,)
UV: W: (290, ,)
SOL-GOOD: W
QUAL: (EHRL., +)(PAULY, +)(NINH., +)
TO: (G.POS.,)(G.NEG.,)(MYCOB.SP.,)
TV: ANTITUMOR
REFERENCES:
DT 3027326

43320-8199

NAME: CLEOMYCIN-11, 3-3-N-BUTYLAMINO-PROPYLAMINO-PROPYLAMINOCLEOMYCIN
PO: S.VERTICILLUS+N-N-BUTYL-N′-3-AMINOPROPYL-1.3-DIAMINOPROPAN
CT: GLYCOPEPTIDE, BLEOMYCIN T., BASIC
FORMULA: C61H93N19O21S2
EA: (N, 17)(S, 4)
MW: 1491
PC: WH., POW.
UV: W: (290, 103,)
SOL-GOOD: W
QUAL: (EHRL., +)(PAULY, +)
TO: (G.POS.,)(G.NEG.,)(MYCOB.SP.,)
TV: ANTITUMOR
REFERENCES:
DT 3027326

43320-8200

NAME: CLEOMYCIN-20, 3-N-METHYL-N-AMINOPROPYLAMINOCLEOMYCIN
PO: S.VERTICILLUS+N.N-BIS-3-AMINOPROPYL-METHYLAMINE
CT: GLYCOPEPTIDE, BASIC, BLEOMYCIN T.
FORMULA: C58H87N19021S2
EA: (N, 18)(S, 4)
MW: 1449
PC: WH., POW.
UV: W: (289, 108,)
SOL-GOOD: W
QUAL: (EHRL., +)(PAULY, +)
TO: (G.POS.,)(G.NEG.,)(MYCOB.SP.,)
TV: ANTITUMOR
REFERENCES:
DT 3027326

43320-8201

NAME: CLEOMYCIN-33, 3-S-R-PHENYLETHYAMINO-PROPYLAMINOCLEOMYCIN
PO: S.VERTICILLUS+N-S-1′-PHENYLETHYL-1.3-DIAMINOPROPAN
CT: GLYCOPEPTIDE, BLEOMYCIN T., BASIC
FORMULA: C62H88N18021S2
EA: (N, 16)(S, 4)
MW: 1606
PC: WH., POW.
UV: W: (290, ,)
SOL-GOOD: W
QUAL: (EHRL., +)(PAULY, +)
TO: (G.POS.,)(G.NEG.,)(MYCOB.SP.,)
TV: ANTITUMOR
REFERENCES:
DT 3027326

43320-8563

NAME:	TALLYSOMYCIN-S1A
PO:	STREPTOALLOTEICHUS HINDUSTANUS+1.3-DIAMINOPROPANE
CT:	GLYCOPEPTIDE, BLEOMYCIN T., BASIC
FORMULA:	C64H101N21O27S2CU
EA:	(N, 15)(S, 4)(CU,)
MW:	1659
PC:	BLUE, POW.
UV:	W: (242, 126,)(292, 102,)
SOL-GOOD:	W, MEOH, DMFA
SOL-POOR:	BUOH, HEX
QUAL:	(NINH., +)(FEHL., -)(SAKA., -)
TO:	(S.AUREUS, .4)(S.LUTEA, .1)(B.SUBT., .003) (E.COLI, .1)(P.VULG., .4)(MYCOB.SP., .012) (C.ALB., 3.1)(FUNGI, .8)
LD50:	(25, IP)
TV:	S-180, P-388, LEWIS, MELANOMA
IS-ION:	(IRC-50-NH4, HCL)(DIAION-HP-20, HCL)
IS-CHR:	(CM-SEPHADEX C-25-HCOO, NH4COOH)

REFERENCES:

JA, 34, 658, 1981; Belg. P 884291; DT 3026425

43320-8564

NAME:	TALLYSOMYCIN-S2A
PO:	STREPTOALLOTEICHUS HINDUSTANUS+3-AMINOPROPYL-DIMETHYLSULFONIUM CHLORIDE
CT:	GLYOPEPTIDE, BLEOMYCIN T., BASIC
EA:	(N, 15)(S, 5)(CU,)
MW:	1705
PC:	BLUE, POW.
UV:	W: (240, 120,)(290, 108,)
SOL-GOOD:	W, MEOH, DMFA
SOL-POOR:	BUOH, HEX
QUAL:	(NINH., +)(FEHL., -)(SAKA., -)
TO:	(S.AUREUS, .2)(S.LUTEA, .8)(B.SUBT., .05) (E.COLI, .05)(P.VULG., .2)(MYCOB.SP., .05) (C.ALB., 6.3)(FUNGI, .4)
LD50:	(25, IP)
TV:	S-180, P-388, LEWIS, MELANOMA
IS-ION:	(IRC-50-NH4, HCL)(DIAION-HP-20, HCL)
IS-CHR:	(CM-SEPHADEX C-25-HCOO, NH4COOH)

REFERENCES:

JA, 34, 658, 1981; Belg. P 884291; DT 3026425

43320-8565

NAME: TALLYSOMYCIN-S6A
PO: STREPTOALLOTEICHUS HINDUSTANUS+DIAMINOPROPANE
CT: GLYCOPEPTIDE, BLEOMYCIN T., BASIC
FORMULA: C66H105N21O28S2CU
EA: (N, 15)(S, 4)(CU,)
MW: 1703
PC: BLUE, POW.
UV: W: (243, , 105)(291, , 83)
SOL-GOOD: W, MEOH, DMFA
SOL-POOR: BUOH, HEX
QUAL: (NINH., +)(FEHL., -)(SAKA., -)
TO: (S.AUREUS, .8)(S.LUTEA, 1.6)(B.SUBT., .03)
(E.COLI, .4)(P.VULG., 1.6)(MYCOB.SP., .4)
(C.ALB., 6.3)(FUNGI, .2)
LD50: (15, IP)
TV: S-180, P-388, LEWIS, MELANOMA
IS-ION: (IRC-50-NH4, HCL)(DIAION-HP-20, HCL)
IS-CHR: (CM-SEPHADEX C-25-HCOO, NH4COOH)
REFERENCES:

JA, 34, 658, 1981; Belg. P 884291; DT 3026425

43320-8566

NAME: TALLYSOMYCIN-S8A
PO: STREPTOALLOTEICHUS HINDUSTANUS+N.N-DI-B''-
HYDROXYETHYL-1.3-DIAMINOPROPANE
CT: GLYCOPEPTIDE, BLEOMYCIN T., BASIC
FORMULA: C68H109N21O29S2CU
EA: (N, 15)(S, 3)(CU,)
MW: 1767
PC: BLUE, POW.
UV: W: (243, , 107)(291, 85,)
SOL-GOOD: W, MEOH, DMFA
SOL-POOR: BUOH, HEX
QUAL: (NINH., +)(FEHL., -)(SAKA., -)
IS-ION: (IRC-50-NH4, HCL)(DIAION-HP-20, HCL)
IS-CHR: (CM-SEPHADEX C-25-HCOO, NH4COOH)
REFERENCES:

JA, 34, 658, 1981; Belg. P 884291; DT 3026425

43320-8567

NAME: TALLYSOMYCIN-S10A
PO: STREPTOALLOTEICHUS HINDUSTANUS+1.4-DIAMINOBUTANE
CT: GLYCOPEPTIDE, BLEOMYCIN T., BASIC
FORMULA: C65H103N21O27S2CU
EA: (N, 16)(S, 4)(CU,)
MW: 1673
PC: BLUE, POW.
UV: W: (244, 242,)(292, 112,)
SOL-GOOD: W, MEOH, DMFA
SOL-POOR: BUOH, HEX
QUAL: (NINH., +)(FEHL., -)(SAKA., -)
TO: (S.AUREUS, 1.6)(S.LUTEA, .4)(B.SUBT., .04)
(E.COLI, .4)(P.VULG., .8)(MYCOB.SP., .2)
(C.ALB., 3.1)(FUNGI, 1.6)
LD50: (27, IP)
TV: S-180, P-388, LEWIS, MELANOMA
IS-ION: (IRC-50-NH4, HCL)(DIAION-HP-20, HCL)
IS-CHR: (CM-SEPHADEX C-25-HCOO, NH4COOH)
REFERENCES:

JA, 34, 658, 1981; Belg. P 884291; DT 3026425

43320-8568

NAME: TALLYSOMYCIN-S11A
PO: STREPTOALLOTEICHUS HINDUSTANUS+N-METHYL-1.3-DIAMINOPROPANE
CT: GLYCOPEPTIDE, BLEOMYCIN T., BASIC
FORMULA: C65H103N21O27S2CU
EA: (N, 16)(S, 4)(CU,)
MW: 1673
PC: BLUE, POW.
UV: W: (244, 62,)(292, 53,)
SOL-GOOD: W, MEOH, DMFA
SOL-POOR: BUOH, HEX
QUAL: (NINH., +)(FEHL., -)(SAKA., -)
TO: (S.AUREUS, 1.6)(S.LUTEA, .2)(B.SUBT., .025)
(E.COLI, .1)(P.VULG., .2)(MYCOB.SP., .2)
(C.ALB., 1.6)(FUNGI, .8)
LD50: (18, IP)
TV: S-180, P-388, LEWIS, MELANOMA
IS-ION: (IRC-50-NH4, HCL)(DIAION-HP-20, HCL)
IS-CHR: (CM-SEPHADEX C-25-HCOO, NH4COOH)
REFERENCES:

JA, 34, 658, 1981; Belg. P 884291; DT 3026425

43320-8569

NAME: TALLYSOMYCIN-S12A
PO: STREPTOALLOTEICHUS HINDUSTANUS+N-A"-PHENYLETHYL-1.3-DIAMINOPROPANE
CT: GLYCOPEPTIDE, BLEOMYCIN T., BASIC
FORMULA: C71H109N21O27S2CU
EA: (N, 15)(S, 3)(CU,)
MW: 1751
PC: BLUE, POW.
UV: W: (243, 102,)(292, 172,)
SOL-GOOD: W, MEOH, DMFA
SOL-POOR: BUOH, HEX
QUAL: (NINH., +)(FEHL., -)(SAKA., -)
TO: (S.AUREUS, .8)(S.LUTEA, .8)(B.SUBT., .1)
(E.COLI, .2)(P.VULG., 3.1)(MYCOB.SP., .1)
(C.ALB., 50)(FUNGI, 3.1)
TV: S-180, P-388, LEWIS, MELANOMA
IS-ION: (IRC-50-NH4, HCL)(DIAION-HP-20, HCL)
IS-CHR: (CM-SEPHADEX C-25-HCOO, NH4COOH)
REFERENCES:

JA, 34, 658, 1981; Belg. P 884291; DT 3026425

43320-8570

NAME: TALLYSOMYCIN-S1B
PO: STREPTOALLOTEICHUS HINDUSTANUS+1.3-DIAMINOPROPANE
CT: GLYCOPEPTIDE, BLEOMYCIN T., BASIC
FORMULA: C58H89N19O26S2CU
EA: (N, 16)(S, 4)(CU,)
MW: 1531
PC: BLUE, POW.
UV: W: (242, 108,)(290, 94,)
SOL-GOOD: W, MEOH, DMFA
SOL-POOR: BUOH, HEX
QUAL: (NINH., +)(FEHL., -)(SAKA., -)
TO: (S.AUREUS, 1)(S.LUTEA, 1.6)(B.SUBT., .05)
(E.COLI, .4)(P.VULG., 1.6)(MYCOB.SP., .05)
(C.ALB., 12)(FUNGI, 1.6)
LD50: (19, IP)
TV: S-180, P-388, LEWIS, MELANOMA
IS-ION: (IRC-50-NH4, HCL)(DIAION-HP-20, HCL)
IS-CHR: (CM-SEPHADEX C-25-HCOO, NH4COOH)
REFERENCES:

JA, 34, 658, 1981; Belg. P 884291; DT 3026425

43320-8571

NAME: TALLYSOMYCIN-S2B
PO: STREPTOALLOTEICHUS HINDUSTANUS+3-AMINOPROPYL-DIMETHYLSULFONIUM CHLORIDE
CT: GLYCOPEPTIDE, BLEOMYCIN T., BASIC
FORMULA: C60H93N18026S3CU
EA: (N, 16)(CU,)(S, 6)
MW: 1577
PC: BLUE, POW.
UV: W: (244.5, 129,)(292, 104,)
SOL-GOOD: W, MEOH, DMFA
SOL-POOR: BUOH, HEX
QUAL: (NINH., +)(FEHL., -)(SAKA., -)
TO: (S.AUREUS, 1.6)(S.LUTEA, 6.3)(B.SUBT., .013)
(E.COLI, .4)(P.VULG., 1.6)(MYCOB.SP., .05)
(C.ALB., 25)(FUNGI, .8)
LD50: (32, IP)
TV: S-180, P-388, LEWIS, MELANOMA
IS-ION: (IRC-50-NH4, HCL)(DIAION-HP-20, HCL)
IS-CHR: (CM-SEPHADEX C-25-HCOO, NH4COOH)
REFERENCES:
JA, 34, 658, 1981; Belg. P 884291; DT 3026425

43320-8572

NAME: TALLYSOMYCIN-S3B
PO: STREPTOALLOTEICHUS HINDUSTANUS+ETHYLENEDIAMINE
CT: GLYCOPEPTIDE, BLEOMYCIN T., BASIC
FORMULA: C57H87N19026S2CU
EA: (N, 16)(S, 4)(CU,)
MW: 1577
PC: BLUE, POW.
UV: W: (240, 123,)(291, 112,)
SOL-GOOD: W, MEOH, DMFA
SOL-POOR: BUOH, HEX
QUAL: (NINH., +)(FEHL., -)(SAKA., -)
TO: (S.AUREUS, .8)(S.LUTEA, 1.6)(B.SUBT., .013)
(E.COLI, .1)(P.VULG., .2)
LD50: (19, IP)
TV: S-180, P-388, LEWIS, MELANOMA
IS-ION: (IRC-50-NH4, HCL)(DIAION-HP-20, HCL)
IS-CHR: (CM-SEPHADEX C-25-HCOO, NH4COOH)
REFERENCES:
JA, 34, 658, 1981; Belg. P 884291; DT 3026425

43320-8573

NAME: TALLYSOMYCIN-S4B
PO: STREPTOALLOTEICHUS HINDUSTANUS+1.3-DIAMINO-2-HYDROXYPROPANE
CT: GLYCOPEPTIDE, BLEOMYCIN T., BASIC
EA: (N, 16)(S, 4)(CU,)
MW: 1547
PC: BLUE, POW.
UV: W: (242, 120,)(290, 103,)
SOL-GOOD: W, MEOH, DMFA
SOL-POOR: BUOH, HEX
QUAL: (NINH., +)(FEHL., -)(SAKA., -)
TO: (S.AUREUS, .8)(S.LUTEA, 3.1)(B.SUBT., .006)
(E.COLI., .4)(P.VULG., .4)(MYCOB.SP., .003)
(NC.ALB., 25)(FUNGI, .8)
LD50: (13, IP)
TV: S-180, P-388, LEWIS, MELANOMA
IS-ION: (IRC-50-NH4, HCL)(DIAION-HP-20, HCL)
IS-CHR: (CM-SEPHADEX C-25-HCOO, NH4COOH)
REFERENCES:

JA, 34, 658, 1981; Belg. P 884291; DT 3026425

43320-8574

NAME: TALLYSOMYCIN-S5B
PO: STREPTOALLOTEICHUS HINDUSTANUS+N-B"-HYDROXYPROPYL-1.2-DIAMINOBUTANE
CT: GLYCOPEPTIDE, BLEOMYCIN T., BASIC
FORMULA: C60H93N19O27S2CU
EA: (N, 16)(S, 4)(CU,)
MW: 1575
PC: BLUE, POW.
UV: W: (244, 124,)(290, 102,)
SOL-GOOD: W, MEOH, DMFA
SOL-POOR: BUOH, HEX
QUAL: (NINH., +)(FEHL., -)(SAKA., -)
TO: (S.AUREUS, 6.3)(S.LUTEA, 12.5)(B.SUBT., .025)
(E.COLI, .4)(P.VULG., 3.1)(MYCOB.SP., .2)
(C.ALB., 25)(FUNGI, .4)
LD50: (46, IP)
TV: S-180, P-388, LEWIS, MELANOMA
IS-ION: (IRC-50-NH4, HCL)(DIAION-HP-20, HCL)
IS-CHR: (CM-SEPHADEX C-25-HCOO, NH4COOH)
REFERENCES:

JA, 34, 658, 1981; Belg. P 884291; DT 3026425

43320-8575

NAME: TALLYSOMYCIN-S6B
PO: STREPTOALLOTEICHUS HINDUSTANUS+N-B"-HYDROXYETHYL-1.3-DIAMINOPROPANE
CT: GLYCOPEPTIDE, BLEOMYCIN T., BASIC
FORMULA: C60H93N19O27S2CU
EA: (N, 16)(S, 4)(CU,)
MW: 1575
PC: BLUE, POW.
UV: W: (243, 128,)(291, 103,)
SOL-GOOD: W, MEOH, DMFA
SOL-POOR: BUOH, HEX
QUAL: (NINH., +)(FEHL., -)(SAKA., -)
TO: (S.AUREUS, 6.3)(S.LUTEA, 12.5)(B.SUBT., .02) (E.COLI, .4)(P.VULG., 3.1)(MYCOB.SP., .2) (C.ALB., 25)(FUNGI, .8)
LD50: (30, IP)
TV: S-180, P-388, LEWIS, MELANOMA
IS-ION: (IRC-50-NH4, HCL)(DIAION-HP-20, HCL)
IS-CHR: (CM-SEPHADEX C-25-HCOO, NH4COOH)
REFERENCES:
JA, 34, 658, 1981; Belg. P 884291; DT 3026425

43320-8576

NAME: TALLYSOMYCIN-S7B
PO: STREPTOALLOTEICHUS HINDUSTANUS+N.N-DIMETHYL-1.3-DIAMINOPROPANE
CT: GLYCOPEPTIDE, BLEOMYCIN T., BASIC
FORMULA: C60H93N19O26S2CU
EA: (N, 16)(S, 4)(CU,)
MW: 1559
PC: BLUE, POW.
UV: W: (245, 120,)(290, 96,)
SOL-GOOD: W, MEOH, DMFA
SOL-POOR: BUOH, HEX
QUAL: (NINH., +)(FEHL., -)(SAKA., -)
TO: (S.AUREUS, 6.3)(S.LUTEA., 12.5)(B.SUBT., .1) (E.COLI, .4)(P.VULG.,)(MYCOB.SP., 3.1)(FUNGI, .4)
LD50: (30, IP)
TV: S-180, P-388, LEWIS, MELANOMA
IS-ION: (IRC-50-NH4, HCL)(DIAION-HP-20, HCL)
IS-CHR: (CM-SEPHADEX C-25-HCOO, NH4COOH)
REFERENCES:
JA, 34, 658, 1981; Belg. P 884291; DT 3026425

43320-8577

NAME: TALLYSOMYCIN-S8B
PO: STREPTOALLOTEICHUS HINDUSTANUS+N.N-DI-B"-HYDROXYETHYL-1.3-DIAMINOPROPANE
CT: GLYCOPEPTIDE, BLEOMYCIN T., BASIC
FORMULA: C62H97N19O28S2CU
EA: (N, 15)(S, 4)(CU,)
MW: 1619
PC: BLUE, POW.
UV: W: (243, 114,)(290, 92,)
SOL-GOOD: W, MEOH, DMFA
SOL-POOR: BUOH, HEX
QUAL: (NINH., +)(FEHL., -)(SAKA., -)
TO: (S.AUREUS, .4)(S.LUTEA, .8)(B.SUBT., .013)(E.COLI, .2)(P.VULG., .5)(MYCOB.SP., .1)(FUNGI, 1.6)
LD50: (30, IP)
TV: S-180, P-388, LEWIS, MELANOMA
IS-ION: (IRC-50-NH4, HCL)(DIAION-HP-20, HCL)
IS-CHR: (CM-SEPHADEX C-25-HCOO, NH4COOH)
REFERENCES:

JA, 34, 658, 1981; Belg. P 884291; DT 3026425

43320-8578

NAME: TALLYSOMYCIN-S9B
PO: STREPTOALLOTEICHUS HINDUSTANUS+N-B"-HYDROXYETHYL-1.2-DIAMINOETHANE
CT: GLYCOPEPTIDE, BLEOMYCIN T., BASIC
FORMULA: C59H91N19O27S2CU
EA: (N, 16)(S, 4)(CU,)
MW: 1561
PC: BLUE, POW.
UV: W: (243, 104,)(290, 87,)
SOL-GOOD: W, MEOH, DMFA
SOL-POOR: BUOH, HEX
QUAL: (NINH., +)(FEHL., -)(SAKA., -)
TO: (S.AUREUS, .4)(S.LUTEA, .8)(B.SUBT., .013)(E.COLI, .1)(P.VULG., .8)(MYCOB.SP., .1)(C.ALB., 25)(FUNGI, 1.6)
LD50: (32, IP)
TV: S-180, P-388, LEWIS, MELANOMA
IS-ION: (IRC-50-NH4, HCL)(DIAION-HP-20, HCL)
IS-CHR: (CM-SEPHADEX C-25-HCOO, NH4COOH)
REFERENCES:

JA, 34, 658, 1981; Belg. P 884291; DT 3026425

43320-8579

NAME: TALLYSOMYCIN-S11B
PO: STREPTOALLOTEICHUS HINDUSTANUS+N-METHYL-1.3-DIAMINOPROPANE
CT: GLYCOPEPTIDE, BLEOMYCIN T., BASIC
FORMULA: C59H91N19O26S2CU
EA: (N, 16)(S, 4)(CU,)
MW: 1545
PC: BLUE, POW.
UV: W: (244, 112,)(292, 91,)
SOL-GOOD: W, MEOH, DMFA
SOL-POOR: BUOH, HEX
QUAL: (NINH., +)(FEHL., -)(SAKA., -)
TO: (S.AUREUS, 3.1)(S.LUTEA, .8)(B.SUBT., .1)
(E.COLI, .2)(P.VULG., 1.6)(MYCOB.SP., .1)
(C.ALB., 12.5)(FUNGI, 6.3)
LD50: (21, IP)
TV: S-180, P-388, LEWIS, MELANOMA
IS-ION: (IRC-50-NH4, HCL)(DIAION-HP-20, HCL)
IS-CHR: (CM-SEPHADEX C-25-HCOO, NH4COOH)
REFERENCES:
JA, 34, 658, 1981; Belg. P 884291; DT 3026425

43320-8580

NAME: TALLYSOMYCIN-S12B
PO: STREPTOALLOTEICHUS HINDUSTANUS+N-3-AMINOPROPYL-MORPHOLINE
CT: GLYCOPEPTIDE, BLEOMYCIN T., BASIC
FORMULA: C62H95N19O27S2CU
EA: (N, 16)(S, 4)(CU,)
MW: 1601
PC: BLUE, POW.
UV: W: (245, 74,)(292, 59,)
SOL-GOOD: W, MEOH, DMFA
SOL-POOR: BUOH, HEX
QUAL: (NINH., +)(FEHL., -)(SAKA., -)
TO: (S.AUREUS, 12.5)(S.LUTEA, 12.5)(B.SUBT., .4)
(E.COLI, 1.6)(P.VULG., 50)(MYCOB.SP., .2)
(FUNGI, 25)
LD50: (100|50, IP)
TV: S-180, P-388, LEWIS, MELANOMA
IS-ION: (IRC-50-NH4, HCL)(DIAION-HP-20, HCL)
IS-CHR: (CM-SEPHADEX C-25-HCOO, NH4COOH)
REFERENCES:
JA, 34, 658, 1981; Belg. P 884291; DT 3026425

43320-8581

NAME: TALLYSOMYCIN-S13B
PO: STREPTOALLOTEICHUS HINDUSTANUS+N-3-AMINOPROPYL-2-PIPECOLINE
CT: GLYCOPEPTIDE, BLEOMYCIN T., BASIC
FORMULA: C64H99N19O26S2CU
EA: (N, 15)(S, 4)(CU,)
MW: 1613
PC: BLUE, POW.
UV: W: (243, 142,)(292, 114,)
SOL-GOOD: W, MEOH, DMFA
SOL-POOR: BUOH, HEX
QUAL: (NINH., +)(FEHL., -)(SAKA., -)
TO: (S.AUREUS, 6.3)(S.LUTEA, 6.3)(B.SUBT., .8)(E.COLI, .4)(P.VULG., 12.5)(MYCOB.SP., .2)(FUNGI, 12.5)
LD50: (35, IP)
TV: S-180, P-388, LEWIS, MELANOMA
IS-ION: (IRC-50-NH4, HCL)(DIAION-HP-20, HCL)
IS-CHR: (CM-SEPHADEX C-25-HCOO, NH4COOH)
REFERENCES:

JA, 34, 658, 1981; Belg. P 884291; DT 3026425

43320-8582

NAME: TALLYSOMYCIN-S14B
PO: STREPTOALLOTEICHUS HINDUSTANUS+N-A"-PHENYLETHYL-1.3-DIAMINOPROPANE
CT: GLYCOPEPTIDE, BLEOMYCIN T., BASIC
FORMULA: C65H97N19O26S2CU
EA: (N, 15)(S, 4)(CU,)
MW: 1626
PC: BLUE, POW.
UV: W: (243, 143,)(292, 109,)
SOL-GOOD: W, MEOH, DMFA
SOL-POOR: BUOH, HEX
QUAL: (NINH., +)(FEHL., -)(SAKA., -)
TO: (S.AUREUS, 3.1)(S.LUTEA, 3.1)(B.SUBT., .4)(E.COLI, .4)(P.VULG., 25)(MYCOB.SP., .2)(FUNGI, 25)
TV: S-180, P-388, LEWIS, MELANOMA
IS-ION: (IRC-50-NH4, HCL)(DIAION-HP-20, HCL)
IS-CHR: (CM-SEPHADEX C-25-HCOO, NH4COOH)
REFERENCES:

JA, 34, 658, 1981; Belg. P 884291; DT 3026425

43320-8583

NAME: <u>TALLYSOMYCIN-S10B</u>
PO: STREPTOALLOTEICHUS HINDUSTANUS+1.4-DIAMINOBUTANE
CT: GLYCOPEPTIDE, BLEOMYCIN T., BASIC
FORMULA: C59H91N19O26S2
EA: (N, 16)(S, 4)
MW: 1545
PC: WH., POW.
OR: (-38, W)
UV: W: (290, 87,)
UV: W-HCL: (243.5, 141,)(292, 114,)
SOL-GOOD: W, MEOH, DMFA
SOL-POOR: BUOH, HEX
QUAL: (NINH., +)(FEHL., -)(SAKA., -)
TO: (S.AUREUS, 3.1)(S.LUTEA, 3.1)(B.SUBT., .2)
(E.COLI, .8)(P.VULG., .4)(K.PNEUM., .2)
(PS.AER., 1.6)(MYCOB.SP., .2)(C.ALB., 12.5)
(FUNGI, 6.3)
LD50: (123, IP)
TV: S-180, P-388, LEWIS, MELANOMA
IS-ION: (IRC-50-NH4, HCL)(DIAION-HP-20, HCL)
IS-CHR: (CM-SEPHADEX C-25-HCOO, NH4COOH)
REFERENCES:

JA, 34, 658, 665, 1981; DT 3026425; Belg. P 884291

43330-8202

NAME: <u>U-60394</u>
PO: S.WOOLENSIS
CT: CHELATE-FORMING, ACIDIC, AMPHOTER
FORMULA: C20H13N3O6
EA: (N, 11)
MW: 391
PC: YELLOW, GREEN, CRYST.
UV: MEOH: (312, ,)
UV: MEOH-HCL: (312, ,)
UV: MEOH-NAOH: (303, ,)(347.5|2.5, ,)
SOL-GOOD: MEOH, CHL, CH2CL2
SOL-POOR: W
QUAL: (FECL3, +)(NINH., -)
TO: (S.LUTEA,)(S.AUREUS, 2)
LD50: (200|100, SC)
IS-FIL: ORIG.
IS-EXT: (CH2CL2, 3, FILT.)
IS-CHR: (SILG, CHL-MEOH)
REFERENCES:

JA, 33, 1391, 1980

40
OTHER LESS KNOWN PEPTIDE ANTIBIOTICS

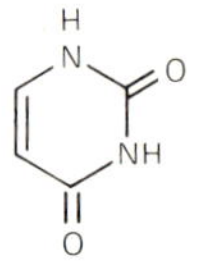

Uracil

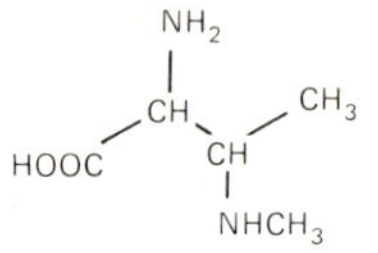

2-Amino-3-N-methylaminobutyric acid

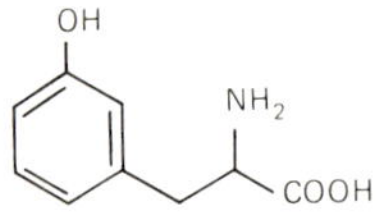

m-Tyrosine

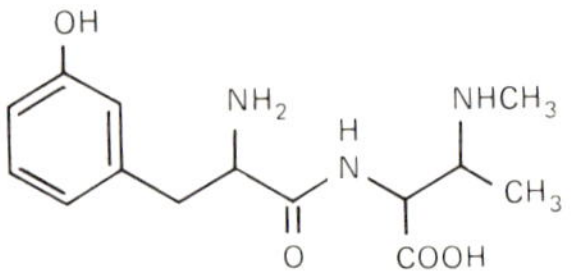

m-Tyrosyl-2-amino-3-N-methylaminobutyric acid

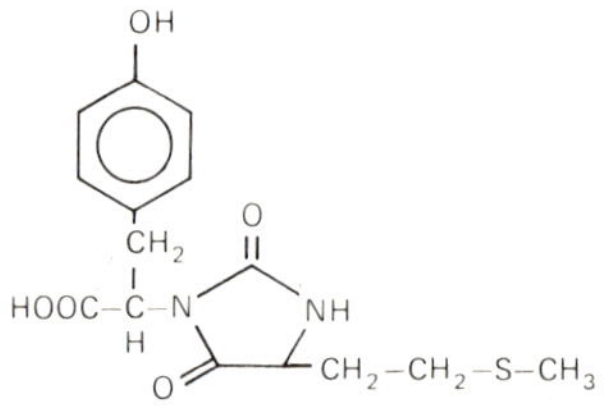

α-[(4-Hydroxyphenyl)methyl]-4-[2-(methylthioethyl)]-2,5-dioxo-1-imidazolidineacetic acid (hydantoin cpd)

A- 38533 (structural elements)

H-Try-pro-glu-ile-ser-trp-thr-arg-asn-gly-cys-S-[Farnesyl]-OH

Rhodotourucin A

40000-6798

NAME: ALTERICIDIN-A
PO: PS.CEPACIA
CT: PEPTIDE
EA: (C, 45)(H, 7)(N, 10)
MW: 1200
PC: WH., POW.
UV: HCL: (278, 55,)
UV: NAOH: (238, 83,)(330, 176,)(244, 115,)(291, 27,)(328, 28,)
SOL-GOOD: DMFA, DMSO, BUOH-W, I.PROH, PYR, BASE
SOL-POOR: MEOH, HEX
QUAL: (DNPH, +)(BIURET, +)(NINH., -)
STAB: (ACID, +)(BASE, -)
TO: (S.CEREV., .39)(C.ALB., .78)(FUNGI, .39)(PHYT.FUNGI, .39)
IS-FIL: 3
IS-EXT: (BUOH, 7, FILT.)
IS-CHR: (SEPHADEX LH-20, PYR-W)(SEPHADEX G-15, W)
IS-CRY: (PREC., NACL, FILT.)(PREC., I.AMOAC, PYR-W)
REFERENCES:
JP 78/127894; *CA*, 90, 184893

40000-6799

NAME: ALTERICIDIN-B
PO: PS.CEPACIA
CT: PEPTIDE
EA: (C, 46)(H, 6)(N, 11)
MW: 1200
PC: WH., POW.
UV: HCL: (278, 67,)
UV: NAOH: (240, 166,)(328, 322,)(243, 217,)(292, 52,)(328, 69,)
SOL-GOOD: DMFA, DMSO, BUOH-W, I.PROH, PYR, BASE
SOL-POOR: MEOH, HEX
QUAL: (NINH., -)(DNPH, +)(BIURET, +)
STAB: (ACID, +)(BASE, -)
TO: (C.ALB.,)(S.CEREV.,)(FUNGI,)(PHYT.FUNGI,)
REFERENCES:
JP 78/127894; *CA*, 90, 184893

40000-6800

NAME: ALTERICIDIN-C
PO: PS.CEPACIA
CT: PEPTIDE
EA: (C, 49)(H, 7)(N, 12)
MW: 1200
PC: WH., POW.
UV: HCL: (278, 129,)
UV: NAOH: (238, 263,)(328, 451,)(243, 236,)
(292, 66,)(328, 103,)
SOL-GOOD: DMFA, DMSO, BUOH-W, I.PROH, PYR, BASE
SOL-POOR: MEOH, HEX
QUAL: (DNPH, +)(BIURET, +)(NINH., -)
STAB: (ACID, +)(BASE, -)
TO: (C.ALB.,)(S.CEREV.,)(FUNGI,)(PHYT.FUNGI,)
REFERENCES:

JP 78/127894; *CA*, 90, 184893

40000-7130

NAME: MICROCIN-140
PO: ESCHERICHIA COLI
CT: PEPTIDE
EA: (N,)
MW: 1000
STAB: (ACID, -)
TO: (E.COLI,)
REFERENCES:

BBRC, 88, 297, 1979; 69, 6, 1976; *J. Bact.*, 135, 342, 1978

40000-7158

NAME: A-38533-B
PO: S.SP.
CT: PEPTIDE, BASIC, AMPHOTER
FORMULA: C39H45N9O12S
EA: (N, 14)(S, 3)
MW: 863
PC: WH., POW.
OR: (+49.5, DMSO)(+57.1, DMSO)
UV: HCL: (222, ,)(258, ,)(290, ,)
UV: NAOH: (243, , 25000)(290, , 8600)
UV: W: (222, , 48000)(258, , 22600)(290, , 6150)
SOL-GOOD: W, MEOH, DMSO, DMFA
SOL-POOR: ACET, HEX
QUAL: (EHRL., +)(PAULY, +)(BIURET, -)(SAKA., -)
(NINH., -)(FECL3, -)
STAB: (ACID, +)(BASE, +)
TO: (PS.AER., 4)
LD50: (2000, IV)
IS-ION: (XAD-4, MEOH-W)
IS-CHR: (FLORISIL, MEOH-W)(SILG, ACCN-W)
IS-CRY: (PREC., MEOH-HCL, ACET)
REFERENCES:

Abst. AAC, 19, 1032, 1033, 1034, 1979; USP 4180564; EP 5956; *CA*, 93, 38347, 166022

40000-7159

NAME: A-38533-A1
PO: S.SP.
CT: PEPTIDE, BASIC, AMPHOTER
FORMULA: C37H44N8O13S
EA: (N, 12)(S, 3)
MW: 840
PC: WH., POW.
OR: (+41.3, DMSO)
UV: HCL: (222, ,)(256, ,)
UV: NAOH: (242, , 35500)(290, , 7500)
UV: W: (222, , 28500)(256, , 21500)
SOL-GOOD: DMSO, W, DMFA
SOL-FAIR: MEOH
SOL-POOR: ACET, HEX
QUAL: (PAULY, +)(EHRL., -)(BIURET, -)(NINH., -)
(SAKA., -)(FECL3, -)
STAB: (ACID, +)(BASE, +)
TO: (PS.AER., 2)
LD50: (2500|500, IV)(3000|400, IP)
REFERENCES:

Abst. AAC, 19, 1032, 1033, 1034, 1979; USP 4180564; EP 5956; *CA*, 93, 38347, 166022

40000-7201

NAME:	AB-102
PO:	ACTINOPLANES JAPANENSIS
CT:	PEPTIDE
EA:	(C,)(H,)(N, 18)(S, 8)
PC:	WH., YELLOW
OR:	(0, DMSO)
SOL-GOOD:	ACET
QUAL:	(BIURET, +)(NINH., -)
TO:	(S.AUREUS,)(MYCOB.TUB.,)
IS-EXT:	(ACET, , MIC.)
IS-CHR:	(SILG,)

REFERENCES:

JP 79/135703; *CA*, 92, 109111

40000-7431

NAME:	A-38533-A2
PO:	S.SP.
CT:	PEPTIDE, BASIC, AMPHOTER
FORMULA:	C38H46N8013
EA:	(C, 54.5|.5)(H, 5.5|.5)(N, 13|.5)
MW:	822
PC:	WH., POW.
OR:	(+46.3, DMSO)
UV:	HCL: (222, , 26500)(256, , 20500)
UV:	NAOH: (242, , 32000)(290, , 6400)
UV:	W: (222, , 26500)(256, , 20500)
SOL-GOOD:	W, MEOH-HCL, DMFA, DMSO
SOL-FAIR:	MEOH
SOL-POOR:	ACET, HEX
QUAL:	(PAULY, +)(EHRL., -)(BIURET, -)(NINH., -) (SAKA., -)(FECL3, -)
STAB:	(ACID, +)(BASE, +)
TO:	(PS.AER., 2)
IS-ION:	(XAD-4, MEOH-W)
IS-CHR:	(FLORISIL, MEOH-W)(SILG, ACCN-W)

REFERENCES:

USP 4180564; EP 5956

40000-7432

NAME: A-38533-C
PO: S.SP.
CT: PEPTIDE, BASIC, AMPHOTER
FORMULA: C40H47N9012
EA: (C, 55)(H, 5)(N, 14)
MW: 846
PC: WH., POW.
OR: (+40, DMSO)
UV: HCL: (222, , 46000)(259, , 21000)(290, , 6000)
UV: NAOH: (243, , 25400)(290, , 8500)
UV: W: (222, , 46000)(259, , 21000)(290, , 6000)
SOL-GOOD: MEOH, MEOH-HCL, DMSO, DMFA
SOL-FAIR: W
SOL-POOR: ACET, HEX
QUAL: (PAULY, +)(EHRL., +)(BIURET, -)(NINH., -)
(SAKA., -)(FECL3, -)
STAB: (ACID, +)(BASE, +)
TO: (PS.AER., 4)
REFERENCES:
USP 4180564; EP 5956

40000-7675

NAME: ANSLIMIN-A
PO: S.OGANOENSIS
CT: PEPTIDE
EA: (C, 58)(H, 8)(N, 14)
MW: 658
PC: YELLOW, CRYST.
OR: (0, MEOH)
UV: MEOH: (200, ,)
SOL-GOOD: MEOH
SOL-FAIR: ETOH
SOL-POOR: W, ACET
QUAL: (NINH., -)
STAB: (ACID, +)(BASE, -)
TO: (SPHAEROTILUS NATANS, .78)
IS-EXT: (BUOH, , W+NACL)
IS-ION: (DIAION HP-20, ACET-W)
IS-CHR: (CEL, ACCN-W)
IS-CRY: (CRYST., MEOH)
REFERENCES:
Appl. Micr., 39, 1123, 1980; JP 81/47297; *CA*, 93, 38190

40000-7676

NAME: ANSLIMIN-B
PO: S.OGANOENSIS
CT: PEPTIDE
EA: (C, 50)(H, 8)(N, 14)
MW: 341
PC: YELLOW, CRYST.
OR: (+1.5, W)
UV: MEOH: (200, ,)
SOL-GOOD: W, MEOH, ETOH
SOL-POOR: ACET
QUAL: (NINH., +)
STAB: (ACID, +)(BASE, -)
TO: (SPHAEROTILUS NATANS, .39)
REFERENCES:

Appl. Micr., 39, 1123, 1980; JP 81/47297; *CA*, 93, 38190

40000-7677

NAME: ALBOLEUTIN
PO: B.SUBTILIS
CT: PEPTIDE, ACIDIC, AMPHOTER
EA: (C, 59)(H, 9)(N, 9)
PC: WH., POW.
OR: (+12, CHL)
UV: MEOH: (200, ,)
SOL-GOOD: ETOAC, BENZ, CHL
SOL-FAIR: ET2O, MEOH, ETOH
SOL-POOR: W, HEX, ACET
TO: (PHYT.FUNGI,)(S.LUTEA,)(PIRICULARIA ORYZAE,)
LD50: (200|100, IP)
IS-EXT: (BUOH, 7, W)
IS-ABS: (CARBON, ACET-W)
IS-CHR: (SILG, CHL-MEOH)(SEPHADEX LH-20, ETOH)
IS-CRY: (DRY, ETOH)
REFERENCES:

40000-7680

NAME: RHODOTORUCIN-A
PO: RHODOSPORIUM TORULOIDES
CT: PEPTIDE, AMPHOTER
EA: (C,)(H,)(N,)(N, 15)
UV: W: (280, ,)
SOL-GOOD: MEOH
SOL-FAIR: W
TO: (C.ALB.,)(S.CEREV.,)
IS-EXT: (BUOH, , EVAP.FILT.)
IS-CHR: (SEPHADEX LH-20, MEOH-AMMONIUM ACETATE)
REFERENCES:

J. Gen. Appl. Microb., 13, 167, 1967; *Agr. Biol. Ch.*, 42, 209, 1978; JP 80/39713; *CA*, 93, 43929

40000-8071

NAME: SF-2068
PO: S.SP.
CT: PEPTIDE, ACIDIC
EA: (C, 48)(H, 7)(N, 13)(O, 32)
MW: 825|25
PC: WH., CRYST.
OR: (+16.3, MEOH)
UV: MEOH: (200, ,)
SOL-GOOD: MEOH, CHL
SOL-POOR: ET2O, HEX, W
QUAL: (NINH., -)
TO: (MYCOPLASMA SP.,)(S.AUREUS, 1.56)(B.SUBT., .19)(S.LUTEA, 6.25)
IS-FIL: 2
IS-EXT: (ETOAC, 2, FILT.)
IS-CHR: (SEPHADEX LH-20, MEOH)(SILG, CHL-MEOH)(SILG, ETOAC-MEOH)
IS-CRY: (CRYST., CHL-ET2O)
REFERENCES:

JP 80/98197; *CA*, 93, 236931

40000-8072

NAME:	CYTOPHAGIN
PO:	CYTOPHAGA SP.
CT:	PEPTIDE
EA:	(C, 47)(H, 7)(N, 13)
MW:	1250|250
PC:	WH., POW.
OR:	(+5, DMSO)
UV:	MEOH: (275, ,)(282, ,)(290, ,)
SOL-GOOD:	ACOH, DMSO
SOL-FAIR:	MEOH
SOL-POOR:	ACET, HEX, W
QUAL:	(NINH., +)(SAKA., -)(EHRL., +)(PAULY, -)
TO:	(S.AUREUS, 1.56)(B.SUBT., 1.56)(S.LUTEA, .1)
REFERENCES:	

JP 80/124794; *CA*, 94, 82162

40000-8147

PO:	MYXOCOCCUS FULVUS
CT:	PEPTIDE
EA:	(N,)
MW:	1000
PC:	WH., POW.
UV:	MEOH: (222, ,)(275, ,)
SOL-GOOD:	MEOH, BUOH
SOL-FAIR:	W
TO:	(B.SUBT., 1)(S.AUREUS, 1)(E.COLI, 30) (PHYT.BACT., 30)
IS-EXT:	(BUOH, , FILT.)
IS-CHR:	(AL, BUOH-MEOH)
IS-CRY:	(DRY, BUOH)
REFERENCES:	

DT 2924006

40000-8272

NAME:	F-16-2
PO:	B.SUBTILIS
CT:	OLIGOPEPTIDE
UV:	(200, ,)
TO:	(G.POS.,)(G.NEG.,)(S.CEREV.,)
REFERENCES:	

40000-8273

NAME: 02-8, 0-28
PO: ARTHROBACTER SP.
CT: BASIC, PEPTIDE
EA: (N,)(S,)
MW: 900
PC: WH., POW.
OR: (+6.8, W)
UV: MEOH: (274, 4.6,)(310, 1,)
UV: MEOH-HCL: (274, ,)(310, ,)
UV: MEOH-NAOH: (259, 4.2,)(345, .9,)(358.45,)
SOL-GOOD: W
SOL-FAIR: MEOH
SOL-POOR: ETOH, HEX
QUAL: (NINH., +)(SAKA., +)(BIURET, +)(PAULY, +)
(FECL3, -)
STAB: (ACID, +)(BASE, +)
TO: (S.AUREUS, 12.5)(B.SUBT., 12.5)(E.COLI, 12.5)
(P.VULG., 50)
LD50: (225|25, IP)
IS-ION: (IRC-50-NH4, NH4OH)(DIAION HP-10, ETOH-W)
IS-CHR: (SEPHADEX G-25, W)
IS-CRY: (LIOF.,)
REFERENCES:

40000-8584

NAME: BN-240-B
PO: ERWINIA SP.
CT: PEPTIDE, BASIC
EA: (C, 44)(H, 6)(N, 17)(O, 32)
PC: WH., POW.
SOL-GOOD: W, MEOH
SOL-POOR: ACET, HEX
QUAL: (NINH., +)(BIURET, +)(FEHL., -)(FECL3, -)
TO: (E.COLI, .78)(PS.AER., 6.25)
IS-ION: (DIAION-HP-25, ACET-W)
IS-CHR: (CH-SEPHADEX, NACL)
IS-CRY: LIOF.
REFERENCES:

JP 81/5495; *CA*, 95, 22943

40000-8639

NAME:	AF-8
PO:	B.SUBTILIS
CT:	PEPTIDE, ACIDIC
EA:	(C, 59)(H, 9)(N, 9)
PC:	WH., POW.
OR:	(+12, CHL)
SOL-GOOD:	BENZ, ETOAC, CHL, BUOH
SOL-FAIR:	ET2O, ETOH, MEOH
SOL-POOR:	HEX, ACET, W
QUAL:	(NINH., -)
TO:	(FUNGI, 6)
IS-ABS:	(CARBON,)

REFERENCES:

JP 80/160788; *CA*, 94, 190311

40000-8839

NAME:	SH-50
PO:	S.GRISEOLUS
CT:	PEPTIDE
FORMULA:	C26H47N7O6
EA:	(N, 17)
MW:	553
TO:	(S.AUREUS,)

REFERENCES:

Egypt J. Bot., 22, 215, 1981; *CA*,, 95, 167032

44
PEPTOLIDES

441 Chromopeptolides
442 Lipopeptolides
443 Heteropeptolides
444 Simple Peptolides
445 Depsipeptide Antibiotics

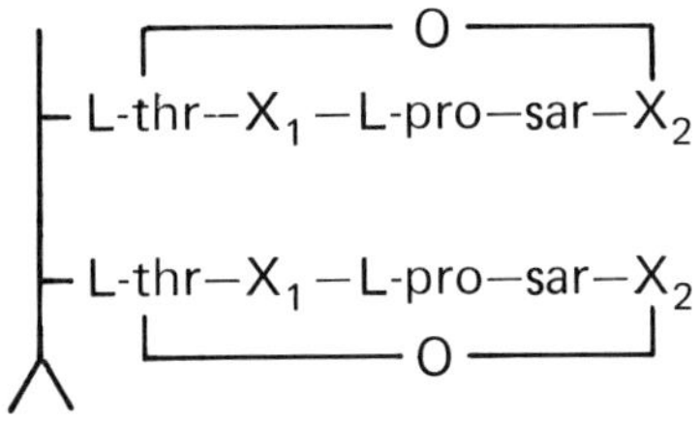

	X_1	X_2
Actinoleucin AY-5	D-leu	L-meval
AU-3B(didemethylactinomycin D)	D-val	L-val

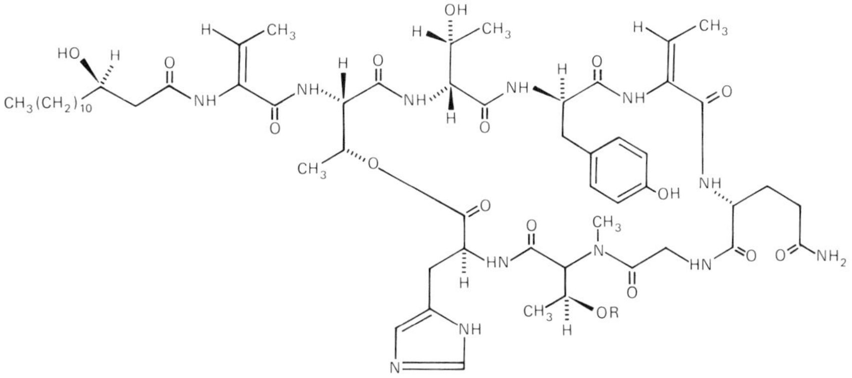

	R
20561	H
20562	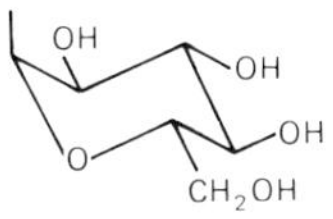

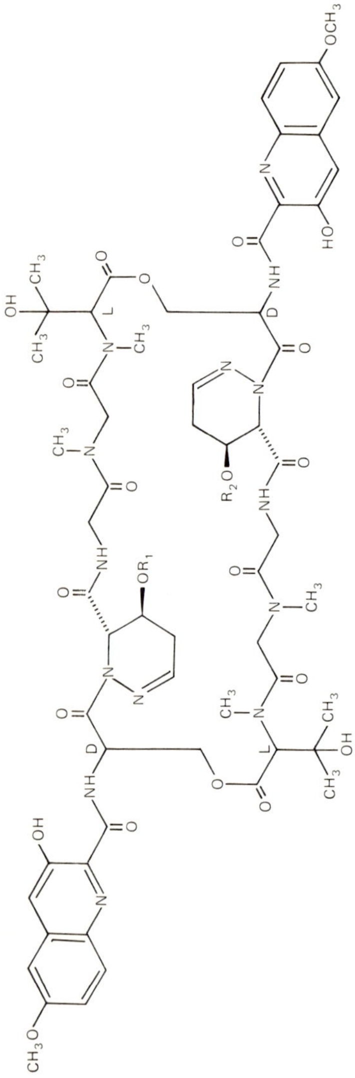

	R_1	R_2
Luzopeptin A	$COCH_3$	$COCH_3$
Luzopeptin B	H	$COCH_3$
Luzopeptin C	H	H

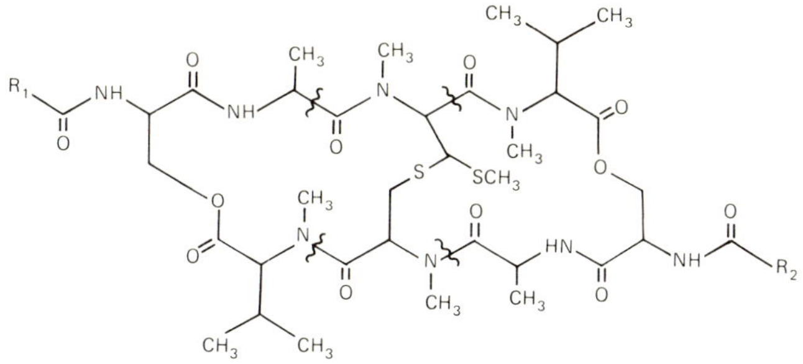

	R_1	R_2	
1QCl	QA	ACl	
2QCl	QCl	QCl	
1TP	QA	TP	
2TP	TP	TP	Quinomycin
1QnM	QA	QnM	series
2QnM	QnM	QnM	
2QnMB*	QnM	QnM	Triostin
2QClB*	QCl	QCl	series

QA:

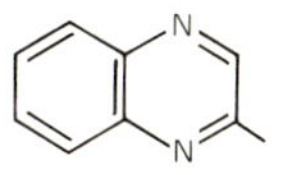

QCl:

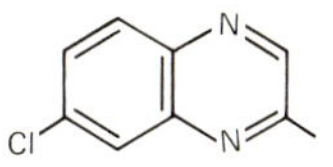

TP:

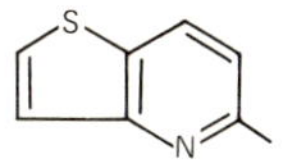

QnM:

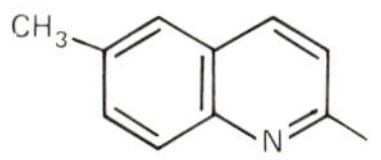

* structural fragment in QnMB and QClB:

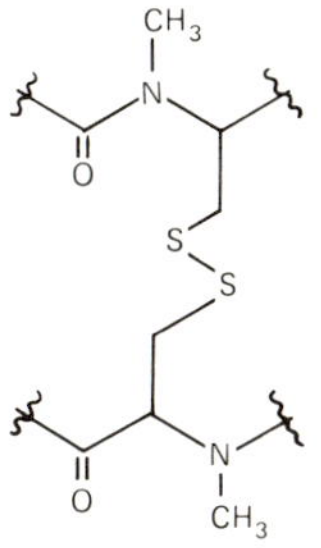

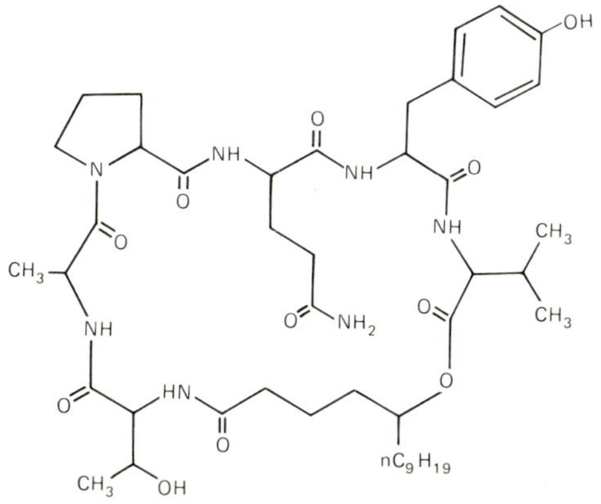

A-43-F

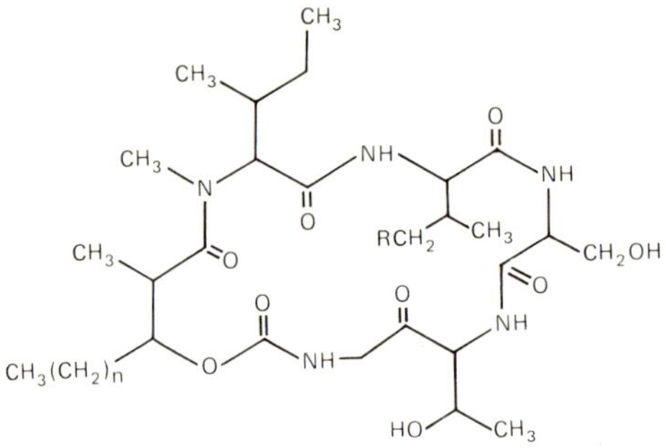

	R	n
SF-1902 A_1 (Globomycin)	CH_3	5
SF-1902 A_2	CH_3	3
SF-1902 A_3	H	5
SF-1902 A_{4a}	CH_3	6
SF-1902 A_{4b}	H	7
SF-1902 A_5	CH_3	7

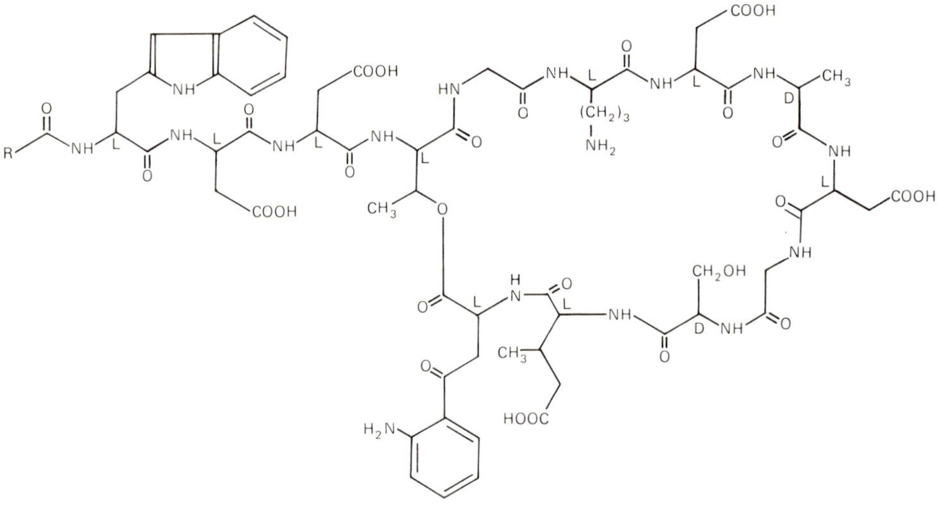

	R
A-21979 C_1	$CO(CH_2)_7CH(CH_3)_2$
A-21979 C_2	$CO(CH_2)_8CH(CH_3)_2$
A-21979 C_3	$CO(CH_2)_9CH(CH_3)_2$
A-21979 C_0	$COC_{10}H_{21}$
A-21979 C_4	$COC_{12}H_{25}$
A-21979 C_5	$COC_{13}H_{27}$

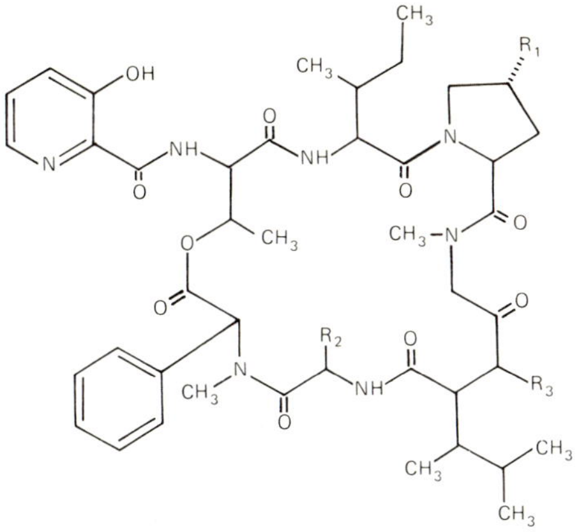

	R_1	R_2	R_3
Etamycin B	H	CH_3	CH_3
Etamycin VII	OH	H	CH_3
Etamycin VI-2	OH	CH_3	H

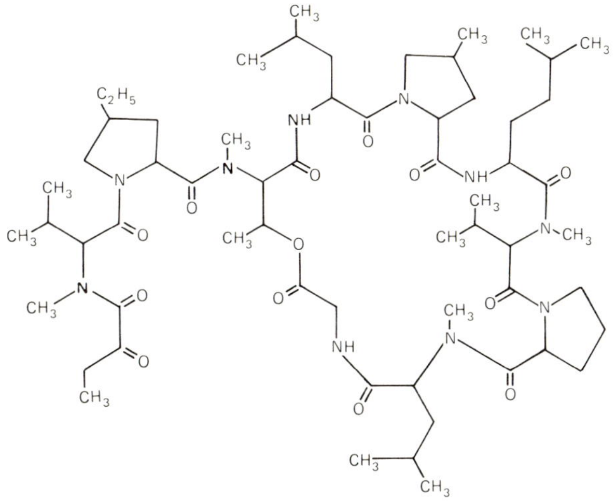

Mycoplanecin

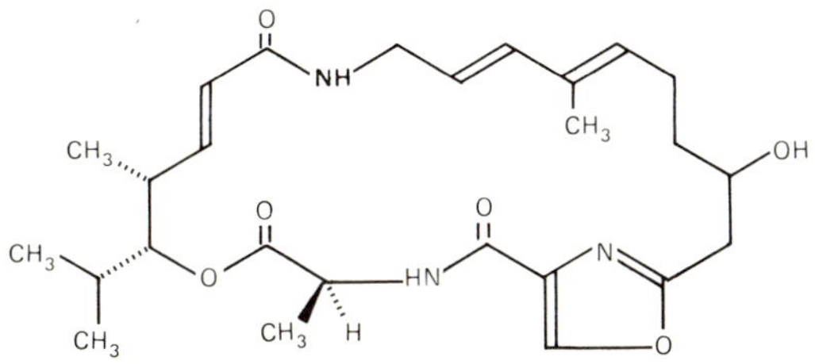

A-17002 C

	R
Destruxin A	CH_2–CH=CH_2
Destruxin B	CH_2–CH(CH_3)$_2$
Destruxin E	CH_2–CH–CH_2 (epoxide, O bridging CH–CH_2)

	R
Lipopeptin A	C_2H_5
Lipopeptin B	CH_3

44110-6802

NAME: ACTINOMYCIN COMPLEX 41156
PO: ACTINOPLANES NIRASAKIENSIS
CT: CHROMOPEPTOLIDE, ACTINOMYCIN T., NEUTRAL
EA: (N,)
PC: RED, CRYST.
UV: MEOH: (242, ,)(442|2, ,)
TO: (S.AUREUS,)
IS-EXT: (ACET, , MIC.)(ETOAC, 8, FILT.)
IS-CHR: (AL, ETOAC-DI.BU.ETHER-W)
REFERENCES:
An. Rep. Sankyo, 30, 84, 1978

44110-6803

NAME: ACTINOMYCIN COMPLEX
PO: ACT.SP., S.CHRYSOMALLUS-CAROTENOIDES
CT: CHROMOPEPTOLIDE, ACTINOMYCIN T., NEUTRAL
EA: (N, 13)
PC: RED, POW.
UV: MEOH: (440, ,)
TO: (B.SUBT.,)
IS-EXT: (ACET, , MIC.)
REFERENCES:
Antib., 7, 1979

44110-7960

NAME: ACTINOLEUCIN AY-5, AY-5, AU-5
PO: ACT.OLIVOBRUNEUS, S.OLIVOBRUNEUS
CT: CHROMOPEPTOLIDE, ACTINOMYCIN T., NEUTRAL
FORMULA: C64H90N12O16
EA: (N, 13)
MW: 1282
PC: RED, CRYST.
UV: MEOH: (240, ,)(440, ,)
TO: (G.POS.,)
TV: ANTITUMOR
REFERENCES:
Antib., 18, 1971

44110-8585

NAME:	AU-3B, N.N-DIDEMETHYLACTINOMYCIN-D, AY-3B
PO:	ACT.SP.
CT:	CHROMOPEPTIDE, ACTINOMYCIN T., NEUTRAL
FORMULA:	C60H82N12016
EA:	(N, 13)
MW:	1226
PC:	ORANGE, YELLOW, POW.
UV:	MEOH: (240, ,)(442, ,)
SOL-GOOD:	MEOH, BENZ
SOL-POOR:	W, HEX
TO:	(G.POS.,)
IS-EXT:	(ETOAC, , WB.)
IS-CHR:	(CEL,)
IS-CRY:	(CRYST., ETOAC)

REFERENCES:

Antib., 5, 1981

44120-7695

NAME:	BBM-928-A, LUZOPEPTIN-A
PO:	ACTINOMADURA LUZONENSIS
CT:	QUINOXALINE-PEPTIDE, ECHINOMYCIN T., NEUTRAL
FORMULA:	C64H78N14O24
EA:	(N, 13)
MW:	1450, 1426
PC:	WH., CRYST.
OR:	(-27, CHL)
UV:	ETOH: (235, 586,)(264, 415,)(345, 165,)
UV:	ETOH-HCL: (234, 610,)(264, 410,)(345, 165,)
UV:	ETOH-NAOH: (230, 569,)(256, 763,)(330, 180,)(383, 170,)
SOL-GOOD:	CHL, CH2CL2
SOL-FAIR:	BENZ, MEOH, BUOH
SOL-POOR:	W, HEX
QUAL:	(FECL3, +)(EHRL., +)(SAKA., -)(NINH., -)
TO:	(S.AUREUS, .2)(S.LUTEA, .2)(B.SUBT., .4)(MYCOB.SP., .4)
LD50:	(.13, IP)
TV:	P-388, L-1210, LEWIS, S-180, MELANOMA
IS-FIL:	8.5
IS-EXT:	(ACET-MEOH, , MIC.)(BUOH, , W)(BUOH, , FILT.)
IS-CHR:	(SILG, ETOAC-MEOH)
IS-CRY:	(CRYST., CHL-MEOH)

REFERENCES:

JA, 33, 1087, 1098, 1980; 34, 148, 1981; *JACS*, 103, 1241, 1243, 1981; *Biochem.*, 19, 5537, 1980; Belg. P 882574

44120-7696

NAME:	LUZOPEPTIN-B, BBM-928-B
PO:	ACTINOMADURA LUZONENSIS
CT:	QUINOXALINE-PEPTIDE, ECHINOMYCIN T., NEUTRAL
FORMULA:	C62H76N14O23
EA:	(N, 12)
MW:	1384
PC:	WH., CRYST.
OR:	(-74, CHL)
UV:	ETOH: (235, 570,)(264, 400,)(345, 163,)
UV:	ETOH-HCL: (234, 556,)(264, 446,)(345, 188,)
UV:	ETOH-NAOH: (230, 530,)(256, 775,)(330, 116,)(383, 112,)
SOL-GOOD:	CHL, CH2CL2
SOL-FAIR:	BENZ, MEOH, BUOH
SOL-POOR:	W, HEX
QUAL:	(FECL3, +)(EHRL., +)(SAKA., -)(NINH., -)
TO:	(S.AUREUS, .4)(B.SUBT., 1.6)(S.LUTEA, .2)(MYCOB.SP., .4)
LD50:	(.18, IP)
TV:	P-388
IS-FIL:	8.5
IS-EXT:	(ACET-MEOH, , MIC.)(BUOH, , W)(BUOH, , FILT.)
IS-CRY:	(CRYST., CHL-MEOH)

REFERENCES:

JA, 33, 1087, 1098, 1980; 34, 148, 1981; *JACS*, 103, 1241, 1243, 1981; *Biochem.*, 19, 5537, 1980; Belg. P 882574

44120-7697

NAME: LUZOPEPTIN-C, BBM-928-C
PO: ACTINOMADURA LUZONENSIS
CT: QUINOXALINE-PEPTIDE, ECHINOMYCIN T., NEUTRAL
FORMULA: C60H70N14O22
EA: (N, 14)
MW: 1470, 1342
PC: WH., CRYST.
OR: (-91, CHL)
UV: ETOH: (235, 638,)(264, 442,)(345, 173,)
UV: ETOH-HCL: (234, 650,)(264, 442,)(345, 173,)
UV: ETOH-NAOH: (230, 580,)(256, 704,)(330, 117,)(383, 122,)
SOL-GOOD: CHL, CH2CL2
SOL-FAIR: BENZ, MEOH, BUOH
SOL-POOR: W, HEX
QUAL: (FECL3, +)(EHRL., +)(SAKA., -)(NINH., -)
TO: (S.AUREUS, 6.3)(S.LUTEA, 6.3)(B.SUBT., 6.3)(MYCOB.SP., .4)
LD50: (.81, IP)
TV: P-388
IS-FIL: 8.5
IS-EXT: (ACET-MEOH, , MIC.)(BUOH, , W)(BUOH, , FILT.)
IS-CHR: (SILG, ETOAC-MEOH)
IS-CRY: (CRYST., CHL-MEOH)
REFERENCES:

JA, 33, 1087, 1098, 1980; 34, 148, 1981; *JACS*, 103, 1241, 1243, 1981; *Biochem.*, 19, 5537, 1980; Belg. P 882574

44120-7698

NAME: LUZOPEPTIN-D, BBM-928-D
PO: ACTINOMADURA LUZONENSIS
CT: QUINOXALINE-PEPTIDE, ECHINOMYCIN T., NEUTRAL
EA: (C, 51)(H, 5)(N, 13)(O, 31)
PC: WH., CRYST.
OR: (-13, CHL)
UV: ETOH: (235, 550,)(264, 380,)(345, 155,)
UV: ETOH-HCL: (234, 565,)(264, 405,)(345, 165,)
UV: ETOH-NAOH: (230, 650,)(256, 930,)(330, 140,)(383, 145,)
SOL-GOOD: CHL, CH2CL2
SOL-FAIR: BENZ, MEOH, BUOH
SOL-POOR: W, HEX
QUAL: (FECL3, +)(EHRL., +)(SAKA., -)(NINH., -)
TO: (S.AUREUS, .4)(B.SUBT., 6.3)(S.LUTEA, .2)
LD50: (.053, IP)
TV: P-388
REFERENCES:

Belg. P 882574

44120-8204

NAME: 1QCL
PO: S.ECHINATUS+7-CHLOROQUINOXALINE-2-CARBOXYLIC ACID
CT: QUINOXALINE-PEPTIDE, ECHINOMYCIN T., NEUTRAL
FORMULA: C51H63N12O12S2CL
EA: (N, 15)(S, 6)(CL, 3)
MW: 1133
PC: WH.
SOL-GOOD: MEOH, BENZ
SOL-POOR: W
TO: (G.POS.,)
TV: CYTOTOXIC
IS-EXT: (CHL-ACET, , MIC.)
REFERENCES:
CC, 46, 1981

44120-8205

NAME: 2QCL
PO: S.ECHINATUS+7-CHLOROQUINOXALINE-2-CARBOXYLIC ACID
CT: QUINOXALINE-PEPTIDE, ECHINOMYCIN T., NEUTRAL
EA: (N, 14)(S, 5)(CL, 6)
MW: 1168
PC: WH.
SOL-GOOD: MEOH, BENZ
SOL-POOR: W
TO: (G.POS.,)
TV: CYTOTOXIC
IS-EXT: (CHL-ACET, , MIC.)
REFERENCES:
CC, 46, 1981

44120-8206

NAME: 1TP
PO: S.ECHINATUS+THIOPENYL-PIRIDINE-2-CARBOXYLIC ACID
CT: QUINOXALINE-PEPTIDE, ECHINOMYCIN T., NEUTRAL
FORMULA: C50H63N11012S3
EA: (N, 14)(S, 9)
MW: 1105
PC: WH.
SOL-GOOD: MEOH, BENZ
SOL-POOR: W
TO: (G.POS.,)
TV: CYTOTOXIC
IS-EXT: (CHL-ACET, , MIC.)
REFERENCES:
CC, 46, 1981

44120-8207

NAME: 2TP
PO: S.ECHINATUS+THIOPENYL-PIRIDINE-2-CARBOXYLIC ACID
CT: QUINOXALINE-PEPTIDE, ECHINOMYCIN T., NEUTRAL
FORMULA: C49H62N10012S4
EA: (N, 13)(S, 11)
MW: 1110
PC: WH.
SOL-GOOD: MEOH, BENZ
SOL-POOR: W
TO: (G.POS.,)
TV: CYTOTOXIC
IS-EXT: (CHL-ACET, , MIC.)
REFERENCES:
CC, 46, 1981

44120-8208

NAME: <u>1QNM</u>
PO: S.ECHINATUS+6-METHYLQUINOLIN-2-CARBOXYLIC ACID
CT: QUINOXALINE-PEPTIDE, ECHINOMYCIN T., NEUTRAL
FORMULA: C53H67N11O12S2
EA: (N, 14)(S, 7)
MW: 1113
PC: WH.
SOL-GOOD: MEOH, BENZ
SOL-POOR: W
TO: (G.POS.,)
TV: CYTOTOXIC
IS-EXT: (CHL-ACET, , MIC.)
REFERENCES:
CC, 46, 1981

44120-8209

NAME: <u>2QNM</u>
PO: S.ECHINATUS+6-METHYLQUINOLIN-2-CARBOXYLIC ACID
CT: QUINOXALINE-PEPTIDE, ECHINOMYCIN T., NEUTRAL
FORMULA: C55H70N10O12S2
EA: (N, 12)(S, 7)
MW: 1126
PC: WH.
SOL-GOOD: MEOH, BENZ
SOL-POOR: W
TO: (G.POS.,)
TV: CYTOTOXIC
IS-EXT: (CHL-ACET, , MIC.)
REFERENCES:
CC, 46, 1981

44120-8586

NAME: <u>2QNMB</u>
PO: S.ECHINATUS+6-METHYLQUINOLINE-2-CARBOXYLIC ACID
CT: QUINOXALINE-PEPTIDE, ECHINOMYCIN T., NEUTRAL
FORMULA: C54H68N10O12S2
EA: (N, 12)(S, 6)
MW: 1112
TO: (G.POS.,)
REFERENCES:
CC, 46, 1981; *JA*, 35, 62, 1982

44120-8587

NAME: 2QCLB
PO: S.ECHINATUS+7-CHLOROQUINOXALINE-2-CARBOXYLIC ACID
CT: QUINOXALINE-PEPTIDE, ECHINOMYCIN T., NEUTRAL
EA: (N, 14)(S, 5)(CL, 6)
MW: 1155
TO: (G.POS.,)
IS-EXT: (CHL, , FILT.)(ACET, , MIC.)
REFERENCES:
CC, 46, 1981; *JA*, 35, 62, 1982

44120-8588

NAME: QUINOMYCIN-X
CT: QUINOXALINE-PEPTIDE, ECHINOMYCIN T., NEUTRAL
PC: YELLOW, CRYST.
UV: MEOH: (242, ,)(320, ,)
TO: (G.POS.,)
REFERENCES:
Egypt J. Micr., 15, 99, 1980/81; *CA*, 95, 40765

44210-7961

NAME: ENDURACIDIN-COMPLEX, B-5477-M
PO: S.FUNGICIDICUS
CT: PEPTOLIDE, ENDURACIDIN T., BASIC
EA: (C, 53)(H, 7)(N, 14)(CL, 3)
MW: 1050|50
PC: WH., POW.
OR: (+87, DMFA)
UV: HCL: (272, 112,)
UV: MEOH: (230, 213,)(263, 150,)
UV: NAOH: (250, 335,)
SOL-GOOD: MEOH
SOL-POOR: ETOAC, HEX
TO: (G.POS.,)
LD50: (100, IV)(400, IP)
IS-EXT: (ACET-W, , MIC.)
REFERENCES:
JP 70/114

44220-6902

NAME:	PANTOMYCIN
PO:	S.HYGROSCOPICUS
CT:	PEPTOLIDE, STENDOMYCIN T., AMPHOTER
FORMULA:	C97H172N20033
EA:	(N,)(N, 13)
MW:	2146
PC:	WH., POW., YELLOW
SOL-GOOD:	MEOH, CHL
SOL-FAIR:	W, BENZ
SOL-POOR:	HEX, ACID
QUAL:	(NINH., +)
STAB:	(HEAT, +)
TO:	(B.SUBT., 100)(S.LUTEA, 1)(S.AUREUS, 10) (K.PNEUM., 100)(P.VULG., 100)(E.COLI, 200) (SHYG., 10)(FUNGI, 10)(PHYT.FUNGI, 1)
TV:	NDV, VACCINIA
IS-FIL:	2
IS-EXT:	(I.PROH, 2, MIC.)(ACET, ,)
IS-CHR:	(SEPHADEX LH-20, ETOH-W)
IS-CRY:	(PREC., ACET, HEX)

REFERENCES:

Mycol., 71, 103, 1979; *AAC*, 17, 980, 1980; USP 3992528

44230-6804

NAME:	20561, W-10
PO:	AEROMONAS SP.
CT:	PEPTOLIDE, PEPTIDOLIPID
FORMULA:	C57H86N12016
EA:	(N, 11)
MW:	1194
PC:	WH., POW.
OR:	(-88, PYR-W)(-66, DMSO-W)
UV:	MEOH: (220, 313,)(280, 20,)
UV:	MEOH-NAOH: (240, 230,)(295, 28,)
SOL-GOOD:	MEOH, BUOH, PYR
SOL-POOR:	ET2O, HEX
TO:	(S.CEREV.,)(C.ALB., .1)
LD50:	(100, PEROS)(1500|500, SC)
IS-EXT:	(BUOH, 7, WB.)
IS-CHR:	(SILG, CHL-MEOH)
IS-CRY:	(PREC., MEOH, ET2O-HEX)

REFERENCES:

USP 4137224, 4232006, 4230799

44230-6805

NAME: 20562, W-10
PO: AEROMONAS SP.
CT: PEPTOLIDE, PEPTIDOLIPID
FORMULA: C63H96N12021
EA: (N, 10)
MW: 1356
PC: WH., POW.
OR: (-60, PYR-W)(-44, DMSO-W)
UV: MEOH: (222, 268,)(280, 14.8,)
UV: MEOH-NAOH: (240, 190,)(292, 194,)
SOL-GOOD: MEOH, BUOH-PYR
SOL-POOR: ET20, HEX
TO: (C.ALB., .075)(S.CEREV.,)
LD50: (800, PEROS)(1000|400, SC)
REFERENCES:
USP 4137224, 4232006, 4230799

44250-7131

NAME: A-43-F, MSD-A43F
PO: VERTICILLIUM LAMELLICOLA
CT: PEPTOLIDE, NEUTRAL
FORMULA: C45H71N7011
EA: (N, 11)
MW: 885
PC: WH., POW.
UV: MEOH: (225, 230,)(278, 38,)
UV: MEOH-HCL: (225, 230,)(278, 38,)
UV: MEOH-NAOH: (245, 200,)(294, 50,)
SOL-GOOD: MEOH, ETOAC
SOL-POOR: W
TO: (FUNGI,)(PHYT.FUNGI,)
IS-EXT: (ETOAC, 7, W)(MEOH, , WB.)
IS-CHR: (SILG, ETOAC-MEOH)(SEPHADEX LH-20, CHL-HEX-MEOH)
IS-CRY: (DRY, DIOXAN)
REFERENCES:
Abst. AAC, 19, 151, 1979; USP 4201771, 4254224

44250-8210

NAME: SF-1902-A2
PO: S.HYGROSCOPICUS
CT: PEPTOLIDE, NEUTRAL
EA: (C, 58)(H, 8)(N, 12)
MW: 630
PC: WH.
OR: (-18, CHL)
UV: MEOH: (200, ,)
SOL-GOOD: MEOH, BENZ
SOL-POOR: W, HEX
QUAL: (NINH., -)(FECL3, -)
TO: (E.COLI,)
IS-FIL: ORIG.
IS-EXT: (ETOAC, , FILT.)(ACET, , MIC.)
IS-CHR: (SILG, ACCN-W)
REFERENCES:

JP 80/47644; *CA*, 93, 112304; *JA*, 34,1416, 1981

44250-8211

NAME: SF-1902-A3
PO: S.HYGROSCOPICUS
CT: PEPTOLIDE, NEUTRAL
EA: (C, 58)(H, 9)(N, 11)
MW: 640
PC: WH.
OR: (-20, CHL)
UV: MEOH: (200, ,)
SOL-GOOD: MEOH, BENZ
SOL-POOR: W, HEX
QUAL: (NINH., -)(FECL3, -)
TO: (E.COLI,)
IS-FIL: ORIG.
IS-EXT: (ETOAC, , FILT.)(ACET, , MIC.)
IS-CHR: (SILG, ACCN-W)
REFERENCES:

JP 80/47644; *CA*, 93, 112304; *JA*, 34, 1416, 1981

44250-8212

NAME: SF-1902-A4
PO: S.HYGROSCOPICUS
CT: PEPTOLIDE, NEUTRAL
EA: (N, 10)(C, 59)(H, 9)
MW: 650
PC: WH.
OR: (-15, CHL)
UV: MEOH: (200, ,)
SOL-GOOD: MEOH, BENZ
SOL-POOR: W, HEX
QUAL: (NINH., -)(FECL3, -)
TO: (E.COLI,)
IS-FIL: ORIG.
IS-EXT: (ETOAC, , FILT.)(ACET, , MIC.)
IS-CHR: (SILG, ACCN-W)
REFERENCES:
JP 80/47644; *CA*, 93, 112304; *JA*, 34, 1416, 1981

44260-7699

NAME: A-21978 CO
PO: S.ROSEOSPORUS
CT: PEPTOLIDE, ACIDIC, AMPHOTER
FORMULA: C72H100N16O27
EA: (N, 13)
MW: 1622
PC: WH., POW.
OR: (+11.9, W)
UV: ETOH: (223, ,)(260, ,)(280, ,)(290, ,)
(360, ,)
SOL-GOOD: W, MEOH, BUOH, THF, DMSO, DIOXAN
SOL-POOR: ACID, ACET, HEX
QUAL: (NINH., +)
STAB: (ACID, +)(BASE, -)
TO: (S.AUREUS, 1)
IS-EXT: (MEOH, ,)(BUOH, 3, W)
IS-ION: (IRA-68-AC, ACOH)
IS-CHR: (SILG, W-MEOH-ACCN)
IS-CRY: (PREC., FILT., ACID)
REFERENCES:
BP 2031437

44260-7700

NAME:	<u>A-21978-C1</u>
PO:	S.ROSEOSPORUS
CT:	PEPTOLIDE, ACIDIC, AMPHOTER
FORMULA:	C73H102N16027
EA:	(N, 13)
MW:	1636
PC:	WH., POW.
OR:	(+16.9, W)
UV:	ETOH: (223, 307,)(260, 62,)(280, 39,)(290, 35,)(360, 33,)
SOL-GOOD:	W, MEOH, BUOH, THF, DMSO, DIOXAN
SOL-POOR:	ACID, ACET, HEX
STAB:	(ACID, +)(BASE, -)
TO:	(S.AUREUS, .5)
LD50:	(400|150, IV)
IS-EXT:	(MEOH, ,)(BUOH, 3, W)
IS-ION:	(IRA-68-AC, ACOH)
IS-CHR:	(SILG, ACCN-W)
IS-CRY:	(PREC., FILT., ACID)

REFERENCES:

BP 2031437

44260-7701

NAME:	<u>A-21978-C2</u>
PO:	S.ROSEOSPORUS
CT:	PEPTOLIDE, ACIDIC, AMPHOTER
FORMULA:	C74H104N16027
EA:	(N, 14)
MW:	1650
PC:	WH., POW.
OR:	(+18.6, W)
UV:	ETOH: (273, 303,)(260, 62,)(280, 41,)(290, 36,)(360, 33,)
SOL-GOOD:	W, MEOH, BUOH, THF, DIOXAN, DMSO
SOL-POOR:	ACID, ACET, HEX
STAB:	(ACID, +)(BASE, -)
TO:	(S.AUREUS, .13)
LD50:	(200|50, IV)(175, SC)

REFERENCES:

BP 2031437

44260-7702

NAME: A-21978-C3
PO: S.ROSEOSPORUS
CT: PEPTOLIDE, ACIDIC, AMPHOTER
FORMULA: C75H106N16027
EA: (N, 14)
MW: 1664
PC: WH., POW.
OR: (+20.9, W)
UV: ETOH: (223, 300,)(260, 63,)(280, 42,)(290, 38,)(360, 32,)
SOL-GOOD: W, MEOH, BUOH, THF, DMSO, DIOXAN
SOL-POOR: ACID, ACET, HEX
STAB: (ACID, +)(BASE, -)
TO: (S.AUREUS, .06)
LD50: (50, IV)(75, SC)
REFERENCES:
BP 2031437

44260-7703

NAME: A-21978-C4
PO: S.ROSEOSPORUS
CT: PEPTOLIDE, ACIDIC, AMPHOTER
FORMULA: C74H104N16027
EA: (N, 13)
MW: 1650
PC: WH., POW.
OR: (+14.8, W)
UV: ETOH: (223, ,)(260, ,)(280, ,)(290, ,)(360, ,)
SOL-GOOD: W, MEOH, BUOH, THF, DMSO, DIOXAN
SOL-POOR: ACID, ACET, HEX
STAB: (ACID, +)(BASE, -)
TO: (S.AUREUS, .25)
REFERENCES:
BP 2031437

44260-7704

NAME: A-21978-C5
PO: S.ROSEOSPORUS
CT: PEPTOLIDE, ACIDIC, AMPHOTER
FORMULA: C75H106N16O27
EA: (N, 16)
MW: 1664
PC: WH., POW.
OR: (+17.9, W)
UV: ETOH: (223, ,)(260, ,)(280, ,)(290, ,)
(360, ,)
SOL-GOOD: W, MEOH, BUOH
SOL-POOR: ACID, ACET, HEX
STAB: (ACID, +)(BASE, -)
TO: (S.AUREUS, .13)
REFERENCES:
BP 2031437

44312-6806

NAME: ETAMYCIN-B, DEOXYETAMYCIN
IDENTICAL: NEOVIRIDOGRISEIN-II
PO: S.GRISEOVIRIDIS
CT: PEPTOLIDE, ETAMYCIN T., AMPHOTER
FORMULA: C44H62N8O10
EA: (N, 13)
MW: 862
PC: WH., CRYST.
OR: (+20, CHL)
UV: ETOH: (304, , 7700)
SOL-GOOD: MEOH, CHL, ACID, BASE
SOL-FAIR: W
SOL-POOR: HEX
QUAL: (FECL3, +)
TO: (G.POS.,)(S.AUREUS,)
REFERENCES:
JA, 32, 392, 1979

44312-6807

NAME: ETAMYCIN COMP.VII
PO: S.GRISEOVIRIDIS
CT: PEPTOLIDE, ETAMYCIN T., AMPHOTER
FORMULA: C43H60N8011
EA: (N, 12)
MW: 864
PC: WH.
TO: (G.POS.,)
REFERENCES:
JA, 32, 392, 1979

44312-7132

NAME: NEOVIRIDOGRISEIN-VII
PO: S.GRISEOVIRIDIS+L-METHIONINE
CT: PEPTOLIDE, ETAMYCIN T., AMPHOTER
FORMULA: C44H62N8011
EA: (N, 12)
MW: 888
TO: (S.AUREUS, 10)
IS-CHR: (SILG, CHL-MEOH)
REFERENCES:
JA, 32, 1002, 1979

44410-8088

NAME: 2928
PO: S.SP.
CT: PEPTOLIDE
EA: (C, 54)(H, 7)(N, 10)
PC: WH., YELLOW, CRYST.
OR: (0,)
UV: ETOH: (245, 188,)(297, 68,)(308, 66,)(353, 60,)
UV: ETOH-HCL: (245, ,)(297, ,)(308, ,)(353, ,)
UV: ETOH-NAOH: (250, ,)(395, ,)
SOL-GOOD: CHL, ETOAC, PYR
SOL-FAIR: MEOH, ETOH
SOL-POOR: W, ET2O, HEX
QUAL: (FECL3, -)(NINH., -)
TO: (S.AUREUS, 1)(B.SUBT., 1)(S.LUTEA, .1)
LD50: (45, SC)
TV: S-180, NK-LY
IS-EXT: (ACET, , MIC.)(CHL, , W)
IS-CHR: (SILG, CCL4-CHL)
IS-CRY: (PREC., CHL, HEX)
REFERENCES:
Antib., 483, 1981

44410-8089

NAME: 5590
PO: S.SP.
CT: PEPTOLIDE, AMPHOTER
EA: (C, 56)(H, 7)(N, 11)
PC: WH., YELLOW, CRYST.
OR: (-25, ETOH)
UV: ETOH: (240, 232,)(297, 69,)(307, 68,)(355, 73,)
UV: ETOH-HCL: (240, ,)(295, ,)(307, ,)(355, ,)
UV: ETOH-NAOH: (245, ,)(305, ,)(395, ,)
SOL-GOOD: CHL, ETOAC, PYR
SOL-FAIR: MEOH, ETOH
SOL-POOR: ET2O, HEX, W
QUAL: (FECL3, +)(NINH., -)
TO: (S.AUREUS, .4)(B.SUBT., .4)(S.LUTEA, .2)(E.COLI, 10)(S.CEREV., 10)
LD50: (39, SC)
TV: S-180, NK-LY
IS-EXT: (ACET, , MIC.)(CHL, , W)
IS-CHR: (SILG, CHL-MEOH)
IS-CRY: (CRYST., ETOH)
REFERENCES:

Antib., 483, 1981

44420-7433

NAME: MYCOPLANECIN-A, MYCOPLANECIN
PO: ACTINOPLANES SP.
CT: AMPHOTER, PEPTOLIDE, GRISELLIMYCIN T.
FORMULA: C61H104N10O13
EA: (C, 62)(H, 8)(N, 12)
MW: 1184
PC: WH., CRYST.
OR: (-69.1, CHL)
UV: MEOH-W: (200, ,)
SOL-GOOD: MEOH, CHL, BENZ
SOL-FAIR: BENZ
SOL-POOR: W, HEX
QUAL: (NINH., -)(DNPH, -)
TO: (MYCOB.TUB., .39)(MYCOB.SP., .01)(S.LUTEA, .025)
IS-EXT: (ETOAC, , FILT.)(ACET, , MIC.)
IS-CHR: (SILG, CHL-ETOAC)(SEPHADEX LH-20, CHL)
IS-CRY: (DRY, CHL)
REFERENCES:

DT 2921085, 3041130; BP 2021558; *CA*, 95, 185545

44440-8213

NAME: DESTRUXIN-E
PO: METARHIZIUM ANISOPLIAE
CT: PEPTOLIDE, NEUTRAL
FORMULA: C29H47N5O8
EA: (N, 12)
MW: 593
PC: CRYST., WH.
OR: (-253, CHL)
TO: INSECTICID
TV: ANTIVIRAL
REFERENCES:
CR Ser. D, 291, 763, 1980; *Phytoch.*, 20, 715, 1981; *CA*, 94, 114300

44450-7519

NAME: LIPOPEPTIN-A, AC-69
PO: S.VIOLACEOCHROMOGENES
CT: ACIDIC, AMPHOTER, PEPTOLIDE
FORMULA: C54H84N10O19
EA: (N, 12)
MW: 1176
EW: 587, 1076
PC: WH., POW.
OR: (-45.4, MEOH)
UV: MEOH: (200, ,)
SOL-GOOD: MEOH, BUOH
SOL-FAIR: ACET, CHL
SOL-POOR: BENZ, HEX
QUAL: (NINH., -)
TO: (PHYT.FUNGI, 150)
IS-FIL: 2
IS-EXT: (ETOAC, 2, FILT.)(ACET-W, 2, MIC.)
IS-CHR: (SILG, I.PROH-NH4OH)(SEPHADEX LH-20, MEOH)
(DEAE-SEPHAROSE-CO3, MEOH-W)
REFERENCES:
JA, 33, 247, 1980; *TL*, 4627, 1980; *Abst. AAC*, 20, 468, 1980; *Agr. Biol. Ch.*, 45, 895, 1981; JP 80/33445; *CA*, 93, 130557

44450-8073

NAME: LIPOPEPTIN-B
PO: S.SP.
CT: PEPTOLIDE
FORMULA: C53H82N10O19
EA: (N, 12)
MW: 1165
PC: WH.
UV: MEOH: (200, ,)
SOL-GOOD: MEOH, BUOH, BASE
SOL-FAIR: W, ACET, ETOAC
SOL-POOR: BENZ, HEX
QUAL: (NINH., -)
TO: (PHYT.FUNGI,)
REFERENCES:

TL, 4627, 1980

44510-6808

PO: FUS.EQUISETI
CT: DEPSIPEPTIDE, VALINOMYCIN T., NEUTRAL
EA: (N, 6)
PC: WH., CRYST.
OR: (-80, MEOH)
SOL-GOOD: MEOH, CHL
SOL-POOR: W
TO: ANTIMICROBIAL
LD50: (41, IV)(175|5, IP)
TV: ANTIVIRAL, CYTOTOXIC
IS-EXT: (TRICHLOROETILEN, , MIC.)
IS-CHR: (AL, CH2CL2)
IS-CRY: (CRYST., ETOH)
REFERENCES:

BP 2010848; DT 2851629; Belg. P 872385

44530-6809

NAME: A-17002-C, A-15104-II
PO: ACTINOPLANES SP.
CT: DEPSIPEPTIDE, NEUTRAL, OSTREOGRICIN-A T.
FORMULA: C26H37N3O6
EA: (N, 9)
MW: 487
PC: WH., CRYST.
OR: (-21, MEOH)
UV: ETOH: (214, , 13200)
UV: HCL: (214, ,)
UV: NAOH: (214, ,)
SOL-GOOD: MEOH, CHL
SOL-POOR: HEX
QUAL: (FECL3, -)(FEHL., -)(DNPH, -)
TO: (S.AUREUS,)(G.POS.,)
IS-EXT: (ETOAC, 3.5, FILT.)
IS-CHR: (SILG, CHL-MEOH)(SILG, ETOAC)
IS-CRY: (PREC., ET2O, HEX)(CRYST., ETOAC-HEX)
REFERENCES:
JA, 32, 108, 1979

44530-6810

NAME: A-17002-A, A-15104-I
PO: ACTINOPLANES SP.
CT: DEPSIPEPTIDE, NEUTRAL, OSTREOGRICIN-A T.
EA: (N,)
TO: (S.AUREUS,)
REFERENCES:
JA, 32, 108, 1979

44530-6811

NAME: A-17002-B, A-15104-X
PO: ACTINOPLANES SP.
CT: DEPSIPEPTIDE, OSTREOGRICIN-A T.
EA: (N,)
TO: (S.AUREUS,)
REFERENCES:
JA, 32, 108, 1979

45
MACROMOLECULAR (PEPTIDE) ANTIBIOTICS

451 Polypeptide Antibiotics
452 Protein Antibiotics
453 Proteide Antibiotics

45110-8216

NAME: "PHYTOACTIN"
PO: S.HYGROSCOPICUS, ACT.HYGROSCOPICUS
CT: POLYPEPTIDE
EA: (N,)
STAB: (ACID, +)(BASE, -)
TO: ANTIMICROBIAL
LD50: PHYTOTOXIC
IS-EXT: (, 3.5, FILT.)
REFERENCES:

Trudy Vsesoj. Nauch. Issl. Inst. Selskohozj. Mikr., 1979, 16; *CA*, 94, 170744

45120-6812

NAME: C-3603
PO: STREPTOCOCCUS MUTANS
CT: POLYPEPTIDE, AMPHOTER, BASIC
EA: (C, 50)(H, 6)(N, 13)(S, 1)(O, 31)
MW: 8000|2000
PC: WH., POW.
UV: W: (280, 40,)(289, 33,)
SOL-GOOD: ACID
SOL-POOR: W, MEOH, HEX
QUAL: (NINH., +)(BIURET, +)
STAB: (ACID, +)(BASE, +)
TO: (B.SUBT., 6.25)(S.AUREUS, 6.25)
IS-CHR: (CM-SEPHADEX C-25, PH7 PUFF)
IS-CRY: (PREC., AMMONIUM SULPHATE, FILT.)
REFERENCES:

DT 2819035

45120-8589

NAME: BOTRYTICIDIN-A
PO: B.SUBTILIS
CT: POLYPEPTIDE, BASIC, AMPHOTER
EA: (C, 53)(H, 8)(N, 24)(O, 15)
MW: 6400
PC: WH., POW.
OR: (+1.1, NACL-H2O)
UV: W: (207, ,)
SOL-GOOD: W
SOL-FAIR: MEOH
SOL-POOR: ETOH, HEX
QUAL: (BIURET, +)(NINH., +)(FEHL., -)
STAB: (HEAT, +)
TO: (PHYT.FUNGI,)(S.AUREUS,)(S.LUTEA,)
LD50: NONTOXIC
IS-ION: (DX-50-H, W)
IS-CHR: (CM-SEPHADEX C-25, NACL)
IS-CRY: LIOF.
REFERENCES:

BP 2061284; Fr. P 2467215; JP 81/53694

45130-8075

NAME: AH-2589-I
PO: S.GLOBISPORUS
CT: POLYPEPTIDE, AMPHOTER
EA: (C, 48)(H, 7)(N, 14)
MW: 6000|4000
OR: (-24, W)
UV: HCL: (262, 42,)
UV: NAOH: (200, ,)
UV: W: (262, 42,)
SOL-GOOD: W
SOL-POOR: MEOH, HEX
QUAL: (NINH., -)
TO: (S.AUREUS, .8)(B.SUBT., 50)(E.COLI, 3.2)
IS-ION: (XAD-2, MEOH-W)
REFERENCES:

JP 80/111498; *CA*, 94, 28880

45130-8215

NAME:	AH-2589-II
PO:	S.GLOBISPORUS
CT:	POLYPEPTIDE, AMPHOTER
EA:	(C, 48)(H, 6)(N, 14)
MW:	6000|4000
OR:	(-68, W)
UV:	W: (200, ,)
SOL-GOOD:	W
SOL-POOR:	MEOH, HEX
QUAL:	(NINH., -)
TO:	(B.SUBT., 1.6)(S.AUREUS, .1)(E.COLI, .4) (K.PNEUM., 50)

REFERENCES:

JP 80/111498; *CA*, 94, 28880

45130-8590

PO:	STAPHYLOCOCCUS EPIDERMIDIS
CT:	POLYPEPTIDE, AMPHOTER
EA:	(N,)
MW:	950
PC:	WH., POW.
UV:	W: (270|5, ,)
SOL-GOOD:	W
STAB:	(HEAT, +)(ACID, +)(BASE, -)
TO:	(S.AUREUS,)(G.POS.,)
IS-CHR:	(SEPHADEX G-50, PH6 BUFF.)
IS-CRY:	(PREC., AMMONIUM SULPHATE, FILT.)

REFERENCES:

WO P 81/962; EP 27710

45211-6813

NAME:	AUROMYCIN, AUROMOMYCIN
PO:	S.MACROMOMYCETICUS
CT:	ACIDIC PROTEIN, AMPHOTER, ACIDIC
FORMULA:	C507|39H845|65N130|130208|16S13|1
EA:	(C, 47)(H, 7)(N, 15)(O, 26)(S, 3)
MW:	12500
EW:	12000
PC:	YELLOW, CRYST.
OR:	(-280, W)
UV:	HCL: (272, 13.2,)(356, 4.4,)
UV:	NAOH: (272|2, 14,)(340, 5.9,)
UV:	W: (273, 13.3,)(357, 4.6,)
SOL-GOOD:	W
SOL-POOR:	MEOH, HEX
QUAL:	(BIURET, +)(NINH., +)(EHRL., +)
STAB:	(LIGHT, -)(ACID, -)
TO:	(S.AUREUS, .1)(S.LUTEA, .1)(B.SUBT., .1) (E.COLI, 3.12)(K.PNEUM., 3.1)(SHYG., 1.56) (P.VULG., 3.12)
LD50:	(3.0|.5, IV)
TV:	EHRLICH, S-180, L-1210
IS-ION:	(IRA-93-CL, W)
IS-CHR:	(SEPHADEX G-50, W)
IS-CRY:	(PREC., AMMONIUM SULPHATE+NA2CO3+FECL3, FILT.)

REFERENCES:

JA, 32, 330, 706, 1979; 33, 1545, 1981; *BBRC*, 89, 1281, 1979; 94, 255, 1980; *Cancer Chem. Pharm.*, 7, 41, 1981; DT 2836821; JP 79/36201, 79/37890; *CA*, 90, 202220

45211-6814

NAME:	DECHROMONEOCARZINOSTATIN, PC
PO:	S.CARZINOSTATICUS
CT:	ACIDIC PROTEIN
FORMULA:	C126H229N20030S.3-4
EA:	(C, 49)(H, 7)(N, 15)(S, 2)
MW:	12000|1000
PC:	WH., POW.
UV:	5: (280, 10.2,)
TO:	(G.POS.,)
LD50:	(100|40, IV)
TV:	EHRLICH, S-180, L-1210
IS-ION:	(XAD-7,)
IS-CHR:	(DEAE-CEL.,)(SEPHADEX G-50, W)

REFERENCES:

JP 79/36202; *CA*, 91, 73136

45211-7137

NAME: RINFOMYCIN
PO: S.GRISEOCHROMOGENES
CT: ACIDIC PROTEIN, ACIDIC, AMPHOTER
EA: (C, 45)(H, 6)(N, 11)
MW: 12000, 11750|2750
PC: BLACK, BROWN, POW.
UV: W: (200, ,)
SOL-GOOD: W, MEOH, ETOH
SOL-POOR: CHL, HEX
QUAL: (NINH., +)(FECL3, -)(SAKA., +)(BIURET, +)
STAB: (ACID, -)(BASE, -)
TV: SN-36, S-180, EHRLICH
IS-CHR: (CG-50-NA, PH6.8 BUFF)
REFERENCES:
JP 79/3957

45211-7964

NAME: NEOCARZINOSTATIN-N1
PO: S.CARZINOSTATICUS
CT: ACIDIC PROTEIN, AMPHOTER, ACIDIC
EA: (C, 47)(H, 7)(N, 12)
MW: 9300
PC: WH., POW.
OR: (-38.7, W)
UV: W: (272.5|2.5, ,)
SOL-GOOD: W
TO: (G.POS.,)(MYCOPLASMA SP.,)
TV: ANTITUMOR
IS-CHR: (BIOREX-70, ACET-W-ACOH)
REFERENCES:
DT 2443560; JP 72/680; 75/52289

45211-7965

NAME: NEOCARZINOSTATIN-N2
PO: S.CARZINOSTATICUS
CT: ACIDIC PROTEIN, AMPHOTER, ACIDIC
EA: (C, 46)(H, 8)(N, 14)
MW: 9000
PC: WH., POW.
OR: (-33.87, W)
UV: W: (272.5|2.5, ,)
SOL-GOOD: W
TO: (G.POS.,)(MYCOPLASMA SP.,)
TV: ANTITUMOR
REFERENCES:
DT 2443560; JP 72/680; 75/52289

45211-8217

NAME: PRENEOCARZINOSTATIN
PO: S.CARZINOSTATICUS
CT: ACIDIC PROTEIN, ACIDIC, AMPHOTER
EA: (N,)
MW: 9000
PC: WH., POW.
UV: W: (280, 10,)
SOL-GOOD: W
TV: HELA
IS-CRY: (PREC., FILT., AMMONIUM SULPHATE)
REFERENCES:
JA, 27, 766, 1974; 33, 110, 1586, 1980; *J. Bioch. (Tokyo)*, 81, 25, 1977; *CA*, 95, 73697

45213-8591

NAME: CI-782
PO: S.PLURICOLORESCENS
CT: AMPHOTER PROTEIN
EA: (C, 33)(H, 5)(N, 7)
PC: YELLOW, BROWN, POW.
UV: W: (258, ,)
SOL-GOOD: W
SOL-POOR: MEOH, HEX
QUAL: (NINH., +)
LD50: (27, IP)
TV: S-180
IS-ABS: (ZEOLIT, NH4OH)
IS-CHR: (SEPHADEX G-100, W)
IS-CRY: LIOF.
REFERENCES:
JP 81/5096; *CA*, 95, 22939

45220-6815

NAME: "CYTOTOXIN"
PO: CLOSTRIDIUM DIFFICILE
CT: PROTEIN, ACIDIC
EA: (N,)
MW: 240000
PC: WH.
UV: W: (280, ,)
STAB: (HEAT, -)(ACID, -)(BASE, -)
TV: CYTOTOXIC
IS-CHR: (SEPHADEX G-200, W)
REFERENCES:
Rev. Inf. Diseases, 1, 379, 1979; *CA*, 91, 16379

45220-7435

NAME: MIEHEIN-21
PO: MUCOR MIENEI
CT: PROTEIN, AMPHOTER
EA: (N,)
MW: 4500|1500
PC: WH., POW.
UV: HCL: (276, ,)
UV: NAOH: (282, ,)(288, ,)
SOL-GOOD: W
STAB: (HEAT, +)(ACID, +)
TO: (B.SUBT.,)
IS-CHR: (CM-SEPHADEX C-25-NA, PH6 BUFF-NACL)(SEPHADEX G-10, ACOH)
IS-CRY: (LIOF.,)
REFERENCES:
Dev. Ind. Microb., 20, 661, 1979; *J. Gen. Micr.*, 81, 1, 1979; *J. Dairy Sci.*, 56, 639, 1973

45220-7709

NAME: CANECELUNIN
PO: S.SP.
CT: PROTEIN, AMPHOTER
EA: (N,)
MW: 62000
PC: WH.
UV: 8: (286, ,)
STAB: (HEAT, -)
TV: S-180, EHRLICH, LEWIS, SA
IS-CHR: (SEPHADEX G-50, W)(DEAE-SEPHADEX A-50, NACL)
IS-CRY: (PREC., FILT., ETOH)
REFERENCES:
JA, 33, 776, 1980

45220-7966

NAME: "LEUCOCIDIN", "CYTOTOXIN"
PO: PS.AERUGINOSA
CT: PROTEIN, ACIDIC, AMPHOTER
EA: (N,)
MW: 25100
PC: WH., POW.
LD50: (.25, IP)
TV: CYTOTOXIC
REFERENCES:

45220-8218

NAME: SML-91 LECTIN
PO: STREPTOCOCCUS MITIS
CT: PROTEIN
EA: (N,)
PC: WH., POW.
TV: ANTITUMOR
IS-CHR: (SEPHADEX G-200, W)
REFERENCES:
JP 80/144894

45220-8219

NAME: SSL-91 LECTIN
PO: STREPTOCOCCUS SANGUIS
CT: PROTEIN
EA: (N,)
PC: WH., POW.
TV: ANTITUMOR
IS-CHR: (SEPHADEX G-200, W)
REFERENCES:
JP 80/144895

45320-6816

NAME: MICHICARCIN, NSC-191595
PO: S.SP.
CT: GLYCOPROTEID, AMPHOTER
EA: (N,)
MW: 20000|10000
TV: P-388, MELANOMA
REFERENCES:
Recent Results Cancer Res., 63, 49, 1978

45320-7476

NAME: A-1
PO: MYCOBACTERIUM BOVIS
CT: GLYCOPROTEID
TV: ANTITUMOR
REFERENCES:
Inf. & Immun., 21, 914, 1978; *CA*, 90, 413

45320-7489

PO: CANDIDA ALBICANS
CT: GLYCOPROTEID
EA: (N,)
LD50: NONTOXIC
TV: S-180
REFERENCES:
J. Surgical Oncology, 15, 99, 1980; *Folia Micr.*, 11, 372, 1966; USP 4182753; *CA*, 93, 230998

45320-7967

PO: STREPTOCOCCUS FAECIUM
CT: GLYCOPROTEID
EA: (N,)
MW: 50000
STAB: (HEAT, +)
TO: (BIFIDOBACTERIUM SP.,)
REFERENCES:
AAC, 18, 58, 1980

45320-7968

PO: AGARICUS HETEROSITES
CT: GLYCOPROTEID, AMPHOTER
EA: (N,)
TV: S-180
REFERENCES:
JP 80/74797; *CA*, 93, 129002

45320-7982

NAME: CI-783
PO: S.AUREOFACIENS
CT: ACIDIC, GLYCOPROTEID
EA: (C, 32)(H, 11)(N, 6)
MW: 20000
PC: WH., YELLOW, CRYST.
OR: (-94, W)
UV: W: (255, ,)
SOL-GOOD: W
SOL-POOR: MEOH, HEX, TOL
QUAL: (NINH., +)
TV: S-180
IS-ION: (ZEOLITE, NH4OH)
IS-CRY: (LIOF.,)
REFERENCES:
JP 80/118393; *CA*, 94, 82163

45320-8220

PO: CANDIDA ALBICANS
CT: GLYCOPROTEID
EA: (N,)
SOL-GOOD: W
SOL-POOR: ETOH
LD50: NONTOXIC
TV: S-180
IS-EXT: (W, , MIC.)
IS-CRY: (PREC., W, ETOH)
REFERENCES:

45320-8221

NAME: C-1740
PO: KLEBSIELLA PNEUMONIAE
CT: GLYCOPROTEID, AMPHOTER
EA: (N,)
MW: 100000|20000
SOL-GOOD: W
TO: (G.POS.,)(G.NEG.,)
TV: ANTITUMOR
REFERENCES:
Int. J. Immunopharmacol., 2, 161, 1980; Holl. P 80/4406

45320-8222

PO: CORYNEBACTERIUM EQUI, CORYNEBACTERIUM PYOGENES, CORYNEBACTERIUM XEROSIS, CORYNEBACTERIUM RENALE
CT: GLYCOPROTEID, AMPHOTER, ACIDIC
EA: (N,)
PC: YELLOW, BROWN, POW.
SOL-GOOD: W
LD50: NONTOXIC
TV: ANTITUMOR
IS-EXT: (W, , MIC.)
REFERENCES:
EP 24941

45320-8592

NAME: TF-130
PO: FUSOBACTERIUM NUCLEATUM
CT: GLYCOPROTEID
EA: (C, 34|6)(H, 5|1)(N, 5|1)
PC: WH., BROWN, POW.
UV: W: (268|12, ,)
SOL-GOOD: W
SOL-POOR: MEOH, HEX
QUAL: (NINH., -)
TV: EHRLICH, S-180
IS-CHR: (SEPHADEX G-200, W)
REFERENCES:
JP 81/45496; Belg. P 884864

45330-6911

NAME: AVS, "AVS"
PO: STREPTOCOCCUS FAECALIS
CT: NUCLEOPROTEID
EA: (N,)
UV: W: (260, ,)
STAB: (HEAT, +)
TV: HERPES
IS-CHR: (SEPHADEX G-100, W)(SEPHADEX G-100, PH7 BUFF.)
REFERENCES:
Kobe J. Med. Sci., 24, 99, 115, 1978; *CA*, 90, 115364

45330-8593

NAME: "VIRUS LIKE PARTICLES"
PO: LEUTINUS EDODES
CT: NUCLEOPROTEID
EA: (N,)
TV: EHRLICH
REFERENCES:
Arch. Virol., 68, 297, 1981; *CA*, 95, 73791

45340-6820

NAME: METHIONINE-G"-LYASE
PO: CLOSTRIDIUM SPOROGENES
CT: ENZYME L.
EA: (N,)
TV: L-1210, P-815
REFERENCES:
Cancer Res., 33, 1866, 1973

45340-6821

NAME: L-LYSINE-A"-OXYDASE
PO: TRICHODERMA VIRIDAE
CT: ENZYME L.
EA: (N,)
STAB: (HEAT, -)(ACID, -)(BASE, +)
TO: (B.SUBT.,)
TV: L-1210, L-5178
REFERENCES:
Agr. Biol. Ch., 43, 337, 1371, 1979; 44, 387, 1980; *Seikagaku*, 50, 923, 1978; *Bioferm.*, 55, 57, 1981; *CA*, 94, 116487

45340-6822

NAME: PHENYLALANINE-AMMONIA LYASE
PO: RHODOTORULA GLUTINIS
CT: ENZYME L.
EA: (N,)
TV: L-5178, L-1210
REFERENCES:
Cancer Res., 32, 258, 1972; 33, 2529, 1973

45340-6823

PO: B.PUMILUS
CT: ENZYME L.
EA: (N,)
MW: 21000
PC: WH.
UV: W: (280, 21,)
SOL-GOOD: W
STAB: (ACID, +)(BASE, -)(HEAT, -)
TO: (RHODOTORULA SP.,)
REFERENCES:
J. Ferm. Techn., 57, 32, 1979

45340-7139

PO: CORTICIUM CENTRIFUGUM
CT: ENZYME L.
EA: (N,)
TO: (LIPOMYCES STARKEYI,)
IS-CHR: (SEPHADEX G-25, W)
IS-CRY: (LIOF.,)(PREC., FILT., AMMONIUM SULPHATE)
REFERENCES:
Agr. Biol. Ch., 43, 1991, 1979

45340-7140

NAME: KU-1
PO: S.SINDENENSIS
CT: ENZYME L.
EA: (N,)
TO: (S.AUREUS,)(B.SUBT.,)
REFERENCES:
J. Antib. & Antifungal Agents, 6, 5, 1978; *CA*, 91, 122134

45340-7141

NAME: "B"-1.3-GLUCANASE+PROTEASE+LYSOZYME"
PO: B.CIRCULANS
CT: ENZYME L.
EA: (N,)
TO: ANTIBACTERIAL, ANTIFUNGAL
REFERENCES:
JP 79/73182;

45340-7142

NAME: N-ACETYLNEURAMINASE-GLYCOHYDROLASE
PO: STREPTOCOCCUS PNEUMONIAE
CT: ENZYME L., ACIDIC, AMPHOTER
EA: (N,)
TV: INFL, MELANOMA
REFERENCES:
CR Ser. D, 285, 837, 1977; EP 4214

45340-7143

NAME: AGA, L-GLUTAMINASE-ASPARAGINASE
PO: ACINETOBACTER SP., ACHROMOBACTER SP.
CT: ENZYME L., ACIDIC, AMPHOTER
EA: (N,)
TV: ANTITUMOR, L-1210, MYELO
REFERENCES:
Cancer Res., 34, 429, 1974; *J. Biol. Chem.*, 250, 4165, 1975; *Life Sci.*, 10, 251, 1971; *Brit. Med. J.*, 1317, 1976; *Canc. Tmt. Rep.*, 63, 1025, 1979

45340-7563

NAME: TYROSINE PHENOL-LYASE
PO: ERWINIA HERBICOLA
CT: ENZYME L.
EA: (N,)
TV: MELANOMA
REFERENCES:
Cancer Res., 36, 167, 1976

45340-7711

PO: S.SP.
CT: ENZYME L.
EA: (N,)
STAB: (HEAT, -)(ACID, -)(BASE, +)
TO: (S.CEREV.,)(FUNGI,)(B.SUBT.,)
REFERENCES:
JP 80/68289

45340-7712

NAME: DEAMIDASE AG
PO: PS.FLUORESCENS
CT: ENZYME L.
EA: (N,)
TV: ANTITUMOR
REFERENCES:
Prob. Zlokach. Rosta, 73, 1977; *CA*, 93, 88530

45340-8223

PO: STREPTOCOCCUS EQUINUS
CT: ENZYME L.
EA: (N,)
MW: 600000
STAB: (HEAT, -)
TO: (STREPTOCOCCUS MUTANS,)
IS-CHR: (SEPHADEX G-200, W)
REFERENCES:
Kurume Igakki Zasshi, 43, 1138, 1980; *CA*, 94, 170752

45340-8594

NAME: PR1-LYSOZYME
PO: PS.AERUGINOSA+MITOMYCIN-C
CT: ENZYME-LIKE
EA: (N,)
TO: (MICROCOCCUS SP.,)
REFERENCES:
J. Bioch. (Tokyo), 83, 727, 1978; *Microbios*, 23, 73, 1979

45350-6817

NAME: K-FACTOR, "KILLER FACTOR"
PO: B.THURINGIENSIS-GALLERIAE
CT: BACTERIOCIN
TO: (B.SP.,)
REFERENCES:
Mikrob., 48, 716, 1979

45350-6818

PO: CLOSTRIDIUM ACETOBUTYLICUM
CT: BACTERIOCIN
EA: (N,)
STAB: (HEAT, -)(ACID, -)(BASE, -)
TO: (CLOSTRIDIUM SP.,)
REFERENCES:
Appl. Micr., 37, 433, 1979; *CA*, 90, 184846

45350-6819

NAME: BACTERIOCIN-14468
PO: MYCOBACTERIUM SMEGMATIS
CT: BACTERIOCIN
EA: (N,)
MW: 75000
STAB: (HEAT, -)
TO: (MYCOB.SP.,)
TV: CYTOTOXIC, HELA
REFERENCES:
AAC, 15, 504, 1979; *Cancer Res.*, 39, 5114, 1979

45350-6824

NAME: THERMOCIN-93
PO: B.STEAROTHERMOPHYLUS
CT: BACTERIOCIN
EA: (N,)
MW: 13500
TO: (B.SUBT.,)
REFERENCES:
J. Gen. Micr., 111, 449, 1979

45350-6825

PO: PS.LIQUEFACIENS
CT: BACTERIOCIN
EA: (N,)
TO: (PS.SP.,)
REFERENCES:
CA, 90, 164443, 169443

45350-7144

NAME: PYOCIN, BACTERIOCIN
PO: PS.SP.+MITOMYCIN-C
CT: BACTERIOCIN, ACIDIC, AMPHOTER
EA: (N,)
STAB: (ACID, -)(BASE, +)(HEAT, -)
TO: (PS.SP.,)
REFERENCES:
JP 79/107596; *CA*, 92, 4702

45350-7145

NAME: SANGUICIN
PO: STREPTOCOCCUS SANGUIS
CT: BACTERIOCIN
EA: (N,)
MW: 280000
STAB: (HEAT, -)
TO: (BACTEROIDES MELANINOGENICUS,)(CORYNEBACTERIUM SP.,)
IS-CHR: (SEPHADEX G-200, W)
REFERENCES:
AAC, 16, 262, 1979; *Bull. Tokyo Dental Coll.*, 18, 217, 1977

45350-7479

PO: BACTEROIDES FRAGILIS
CT: BACTERIOCIN
EA: (N,)
MW: 13500, 18700
STAB: (HEAT, -)(ACID, -)(BASE, -)
TO: (BACTEROIDES FRAGILIS,)
IS-CHR: (SEPHADEX G-100, W)
REFERENCES:

AAC, 16, 724, 1979

45350-7564

NAME: <u>ERWINICIN</u>
PO: ERWINIA AROIDEA
CT: BACTERIOCIN
EA: (N,)
MW: 150000
STAB: (HEAT, -)
TO: (ERWINIA CAROTOVORA,)
REFERENCES:

Zh. Micr. Epid. Imm., (2), 77, 1980; *CA*, 92, 160192

45350-7565

NAME: <u>S34-BACTERIOCIN</u>
PO: STREPTOCOCCUS FAECIUM
CT: BACTERIOCIN
EA: (N,)
MW: 290000
STAB: (ACID, +)(BASE, +)
TO: (STREPTOCOCCUS SP.,)
REFERENCES:

CA, 92, 160187, 160188

45350-7566

PO: ERWINIA UREDOVORA
CT: BACTERIOCIN
EA: (N,)
MW: 150000
STAB: (HEAT, -)
TO: (ERWINIA CAROTOVORA,)
REFERENCES:

Arch. Int. Physiol. Biochim., 87, 845, 1979; *CA*, 92, 90547

45350-7713

PO: STREPTOCOCCUS SP.
CT: BACTERIOCIN
EA: (N,)
STAB: (HEAT, -)
TO: ANTIBACTERIAL
REFERENCES:
Nippon Saikingaku Zasshi, 35, 358, 1980; *CA*, 93, 22246

45350-7714

PO: ERWINIA AROIDEA, ERWINIA CHRYSANTHEMI, ERWINIA QUERSINIA
CT: BACTERIOCIN
EA: (N,)
MW: 25000|8000
STAB: (HEAT, -)
TO: (ERWINIA SP.,)
REFERENCES:
Prikl. Biokh. Mikr., 16, 372, 1980; *CA*, 93, 63990

45350-7970

NAME: STAPHYLOCOCCIN
PO: STAPHYLOCOCCUS EPIDERMIDIS
CT: BACTERIOCIN
EA: (N,)
TO: (STAPHYLOCOCCUS SP.,)
TV: CYTOTOXIC
REFERENCES:
Pol. J. Pharm. Pharmacol., 33, 337, 1981; Pol. P 105748; *CA*, 92, 213535

45350-8224

NAME: MUTACIN GS-5
PO: STREPTOCOCCUS MUTANS
CT: BACTERIOCIN
EA: (N,)
MW: 20000
PC: WH., POW.
STAB: (HEAT, +)
TO: (STREPTOCOCCUS PYOGENES,)
REFERENCES:
AAC, 19, 166, 1981; *Inf. & Immun.*, 12, 1375, 1975

45350-8595

PO: CORYNEBACTERIUM, NEBRASCENSE, CORYNEBACTERIUM MICHIGINENSE, CORYNEBACTERIUM SP.
CT: BACTERIOCIN
EA: (N,)
TO: (CORYNEBACTERIUM SP.,)
REFERENCES:
Can. J. Micr., 25, 367, 1979

45350-8596

NAME: LACTOCIN C-183, LACTOCIN C-290, LACTOCIN C-308
PO: LACTOBACTERIUM CASEI
CT: BACTERIOCIN
TO: (LACTOBACILLUS SP.,)
REFERENCES:
Antib., 837, 1980

45350-8597

NAME: LACTOCIN P-109, LACTOCIN P-186
PO: LACTOBACILLUS PLANTARUM
CT: BACTERIOCIN
TO: (LACTOBACILLUS SP.,)
REFERENCES:
Antib., 837, 1980

45350-8598

NAME: THOMYCIDE
PO: STREPTOCOCCUS SP.
CT: BACTERIOCIN
TO: (G.POS.,)
UTILITY: ON CLINICAL TRIAL
REFERENCES:
Antib., 528, 1980

45350-8860

NAME: LACTOCIN-A1
PO: LACTOBACILLUS ACIDOPHYLLUS
CT: BACTERIOCIN
TO: (S.AUREUS,)(B.SUBT.,)(P.VULG.,)
REFERENCES:
Antib., 843, 1981

45360-7715

NAME: INTERFEROIDE
PO: BACTERIUM SP.+DNA RECOMBINANT
TV: ANTIVIRAL
REFERENCES:
Chem. Engn. News, 58(4), 38, 1980

45370-6702

PO: STV.HACHIJOENSE, STV.GRISEOCARNEUM, STV.CINNAMONEUM
CT: CELL-WALL COMPONENT
LD50: (140, IP)(400, SC)(900, PEROS)
TV: S-180, L-1210
REFERENCES:
JP 78/99315; *CA*, 90, 101949

45370-6705

PO: ACTINOPYCNIDIUM COERULEUM
CT: CELL-WALL COMPONENT
LD50: (150, IP)(420, SC)
TV: ANTITUMOR, EHRLICH, INFL
REFERENCES:
JP 78/118511

45370-6737

NAME: "CELL-WALL COMPONENT"
PO: PS.AERUGINOSA, SALMONELLA TYPHIMURIUM, ESCHERICHIA COLI
CT: LIPOPOLYSACCHARIDE, CELL-WALL COMPONENT
EA: (N,)
TV: S-180
REFERENCES:
Gann, 63, 503, 1972; 64, 523, 1973; *J. Bioch. (Tokyo)*, 83, 711, 1978; *Jap. J. Exp. Med.*, 41, 493, 1971; 47, 393, 1977; *Eur. J. Bioch.*, 97, 623, 1979

45370-7134

PO: B.MEGATHERIUM, B.LICHENIFORMIS, CORYNEBACTERIUM POINSETTIAE
CT: MACROMOLECULAR, GLYCOPROTEID, CELL-WALL COMPONENT
EA: (N,)
TV: ANTITUMOR
REFERENCES:
Fr. P. 2410476

45370-7135

NAME: N-T-1, N-A-1, N-S-1, N-B-1
PO: MYCOBACTERIUM TUBERCULOSIS, MYCOBACTERIUM AVIUM, MYCOBACTERIUM SMEGMATIS, MYCOBACTERIUM BOVIS
CT: MACROMOLECULAR, CELL-WALL COMPONENT
EA: (N,)
LD50: (950|250, IP)
TV: S-180, EHRLICH
IS-CRY: (LIOF.,)
REFERENCES:
DT 2908241; *CA*, 92, 20560

45370-7136

NAME: MER, "METANOL EXTRACTABLE RESIDUE"
PO: MYCOBACTERIUM TUBERCULOSIS BCG
CT: MACROMOLECULAR, CELL-WALL COMPONENT
EA: (N, 8)
TV: ANTITUMOR
IS-EXT: (MEOH, , MIC.)
REFERENCES:
JA, 32, 1011, 1979; *Nat. Cancer Inst. Mtg.*, 35, 157, 1972; *Brit. J. Cancer*, 31, 176, 1975

45370-7138

PO: ESCHERICHIA COLI
CT: GLYCOPROTEID, CELL-WALL COMPONENT
EA: (N,)(P, .8)
SOL-GOOD: W
REFERENCES:

Fr. P. 2396018; *CA*, 91, 122162

45370-7146

PO: VIBRIO ANGUILLARUM
CT: CELL-WALL COMPONENT
EA: (N,)
TV: EHRLICH
REFERENCES:

Gann, 70, 429, 1979

45370-7147

NAME: CP, NSC-220537
PO: CORYNEBACTERIUM PARVUM
CT: CELL-WALL COMPONENT
TV: ANTITUMOR
UTILITY: ANTITUMOR DRUG
REFERENCES:

Canc. Tmt. Rep., 62, 1919, 1978; *Cancer Res.*, 39, 3554, 1979; *J. Immunopharm.*, 3, 49, 1981; USP 4069314; *CA*, 95, 22826

45370-7199

PO: CLOSTRIDIUM SACCHAROPERBUTYLACETONICUM, CLOSTRIDIUM ACETOBUTYLICUM, CLOSTRIDIUM BUTYRICUM
EA: (C, 42)(H, 7)(N, 8)(P, 1)(ASH, 1)(C, 46)(N, 12)
PC: YELLOW
OR: (-42, HCL)
SOL-GOOD: ACID
SOL-FAIR: W
SOL-POOR: ETOH, ET2O
TV: S-180
REFERENCES:

JP 79/76817; JP 80/118417;

45370-7434

PO: STREPTOCOCCUS EQUISIMILIS
CT: MACROMOLECULAR
TV: EHRLICH
REFERENCES:
DT 2921829

45370-7490

NAME: 60-F
PO: STREPTOCOCCUS PYOGENES, STREPTOCOCCUS EQUISIMILIS
CT: POLYSACCHARIDE-PROTEIN COMPLEX
EA: (C, 43)(H, 6)(N, 7)
PC: WH., POW.
UV: (200, ,)
SOL-POOR: W
TV: S-180, EHRLICH, YOSHIDA
REFERENCES:
Juzen Med. J., 87, 609, 1978; DT 2926406

45370-7491

NAME: "CELL-WALL SKELETON", CWS
PO: PROPIONIBACTERIUM ACNES
CT: MACROMOLECULAR
TV: ANTITUMOR
REFERENCES:
Gann, 73, 73, 1979

45370-7705

PO: LACTOBACILLUS P.
CT: CELL-WALL COMPONENT
EA: (N,)
TO: (E.COLI,)(PS.AER.,)
REFERENCES:
Zbl. Bakt. Parasit., 170B, 133, 1980; *CA*, 92, 211535

45370-7706

PO: BACTEROIDES FRAGILIS-VULGATUS, MEGASPHAERA SP., PASTEURELLA SP.
CT: CELL-WALL COMPONENT
EA: (N,)
PC: WH., POW.
SOL-GOOD: W
TV: EHRLICH
REFERENCES:
JP 80/57520; *CA*, 93, 236925

45370-8225

NAME: LC-9018
PO: LACTOBACILLUS CASEI
CT: CELL-WALL COMPONENT
EA: (N,)
LD50: NONTOXIC
TV: S-180, L-1210, P-388
REFERENCES:
39th Meet. Jap. Canc. Ass., Tokyo, 1980, Abst. 795; Holl. P 80/1045

45370-8226

PO: PROPIONIBACTERIUM GRANULOSUM, PROPIONIBACTERIUM ACNES, PROPIONIBACTERIUM AVIDUM
CT: CELL-WALL COMPONENT
EA: (N,)
TV: ANTITUMOR
REFERENCES:
Canc. Chemother. Res., 167, 1980

45370-8227

NAME: RP-41200, RP-32919
PO: MICROCOCCUS SEDOGENES
CT: CELL-WALL COMPONENT
EA: (C, 46)(H, 7)(N, 5)(O, 37)(P, 1)(S, .5)(NA, 2)
MW: 500000
PC: WH., POW.
UV: W: (200, ,)
SOL-GOOD: W
LD50: (23, IV)
TV: S-180, ANTIVIRAL
REFERENCES:
Belg. P 882841; Holl. P 80/2108; *CA*, 94, 190309

45370-8844

NAME: CW-5
PO: BIFIDOBACTERIUM LONGUM
CT: CELL-WALL COMPONENT, GLYCOPROTEIN
EA: (N,)
TV: ANTITUMOR
REFERENCES:
JP 81/103194

45370-8861

NAME: NS, NS-B-1
PO: MYCOBACTERIUM BOVIS, MYCOBACTERIUM AOYAMA, MYCOBACTERIUM TUBERCULOSIS
CT: CELL-WALL COMPONENT
EA: (N,)(P,)
PC: WH., POW.
TV: EHRLICH, S-180, MELANOMA
REFERENCES:
JP 80/92689; *CA*, 93, 148126

45370-8862

NAME: NCWS
PO: NOC.RUBRA, NOC.OPACA, NOC.ASTEROIDES, NOC.CORALLINA
CT: CELL-WALL COMPONENT
EA: (N,)
TO: (E.COLI,)
TV: ANTITUMOR
REFERENCES:
Abst. 12th Int. Congr. Chemother., Florence, 1981, 379; JP 81/32493

450
OTHER LESS KNOWN MACROMOLECULAR COMPOUNDS

45000-6826

PO: LACTOBACILLUS SP.
CT: PROTEIN
EA: (N,)
MW: 10000
STAB: (HEAT, +)
TO: (S.AUREUS,)(PS.AER.,)(B.SP.,)
REFERENCES:

Swed. J. Agric. Res., 8, 61, 67, 1978; *CA*, 90, 4763, 4764

45000-7133

NAME: LEUCODIDIN, "CYTOTOXIC PROTEIN"
PO: PS.AERUGINOSA
CT: PROTEIN, ACIDIC, AMPHOTER
EA: (N,)
MW: 25100, 22550|250
EW: 23100
STAB: (HEAT, -)
LD50: (.025, IV)
TV: CYTOTOXIC
IS-CRY: (PREC., FILT., AMMONIUM SULPHATE)
REFERENCES:

Toxicon, 17, 467, 1979; *J. Gen. Micr.*, 93, 292, 1976

45000-7707

PO: STREPTOCOCCUS SP.
CT: MACROMOLECULAR
EA: (N,)
MW: 8500|500
STAB: (HEAT, +)(ACID, +)(BASE, -)
IS-CHR: (SEPHADEX G-100, W)
REFERENCES:

J. Med. Microbiol., 12, 413, 1979

45000-7708

PO: S.ALBUS
CT: MACROMOLECULAR
EA: (N,)
TV: ANTITUMOR
REFERENCES:

JP 80/15741

45000-7710

NAME: BIOFLORIN
PO: STREPTOCOCCUS FAECIUM
CT: MACROMOLECULAR
EA: (N,)
TO: ANTIMICROBIAL
UTILITY: ON CLINICAL TRIAL
REFERENCES:
Curr. Ther. Res., 26, 967, 978, 1979

45000-7962

PO: S.SP.
CT: PROTEIN
EA: (N,)
TV: P-388
REFERENCES:
Ind. J. Exp. Biol., 18, 935, 1980; *CA*, 93, 202647

45000-7963

PO: DALDINIA CONCENTRICA
CT: PROTEIN, MACROMOLECULAR
EA: (N,)
TO: (S.CEREV.,)(FUNGI,)
REFERENCES:
Fr. P 243836

45000-8214

PO: S.SP.
CT: MACROMOLECULAR, PROTEIN
EA: (N,)
TV: P-388
REFERENCES:

Indices

SEQUENCE OF ALPHABETIZING

These indices are computer-generated, and the following order represents the sequence used in alphabetizing characters within an entry:

SPACE	
"	(quote)
'	(apostrophe)
(	(open paren)
)	(close paren)
+	(plus)
,	(comma)
−	(minus or hyphen)
.	(period)
/	(slash)
:	(colon)
		(vertical bar)
A to Z	(letters)
0 to 9	(numbers)

Numbers are sequenced by digit from left to right. For example:

1		
1	2	
1	3	0
2		
2	1	
2	2	0
3		

Also, please note that unnamed compounds do not appear in the *Index of Names of Antibiotics* or the *Index of Antibiotic Numbers and Names.*

This volume contains new and revised information on compounds listed in Volumes I through IX of the *CRC Handbook of Antibiotic Compounds* as well as information on new compounds that have been developed since the publication of Volume IX. In these indices, antibiotics with compound numbers less than 6700 (e.g., 11200-2750) are to be found in Section I of this volume, while those with compound numbers greater than or equal to 6700 are to be found in Section II.

INDEX OF NAMES OF ANTIBIOTICS

INDEX OF ANTIBIOTIC NUMBERS AND NAMES

10000-3594 SPORAVIRIDIN
10000-7885 460-B
10000-7886 OS-469
10000-7887 K-7

11 PURE SACCHARIDES

11121-5564 NOJIRIMYCIN-B
SG-1
11121-6731 N-CARBAMOYL-B"-D-GLUCOSAMINIDE
N-CARBAMOYL-D-GLUCOSAMINE
SF-1993
11121-7553 IA-II
N-CARBAMOYL-B"-D-GLUCOSAMINIDE
SF-1993
11122-7601 U-59834
3-TREHALOSAMINE
11132-5843 K-52-B
11132-6178 OLIGOSTATIN-E
11132-6179 OLIGOSTATIN-D
11132-7602 TAI-A
11132-7603 TAI-B
11132-8465 OLIGOSTATIN-C
SF-1130-X3
11140-0012 STREPTOZOTOCIN
ZANOSAR
11140-0013 PRUMYCIN
11200-7524 LEVAN
11200-8859 PK-1
11210-0019 LENTINAN
11210-0020 G-2
11210-0023 NO NAME
11210-0025 SCHYSOPHYLLAN
SCHYZOPHYLLAN
11210-6181 GU-1 (GU-2
GU-3)
11210-6182 GU-4
11210-6732 G-I-2A"B"
11210-6734 "GLUCAN"
CORIOLAN
D-II
11210-6735 B"-1.3-GLUCAN
11210-7062 HELVELLAN
11210-7063 A-5 MANNAN
11210-7509 "YEAST GLUCAN"
"YEAST-B"-GLUCAN"
11210-7604 NEOSCHYZOPHYLLAN
TSH
11210-7605 "FRUCTAN"
"GLUCAN"
11210-7606 "MANNAN"
11210-8093 "HYDROLYSED CORTICANE"
11210-8162 "GLUCAN-I"
11210-8466 B"-1.3-GLUCAN-POLYOL
11210-8857 "GLUCAN"
11220-0044 KRESTIN
POLYSACCHARIDE-K
11220-6485 "PEPTIDOMANNAN"
KS-2-A
11220-7065 C-45
11220-7413 KS-SUBSTANE
11220-7414 KS-2B
11220-7609 DH-6665
11220-7612 SP-MT-1
TO-B1
T1-B1
T2-B1
11220-8094 "PROTEOGLYCAN"
11220-8095 D-MANNAN
JA-AS-15-712
11220-8468 KS-2D
11220-8783 IES-1638
11230-8784 "ARABINOMANNAN"
11240-4084 H-2075
H-2609
SOEDOMYCIN
11240-6736 PS-9-5H
11240-8163 PH-2
11240-8164 AP

12 AMINOGLYCOSIDE ANTIBIOTICS

12000-4105 CORALLINOMYCIN
12000-5306 FA-252-C
12000-7067 DESOMYCIN
12000-7613 SF-1999
12110-0076 DIHYDROSTREPTOMYCIN
12110-6740 741
12130-0090 SPECTAM
SPECTINOMYCIN
SPECTOGARD
12130-4043 B-2847-A"
12211-0096 NEOMYCIN-B
12211-0097 NEOMYCIN-C
12211-8470 COMP.X1
6'''-DEAMINO-6'''-HYDROXYPAROMOMYCIN-I
12211-8471 COMP.X2
6'''-DEAMINO-6'''-HYDROXYPAROMOMYCIN-II
12211-8472 COMP.Y2
6'''-DEAMINO-6'''-HYDROXYNEOMYCIN-C
12211-8473 NEOMYCIN-G
12211-8474 NEOMYCIN-K
12212-0119 H-60
IBISTACIN
RIBAMICINA
RIBOMYCINE
RIBOSTAMIN
RIBOSTAMYCIN
12212-0124 XYLOSTACIN
XYLOSTAMYCIN
XYLOSTATIN
12212-0126 BUTIROSIN-A
12212-0127 BUTIROSIN-B
12212-6129 6'-N-METHYLBUTIROSIN-A
12212-6346 DMB-A
3'.4'-DIDEOXY-6'-C-

	METHYLBUTIROSIN
12212-6347	6'-N-METHYLBUTIROSIN-B
12212-6739	H-60
	RIBOSTAMYCIN
12221-0149	NEBRAMYCIN COMPLEX
12221-0154	GERNEBCIN
	OBRACIN
	TOBRACIN
	TOBRAMYCIN
	3'-DEOXYKANAMYCIN-B
12221-6595	5-DEOXYKANAMYCIN
	5-DEOXYKANAMYCIN-A
12221-6596	NO NAME
12222-0156	GENTAMICIN-C1
	REFOBACIN
	SULMYCIN
12222-0157	GENTAMICIN-C2
	REFOBACIN
	SULMYCIN
12222-0158	GENTAMICIN-C1A
	REFOBACIN
	SULMYCIN
12222-0159	MICRONOMYCIN
	SAGAMICIN
12222-0160	GENTAMICIN-A
12222-0163	GENTAMYCIN-B
12222-0165	GENTMICIN-X2
12222-0166	G-418
12222-0167	BAYMICINE
	G-52-2
	SISOLINE
	SISOMICIN
	SISOMINE
12222-0168	66-40-B
12222-0169	66-40-D
12222-0170	VERDAMICIN-I
12222-0171	G-52
12222-3875	MUTAMICIN-6
12222-5309	IPROMICIN
	WIN-42122-2
	2-HYDROXYGENTAMICIN-C1
12222-5394	IPROMICIN
	WIN-42122-2
	2-HYDROXYGENTAMICIN-C2
12222-6196	3"-N-DEMETHYLSISOMICIN
	66-40-G
12222-6597	XK-62-4
	6'-N-METHYLSAGAMICIN
12222-6598	XK-62-3
	1-N-METHYLGENTAMICIN-C1A
12222-7071	477-2H
	6'-O-GLUCOSYLGENTAMICIN-A
12222-7555	MU-7
	MUTAMICIN-7
	3"-N-METHYLSISOMICIN
12222-7556	MU-8
	MUTAMICIN-8
	1-N-METHYL-5-DEOXYSISOMICIN
12222-7557	MU-X
	5-EPIFLUORO-5-DEOXYSISOMICIN
12222-7614	XK-62-5
	3"-N-DEMETHYLSAGAMICIN
12222-7615	XK-62-6
12222-7616	XK-62-7
12222-7617	XK-62-8
12222-7618	UAA-3
12222-8064	TPJ-B
12222-8065	K-26-2
	6'-N-METHYLVERDAMICIN
12222-8083	2-HYDROXYSAGAMICIN
12222-8096	G-367-1
	2'-N-FORMYLSISOMICIN
12222-8097	EPI-SISOMICIN
	G-367-2
12222-8098	D-53
	3"-N-DEMETHYL-G-52
12222-8475	GENTOXIMICIN-A
	2'-DEAMINO-2'-HYDROXYGENTAMICIN-C1A
	3'.4'-DIDEOXYGENTAMICIN-B
12222-8476	GENTOXIMICIN-B
	2'-DEAMINO-2'-HYDROXYGENTAMICIN-C2
	3'.4'-DIDEOXYGENTAMICIN-B1
12222-8477	3"-N-DEMETHYLGENTAMICIN-C1A
12222-8478	3"-N-DEMETHYLSISOMICIN
	66-40-G
12222-8479	2-HYDROXYGENTAMICIN-A3
12222-8480	2-HYDROXYGENTAMICIN-B
12222-8481	2-HYDROXYGENTAMICIN-B1
12222-8482	2-HYDROXY-JI-20A
12222-8483	2-HYDROXY-JI-20B
12223-4756	SELDOMYCIN-5
12224-6738	SU-2
	1-DEAMINO-1-HYDROXYGENTAMICIN-C1A
12224-7068	SU-1
	1-DEAMINO-1-HYDROXYGENTAMICIN-C2
12224-7069	SU-3
	SUM-3
	1-DEAMINO-1-HYDROXYSAGAMICIN
12224-7070	HYDROXY-SU-2
	SU-4
	SUM-4
12224-7889	S-11-A
	1-DEAMINO-1-HYDROXYXYLOSTATIN
12224-8165	K-114-E
12224-8166	K-114-G
12225-7415	COMBIMICIN-A2
	I-A2
	3"-N-METHYL-4"-C-METHYL-3'.4'-DI DEOXYKANAMYCIN-A
12225-7416	COMBIMICIN-B1
	COMBIMICIN-B1
	I-B1
	3"-N-METHYL-4"-C-METHYL-3'.4'-DIDEOXY-6-N'-METHYLKANAMICIN-B

12225-7417 COMBIMICIN-B2
3"-N-METHYL-4"-C-METHYL-
3'.4'-DI
DEOXYKANAMYCIN-B
12225-8099 I-A1
I-SK-A1
2'-DEAMINO-2'-HYDROXY-
5"-HYDRO
XYMETHYLSISOICIN
12225-8100 I-B1
I-SK-B1
5"-HYDROXY
METHYLSISOMICIN
12225-8101 I-B2
I-SK-B2
5"-HYDROXYMETHYL-4"-
DEMETHYLSISOMICIN
12225-8102 COMBIMICIN-A4
3"-N-METHYLKANAMYCIN-A
12225-8103 COMBIMICIN-A3
3"-N-METHYL-4"-C-
METHYLKANAMYCIN-A
12225-8104 COMBIMICIN-B5
3"-N-METHYLKANAMYCIN-B
12225-8105 3"-N-METHYL-4"-C-
METHYLKANAMYCIN-B
12225-8106 COMBIMICIN-T2
I-T2
3"-N-METHYLTOBRAMYCIN
12225-8107 COMBIMICIN-B4
3"-N-METHYLDIBEKACIN
12225-8108 COMBIMICIN-T1B
I-T1B
3"-N-METHYL-4"-C-
METHYLTOBRAMYCIN
12225-8484 COMBIMICIN-A3
I-A3
3"-N-METHYL-4"-C-
METHYLKANAMYCIN-A
12225-8485 COMBIMICIN-A4
I-A4
3"-N-METHYLKANAMYCIN-A
12225-8486 COMBIMICIN-B3
I-B3
6'-N-METHYL-3'.4'-
DIDEOXY-3"-N-
METHYLKANAMICIN-B
12225-8487 COMBIMICIN-B4
I-B4
3"-N-METHYL-3'.4'-
DIDEOXYKANAMYCIN-B
3"-N-METHYLDIBEKACIN
12225-8488 COMBIMICIN-B5
I-B5
3"-N-METHYLKANAMYCIN-B
12225-8489 COMBIMICIN-T1A
I-T1A
3'-DEOXY-4"-C-METHYL-3"-
N-METHYLKANAMYCIN-A
12225-8490 COMBIMICIN-T2
I-T2
3"-N-METHYLTOBRAMYCIN
3'-DEOXY-3"-N-
METHYLKANAMYCIN-B
12225-8491 COMBIMICIN-T1B
I-T1B
3"-N-METHYL-4"-C-
METHYLTOBRAMYCIN
3'-DEOXY-4"-C-METHYL-3"-
N-METHYLKANAMYCIN-B
12230-0173 DESTOMYCIN-A
XK-33-F1
12230-0176 A-16316-A
A-396-I
12230-8785 A-9594
12240-0178 APRAMYCIN
12240-3877 OXYAPRAMYCIN
12250-5835 GARAMINE
12250-5837 GENTAMINE-C2
12250-7072 FU-10
12250-7558 5.6-DIDEOXYNEAMINE
12310-0179 VALIDACIN
VALIDAMYCIN-A
12310-5830 JINGGANGMYCIN-II
12320-0187 MINOSAMINOMYCIN
12330-0189 MYOMYCIN-B
12330-7073 DC-5-4
12330-7074 BM-782-A"1
LL-BM-782-A"1
12330-7075 BM-782-A"1A
LL-BM-782-A"1A
12330-7076 BM-782-A"2
LL-BM-782-A"2
12330-8492 MYOMYCIN-C
12340-0420 LL-BM-123-A"
12350-8109 X-14847
2-AMINO-2-DEOXY-A"-D-
GLUCOPYRANOSYL-1-O-D-
MYO-INOSITOL
12410-0135 FORTIMICIN-B
12410-3861 ABBOT-44747
ASTROMICIN
FORTIMICIN-A
KW-1070
12410-5435 DACTIMICIN-B
N-FORMYLFORTIMICIN-A
SF-1854
12410-6133 FORTIMICIN-D
6'-DEMETHYLFORTIMICIN-A
12410-6134 FORTIMICIN-KE
6'-DEMETHYLFORTIMICIN-B
12410-6486 KA-6606-I
SPORARICIN-A
12410-6487 SPORARICIN-B
12410-6488 SPORARICIN-C
12410-6489 SPORARICIN-D
12410-6741 FORTIMICIN-KG
12410-6742 FORTIMICIN-KF
12410-6743 FORTIMICIN-E
FORTIMICIN-KH
3.4-DIEPIFORTIMICIN-B
12410-6744 ISOFORTIMICIN
ISOFORTIMICIN-A
12410-7077 ISTAMYCIN-A
KA-7038-I
SANNAMYCIN-A
12410-7078 ISTAMYCIN-A0
KA-7038-III

SANNAMYCIN-B
12410-7079 ISTAMYCIN-A
SANNAMYCIN-A
12410-7080 ISTAMYCIN-B
1-EPI-ISTAMYCIN-A
12410-7081 FORTIMICIN-KG1
12410-7082 FORTIMICIN-KG2
12410-7083 FORTIMICIN-KG3
12410-7084 FORTIMICIN-KO1
3-EPI-FORTIMICIN-B
12410-7418 KA-7038-II
12410-7419 DACTIMICIN
SF-2052
12410-7510 KA-7038-IV
12410-7511 KA-7038-V
12410-7512 KA-7038-VI
SANNAMYCIN-C
12410-7513 KA-7038-VII
12410-7514 FORTIMICIN-AE
12410-7619 ISTAMYCIN-AO
SANNAMYCIN-B
12410-7620 ISTAMYCIN-BO
1-EPI-SANNAMYCIN-B
12410-7621 KA-6606-V
3-EPI-SPORARICIN-B
12410-7622 KA-6606-VI
1-EPI-SPORARICIN-B
12410-7623 FORTIMICIN-AH
12410-7624 FORTIMICIN-AI
12410-7625 FORTIMICIN-AK
12410-7626 FORTIMICIN-AL
3-O-DEMETHYLFORTIMICIN-KG1
12410-7627 FORTIMICIN-AM
3-O-DEMETHYL-4-EPI-FORTIMICIN-B
12410-7628 FORTIMICIN-AN
12410-7629 FORTIMICIN-AO
12410-7630 FORTIMICIN-AP
3-O-DEMETHYLFORTIMICIN-E
12410-7631 FORTIMICIN-AQ
12410-7632 FORTIMICIN-AS
12410-7633 FORTIMICIN-KO
12410-7634 A-49759
3-O-DEMETHYLFORTIMICIN-A
12410-7635 3-O-DEMETHYL-2′-N-GLYCYLFORTIMICIN-B
12410-7888 AX-127-B
LYSINOMYCIN
12410-7890 DEMETHYLFORTIMICIN-B
FORTIMICIN-KR1
12410-7891 FORTIMICIN-KQ
12410-7892 3-O-DEMETHYLFORTIMICIN-KE
12410-7893 DEMETHYL-ISOFORTIMICIN-C
3-O-DEMETHYL-1-N-HYDANTHOYLFORTIMICIN-B
12410-8786 DACTIMICIN-B
SF-1854
SF-2052-B
12510-5156 GL-A1
SORBISTIN-A1
12510-5158 GL-A2
SORBISTIN-B
12510-5439 GL-A2
P-2563-A
12510-5440 GL-A1
P-2563-B
12510-7085 GL-A1
LL-AM-31-B"
SORBISTIN-A1
12510-7086 GL-A2
LL-AM-31-G"
SORBISTIN-B

13 OTHER GLYCOSIDES

13110-0190 STREPTOTHRICIN-F
13110-0205 PRACTOMYCIN-C
RACEMOMYCIN-B
13110-0271 BOSEIMYCIN-III
13110-0278 S-15-1-A
13110-0437 AMYCIN
13110-1884 PRACTOMYCIN-C
RACEMOMYCIN-B
STREPTOTHRICIN-D
13110-4757 Y-U17W-C1
13110-7894 Y-U17W-C2
13110-8167 402
13110-8168 S-15-1-C1
13110-8493 AY-24488
13120-0259 BY-81
13120-0261 BD-12
13120-0262 LL-AC-541
13120-0264 LL-AB-664
13130-0436 24010-B1
13200-6771 AM-3696
AZURENOMYCIN-A
13200-7087 AM-3696-B
AZURENOMYCIN-B
13200-7768 OA-7653
13200-8110 A-16686
13210-0280 RISTOCETIN-A
13210-0281 RISTOCETIN-B
13210-0282 RISTOMYCIN-A
13210-0283 ACTINOIDIN-A
13210-0284 ACTINOIDIN-B
13220-0286 VANCOMYCIN
13220-0289 AVOPARCIN
AVOPARCIN-B"
13220-5378 TEICHOMYCIN-A2
13220-5815 A-35512-B
13220-6745 AVOPARCIN-A"
LL-AV-290-A"
13230-0293 CHROMOMYCIN-A3
13230-0305 OLIVOMYCIN-A
13230-0311 MITHRAMYCIN
13230-0317 VARIAMYCIN
13230-4864 C-2449
S-2449
13230-7895 SR-1768-F
13230-7896 144-3-I
13230-7897 144-3-II
13230-7898 144-3-III

14 SUGAR DERIVATIVES

14110-0325 EVERNINOMYCIN-D

14110-0332 AVILAMYCIN-A
14110-0333 CURAMYCIN
CURAMYCIN-A
14110-0334 FLAMBAMYCIN
14110-4001 LIPIARMYCIN
14110-5312 SE-73B
14110-7088 SF-2033
14110-7636 SE-73
14120-7637 BU-2545
14120-7899 N-ETILLINCOMYCIN
U-24166
14130-5386 PAPULACANDIN-A
14130-5387 PAPULACANDIN-B
14130-5388 PAPULACANDIN-D
14130-5525 PAPULACANDIN-C
14210-0354 MOENOMYCIN-A
14210-4760 ENSANCHOMYCIN
14220-0380 CORD FACTOR
14220-7420 SCHIZONELLIN-A
14220-7421 SCHIZONELLIN-B
14230-0421 BM-123-B"1
CINODINE-B"1
14230-0422 BM-123-G"1
CINODINE-G"1
14230-4070 PULVOMYCIN
1063-Z
14230-4088 84-B-3
14230-5400 BM-123-G"2
CINODINE-G"2
14230-5401 BM-123-B"2
CINODINE-B"2
14230-8084 BU-2349-C
GLYSPERIN-C
14230-8111 BU-2349-A
GLYSPERIN-A
14230-8112 BU-2349-B
GLYSPERIN-B

21 MACROLIDE ANTIBIOTICS

21110-0439 METHYMYCIN
21110-0441 NA-181
21122-0457 ANAMYCIN
EROMERZIN
ERYTHROMYCIN-A
21122-0461 OLEANDOCIN
OLEANDOMYCIN
21122-5027 1745-A-X
21123-0465 MEGALOMYCIN-A
21124-4082 RP-23671
21124-4083 BF-2126208
RP-23672
21211-0475 MAIDIMEISU
MIDECAMINE
MIDICACIN
MYDECAMYCIN
RUBIMYCIN
21211-0478 LEUCOMYCIN-A3
21211-0480 LEUCOMYCIN-A5
21211-0485 DHA
LEUCOMYCIN-A10
LEUCOMYCIN-U
21211-0486 LEUCOMYCIN-A11
LEUCOMYCIN-V

21211-0492 JOSACIN
JOSAMYCIN
21211-0495 PLATENOMYCIN-A1
21211-0503 ESPINOMYCIN-A2
21211-0506 LEUCOMYCIN COMPLEX
TURIMYCIN COMPLEX
21211-0575 DHH
TURIMYCIN-HO
21211-6746 KITASAMYCIN-A12
LEUCOMYCIN-A12
21211-6747 KITASAMYCIN-A13
LEUCOMYCIN-A13
21211-6748 KITASAMYCIN-A14
LEUCOMYCIN-A14
21211-6749 KITASAMYCIN-A15
LEUCOMYCIN-A15
21211-6750 SF-837-MI
4′-DEACYL-MYDECAMYCIN
4′-DEACYL-SF-837-A1
21212-0514 SPIRAMYCIN-I
21221-0524 CARBOMYCIN-A
21221-5317 DELTAMYCIN-A1
21221-5954 DELTAMYCIN-X
21221-7090 DEEPOXY-CARBOMYCIN-A
21222-7422 DESMYCAROSYL-ANGOLAMYCIN
STAPHCOCCOMYCIN
21223-0529 ROSAMYCIN
21223-0535 A-6888-F
CIRRAMYCIN-A1
21223-0536 A-6888-A
CIRRAMYCIN-B
21223-0537 A-6888-A
ACUMYCIN
21223-6751 SCH-23831
21223-7091 A-6888-C
21223-7092 A-6888-X
21224-7093 A-11725-I
AR5-1
MYCINAMYCIN-I
21224-7094 A-11725-II
AR5-2
MYCINAMYCIN-II
21231-0539 CARBOMYCIN-B
21231-7900 PLATENOMYCIN-W3
YL-704-W3
21232-0541 RELOMYCIN
S-1
20-DIHYDROTYLOSIN
21232-0549 TYLOSIN
21232-5849 3-ACETYLTYLOSIN
21232-5850 3-PROPYONYLTYLOSIN
21232-5851 4"-BUTYRLTYLOSIN
21232-5852 4"-ISOVALERYLTYLOSIN
21232-7901 S-2
9.20-TETRAHYDROTYLOSIN
21232-8113 3-ACETYL-4"-
BUTYRYLTYLOSIN
21232-8114 3-ACETYL-4"-
ISOVALERYLTYLOSIN
21232-8115 3-PROPIONYL-4"-
BUTYRYLTYLOSIN
21232-8116 3-PROPIONYL-4"-
ISOVALERYLTYLOSIN
21232-8169 N-1

21233-5174 DESOSAMINYL-PROTYLONOLIDE
M-4365-G1
21233-5175 M-4365-G2
21234-7095 A-11725-IV
MYCINAMYCIN-IV
12.13-DEEPOXY-AR5-1
21234-7096 A-11725-V
MYCINAMYCIN-V
12.13-DEEPOXY-AR5-2
21234-7423 A-11725-III
MYCINAMYCIN-III
21241-0555 MARIDOMYCIN-I
21241-0556 MARIDOMYCIN-II
21241-0557 MARIDOMYCIN-III
21241-0558 MARIDOMYCIN-IV
21241-0559 MARIDOMYCIN-V
21241-0561 MARIDOMYCIN-VI

22 POLYENE ANTIBIOTICS

22110-0595 RAPAMYCIN
22110-8494 M-5070
22200-0633 AC2-435
22200-5294 BARODAMYCIN
22200-7904 TETRAENE BG-6
TETRAENE OS-1
22200-7906 HA-236
22210-0603 ETRUSCOMYCIN
3605-5
3608-5
4995-35
22210-0608 TETRAMYCIN
22210-7638 ALBICIDIN
22230-0618 NYSTATIN-A2
22230-0619 NYSTATIN-A3
22230-5853 POLYFUNGIN-B
22300-7905 PENTAENE DG-15
PENTAENE EG-4
22310-0649 FUNGICHROMIN
22310-0650 LAGOSIN
22310-0651 MOLDICIDIN-B
22310-0652 PENTAMYCIN
22310-0653 COGOMYCIN
22310-0659 FILIPIN COMPLEX
U-5956
22310-6204 FCRC-21
MOLDCIDIN-B
NSC-277813
PENTAMYCIN
22310-6602 LYMPHOSARCIN
NSC-208642
22310-8495 ELIZABETHIN
22320-7907 LIA-0371
22320-8496 GANNIBAMYCIN
22330-0703 LIENOMYCIN
2995
22500-0791 NEOHEPTAENE
22511-0726 AUREOFUNGIN
CANDIMYCIN
DJ-400-B1
22511-0727 DJ-400-B2
22511-0728 CANDICIDIN
CANDICIDIN COMPLEX
HEPTAMYCIN
22511-0729 HAMYCIN
22511-0732 CANDICIDIN-D
CANDICIDIN-D1
22511-0733 TRICHOMYCIN-A
22511-0734 TRICHOMYCIN-B
22511-0735 LEVORIN-A0
22511-0736 LEVORIN-A1
22511-0737 LEVORIN-A2
22511-0738 LEVORIN-A3
22511-0741 HEPTAMYCIN
LEVORIN
LEVORIN COMPLEX
22511-0747 HEPTAFUNGIN
22511-0769 AUREOFUNGIN-A
CANDIMYCIN-A
DJ-400-B1
22511-0771 AUREOFUNGIN-A
CANDIMYCIN
22511-0777 ORAFUNGIN
PARTRICIN
PARTRICIN-A
SN-654
TRICANDINE (METHYL ESTER)
22511-0806 ASCOSIN
CANDICIDIN
HEPTAMYCIN
LEVORIN
22511-4767 AUREOFACIN-A
AYF
AYFACTIN
VACIDIN-A
22511-5854 HAMYCIN
NO NAME
67-121 COMPLEX
22511-6205 GLOBOROSEOMYCIN
GLOBORUBERMYCIN
22511-7908 AUREOFACIN
AYFACTIN
PARTRICIN-B
VACIDIN-A
22511-7909 HEPTAFUNGIN-B
TRICHOMYCIN-B
22511-8497 "ANTIBIOTIC A"
22511-8498 67-121-D
22520-0776 AMPHOTERICIN-B
22520-0786 CANDIDININ
22520-0787 MYCOHEPTIN
MYCOHEPTIN-A2
22520-4058 HYDROHEPTIN
22520-7639 MYCOHEPTIN-A1
22520-8499 LIA-0735
22600-5266 EPIRODIN-A
22610-6752 OCTAMYCIN
4041-II
22710-0814 FLAVOFUNGIN
464
22710-0817 FLAVOMYCOIN
ROFLAMYCIN
22710-0818 MYCOTICIN-A
464

22710-0820 BRUNEOFUNGIN
22710-6207 FLAVOPENTIN
22710-7097 LIA-0167
NIGROFUNGIN
22720-0822 DERMOSTATIN-A

23 MACROCYCLIC LACTONE ANTIBIOTICS

23000-2344 MUSARIN
23000-2345 HYGROSTATIN
Y-5443
23000-4028 COPIAMYCIN
23000-7441 FA-2713-I
23000-7442 FA-2713-II
23000-8501 L-681110
23110-0829 OLIGOMYCIN-A
178
23110-4987 560
23110-8787 CYTOVARICIN
H-230
23120-0838 BOTRYCIDIN
VENTURICIDIN-X
23120-0843 AABOMYCIN
AABOMYCIN-A
23120-0844 A-150-A
23120-0845 BOTRYCIDIN
VENTURICIDIN-X
23120-5034 AABOMYCIN-S
23130-0846 AXENOMYCIN-A
23130-0848 AXENOMYCIN-D
23140-0849 PRIMYCIN
23140-4002 MONAZOMYCIN
23140-4003 MONAZOMYCIN
TAKACIDIN
23140-4004 LL-A-491
23150-0952 SCOPAFUNGIN
23150-3835 P-6226
23150-3836 AZALOMYCIN-F
23150-7098 AB-99
TENDAMYCIN
TENDOMYCIN
23150-8078 SF-2077
23150-8171 ENDOMYCIN COMPLEX
23150-8836 NIPHITHRICIN-A
23150-8837 NIPHITHRICIN-B
23160-0956 SCOPAMYCIN-A (TENTATIVE)
SCOPATHRICIN
SCOPATHRICIN-I
23160-5940 SF-1540
SF-1540-A
23160-5958 SF-1540-B
23160-7910 SCOPATHRICIN-II
23160-8500 CONCANAMYCIN-A
23210-0866 RECIFEOLIDE
23210-7100 PYRENOLIDE-A
23210-8117 PYRENOLIDE-B
23210-8118 PYRENOLIDE-C
23220-4774 A-26771-B
23230-0444 ALBOCYCLIN
23240-7426 EM-4615-A
IZUMENOLIDE
23310-0870 ANTIMYCIN-A3
23310-7911 ANTIMYCIN-AOA
23310-7912 ANTIMYCIN-AOC
23310-7913 ANTIMYCIN-AOD
23310-7914 ANTIMYCIN-AOB
23320-0884 PYRENOPHORIN
23320-0885 VERMICULIN
23320-7099 GRAHAMIMYCIN-A1
23320-8119 GRAHAMIMYCIN-A
23320-8120 GRAHAMIMYCIN-B
23330-0953 ELAIOPHYLIN
ELAIOPHYLIN-A
255-E
23330-0954 AZALOMYCIN-B
SAPROMYCETIN-A
255-E
23330-0955 AZALOMYCIN-B
ELAIOPHYLIN-A
SAPROMYCETIN-A
23410-0890 TETRANACTIN
23410-0891 MACROTETROLIDE B
23410-0892 MACROTETROLIDE C
23410-0893 MACROTETROLIDE D
23410-0894 MACROTETROLIDE G
23410-7915 SP-351-D
23420-0883 BOROMYCIN
23420-5322 APLASMOMYCIN-A
23511-0915 CHLOROTHRICIN
23512-0850 MILBEMYCIN-A"10
23512-0851 MILBEMYCIN-A"9
23512-0852 MILBEMYCIN-A"3
23512-0853 MILBEMYCIN-A"6
23512-0854 MILBEMYCIN-B"1
23512-0855 MILBEMYCIN-A"5
23512-0856 MILBEMYCIN-A"1
23512-0857 MILBEMYCIN-A"2
23512-0858 MILBEMYCIN-A"4
23512-0859 MILBEMYCIN-B"2
23512-0860 MILBEMYCIN-B"3
23512-6142 AVERMECTIN-A1A
IVERMECTIN
23512-6143 AVERMECTIN-A1B
IVERMECTIN
23512-6144 AVERMECTIN-A2A
IVERMECTIN
23512-6145 AVERMECTIN-A2B
IVERMECTIN
23512-6146 AVERMECTIN-B1A
IVERMECTIN
23512-6147 AVERMECTIN-B1B
IVERMECTIN
23512-6148 AVERMECTIN-B2A
IVERMECTIN
23512-6149 AVERMECTIN-B2B
IVERMECTIN
23512-8170 B-41-D
MILBEMYCIN-D
23513-7193 ANTLERMICIN-A
DC-11
TETROCARCIN-A
23513-7523 ANTLERMICIN-A
TETROCARCIN-A
23513-7647 DC-11-A2
TETROCARCIN-B
23513-7648 DC-11-A3
TETROCARCIN-C
23513-7649 ANTLERMICIN-B

23513-7650 ANTLERMICIN-C
23513-8517 KIJANIMICIN
KIJANMYCIN
SCH-25663
23513-8788 DC-11-A3
TETROCARCIN-C
23520-0922 CYTOCHALASIN-E
23520-8502 CYTOCHALASIN-L
23520-8503 CYTOCHALASIN-M
23530-0925 BREFELDIN-A
23530-0929 RADICICOL
23530-0930 MONORDEN
23530-6718 F-2 TOXIN
FES
ZEARALENONE
23530-6719 ZEAENOL
23530-7559 HYPOTHEMYCIN
23530-7560 TRANS-RESORCYLIDE
23530-7641 BREFELDIN-C
7-DEOXYBREFELDIN-A
23530-7642 DECHLOROMONORDEN
MONOCILLIN-I
23530-7643 MONOCILLIN-III
23530-7644 MONOCILLIN-II
23530-7645 MONOCILLIN-IV
23530-7646 MONOCILLIN-V
23530-8504 +-DIHYDROHYPOTHEMYCIN
23540-0932 CYTOCHALASIN-D
23540-0940 CHAETOGLOBOSIN-A
23540-0941 CHAETOGLOBOSIN-B
23540-3325 ASPOCHALASIN-B
ASPOSTEROL
23540-5035 CHAETOGLOBOSIN-C
23540-5036 CHAETOGLOBOSIN-D
23540-5037 CHAETOGLOBOSIN-E
23540-5038 CHAETOGLOBOSIN-F
23540-6753 ASPOCHALASIN-A
23540-6754 ASPOCHALASIN-B
ASPOSTEROL
23540-6755 ASPOCHALASIN-C
23540-6756 ASPOCHALASIN-D
23540-7561 CHAETOGLOBOSIN-K
23540-8066 ENGLEROMYCIN
23540-8505 ACETYLCHAETOGLOBOSIN-A
CYTOCHALASIN-K
23550-6893 CP-47444
NARGENICIN-A1
47444
23550-7640 NODUSMICIN
23550-8121 CP-51467
NARGENICIN-B1
23550-8122 CP-52726
NARGENICIN-B2
23550-8123 CP-52748
NARGENICIN-B3

24 MACROLACTAM ANTIBIOTICS

24110-0957 RIFAMYCIN-B
24110-0966 RIFAMYCIN-S
24110-0981 TOLYPOMYCIN-R
24110-0982 TOLYPOMYCIN-RB
24110-4113 B-2847-A"L
24110-5039 RIFAMYCIN-P
24110-5040 RIFAMYCIN-Q
24110-5041 RIFAMYCIN-R
24110-6150 RIFAMYCIN-VERDE
24110-7425 PROANSAMYCIN B-M1
24110-7916 3-HYDROXYRIFAMYCIN-S
24110-7917 3.31-DIHYDROXYRIFAMYCIN-S
24110-7918 1-DEOXY-1-OXARIFAMYCIN-S
16.17-DEHYDRORIFAMYCIN-G
24110-8124 8-DEOXYRIFAMYCIN-S
24110-8125 8-DEOXY-3-HYDROXYRIFAMYCIN-S
24110-8126 8-DEOXYRIFAMYCIN-B
24120-5044 DAMAVARICIN-C
24120-5045 DAMAVARICIN-D
24120-7515 STREPTOVARICIN-U
24120-8127 CP-50833
21-HYDROXY-25-DEMETHYL-25-METHYLTHIOPROTOSTREPTOVARICIN
24120-8128 ACTAMYCIN
24130-0990 GELDANAMYCIN
24130-6493 MACBECIN-I
24130-6494 MACBECIN-II
24130-6757 AM-3672
HERBIMYCIN
HERBIMYCIN-A
24130-7101 17-O-DEMETHYLGELDANAMYCIN
24130-8067 HERBIMYCIN-B
24130-8789 ANSATRIENIN-A
24130-8790 ANSATRIENIN-B
24140-1289 NAPHTOMYCIN
RO-7-7961
24210-5961 ANSAMITOCIN-P2
C-15003-P2
24210-5962 ANSAMITOCIN-P1
C-15003-P1
24210-5963 ANSACRIN
ANSAMITOCIN-P3′
C-15003-P3′
24210-5964 ANSACRIN
ANSAMITOCIN-P3
C-15003-P3
24210-5965 ANSACRIN
ANSAMITOCIN-P4
C-15003-P4
24210-8129 C-15003 PND-1
N-DEMETHYLANSAMITOCIN-P-1
24210-8130 C-15003 PND-2
N-DEMETHYLANSAMITOCIN-P-2
24210-8131 C-15003 PND-3
N-DEMETHYLANSAMITOCIN-P-3
24210-8132 C-15003 PND-4
N-DEMETHYLANSAMITOCIN-P-4
24210-8506 ANSAMITOCIN-PHM-1
C-15003-PHM-1
26-HYDROXYANSAMITOCIN-P1
24210-8507 ANSAMITOCIN-PHM-2

C-15003-PHM-2
24210-8508 ANSAMITOCIN-PHM-3
C-15003-PHM-3
24210-8509 ANSAMITOCIN-PHM-4
C-15003-PHM-4
24210-8510 ANSAMITOCIN-PHO-3
C-15003-PHO-3
DESACETYLMAYTANBUTACIN
15-HYDROXYASAMITOCIN-P3
24210-8511 EPI-ANSAMITOCIN-PHO-3
EPI-C-15003-PHO-3
24210-8512 ANSAMITOCIN-PND-4B"HY
C-15003-PND-4-B"HY
24210-8513 ANSAMITOCIN-P-4-B"HY
C-15003-P-4-B"HY
24210-8514 ANSAMITOCIN-P-4-G"HY
C-15003-P-G"HY
24210-8515 C-15003-QND-0
24210-8516 C-15003-DECL-QND-0
24220-1307 RUBRADIRIN
RUBRADIRIN-A
24220-6351 RUBRADIRIN-B
24220-6720 RUBRADIRIN-C
24220-7102 CP-48926
24220-7103 CP-48927
24300-8172 STUBOMYCIN

30000-1453 768
30000-3892 CRATERIFERMYCIN
CRATERIFERMYCIN COMPLEX
30000-6700 SF-1971
30000-6758 XK-99
30000-7653 SF-2012
30000-7654 C-15462-A
30000-7655 C-15462-V
30000-7656 VETICILLOMYCIN
30000-7922 CRATERIFERMYCIN-A
30000-7923 CRATERIFERMYCIN-B
30000-7924 CRATERIFERMYCIN-C
30000-7925 CRATERIFERMYCIN-D

31 TETRACYCLIC COMPOUNDS AND ANTHRAQUINONES

31111-5979 DOXYCYCLINE
31111-7104 6-DEOXYTETRACYCLINE
31111-7926 4-N-DEMETHYL-4-N-ETHYL-
6-DEMETHYLTETRACYCLINE
31112-1011 CHELOCARDIN
31112-6491 ISOCHELOCARDIN
31120-1012 PILLAROMYCIN-A
31130-7657 VEROTETRONE
31200-1423 54
31200-7105 MYCETIN-C
31211-1035 G"-RHODOMYCIN-I
IREMYCIN
31211-1036 G"-RHODOMYCIN-II
ROSEORUBICIN-B
31211-1039 G"-RHODOMYCIN-
ROA.DEOFUC.ROD
RHODOMYCIN-Y
31211-1041 G"-RHODOMYCIN-ROA2.ROD
31211-1043 RETAMICINA
RETAMYCIN
31211-1047 B-5794
G"-RHODOMYCIN COMPLEX
31211-1048 ISORHODOMYCIN-B
VIOLAMYCIN-A
VIOLAMYCIN-A1
31211-1049 VIOLAMYCIN-B1
31211-6721 A-43615
G"-RHODOMYCIN-I
IREMYCIN
31211-6759 ROSEORUBICIN-A
31211-6760 G"-RHODOMYCIN-II
ROSEORUBICIN-B
31211-7927 RHODOMYCIN-COMPLEX
31211-7928 VIOLAMYCIN-A2
31211-7929 VIOLAMYCIN-A3
31211-7930 VIOLAMYCIN-A4
31211-7931 VIOLAMYCIN-A5
31211-7932 VIOLAMYCIN-A6
31211-7933 VIOLAMYCIN-B-I2
VIOLAMYCIN-B2
31211-7934 VIOLAMYCIN-B3
31211-7935 VIOLAMYCIN-B4
31211-7936 VIOLAMYCIN-B5
31211-7937 VIOLAMYCIN-B6
31211-8133 G-6
11-HYDROXYCINERUBIN-A
31211-8134 G-7
10-DECARBOMETHOXY-10.11-
DIHYDROX
YACLACINOMYCIN-A
31211-8135 G-7
4"-DEHYDRORHODOMYCIN-Y
31211-8136 G-9
10-DECARBOMETHOXY-10-
HYDROXYCINERUBIN-A
31212-1058 CINERUBIN-A
G-4
5888-III
31212-5668 C-36145
MUSETTAMYCIN
NSC-219941
NSC-284671
PYRROCYCLINE-A
31212-5669 MARCELLOMYCIN
PYRROCYCLINE-B
31212-5816 RHODIRUBIN-A
31212-5876 RHODIRUBIN-B
31212-5975 MA-144-S2
MUSETTAMYCIN
RHODIRUBIN-D
31212-6603 NSC-267695
NSC-293858
PYRROCYCLINE-C
RUDOLPHOMYCIN
31212-7106 ALCINDOROMYCIN
N-MONODEME
THYLMARCELLOMYCIN
31212-7107 COLLINEMYCIN
10-EPI-MUSETTAMYCIN
31212-7108 MIMIMYCIN
10-EPI-MARCELLOMYCIN
31212-7938 TRYPANOMYCIN COMPLEX
31212-8137 G-5
11-HYDROXYACLACINOMYCIN-

	A
31212-8518	1-HYDROXYAURAMYCIN-A
31212-8519	1-HYDROXYAURAMYCIN-B
31212-8520	1-HYDROXYSULFURMYCIN-A
31212-8521	1-HYDROXYSULFURMYCIN-B
31213-1064	AKLAVIN
	5888-II
31213-4778	ACLACINOMYCIN-A
	ACLACINON
	ACLARUBICIN
	G-1
	NSC-208734
31213-5855	MA-144-T1
31213-5857	MA-144-L1
31213-5858	MA-144-M1
31213-5860	MA-144-G1
31213-5861	MA-144-N1
31213-5862	MA-144-S1
31213-6062	ACLACINOMYCIN-Y
31213-8068	MA-144-KH
	7-O-DAUNOS AMINYLAKLAVINONE
31213-8138	G-2
	10-DECARBO METHOXYACLACINOMYCIN-A
31213-8139	G-3
	4-O-METHYLACLACINOMYCIN-A
31213-8140	AURAMYCIN-A
31213-8141	AURAMYCIN-B
31213-8142	SULFURMYCIN-A
31213-8143	SULFURMYCIN-B
31213-8522	13-METHYLACLACINOMYCIN-A
31213-8523	2-HYDROXYACLACINOMYCIN-A
31214-1069	DAUNOMYCIN
31214-1070	ADRIAMYCIN
31214-1085	CARMINOMYCIN
	CARMINOMYCIN-1
31214-1086	CARMINOMYCIN-II
	CARMINOMYCIN-2
	D-326-III
	DF-4466-A
	RUBEOMYCIN-A
	4-HYDROXYBAUMYCIN-A2
31214-1087	CARMINOMYCIN-III
	CARMINOMYCIN-3
	D-326-IV
	DF-4466-B
	RUBEOMYCIN-A1
	4-HYDROXYBAUMYCIN-A1
31214-5323	DIHYDROCARMINOMYCIN
31214-5818	BAUMYCIN-A1
	BAUNOMYCIN
31214-6722	11-DEOXYDAUNOMYCIN
	11-DEOXYDAUNORUBICIN
31214-6723	11-DEOXYADRIAMYCIN
	11-DEOXYDOXORUBICIN
31214-6724	11-DEOXY-RP-20798
	11-DEOXY-13-DIHYDRODAUNOMYCIN
31214-6725	11-DEOXY-13-DEOXO-DAUNORUBIN
	11-DEOXY-13-DEOXODAUNOMYCIN
31214-7658	GP-I
	OODM
	1-HYDROXY-13-DIHYDRODAUNOMYCIN
31214-7659	FOODM
	GP-II
	N-FORMYL-1-HYDROXY-13-DIHYDRODAUNOMYCIN
31214-7660	COMP Y
	11-DEOXYCARMINOMYCIN
31214-7661	COMP Z
	11-DEOXY-13-DEOXOCARMINOMYCIN
31214-7662	COMP W
	11-DEOXY-14-HYDROXYCARMINOMYCIN
31214-7663	COMP X
31214-8069	GP-III
31214-8144	DOC
	13-DEOXYCARMINOMYCIN
31214-8173	CARMINOMYCIN-2
	D-326-III
	DF-4466-A
	FA-1180-A
	RUBEOMYCIN-A
31214-8174	CARMINOMYCIN-3
	D-326-IV
	DF-4466-B
	FA-1180-A1
	RUBEOMYCIN-A1
31214-8175	FA-1180-B
	RUBEOMYCIN-B
31214-8176	FA-1180-B1
	RUBEOMYCIN-B1
31214-8524	FEUDOMYCIN-A
	13-DEOXYDAUNOMYCIN
31214-8525	FEUDOMYCIN-B
31214-8526	D-326-I
	4-HYDROXYBAUMYCINOL-A2
31214-8527	D-326-II
	4-HYDROXYBAUMYCINOL-A1
31214-8528	CARMINOMYCIN-2
	D-326-III
	DF-4466-A
	RUBEOMYCIN-A
	4-HYDROXYBAUMYCIN-A2
31214-8529	CARMINOMYCIN-3
	D-326-IV
	DF-4466-B
	RUBEOMYCIN-A1
	4-HYDROXYBAUMYCIN-A1
31214-8791	CARMINOMYCIN-2
	DF-4466-A
	RUBOMYCIN-A
	4466-A
31214-8792	CARMINOMYCIN-3
	DF-4466-B
	RUBOMYCIN-A1
	4466-B
31221-1090	NOGALAMYCIN
31221-1098	BEROMYCIN-B
31221-1099	BEROMYCIN-C
31221-8530	VIRIPLANIN
31222-1121	ARANCIAMYCIN

SM-173-A
31222-1123 STEFFIMYCIN
31222-6211 SM-173-B
31222-7516 ARANCIAMYCIN
SM-173-A
31222-7939 U-58875
10-DIHYDROSTEFFIMYCIN-A
31222-7940 U-58874
10-DIHYDROSTEFFIMYCIN-B
31223-8842 A-12918
31230-6761 STEFFIMYCINOL
31230-7941 BISANHYDROAKLAVINONE
GALIRUBINONE-B1
MA-144-F
31230-8793 DC-44-A
31230-8794 DC-44-B
31240-6727 TETRACENOMYCIN-C
31240-8795 3-DEMETHYOXY-3-
ETHOXYTETRACENOMYCIN-C
31240-8796 TETRACENOMYCIN-A2
31240-8797 TETRACENOMYCIN-B1
31240-8798 TETRACENOMYCIN-B2
31240-8799 TETRACENOMYCIN-D
31300-4685 MADURAMYCIN
31310-1141 CLAVOXANTHIN
ENDOCROCIN
31310-4780 EMODIN
31310-5568 CYNODONTHIN
31310-6063 CATENARIN
31310-6604 CHRYSOROBIN
PARIETHIN
PHYSCION
31310-7110 CATENARIN-5-METHYLETHER
31310-7111 PHYSCION-9-ANTHRONE
31310-7112 CATENARIN-7-METHYLETHER
ERYTHROGLAUCIN
31320-1144 LUTEOSKYRIN
31320-5978 RUGULIN
31330-1149 AQUAYAMYCIN
31330-5329 792
31330-5555 SEN-136-A
YORONOMYCIN
31330-5966 OS-4742-A1
P-1894-B
VINENOMYCIN-A1
31330-5967 OS-4742-A2
VINENOMYCIN-A2
31330-5968 OS-4742-B1
VINENOMYCIN-B1
31330-5969 OS-4742-B2
VINENOMYCIN-B2
31330-7562 OS-4742-A1
P-1894-B
VINENOMYCIN-A1
31350-1157 PLURAMYCIN-A
31350-1160 DESACETYLPLURAMYCIN-A
RUBIFLAVIN
RUBIFLAVIN-A
31350-1162 HEDAMYCIN
31350-1170 KIDAMYCIN
KIDAMYCIN-F
RUBIFLAVIN-B
31350-5234 GRISEORUBIN COMPLEX
31350-7424 DC-14
31350-7942 KIDAMYCIN-E
289-E
31370-7943 VARIACYCLOMYCIN-A
31380-7664 VIOCRISTIN
31380-7665 ISOVIOCRISTIN

32 NAPHTHOQUINONES

32120-1186 JUGLOMYCIN-A
32120-1187 JUGLOMYCIN-B
32120-1199 FUSARUBIN
32120-1200 JAVANICIN
32120-1201 NOVARUBIN
32120-1202 NORJAVANICIN
32120-7944 WS-5995-A
32120-8070 WS-5995-B
32120-8531 TRICHIONE
32120-8532 HOMOTRICHIONE
32211-1213 BOSTRYCIN
32211-3539 DACTYLARIN
32211-6114 ALTERSOLANOL-B
32221-1234 ACTINORHODIN
32221-7945 PROACTINORHODIN
32222-1239 PURPUROMYCIN
4041-I
32222-1242 G"-NAPHTOCYCLINONE
32222-1243 B"-NAPHTOCYCLINONE
32222-1244 A"-NAPHTOCYCLINONE
32222-1247 GRISEORHODIN-A
32222-1248 GRISEORHODIN-B
32222-1249 GRISEORHODIN-C
32222-6115 GRISEORHODIN-C2
32222-6565 FCRC-57G
GRISEORHODIN-G
32222-6566 FCRC-57U
32223-1261 GRANATICIN
GRANATOMYCIN-C
32223-5410 DIHYDROGRANATICIN
GRANATOMYCIN-D
32223-6728 GRANATICINIC ACID
32223-7946 GRANATOMYCIN-A
METHYL-DIHYDROGRANATICIN
32224-1268 BIKAVERIN
32224-1269 NORBIKAVERIN
32224-1277 MARTICIN
32224-1278 ISOMARTICIN
32224-6762 O-ETHYLFUSARUBIN
32224-6763 DIHYDROFUSARUBIN
32224-6764 3-O-ETHYLD
IHYDROFUSARUBIN
32225-1267 KALAFUNGIN
32225-1271 DEOXYFRENOLICIN
DESOXYFRENOLICIN
32225-1272 FRENOLICIN
32225-3849 NANAOMYCIN-A
32225-3850 NANAOMYCIN-B
32225-5048 NANAOMYCIN-D
32225-5049 GRISEUSIN-A
32225-6352 FRENOLICIN-B
32225-6765 NANAOMYCIN-E
32226-1270 XANTHOMEGNIN
32226-1279 LAMBERTELLIN
32226-8177 3.4-DEHYDROXANTHOMEGNIN
32227-7113 GUNACIN

32228-8633 FCRC-A-48-A
FREDERICAMYCIN-A
32230-1284 BOSTRYCOIDIN
32310-1421 MEDERMYCIN
32320-1300 XANTHOMYCIN
32320-5549 SF-1739
32320-6766 A-5945
32320-8533 XANTHOMYCIN-LIKE

33 BENZOQUINONES

33110-6492 490-QUINONE
33140-1334 COMPOUND (NO NAME) 6663
PLEUROTIN
P1
33140-5327 ASTERRIQUINONE
33140-6116 MYCORRHIZIN
33140-6117 CHLOROMYCORRHIZIN
33140-8145 PLEUROTIN
P1
33140-8534 ASTERRIQUINONE A-1
DIMETHYLASTERRIQUINONE
33140-8535 ASTERRIQUINONE A-2
33140-8536 ASTERRIQUINONE A-3
33140-8537 ASTERRIQUINONE A-4
33140-8538 ASTERRIQUINONE B-1
33140-8539 ASTERRIQUINONE B-2
33140-8540 ASTERRIQUINONE B-3
33140-8541 ASTERRIQUINONE B-4
33140-8542 ASTERRIQUINONE C-1
33140-8543 ASTERRIQUINONE-C-2
33140-8544 ASTERRIQUINONE-D
33150-6729 SARCINAMYCIN-A
SARUBICIN-A
U-58431
33150-7947 SARUBICIN-A
U-58431
33211-1336 MITOMYCIN-B
33211-1337 MITOMYCIN-C
33211-1344 PORFIROMYCIN
33211-7114 7-DEMETHOXY-7-
AMINOMITOMYCIN-B
33211-7667 7-DEMETHOXY-7-AMINO-9A-
O-METHYLMITOMYCIN-B
33211-7668 9A-O-METHYLMITOMYCIN-B
33211-7669 1A-N-METHYLMITOMYCIN-A
33211-7670 -DEHYDRO-7-
DEMETHYLMITOMYCIN-B
MITOMYCIN-G
10-DECARBAMOYLOXY-9-
DEHYDROPORFIROMYCIN
7-AMINO-9A-O-METHYL-10-
DECARBAMOYLOXY-9
33211-7671 MITOMYCIN-K
10-DECARBAMOYLOXY-9-
DEHYDRO-N-
METHYLMITOMYCIN-A
9A-O-METHYL-10-
DECARBAMOYLOXY-9-
DEHYDROMITOMYCIN-B
33211-7672 MITOMYCIN-H
10-DECARBAMOYLOXY-9-
DEHYDROMITOMYCIN-B
33212-7115 49-A
33220-1350 STREPTONIGRIN
33220-7517 K-82-A
LAVENDAMYCIN
33220-8545 AB-111
33220-8800 K-82-A
LAVENDAMYCIN
33240-5252 MIMOSAMYCIN
33240-7673 MIMOCIN
33250-5253 CHLOROCARCIN-A
33250-5254 CHLOROCARCIN-B
33250-5255 CHLOROCARCIN-C
33250-5970 SAFRAMYCIN-A
21-CYANOSAFRAMYCIN-B
33250-5971 SAFRAMYCIN-B
33250-5972 SAFRAMYCIN-C
33250-5973 SAFRAMYCIN-D
33250-5974 SAFRAMYCIN-E
33250-7674 DECYANOSAFRAMYCIN-A
SAFRAMYCIN-S
33260-6730 C-14482-A1
DNACTIN-A1
33260-7919 C-14482-B1
DNACTIN-B1
33260-7920 C-14482-B2
DNACTIN-B2
33260-7921 C-14482-B3
DNACTIN-B3

34 QUINONE-LIKE COMPOUNDS

34110-1363 ASCOCHYTIN
34110-1367 CITRININ
34110-8801 DEFLECTIN-1A
34110-8802 DEFLECTIN-A
DEFLECTIN-1B
34110-8803 DEFLECTIN-1C
34110-8804 DEFLECTIN-B
DEFLECTIN-2A
34110-8805 DEFLECTIN-2B
34132-1388 ATROVENTIN
34132-1389 HERQUEICHRYSIN
34133-6767 NEOCERCOSPORIN
NEOSPORIN
34133-6769 CERCOSPORIN
34210-1392 EPOXYDON
34210-1396 DESOXYEPOXYDON
EPOFORMIN
34210-1397 TERREIC ACID
Y-8980
34210-1399 TERREMUTIN
34210-1400 PHYLLOSTIN
PHYLLOSTINE
34210-1401 PANEPOXYDONE
34210-1402 PANEPOXYDIONE
34210-6118 ANTIPHENICOL
34210-6768 +-ISOEPOXYDON
U-III
34210-8146 CORILOXIN

40000-4096 AQUINOMYCIN
40000-5059 GARDIMYCIN
40000-5197 FR-3383
40000-5198 NRC-501
40000-5455 S-19

40000-5536	SC-4
40000-5574	MONOKETOORGANOMYCIN
40000-5993	AM-2504
40000-6125	BN-192
40000-6219	SYRINGOTOXIN
40000-6798	ALTERICIDIN-A
40000-6799	ALTERICIDIN-B
40000-6800	ALTERICIDIN-C
40000-7130	MICROCIN-140
40000-7158	A-38533-B
40000-7159	A-38533-A1
40000-7201	AB-102
40000-7431	A-38533-A2
40000-7432	A-38533-C
40000-7675	ANSLIMIN-A
40000-7676	ANSLIMIN-B
40000-7677	ALBOLEUTIN
40000-7680	RHODOTORUCIN-A
40000-8071	SF-2068
40000-8072	CYTOPHAGIN
40000-8272	F-16-2
40000-8273	O-28 O2-8
40000-8584	BN-240-B
40000-8639	AF-8
40000-8839	SH-50

41 AMINO ACID DERIVATIVES

41110-1454	AZASERIN AZASERINE 40816
41110-1455	DON
41110-1457	DUAZOMYCIN-B
41121-1460	ALANOSINE L-ALANOSINE NSC-153353
41121-1464	ARMENTOMYCIN
41121-1474	L-2-AMINO-4-METHOXY-TRANS-BUT-3-ENOIC ACID L-2-AMINO-4-METHOXY-TRANS-3-BUTENOIC ACID
41121-7117	AMHA L-2-AMINO-5-METHYL-5-HEXENOIC ACID
41121-7518	FR-900130 L-2-AMINO-3-BUTYNOIC ACID
41121-8148	AL-719 G"-CHLORONORVALINE
41122-1466	L-2-AMINO-4-2-AMINOETHOXY-3-BUTENOIC ACID
41122-1477	I-677 L-4-OXALYSINE
41122-8546	L-4-OXALYSIN 1-667
41123-1483	DHPA L-2.5-DIHY DROPHENYLALANINE
41123-6772	B"-AMINOXY-D-ALANINE
41123-8085	L-2-1-METHYLCYCLOPROPYL-GLYCINE PA-4046-I
41123-8149	AL-719-Y G"-HYDROXY NORVALINELACTONE
41124-1486	L-AZETIDINE-2-CARBOXYLIC ACID
41124-1488	ACIVICIN U-42126
41124-1491	CYCLOSERINE
41124-1492	FURANOMYCIN
41124-6221	SF-1835 SF-1836
41125-6119	FORPHENICINE FORPHENICOL
41211-1497	6-AMINOPENICILLANIC ACID
41211-1506	PENICILLIN-N
41211-1507	ISOPENICILLIN-N
41211-7681	"PENICILLIN LIKE FACTOR"
41212-1508	CEPHALOSPORIN-C
41212-1510	DESACETYLCEPHALOSPORIN-C
41212-1511	CEPHAMYCIN-A
41212-1512	CEPHAMYCIN-B
41212-1513	CEPHAMYCIN-C S-3907C-2 SF-1584
41212-1518	WS-3442-D
41212-1519	WS-3442-E
41212-5054	OGANOMYCIN-D3 Y-G19Z-D3 7-METHOXYD EACETYLCEPHALOSPORIN-C
41212-5278	OGANOMYCIN-D3 7-METHOXYD EACETYLCEPHALOSPORIN-C
41212-5671	OGANOMYCIN-F Y-G19Z-F
41212-5672	OGANOMYCIN-G Y-G19Z-G
41212-5673	OGANOMYCIN-H Y-G19Z-H
41212-5674	OGANOMYCIN-I Y-G19Z-I
41212-5874	OGANOMYCIN-D2
41212-7440	XK-201-IV
41212-7949	"COMPOUND A" OGANOMYCIN-A
41212-7950	"COMPOUND-C" OGANOMYCIN-B
41212-8086	PA-32413-I
41212-8547	S-3907C 3
41212-8548	S-3907C 4B
41213-5186	NOCARDICIN-A
41213-5675	NOCARDICIN-E
41213-5676	NOCARDICIN-F
41213-5875	NOCARDICIN-C
41213-5876	NOCARDICIN-D
41213-5980	NOCARDICIN-G
41214-2262	"OLIVANIC ACID" MM-4550
41214-4253	MC696-SY2-A
41214-4800	EPITHIENAMYCIN-E MM-13902
41214-4902	EPITHIENAMYCIN-A MM-22380 MSD-890-A1

	17927-A1
41214-5056	THIENAMYCIN
41214-5370	EPITHIENAMYCIN-F
	MM-17880
41214-5677	N-ACETYL-THIENAMYCIN
41214-5982	EPITHIENAMYCIN-B
	MM-22382
	MSD-890-A2
	17927-A2
41214-5983	EPITHIENAMYCIN-D
	MM-22383
	MSD-980-A5
	PS-4
41214-6080	PS-5
41214-6120	EPITHIENAMYCIN-E
	MSD-890-A9
41214-6121	EPITHIENAMYCIN-F
	MSD-890-A10
41214-6223	EPITHIENAMYCIN-A
	MM-22380
	17927-A1
41214-6224	EPITHIENAMYCIN-C
	MM-22381
	PS-3
41214-6225	EPITHIENAMYCIN-B
	MM-22382
	17927-A2
41214-6226	EPITHIENAMYCIN-D
	MM-22383
	PS-4
41214-6227	EPITHIENAMYCIN-A
	MM-22380
	17927-A1
41214-6228	EPITHIENAMYCIN-B
	MM-22382
	17927-A2
41214-6265	EPITHIENAMYCIN-C
	MM-22381
	MSD-890-A3
	PS-3
41214-6568	N-ACETYL-D EHYDROTHIENAMYCIN
41214-6773	DIHYDRO-PS-8
	PS-6
41214-6774	PS-7
41214-7118	SF-2050
41214-7119	SF-2050-B
41214-7684	17927-D
41214-7685	EPITHIENAMYCIN-C
	MM-22381
	PS-3
	890-A1
	890-A3
41214-7686	EPITHIENAMYCIN-D
	MM-22383
	PS-4
	890-A5
41214-7687	DESACETYL-PS-5
	DESHYDROXYTHIENAMYCIN
	NS-5
41214-7951	C-19393-H2
	CARPETIMYCIN-A
	S-19393-H2
41214-7952	C-19393-S2
	CARPETIMYCIN-B
41214-8150	C-19393-H2
	CARPETIMYCIN-A
	KA-6643-A
41214-8151	C-19393-S2
	CARPETIMYCIN-B
	KA-6643-B
41214-8152	ASPERENOMYCIN-A
	PA-31088-IV
41214-8178	NORTHIENAMYCIN
41214-8549	PS-8
41214-8550	ASPERENOMYCIN-B
	PA-39504-X1
41215-5057	AUGMENTIN (WITH AMPICILLIN)
	CLAVULANIC ACID
41215-6610	FORMYLOXYMETHYLCLAVAM
41215-6611	CLAVAM-2-CARBOXYLIC ACID
41215-7953	2-3-ALANYL-CLAVAM
41215-8179	2-2-HYDROXYETHYL-CLAVAM
41216-6895	G-6302
	MONOBACTAM-I
	SULFAZECIN
41216-7771	ISOSULFAZECIN
	SB-72310
41216-8551	EM-5210
	MONOBACTAM-I
	SQ-26445
	SULFAZECIN
41216-8552	E-5117
	MONOBACTAM-III
	SQ-26180
41216-8553	MONOBACTAM-VII
41216-8554	MONOBACTAM-VIII
41216-8555	MONOBACTAM-IX
41216-8556	MONOBACTAM-X
41216-8557	MONOBACTAM-XI
41220-1523	AUREOTHRICIN
41220-1524	HOLOMYCIN
	MM-21801
41220-1525	THIOLUTIN
41220-1528	ISOBUTYRROPYRROTHIN
41230-1534	LYDIMYCIN
41310-1540	PIPERAZINEDIONE
41310-8558	CAIROMYCIN-A
41321-1543	GLIOTOXIN
41321-5259	EPICORAZINE A
41321-6775	EPICORAZINE-B
41321-7122	BISDETHIO-BIS-METHYLTHIOGLIOTOXIN
41322-1554	CHAETOCIN
41322-1563	CHAETOMIN
	CHETOMIN
41323-6495	SPORIDESMIN-H
41324-1571	HYALODENDRIN
41325-7120	SIRODESMIN-PL
41325-7121	DEACETYLSIRODESMIN-PL
41330-1584	NEOASPERGILLIC ACID
41330-8559	ASTERCHROME

42 HOMOPEPTIDES

42111-1598	DISTAMYCIN-A
42112-1601	AMIDINOMYCIN

MYXOVIROMYCIN
42113-2630 SF-98
42120-1609 NEGAMYCIN
42120-1610 LEUCYLNEGAMYCIN
42120-1612 MALONOMYCIN
42120-1617 TETAINE
42120-1618 BACILYSIN
KM-208
42120-1624 SF-1293
42120-4940 E-64
42120-5989 BESTATIN
42120-6776 PHEGANOMYCIN
42120-6777 PHEGANOMYCIN-D
42120-6778 PHEGANOMYCIN-DR
42120-6779 PHEGANOMYCIN-DGPT
42120-6780 DEOXYPHEGANOMYCIN-D
42120-7439 MALIOXAMYCIN
42120-7688 AJ-9406
N-6-DEOXY-L-TALOSYLOXY-HYDROXYPHOSPHINYL-L-LEUCYL-L-TRYPTOPHANE
42120-7689 FR-900137
42120-7690 FR-900148
42120-8087 N-2.6-DIAMINO-6-HYDROXYMETHYLPIMELYL-ALANINE
42120-8560 CAIROMYCIN-C
42120-8561 FK-156
FR-900156
42130-1627 BICOZAMYCIN
BICYCLOMYCIN
42140-6703 CC-1065
NSC-298223
RACHELMYCIN
42140-6770 CC-1065
NSC-219877
RACHELMYCIN
42210-1632 GRAMICIDIN-A
42220-1639 EDEIN-A1
EDEINE-A
42220-1640 EDEIN-B1
EDEINE-B
42220-1641 EDEIN-D
42220-3898 GALANTIN-II
42220-3899 GALANTIN-I
42220-6122 EDEIN-F
42220-6781 K-582-A
K-582M-A
MYRORIDIN-K
42220-6782 K-582-B
K-582M-B
42220-7691 TATUMINE
42230-8153 "PENTADECAPEPTIDE"
42240-6789 IMACIDIN-B
42240-6790 IMACIDIN-C
42240-6791 IMACIDINIC ACID-C
42250-1697 ALAMETHICIN-I
42250-1869 EMERIMICIN-II
42250-1872 ZERVAMICIN-I
42250-5193 SAMAROSPORIN
42250-6071 ANTIAMOEBIN-II
42250-6930 HYPELCIN-A
42250-6931 HYPELCIN-B
42260-3905 CEREXIN-A
42260-3906 CEREXIN-B
42260-5424 CEREXIN-C
42260-5425 CEREXIN-D
42260-8154 CEREXIN-B2
42260-8155 CEREXIN-B3
CEREXIN-B4
42260-8156 CEREXIN-D2
42260-8157 CEREXIN-D3
CEREXIN-D4
42260-8203 LEUPEPTIN-PRLL
42270-1877 LEUCINOSTATIN-A
42270-2100 ICI-13595
42270-5994 TRICHOPOLIN-A
TRICHPOLYN-I
42270-5995 TRICHOPOLIN-B
TRICHOPOLYN-II
42270-8158 P-168
1907-VIII
42270-8159 1907-II
42270-8160 P-168
1907-VIII
42300-1689 CHLAMYDOCIN
42300-6783 TERNATIN
42310-1669 GRAMICIDIN-S
42310-1706 GRATIZIN
42320-1673 BACITRACIN-A
42320-4393 GALLERIN
26A
42330-1677 VIOMYCIN
XK-33-F-3
42330-1680 CAPREOMYCIN-I-A
42330-1685 TUBERACTINOMYCIN-N
42330-5682 LL-BM-547-B"
42350-4873 CYCLOSPORIN-A
42350-5060 CYCLOSPORIN-C
42350-5575 CYCLOSPORIN-D
42350-7692 CYCLOSPORIN-G

43 HETEROMER PEPTIDES

43100-6784 MANILOSPORIN-A
43100-6785 MANILOSPORIN-B1
43100-6786 MANILOSPORIN-B2
43100-7678 HERBICOLIN-A
43100-7679 HERBICOLIN-B
43110-1714 AMPHOMYCIN
33B
43110-1716 LASPARTOMYCIN
RP-18887
43110-4052 LASPARTOMYCIN
RP-18887
43110-7954 PARVULIN-B
43110-7955 PARVULIN-C
43110-8180 33-A
43121-1722 POLYMIXIN-A
43121-1723 POLYMIXIN-M
43121-1724 POLYMYXIN-B1
43121-1725 POLYMIXIN-B2
43121-1726 POLYMYXIN-C
POLYMYXIN-C COMPLEX
POLYMYXIN-C1
POLYMYXIN-P
43121-1727 POLYMIXIN-D1
43121-1728 COLISTIN-A

	POLYMIXIN-E1
43121-5991	POLYMYXIN-F1
43122-1742	OCTAPEPTIN-A2
43122-1743	OCTAPEPTIN-A1
43122-1744	OCTAPEPTIN-B2
43122-1745	OCTAPEPTIN-B1
43122-6792	AM-157
43122-6793	OCTAPEPTIN-A4
43122-6794	OCTAPEPTIN-B4
43122-7123	OCTAPEPTIN-D
43123-6795	NLF-II
	PERMETIN-A
	PERMYCIN-A
43123-7124	PELMYCIN
	PERMETIN-B
	PERMYCIN-B
43124-5426	TRIDECAPTIN-AA"
43124-6229	TRIDECAPTIN-BA"
43124-6230	TRIDECAPTIN-CA"1
43124-8181	TRIDECAPTIN-CA"2
43124-8182	TRIDECAPTIN-CB"1
43124-8183	TRIDECAPTIN-BB"
	TRIDECAPTIN-BG"
43124-8184	TRIDECAPTIN-BD"
43130-1749	A-30912-A
	ECHINOCANDIN-B
43130-3902	ACULEACIN COMPLEX
43130-5064	A-30912-B
	ECHINOCANDIN-C
43130-5065	A-30912-C
	ECHINOCANDIN-D
43130-5066	ACULEACIN-A
43130-5429	S-31794-F-1
43130-5685	A-30912-A
	ECHINOCANDIN-B
	SL-7810-F
43130-5686	A-30912-B
	ECHINOCANDIN-C
	SL-7810-FII
43130-5687	A-30912-C
	ECHINOCANDIN-D
	SL-7810-FIII
43130-8185	A-30912-H
	A-42355
43140-1650	"RAUBITSCHEK SUBSTANCE"
	BACILLOMYCIN-D
43140-1663	MYCOSUBTILIN
43140-1750	ITURIN-A
43140-5639	BACILLOMYCIN-L
43140-6787	MANILOSPORIN-C1
43140-6788	MANILOSPORIN-C2
43140-8074	"RAUBITSCHEK SUBSTANCE"
	BACILLOMYCIN-D
43200-5990	MC-902-I
43200-6127	MC-902-I′
43200-8562	A-7413-D
43200-8601	L-31-A
43200-8602	L-31-B
43211-1752	THIOSTREPTON
43211-1754	SIOMYCIN-A
43211-1757	SIOMYCIN-C
43211-1758	THIOPEPTIN-A1A
43211-1759	THIOPEPTIN-A2A
43211-1760	THIOPEPTIN-A3A
43211-1761	THIOPEPTIN-A4A
43211-1762	THIOPEPTIN-BA
43211-1771	ACTINOTHIOCIN
43211-1980	TAITOMYCIN COMPLEX
43211-1981	TAITOMYCIN-B
43211-1982	3354-1
43211-1983	3354-1
43211-1984	1542-19
43211-1985	RP-9671
43211-5534	NOSIHEPTID
43211-5572	SCH-18640
43211-6235	THIOPEPTIN-BB
43211-6236	THIOPEPTIN-A1B
43211-6237	THIOPEPTIN-A3B
43211-6238	THIOPEPTIN-A4B
43211-6796	RP-35665
43211-6801	A-6984-B
	PEPTIDE-B
43211-7125	A-6984-C
	PEPTIDE-C
43211-7126	A-6984-D
	PEPTIDE-D
43211-7127	A-10947
43211-7129	SYRIAMYCIN
43211-7427	S-54832-A-I
43211-7428	S-54832-A-II
43211-7429	S-54832-A-III
43211-7430	S-54832-A-IV
43211-7693	CINROPEPTIN
	TSINROPEPTIN
43211-8186	THIOSTREPTON-B
43211-8187	SIOMYCIN-D1
43211-8188	PLANOTHIOCIN-A
43211-8189	PLANOTHIOCIN-B
43211-8190	PLANOTHIOCIN-C
43211-8191	PLANOTHIOCIN-D
43211-8192	PLANOTHIOCIN-E
43211-8193	PLANOTHIOCIN-F
43211-8194	PLANOTHIOCIN-G
43212-1767	ALTHIOMYCIN
43213-8195	JINGSIMYCIN
43214-5194	THIOCILLIN-I
43214-5195	THIOCILLIN-II
43214-5196	THIOCILLIN-III
43220-1773	BOTTROMYCIN-A
43220-1774	BOTTROMYCIN-A2
43220-1775	BOTTROMYCIN-B1
43220-1777	BOTTROMYCIN-C2
43230-1782	SULFOMYCIN-I
43230-1785	BERNINAMYCIN-A
43311-6898	216
43311-7959	SF-2012-L
43315-6797	DESFERRITR IACETYLFUSIGEN-LIKE V
43315-7694	TRIORNICINE-A
43320-1826	BLEOMYCIN COMPLEX
43320-1828	BLEOMYCIN-A2
	ZHENGGUANGMYCIN-A2
43320-1835	BLEOMYCETIN
	BLEOMYCIN-A5
43320-1838	BLEOMYCIN-B2
	ZHENGGUANGMYCIN-B2
43320-1848	PHLEOMYCIN-G

43320-5431 TALLYSOMYCIN-A
43320-5432 TALLYSOMYCIN-B
43320-5689 SF-1771
SF-1771-A
43320-5690 CU-FREE
SF-1771
SF-1771-B
43320-6353 PEP-BLEOMYCIN
43320-6618 SF-1961-B
43320-6619 SF-1961-A
43320-7128 3-S-1'-PHENYLETYLAMINO-PROPILAMINO-BLEOMYCIN
43320-7573 BLEOMYCIN-A5
PINGYANMYCIN
43320-7956 BLEOMYCIN-B1
43320-7957 CLEOMYCIN-B2
43320-7958 MC-637-SY-1
43320-8196 CLEOMYCIN-A2
CLEOMYCIN-1
43320-8197 CLEOMYCIN-2
4-GUANIDIN OBUTYLAMINOCLEOMYCIN
43320-8198 CLEOMYCIN-6
3-4-AMINOBUTYLAMINO-PROPYLAMINOCLEOMYCIN
43320-8199 CLEOMYCIN-11
3-3-N-BUTYLAMINO-PROPYLAMINO-PROPYLAMINOCLEOMYCIN
43320-8200 CLEOMYCIN-20
3-N-METHYL-N-AMINOPRO PYLAMINOCLEOMYCIN
43320-8201 CLEOMYCIN-33
3-S-R-PHENYLETHYAMINO-PROPYLAMINOCLEOMYCIN
43320-8563 TALLYSOMYCIN-S1A
43320-8564 TALLYSOMYCIN-S2A
43320-8565 TALLYSOMYCIN-S6A
43320-8566 TALLYSOMYCIN-S8A
43320-8567 TALLYSOMYCIN-S10A
43320-8568 TALLYSOMYCIN-S11A
43320-8569 TALLYSOMYCIN-S12A
43320-8570 TALLYSOMYCIN-S1B
43320-8571 TALLYSOMYCIN-S2B
43320-8572 TALLYSOMYCIN-S3B
43320-8573 TALLYSOMYCIN-S4B
43320-8574 TALLYSOMYCIN-S5B
43320-8575 TALLYSOMYCIN-S6B
43320-8576 TALLYSOMYCIN-S7B
43320-8577 TALLYSOMYCIN-S8B
43320-8578 TALLYSOMYCIN-S9B
43320-8579 TALLYSOMYCIN-S11B
43320-8580 TALLYSOMYCIN-S12B
43320-8581 TALLYSOMYCIN-S13B
43320-8582 TALLYSOMYCIN-S14B
43320-8583 TALLYSOMYCIN-S10B
43330-1963 VIRIDOMYCIN-A
43330-8202 U-60394

44 PEPTOLIDES

44110-1818 GLUCONIMYCIN
44110-1895 ACTINOMYCIN-IV
44110-1904 ACTINOMYCIN-C COMPLEX
KENGSHENGMYCIN
44110-1912 ACTINOLEVALINE
44110-1913 ACTINOLEUCINE
44110-1925 ACTINOMYCIN-PIP-1B"
44110-1940 ETABETACIN
44110-4149 ACTINOMYCIN-C COMPLEX
KENGSHENGMYCIN
44110-6281 TOXIFERTILIN
44110-6802 ACTINOMYCIN COMPLEX 41156
44110-6803 ACTINOMYCIN COMPLEX
44110-7960 ACTINOLEUCIN AY-5
AU-5
AY-5
44110-8585 AU-3B
AY-3B
N.N-DIDEME THYLACTINOMYCIN-D
44120-1963 TRIOSTIN-A
44120-1972 QUINONMYCIN-B
44120-1973 QUINONMYCIN-C
44120-1974 QUINONMYCIN-A
44120-7695 BBM-928-A
LUZOPEPTIN-A
44120-7696 BBM-928-B
LUZOPEPTIN-B
44120-7697 BBM-928-C
LUZOPEPTIN-C
44120-7698 BBM-928-D
LUZOPEPTIN-D
44120-8204 1QCL
44120-8205 2QCL
44120-8206 1TP
44120-8207 2TP
44120-8208 1QNM
44120-8209 2QNM
44120-8586 2QNMB
44120-8587 2QCLB
44120-8588 QUINOMYCIN-X
44210-7961 B-5477-M
ENDURACIDIN-COMPLEX
44220-6902 PANTOMYCIN
44230-1960 SURFACTIN
44230-1988 ISARIIN
44230-6804 W-10
20561
44230-6805 W-10
20562
44250-5941 SF-1902-A1
44250-6081 GLOBOMYCIN
SF-1902-A1
44250-6624 SF-1902-A5
44250-7131 A-43-F
MSD-A43F
44250-8210 SF-1902-A2
44250-8211 SF-1902-A3
44250-8212 SF-1902-A4
44260-7699 A-21978 C0
44260-7700 A-21978-C1
44260-7701 A-21978-C2
44260-7702 A-21978-C3
44260-7703 A-21978-C4
44260-7704 A-21978-C5

44311-2001	STAPHYLOMYCIN-S
	16-5B
44311-5882	PLAURACIN-37277
44311-5883	PLAURACIN-37932
44311-5884	PLAURACIN-40042
44312-2010	ETAMYCIN
44312-6082	NEOVIRIDOGRISEIN-I
44312-6083	ETAMYCIN-B
	NEOVIRIDOGRISEIN-II
44312-6084	NEOVIRIDOGRISEIN-III
44312-6806	DEOXYETAMYCIN
	ETAMYCIN-B
	NEOVIRIDOGRISEIN-II
44312-6807	ETAMYCIN COMP.VII
44312-7132	NEOVIRIDOGRISEIN-VII
44320-2015	MONAMYCIN COMPLEX
44410-8088	2928
44410-8089	5590
44420-7433	MYCOPLANECIN
	MYCOPLANECIN-A
44430-3900	A-3302-B
	TL-119
44440-8213	DESTRUXIN-E
44450-7519	AC-69
	LIPOPEPTIN-A
44450-8073	LIPOPEPTIN-B
44510-2040	ISOLEUCINOMYCIN
	VALINOMYCIN
44510-2043	ENNIATIN-A
44510-2044	ENNIATIN-B
44510-2045	ENNIATIN-C
44510-2046	ENNIATIN-A1
44510-2052	BEAUVERICIN
44510-5447	BASSIANOLIDE
44510-6240	ENNIATIN-B1
44520-2056	SPORIDESMOLIDE-I
44530-1629	GRISEOVIRIDIN
44530-2057	OSTREOGRICIN-A
	16-5A
44530-2059	PRISTINAMYCIN-IIA
44530-2065	OSTREOGRICIN-G
	16-5D
44530-2066	A-17002
	A-2315-A
	PLAURACIN-35763
44530-2067	A-2315-B
	MADUMYCIN-I
	PLAURACIN-36926
44530-2069	A-2315-B
	MADUMYCIN-I
	PLAURACIN-36295
44530-2070	A-15104-V
	A-17002-F
	MADUMYCIN-II
	PLAURACIN-35763
44530-2071	YAKUSIMYCIN-A
44530-5085	1745-Z3-BW
44530-5680	GEMINIMYCINS
44530-5885	A-2315-B
	MADUMYCIN-I
	PLAURACIN-36926
44530-5886	A-17002-F
	A-2315-A
	MADUMYCIN-II
	PLAURACIN-35763
44530-6809	A-15104-II
	A-17002-C
44530-6810	A-15104-I
	A-17002-A
44530-6811	A-15104-X
	A-17002-B
44540-2075	DETOXIN-D1

45 MACROMOLECULAR PEPTIDES

45000-4104	NA-699
45000-7133	"CYTOTOXIC PROTEIN"
	LEUCODIDIN
45000-7710	BIOFLORIN
45110-2078	PHYTOACTIN
45110-6621	BACILEUCINE-A
45110-6622	BACILEUCINE-B
45110-8216	"PHYTOACTIN"
45120-6812	C-3603
45120-8589	BOTRYTICIDIN-A
45130-2105	SYRINGOMYCIN
45130-8075	AH-2589-I
45130-8215	AH-2589-II
45140-2120	NISIN
45140-2200	DIPLOCOCCIN
45211-2137	NEOCARZINOSTATIN
45211-2138	ACTINOXANTHIN
45211-2140	LYMPHOMYCIN COMPLEX
45211-2143	MACROMOMYCIN
45211-2145	MITOMALCIN
45211-6813	AUROMOMYCIN
	AUROMYCIN
45211-6814	DECHROMONEOCARZINOSTATIN
	PC
45211-7137	RINFOMYCIN
45211-7964	NEOCARZINOSTATIN-N1
45211-7965	NEOCARZINOSTATIN-N2
45211-8217	PRENEOCARZINOSTATIN
45212-5343	SPORAMYCIN
45213-2183	PICIBANIL
45213-8591	CI-782
45220-2207	SILLUCIN
45220-6815	"CYTOTOXIN"
45220-7435	MIEHEIN-21
45220-7709	CANECELUNIN
45220-7966	"CYTOTOXIN"
	"LEUCOCIDIN"
45220-8218	SML-91 LECTIN
45220-8219	SSL-91 LECTIN
45320-6294	NO NAME
45320-6816	MICHICARCIN
	NSC-191595
45320-7476	A-1
45320-7982	CI-783
45320-8221	C-1740
45320-8592	TF-130
45330-6911	"AVS"
	AVS
45330-8593	"VIRUS LIKE PARTICLES"
45340-2274	ASPARAGINASE
	L-ASPARAGINASE
45340-2286	PTERIDIN DEAMINASE
	PTERIN DEAMINASE

45340-6288	CARBOXYPEPTIDASE G1 CPD-G1
45340-6820	METHIONINE-G"-LYASE
45340-6821	L-LYSINE-A"-OXYDASE
45340-6822	PHENYLALANINE-AMMONIA LYASE
45340-7140	KU-1
45340-7141	"B"-1.3-GL UCANASE+ PROTEASE+LYSOZYME"
45340-7142	N-ACETYLNEURAMINASE-GLYCOHYDROLASE
45340-7143	AGA L-GLUTAMINASE-ASPARAGINASE
45340-7563	TYROSINE PHENOL-LYASE
45340-7712	DEAMIDASE AG
45340-8594	PR1-LYSOZYME
45350-4802	VIRIDIN-B
45350-4804	HAEMOCIN
45350-5719	CAROTOVORICIN
45350-6094	VIRIDICIN VIRIDICINE
45350-6241	CLOSTOCIN-O
45350-6354	GLAUCESCIN
45350-6496	FULVOCIN-C
45350-6628	COLICIN-E-3
45350-6817	"KILLER FACTOR" K-FACTOR
45350-6819	BACTERIOCIN-14468
45350-6824	THERMOCIN-93
45350-7144	BACTERIOCIN PYOCIN
45350-7145	SANGUICIN
45350-7564	ERWINICIN
45350-7565	S34-BACTERIOCIN
45350-7970	STAPHYLOCOCCIN
45350-8224	MUTACIN GS-5
45350-8596	LACTOCIN C-183 LACTOCIN C-290 LACTOCIN C-308
45350-8597	LACTOCIN P-109 LACTOCIN P-186
45350-8598	THOMYCIDE
45350-8860	LACTOCIN-A1
45360-2330	"ANTI VIRAL FACTOR" AVF
45360-7715	INTERFEROIDE
45370-6737	"CELL-WALL COMPONENT"
45370-7135	N-A-1 N-B-1 N-S-1 N-T-1
45370-7136	"METANOL EXTRACTABLE RESIDUE" MER
45370-7147	CP NSC-220537
45370-7490	60-F
45370-7491	"CELL-WALL SKELETON" CWS
45370-8225	LC-9018
45370-8227	RP-32919 RP-41200
45370-8844	CW-5
45370-8861	NS NS-B-1
45370-8862	NCWS

INDEX OF PRODUCING ORGANISMS

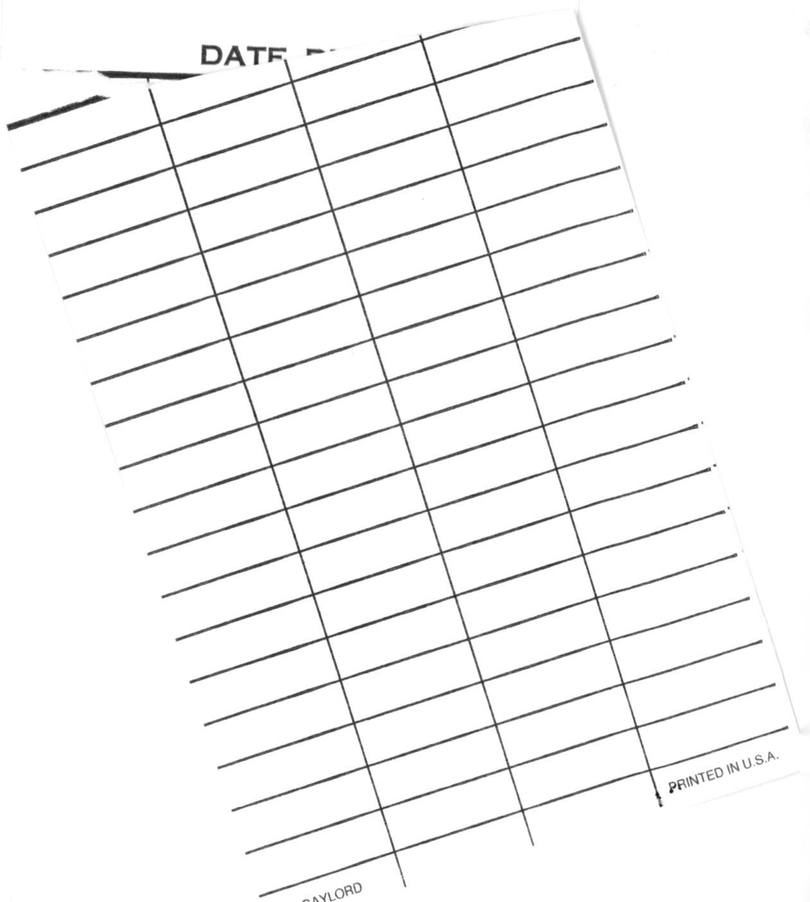
GAYLORD
PRINTED IN U.S.A.

3 1211 01694505 3